AF411442

MOFs for Advanced Applications

MOFs for Advanced Applications

Editors

Jorge Bedia
Carolina Belver

MDPI • Basel • Beijing • Wuhan • Barcelona • Belgrade • Manchester • Tokyo • Cluj • Tianjin

Editors

Jorge Bedia
Departamento de Ingeniería
Química, Universidad
Autónoma de Madrid,
Cantoblanco
Spain

Carolina Belver
Departamento de Ingeniería
Química, Universidad
Autónoma de Madrid,
Cantoblanco
Spain

Editorial Office
MDPI
St. Alban-Anlage 66
4052 Basel, Switzerland

This is a reprint of articles from the Special Issue published online in the open access journal *Catalysts* (ISSN 2073-4344) (available at: https://www.mdpi.com/journal/catalysts/special_issues/ MOF_Metal_Organic_Frameworks).

For citation purposes, cite each article independently as indicated on the article page online and as indicated below:

LastName, A.A.; LastName, B.B.; LastName, C.C. Article Title. *Journal Name* **Year**, *Volume Number*, Page Range.

ISBN 978-3-0365-3541-8 (Hbk)
ISBN 978-3-0365-3542-5 (PDF)

Contents

About the Editors

Jorge Bedia obtained his degree in Industrial Engineering and obtained his PhD in Chemical Engineering at the University of Malaga (Spain). He has been an Assistant Professor at the Chemical Engineering Department of the Autonomous University of Madrid (Spain) since 2013. His research interests include: (i) the synthesis and characterization of carbon-based materials for adsorption and catalysis; (ii) synthesis and applications of MOFs; (iii) water purification via advanced oxidation processes, especially photocatalysis; (iv) the gas-phase hydrodechlorination of chlorinated volatile organic compounds and (v) separation and catalytic processes with ionic liquids and supported ionic liquids. Dr. Bedia is co-author of around 100 refereed journal publications (more than 3500 citations with an H factor of 38), 3 book chapters and 2 Spanish patents. He has presented over 150 works to national and international conferences with over 20 oral presentations. He has been involved in more than 20 research projects from different entities: European, national, integrated or cooperative with other countries (Russia, Mexico, Germany, Peru and the USA) and several research contracts with private companies. He is a member of the Editorial Board of the *Chemical Engineering Journal* and *Associate Editor of Separation and Purification Technology*.

Carolina Belver has been an Associate Professor in Chemical Engineering Section at the Autonomous University of Madrid, Spain since 2011. She received her chemistry degree from the Autonomous University of Madrid in 1997 and obtained her Ph.D. in 2004 from the University of Salamanca (Spain). She performs research in the fields of (i) water purification via photocatalysis, (ii) the design of novel heterostructures for environmental remediation, (iii) water treatment via adsorption, (iv) advanced oxidation technologies for water treatment and (vi) environmental nanotechnology (fundamental and applications of nanomaterials). Dr. Belver is co-author of 90 publications with more than 3600 citations, receiving an H factor of 35, and 11 chapter-books. She has presented over 120 works to national and international conferences. In 2013, Dr. Belver received a Fulbright Fellowship, and she has been involved in more than 25 research projects from different entities: European, national and cooperative with other countries (China, Mexico, Peru and the USA) and several research contracts with private companies. She is a principal investigator of two national research projects and two collaboration projects which are Latin-American. She is Editor of the *Chemical Engineering Journal*.

Editorial

Metal Organic Frameworks for Advanced Applications

Carolina Belver * and Jorge Bedia *

Departamento de Ingeniería Química, Universidad Autónoma de Madrid, Cantoblanco, 28049 Madrid, Spain
* Correspondence: carolina.belver@uam.es (C.B.); jorge.bedia@uam.es (J.B.)

Citation: Belver, C.; Bedia, J. Metal Organic Frameworks for Advanced Applications. *Catalysts* **2021**, *11*, 648. https://doi.org/10.3390/catal11050648

Received: 30 April 2021
Accepted: 30 April 2021
Published: 20 May 2021

Publisher's Note: MDPI stays neutral with regard to jurisdictional claims in published maps and institutional affiliations.

1. Introduction

Metal organic frameworks (MOFs) are a class of porous materials with a modular structure. This allows for a very wide structural diversity and the possibility of synthesizing materials with tailored properties for advanced applications [1,2]. Thus, MOF materials are the subject of intense research, with strong relevance to both science and technology. MOFs are formed by the assembly of two components—cluster or metal ion nodes—which are also called secondary building units (SBUs), and organic linkers between the SBUs, usually giving rise to crystalline structures with an open framework and significant porous texture development. The main aim of this Special Issue of *Catalysts* is to present the most relevant and recent insights in the field of synthesis and characterization of MOFs and MOF-based materials for advanced applications, including adsorption, gas storage/capture, drug delivery, catalysis, photocatalysis, and/or chemical sensing.

2. Metal–Organic Frameworks for Advanced Applications

This Special Issue includes outstanding studies focused on the synthesis of new MOF-based materials with enhanced properties in terms of behavior and/or stability for adsorption and catalytic applications. In this sense, Khraisheh et al. [3] analyzed the potential application of two MOFs (SIFSIX-3-Ni and NbOFFIVE-1-Ni) for the separation of prime olefins from gas streams at different temperatures and pressures. Dynamic adsorption breakthrough tests were also conducted, and the stability and regeneration ability of the MOFs were established after eight consecutive cycles. The results revealed that SIFSIX-3-Ni can be considered an effective adsorbent for the separation of 10/90 v/v C_3H_4/C_3H_6 under the range of experimental conditions analyzed. The separated C_3H_6 was obtained with a 99.98% purity. Following with the stream purification applications, Mirante et al. [4] researched the catalytic efficiency of a layered coordination polymer to remove simultaneous sulfur and nitrogen compounds from fuels. An ionic lamellar coordination polymer based on a flexible triphosphonic acid linker achieved complete desulfurization and denitrogenation after 2 h using single model diesels, an ionic liquid as extraction solvent ([BMIM]PF$_6$) and H_2O_2 as oxidant. The lamellar catalyst showed a high recycle capacity for desulfurization. The reusability of the diesel/H_2O_2/[BMIM]PF$_6$ system catalyzed by a lamellar catalyst was more efficient for denitrogenation than for the desulfurization process when using a multicomponent model diesel. This behavior was not associated with the catalyst performance, but mainly due to the saturation of S/N compounds in the extraction phase. Salahuddin et al. [5] studied the use of MnO_2-doped ZIF-67 MOF for the electrocatalytic oxygen reduction reaction. The material showed an improved performance due to the incorporation of MnO_2 not only by enhancing the surface area, but also the conductivity. Chen et al. [6] reported photocatalytic CO_2 reduction with MIL-100(Fe)-CsPbBr$_3$ composites. The composites, with a high specific surface area, displayed an enhanced solar light response, and an improved charge carrier separation, resulted in an improved photocatalytic performance. Optimization of the relative composition, with the formation of a dual-phase CsPbBr$_3$ to CsPb2Br$_5$ perovskite composite, showed an excellent photocatalytic performance with 20.4 µmol CO produced per gram of photocatalyst during one hour under visible light irradiation. Palani et al. [7] explored the practical advantages and limitations of applying a UiO-66-based MOF catalyst in a flow microreactor for the catalytic hydrolysis of ethyl paraoxon, an organophosphorus chemical agent. It was concluded that tableting and sieving were viable methods to obtain MOF particles of a suitable size to be successfully screened under flow conditions in a microreactor, the optimal condition being catalyst particles with a sieved fraction between 125 and 250 µm. The synthesis of new materials was also addressed in this Special Issue. Li et al. [8] used tetracaboxylic acid, 3-(20,40-dicarboxylphenoxy)phthalic acid (H4dpa), as a multifunctional linker for the hydrothermal synthesis of new coordination polymers. Hence, a simple synthetic procedure led to the formation of a series of copper(II), manganese(II), and zinc(II) coordination polymers, whose catalytic activity was also explored in the mild cyanosilylation of benzaldehyde substrates with trimethylsilyl cyanide. The zinc(II)-based one functions as an effective and reusable heterogeneous catalyst to produce cyanohydrin products in up to 93% yields. Santibáñez et al. [9] described the aerobic oxidation of cyclohexene using heterometallic MOFs as catalysts in solvent-free conditions with oxydiacetic acid. It was concluded that the Langmuir surface area and the redox potentials were more important than the acid strength and acid sites of the studied MOFs, in terms of the referred catalytic performance. In addition, the reaction conditions were also shown to play an important role in the catalytic performance of the studied systems. In particular, the type of oxidant and the way to supply it to the reaction medium influenced the catalytic results. Jiao et al. [10] described the fast immobilization of human carbonic anhydrase II on Ni-based MOF nanorods with high catalytic performance. These results showed that the Ni-MOFs have great potential and high efficiency for the specific binding of immobilized enzymes. Pertiwi et al. [11] used MIL-101(Cr) MOF as a solid–acid catalyst for the solution conversion of biomass-derived glucose to 5-hydroxymethyl furfural (5-HMF). The substitution of Cr^{3+} by Fe^{3+} and Sc^{3+} in the MIL-101 structure resulted in more environmentally benign catalysts. It was observed that MIL-101(Fe) could be prepared, and the inclusion of Sc was possible at low levels (10% of Fe replaced). During extended synthesis times, the polymorphic MIL-88B structure was formed instead. The optimum material was a bimetallic (Fe,Sc) form of MIL-88B, which provides ~70% conversion of glucose with 35% selectivity towards 5-HMF after 3 h at 140 °C, which was a higher conversion compared to other heterogeneous catalysts reported in the same solvent. Padmanaban and Yoon [12] studied the surface modification of an MOF-based catalyst with Lewis metal salts for improved catalytic activity in the fixation of CO_2 into polymers.

The stability of MOFs is a crucial aspect only scarcely analyzed in the literature. In this sense, Gomez-Avilés et al. [13] reported a quantitative study of the water stability of NH_2-MIL-125(Ti), analyzing the ligand release along the contact time in water. This study demonstrates that NH_2-MIL-125(Ti) easily leached out over time while maintaining its structure. The effect of different thermal treatments applied to the MOF was investigated in order to enhance its water stability. Similarly, Wang et al. [14] synthesized Mn-MOF-74, which was further modified via two paths for enhanced water resistance. The results of selective catalytic reduction (SCR) performance tests showed that polyethylene oxide-polypropylene-polyethylene oxide (P123)-modified Mn-MOF-74 exhibited outstanding NO conversion of up to 92.1% in the presence of 5 vol.% water at 250 °C, compared to 52% for Mn-MOF-74 under the same conditions. It was concluded that the water resistance of Mn-MOF-74 was significantly promoted after the introduction of P123 and that the unmodified P123-Mn-MOF-74 was proven to be a potential low-temperature SCR catalyst.

This Special Issue contains several reviews, which analyzed promising approaches for the synthesis and advanced applications of MOFs. The combination of MOFs with other materials constitutes a promising methodology to synthesize materials with tunable and optimized characteristics for specific applications. Thus, Sun et al. [15] analyzed the synthesis of hybrid materials by supporting or incorporating polyoxometalates (POMs) into/onto MOFs. These hybrid materials combine the strong acidity, oxygen-rich surface, and redox capability of POMs while overcoming their drawbacks, such as difficult handling, a low surface area, and a high solubility. MOFs are ideal hosts due to their high surface area, long-range ordered structure, and high tunability in terms of the pore size and channels. In some cases, MOFs add an extra dimension to the functionality of hybrids. A study discussed their synthesis strategies, together with their major applications, such as their use in different catalytic applications. Jang et al. [16] described the use of MOFs as catalysts for air and water pollution environmental remediation, including an interesting approach focused on the synthesis of MOF-based filtering membranes by electrospinning using an eco-friendly polymer. Elhenawy et al. [17] reviewed the applications of MOFs in the carbon capture, focused on the applications of MOFs in the adsorption, membrane separation, catalytic conversion, and electrochemical reduction of CO_2 to provide new practical and efficient techniques for CO_2 mitigation. Finally, Liu et al. [18] described the recent advances in MOF-based nanocatalysts for photo-promoted CO_2 reduction applications. This review described the unique advantages of MOF-based materials for the photocatalytic reduction of CO_2. The catalytic reaction process, conversion efficiency, as well as the product selectivity of photocatalytic CO_2 reduction while using MOF-based materials was thoroughly discussed, with a special focus to the catalytic mechanism of CO_2 reduction with the aid of electronic structure investigations.

It can be concluded that MOFs have undiscovered potential for multiple advanced applications. We will probably witness an explosion of different applications of these versatile and attractive materials in the coming years. It is likely that their use in real applications will be the key for their further development.

Author Contributions: Conceptualization, C.B. and J.B.; writing—original draft preparation, C.B. and J.B.; writing—review and editing, C.B. and J.B. All authors have read and agreed to the published version of the manuscript.

Acknowledgments: We would like to thank the authors, reviewers, and editorial team of *Catalysts* for the work devoted to this Special Issue. We would also like to thank Vivian Niu for all her help, advice, and unvaluable work throughout the process.

Conflicts of Interest: The authors declare no conflict of interest.

References

1. Bedia, J.; Muelas-Ramos, V.; Peñas-Garzón, M.; Gómez-Avilés, A.; Rodríguez, J.J.; Belver, C. A review on the synthesis and characterization of metal organic frameworks for photocatalytic water purification. *Catalysts* **2019**, *9*, 52. [CrossRef]
2. Bedia, J.; Muelas-Ramos, V.; Peñas-Garzón, M.; Gómez-Avilés, A.; Rodriguez, J.J.; Belver, C. Metal–organic frameworks for water purification. In *Nanomaterials for the Detection and Removal of Wastewater Pollutants*; Elsevier: Amsterdam, The Netherlands, 2020; pp. 241–283.
3. Khraisheh, M.; Almomani, F.; Walker, G. Effective Separation of Prime Olefins from Gas Stream Using Anion Pillared Metal Organic Frameworks: Ideal Adsorbed Solution Theory Studies, Cyclic Application and Stability. *Catalysts* **2021**, *11*, 510. [CrossRef]
4. Mirante, F.; Mendes, R.F.; Paz, F.A.A.; Balula, S.S. High Catalytic Efficiency of a Layered Coordination Polymer to Remove Simultaneous Sulfur and Nitrogen Compounds from Fuels. *Catalysts* **2020**, *10*, 731. [CrossRef]
5. Salahuddin, U.; Iqbal, N.; Noor, T.; Hanif, S.; Ejaz, H.; Zaman, N.; Ahmed, S. ZIF-67 derived MnO_2 doped electrocatalyst for oxygen reduction reaction. *Catalysts* **2021**, *11*, 92. [CrossRef]
6. Cheng, R.; Debroye, E.; Hofkens, J.; Roeffaers, M.B.J. Efficient photocatalytic CO_2 reduction with MIL-100(Fe)-cspbbr3 composites. *Catalysts* **2020**, *10*, 1352. [CrossRef]
7. Elumalai, P.; Elrefaei, N.; Chen, W.; Al-Rawashdeh, M.; Madrahimov, S.T. Testing Metal–Organic Framework Catalysts in a Microreactor for Ethyl Paraoxon Hydrolysis. *Catalysts* **2020**, *10*, 1159. [CrossRef]
8. Li, Y.; Liang, C.; Zou, X.; Gu, J.; Kirillova, M.V.; Kirillov, A.M. Metal(II) coordination polymers from tetracarboxylate linkers: Synthesis, structures, and catalytic cyanosilylation of benzaldehydes. *Catalysts* **2021**, *11*, 204. [CrossRef]
9. Santibáñez, L.; Escalona, N.; Torres, J.; Kremer, C.; Cancino, P.; Spodine, E. CuII- and CoII-Based MOFs: {[La₂Cu₃(μ-H₂O)(ODA)₆(H₂O)₃]·3H₂O}ₙ and {[La₂Co₃(ODA)₆(H₂O)₆]·12H₂O}ₙ. The Relevance of Physicochemical Properties on the Catalytic Aerobic Oxidation of Cyclohexene. *Catalysts* **2020**, *10*, 589. [CrossRef]
10. Jiao, M.; He, J.; Sun, S.; Vriesekoop, F.; Yuan, Q.; Liu, Y.; Liang, H. Fast Immobilization of Human Carbonic Anhydrase II on Ni-Based Metal-Organic Framework Nanorods with High Catalytic Performance. *Catalysts* **2020**, *10*, 401. [CrossRef]
11. Pertiwi, R.; Oozeerally, R.; Burnett, D.L.; Chamberlain, T.W.; Cherkasov, N.; Walker, M.; Kashtiban, R.J.; Krisnandi, Y.K.; Degirmenci, V.; Walton, R.I. Replacement of Chromium by Non-Toxic Metals in Lewis-Acid MOFs: Assessment of Stability as Glucose Conversion Catalysts. *Catalysts* **2019**, *9*, 437. [CrossRef]

12. Padmanaban, S.; Yoon, S. Surface Modification of a MOF-based Catalyst with Lewis Metal Salts for Improved Catalytic Activity in the Fixation of CO_2 into Polymers. *Catalysts* **2019**, *9*, 892. [CrossRef]
13. Gómez-Avilés, A.; Muelas-Ramos, V.; Bedia, J.; Rodriguez, J.J.; Belver, C. Thermal Post-Treatments to Enhance the Water Stability of NH_2-MIL-125(Ti). *Catalysts* **2020**, *10*, 603. [CrossRef]
14. Wang, S.; Gao, Q.; Dong, X.; Wang, Q.; Niu, Y.; Chen, Y.; Jiang, H. Enhancing the Water Resistance of Mn-MOF-74 by Modification in Low Temperature NH_3-SCR. *Catalysts* **2019**, *9*, 1004. [CrossRef]
15. Sun, J.; Abednatanzi, S.; Van Der Voort, P.; Liu, Y.Y.; Leus, K. Pom@mof hybrids: Synthesis and applications. *Catalysts* **2020**, *10*, 578. [CrossRef]
16. Jang, S.; Song, S.; Lim, J.H.; Kim, H.S.; Phan, B.T.; Ha, K.T.; Park, S.; Park, K.H. Application of various metal-organic frameworks (Mofs) as catalysts for air and water pollution environmental remediation. *Catalysts* **2020**, *10*, 195. [CrossRef]
17. Elhenawy, S.E.M.; Khraisheh, M.; Almomani, F.; Walker, G. Metal-organic frameworks as a platform for CO_2 capture and chemical processes: Adsorption, membrane separation, catalytic-conversion, and electrochemical reduction of CO_2. *Catalysts* **2020**, *10*, 1293. [CrossRef]
18. Liu, C.; Wang, W.; Liu, B.; Qiao, J.; Lv, L.; Gao, X.; Zhang, X.; Xu, D.; Liu, W.; Liu, J.; et al. Recent advances in MOF-based nanocatalysts for photo-promoted CO_2 reduction applications. *Catalysts* **2019**, *9*, 658. [CrossRef]

Article

Effective Separation of Prime Olefins from Gas Stream Using Anion Pillared Metal Organic Frameworks: Ideal Adsorbed Solution Theory Studies, Cyclic Application and Stability

Majeda Khraisheh [1,*], **Fares Almomani** [1] **and Gavin Walker** [2]

[1] Department of Chemical Engineering, College of Engineering, Qatar University, Doha P.O. Box 2713, Qatar; falmomani@qu.edu.qa

[2] Department of Chemical Sciences, SSPC, Bernal Institute, University of Limerick, V94 T9PX Limerick, Ireland; Gavin.Walker@ul.ie

* Correspondence: m.khraisheh@qu.edu.qa

Abstract: The separation of C_3H_4/C_3H_6 is one of the most energy intensive and challenging operations, requiring up to 100 theoretical stages, in traditional cryogenic distillation. In this investigation, the potential application of two MOFs (SIFSIX-3-Ni and NbOFFIVE-1-Ni) was tested by studying the adsorption-desorption behaviors at a range of operational temperatures (300–360 K) and pressures (1–100 kPa). Dynamic adsorption breakthrough tests were conducted and the stability and regeneration ability of the MOFs were established after eight consecutive cycles. In order to establish the engineering key parameters, the experimental data were fitted to four isotherm models (Langmuir, Freundlich, Sips and Toth) in addition to the estimation of the thermodynamic properties such as the isosteric heats of adsorption. The selectivity of the separation was tested by applying ideal adsorbed solution theory (IAST). The results revealed that SIFSIX-3-Ni is an effective adsorbent for the separation of 10/90 v/v C_3H_4/C_3H_6 under the range of experimental conditions used in this study. The maximum adsorption reported for the same combination was 3.2 mmol g^{-1}. Breakthrough curves confirmed the suitability of this material for the separation with a 10-min gab before the lighter C_3H_4 is eluted from the column. The separated C_3H_6 was obtained with a 99.98% purity.

Keywords: metal organic frame works; olefin paraffin separations; propyne; propylene; adsorption isotherms; dynamic breakthrough

Citation: Khraisheh, M.; Almomani, F.; Walker, G. Effective Separation of Prime Olefins from Gas Stream Using Anion Pillared Metal Organic Frameworks: Ideal Adsorbed Solution Theory Studies, Cyclic Application and Stability. *Catalysts* **2021**, *11*, 510. https://doi.org/10.3390/catal11040510

Academic Editor: Jorge Bedia

Received: 7 March 2021
Accepted: 16 April 2021
Published: 18 April 2021

Publisher's Note: MDPI stays neutral with regard to jurisdictional claims in published maps and institutional affiliations.

1. Introduction

Separation processes in the oil, gas and chemical industries account for up to 15% of their total energy requirements [1–4]. Light hydrocarbons (C_1–C_9) are vital chemical feedstocks and energy resources around the world [5]. In addition, olefin separations are critical for the chemical industry, with the greatest demand on high purity propylene (C_3H_6). The demand for high purity propylene has risen sharply in recent years, and the compound is now the second most widely produced hydrocarbon by volume in the world after ethylene [3,6–9]. In 2019, the production of propylene was around 145 million tons globally [5]. Propylene is an intermediate essential chemical in a large number of important chemical industries such as polypropylene based plastics, propylene oxides, isopropanol, acrylonitrile, and other copolymers [3,6,9–19]. Steam/catalytic cracking of higher chain hydrocarbons is the main method of producing propylene, although the resulting product inevitably contains amounts of propyne. Propyne (C_3H_4) is a common impurity that is known to cause a poisoning effect of the catalyst during the cracking process, with detrimental effect on the production of propylene [3]. To meet polymer grade propylene requirements, the content of propyne must be reduced to less than 5 ppm. It is therefore imperative to remove propyne from the propylene gas streams to produce the required propylene polymer grade gas (>99.99% purity). The separation of propane (C_3H_8)

and propylene is well reported in literature given its demanding energy requirements and the close relative volatilities of both compounds at the temperature range of operation (typically between 244–327 K) [2,7–10,13,14,17]. However, only a few studies have reported the separation of C_3H_4/C_3H_6 mixtures [4,20–22]. This separation is very challenging due to the physical and chemical similarities (molecular size C_3H_4: $4.16 \times 4.01 \times 6.51$ Å [3] and C_3H_6: $5.25 \times 4.16 \times 6.44$ Å [3]) between propyne and propylene compounds; more so than C_2H_2/C_2H_4 separations, for example [4]. Cryogenic distillation is the most commonly used method. Typically, it requires more than 100 theoretical stages to perform the separations of the gas fractions under low temperature and high pressure conditions [6], resulting in very high energy demands. Other technologies involve adsorption-based separations using a variety of materials such as zeolites and carbons in a variety of settings such as pressure, vacuum, or temperature swing settings [2,7,12,14,23–26]. Low separations and selectivity remain a challenge in difficult separations like those related to propyne and C_3H_6 separations. New developments in metal organic frameworks (MOFs) and the development of crystal engineering represent huge potential for previously difficult separations [27–29]. Metal organic frameworks, sometimes called metal coordinated polymers, are crystalline micropore materials with huge potential, owing to the flexibility in their chemical structure and intrinsic properties that can be fine-tuned to suit a given gas separation application [30–32]. Accordingly, the MOF internal aperture can be tuned for selective size or shape separation. They generally comprise organic ligands and inorganic metal clusters. Accordingly, the changes in structure pore size, one of the most appealing features of MOFs, and functional groups lends itself to many gas applications and separation processes. The structural flexibility and dynamic behavior of metal organic frameworks, such as the gate-opening effect due to variations in temperature, pressure, or other eternal stimuli, are another distinct feature of metal organic frameworks compared to traditional adsorbents such as zeolites [5]. A class of MOFs (hybrid ultra-microporous materials) are setting a new benchmark in a number of gas separations [27]. These materials are based on 3D Zn/Cu/Ni coordination networks that contain metal nodes in addition to organic linkers. Anions such as SiF_6^{2-} (SIFSIX), $NbOF_5^{2-}$ (NbOFFIVE) form bridges in the MOF structure and are reported to have potential application in gas separations, such as C_3H_6/C_3H_8 and other lower alkene/alkane separations.

Gas hydrocarbons are adsorbed inside the molecule and held via van der Waals, metal-binding or hydrogen interactions, depending on the type of MOF used. To this end, equilibrium, kinetic, or molecular sieving separations can be the predominant mechanism that aids the use of a particular MOF in a given gas separation. Molecular sieving is based on size/shape exclusion, while thermodynamic equilibrium separation depends on the type and strength of the interactions and the affinity of the host-guest interactions i.e., MOF and the gas species. In equilibrium separations, there are relative thermodynamic affinities between the gas species and the MOF adsorbent via the introduction of strong interaction sites on the MOF frame. Lewis acidity related to an uncoordinated open metal site MOF is used in CO_2 or olefin separations. These interactions form strong bonds which is the major challenge associated with their use in such separations due to the associated extra energy requirements related to the desorption of the gas and its subsequent release and separation, in addition to the MOF regeneration. Kinetic operations are based on the diffusivity of the gas compounds and their relative selectivity.

The use of metal organic frameworks for the separation of propyne/propylene is still in the early stages, with few reported studies in literature [4]. The first study on C_3H_4/C_3H_6 separation was reported using [Cu(4,4′-bipyridine)(trifluo-romethanesulfonate (OTF)$^-_2$](ELM-12) [21]. This material showed good potential for adsorption selectivity through the estimation of the ideal adsorbed solution theory (IAST) selectivity at a trace concentration of propyne (ratio 1/99 C_3H_4/C_3H_6), with good adoption capacity at low-pressure range. The work highlighted the potential of metal organic frameworks of flexible structures for hydrocarbon separation at low concentrations. Anion pillared metal organic frameworks with various structures were also studied, including the $NbOF_5^{2-}$ and SiF_6^{2-}

pyrazine based MOF family (NbOFFIVE-1-Ni (sometimes called KAUST-7 [33]), SIFSIX-2-Cu, SIFSIX-3-Ni, SIFSIX-3-Zn) [18]. The study highlighted that the geometric disposition and pore size of SiF_6^{2-} anions could be varied to best match the propyne molecules, resulting in very promising separation under the applied experimental conditions. The adsorption capacity reportedly reached 2.6 mmol g^{-1} for propyne concentrations of around 300 ppm. The study reported that both molecules were adsorbed, limiting productivity in separation [5]. Li et al. [4] screened 20 different MOF materials with varying structures, pore sizes, preparation methods, and functional groups for propyne/propylene separation. Their study showed that one material, called UTSA-200, exhibited great potential (95 cm^3 cm^{-3} at low pressure and 298 K) at an ultralow concentration (0.1:99.9 v/v C_3H_4/C_3H_6). The high affinity was reported to be related to the suitable pore size resulting from the rotation of the pyridine rings in the MOF with blocking propylene effect [4,5]. Recently, calcium based metal organic frameworks were investigated for C_3H_4/C_3H_6 separation [3]. An uptake gas value of 2.4 mmol g^{-1} was recorded at low pressures and trace concentrations.

In the few reported studies on propyne/propylene separation, very low concentrations were studied at pressures up to 1 bar. To establish a good trade off, high productivity and purity is required from the adsorbent material. In addition, most studies considered the adsorption capacity alone, while a few considered breakthrough analysis for a number of different cycles. There is a lack of systematic studies on this particular separation that take into account the adsorption isotherms and kinetics of the stream under consideration. Targeting materials with well-developed porous structures and stability is required to enhance the separation of propyne and propylene, to ensure the high purity of propylene, which is needed for industrial grade applications, and to reduce the energy requirements of this process. As mentioned earlier, the study of the selectivity is important to establish the type of integration between the gas and the MOF adsorbent to facilitate the understanding of the desorption and the reversibility of the interaction. Accordingly, in this study, a systematic approach was considered to study the effect of two types of pyrazine-based inorganic anions SiF_6^{2-}, $NbOF_5^{2-}$) metal organic frameworks (SIFSIX-3-Ni and NbOFFIVE-1-Ni) under a range of temperatures (300–360 K) and pressures (up to 100 kPa) for the separation of propyne/propylene in various concentrations (10/90 v/v). Adsorption capacities were analyzed using four isotherm models, i.e., single (Langmuir and Freundlich) and multi constant isotherms (Sips and Toth). No other reported studies have addressed the mathematical isotherm fitting of the experimental data for the C_3H_4/C_3H_6 system in combination with the metal organic frameworks used in this study. Such data are important in engineering calculations related to scale up and industrial applications. In addition, the selectivity was analyzed using the ideal adsorbed solution theory (IAST). Dynamic studies were conducted and breakthrough curves were established at varying numbers of cycles.

2. Experimental Materials and Methods

High purity analytical materials were supplied by Sigma-Aldrich and Buzwair Inc. Qatar. Chemical structures and characteristics of propyne and C_3H_6 are shown in Table 1. Gases were of 99.99% purity.

Table 1. Characteristic information for adsorbents and adsorbate gases.

Sorbent	BET ($m^2\ g^{-1}$)	Pore Size (nm)	Pore Volume ($cm^3\ g^{-1}$)	Ref	Size ($Å^3$) *	References
NbOFFIVE-1-Ni	248	0.139	0.095	This study, ref. [34]	D1:4.66, D2:3.21, D3:4.9 D2:3.047	[1,18] [1]
SIFSIX-3-Ni	368	0.36	0.167	This study, ref. [34]	D1:5.03, D2:3.75, D3:4.6 D2:4.2 D1:5.047	[18] [4] [1]
Hydrocarbon	**Structure**				**Size ($Å^3$) ***	**References**
C_3H_4					4.16 × 4.01 × 6.51	[3–18]
C_3H_6					5.25 × 4.16 × 6.44	[3–5,18]

* D1, D2 and D3 refer to distances as illustrated in Figure 1b.

2.1. Adsorbent Synthesis

The two metal organic frameworks (NbOFFIVE-1-Ni and SIFSIX-3-Ni) were prepared as detailed in our earlier work [34]. Both metal organic frameworks were pyrazine based, as detailed in Figure 1. The metal organic frameworks were architected by the bridging of the pyrazine-Ni^{2+} square grid layers with NbO_5^{2-} (Figure 1). The difference between the NbO_5^{2-} and SiF_6^{2-} resulted in pyrazine moieties free rotation, and it affected the pore cavity which is smaller in NbO_5^{2-} compared to that of SiF_6^{2-} (= hexafluorosilicate) (Table 1).

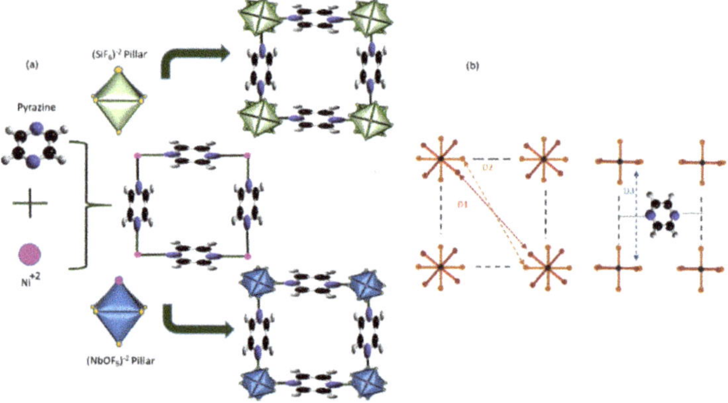

Figure 1. (**a**) Schematic of the basic structure of the pyrazine based metal organic frameworks used in the study. (**b**) Structural representation of the metal organic frameworks and size dimensions. Values of D1, D2 and D3 are given in Table 1 (dimensions are not to scale and for illustration purposes).

2.2. Sample Characterization:

2.2.1. Brunauer-Emmett-Teller (BET) Analysis

Liquid nitrogen was used for the N_2 gas adsorption tests at 77 K. Micromeritics characterization was carried out using ASAP 2420 surface and porosity analyzer (Micromeritics GmbH, Unterschleißheim, Germany). Surface area was established using BET model while pore size distribution was found using the BHJ method [34]. Table 1 presents the main characteristics of the metal organic frameworks used.

2.2.2. Thermogravimetric Analysis (TGA), SEM and FTIR

TGA analysis may be used to establish the weight loss of metal organic frameworks under thermal stress with high temperatures as an indication to the sample thermal stability. Tests were conducted using a Perkin Elmer Pyris 6 analyzer. Details of these tests have been reported by Khraisheh et al. [34].

Furthermore, Fourier-transform infrared spectroscopy (FTIR) (using Bruker Vertex 80) for the adsorbents was conducted in the range of 4000–400 cm^{-1}. In addition, SEM analyses were conducted following standard protocols.

2.3. Gas Uptake

Equilibrium and Kinetic Breakthrough Gas Adsorption Studies

To establish the maximum and equilibrium uptake capacity of the solids towards C_3H_4/C_3H_6 gases, Micromeritics ASAP 2420 (Germany) was used. Samples of around 100 mg were outgassed for at least 12 h under vacuum at 338 K before the adsorption/desorption experiments. Tests were conducted at a range of temperatures between 300–360 K (controlled via water (300–320 K) or oil (340–360 K) jacket) and pressures from 0–100 KPa. An in-house simple setup was used to test the breakthrough experiments (dynamic tests). The setup consisted mainly of a small quartz adsorption column (I.D 4 mm, height: 160 mm), gas flow controllers and gas analyzer. The required amount of adsorbent was used in the columns and then activated inside the column at 338 K for a couple of hours using helium (He) gas to ensure that any unwanted inert gases had been fully purged from the column. To eliminate any errors in the use of the gas mixtures, a (10/90 v/v) C_3H_4/C_3H_6 gas mixture cylinder was purchased (Buzwair gas Inc. Qatar) and used in the experiments. A total flow rate of 5 mL (STP min^{-1}) was used in all experiments. The flow rate was found to offer the best optimum conditions without having to fluidize the bed. The beds were regenerated ahead of a new gas breakthrough cycle using He (99.99 purity) for 30 min at 298 K and a flow rate of 5 mL min^{-1}. The same sample of solid MOF sorbent was used for up to eight successive breakthrough cycles in order to study the stability of the adsorbent during multiple cycle operations of sorption and desorption. The experimental procedure for each cycle was identical to that described above.

3. Theoretical Modeling

3.1. Isotherm and Kinetic Models

The adsorption isotherms of the pure compounds on the NbOFFOVE-1-Ni, SIFSIX-3-Ni were determined. Many models are reported in the literature to describe the adsorbent capacity for a certain species. The most commonly used in the case of gas-solid adsorption system are Langmuir, Freundlich, Sips and Toth isotherms [10]. The first two are known as the two parameter models, while the latter models are a hybrid combination of the two parameter models.

The Langmuir model is one of the most widely used isotherms in many applications including solid-gas and solid-liquid separations. The model is based on the assumption of monolayer interaction between the adsorbent and adsorbate with the assumption that all adsorption sites have similar adsorption energy requirements. Equation (1) represents one of the simplest forms to describe the Langmuir isotherm, where the maximum or saturation adsorption capacity is related to the system pressure and Langmuir constant, as shown in Equation (1):

$$Q = q_{sat}\frac{k_1 P}{1 + k_1 P} \tag{1}$$

where Q (mmol g^{-1}) is the adsorption capacity, q_{sat} (mmol g^{-1}) is the equilibrium uptake capacity of the gas species, P is the system pressure (kPa), and k_1: isotherm constant related to the energy of adsorption

The two main factors of Equation (1) were estimated from experimental data and used to establish the model fit of the data to the isotherm. The applicability of the isotherm is typically related to another factor that is associated with the Langmuir model, as described in Equation (2):

$$R_l = \frac{1}{1 + k_l P} \tag{2}$$

The value of the separation factor R_l is indicative of the ability of the model to fit the experimental data.

The Freundlich isotherm, on the other hand, is based on the assumption of multilayer adsorption with various adsorption energies [35]. In this model (Equation (3)), the increase in the pressure of the system increases the uptake capacity of the adsorbent.

$$Q = k_f P^{\frac{1}{n}} \tag{3}$$

where k_f is the Freundlich constant and n is a constant. A value of n lower than 1 is indicative of chemisorption rather than physisorption.

Although both models are widely used in many applications, their ability to fit the experimental data in a wide range of applications is limited in systems related to adsorption of many hydrocarbons. Accordingly, hybrid multiconstant models that are based on the two models have been used with better representation of the adsorption of hydrocarbons on solid materials.

The Sips model is an isotherm derived from both Langmuir and Freundlich isotherms and is presented in Equation (4) [10]:

$$Q = q_{sat}(k_s P)^{1/m} / (1 + k_s P)^{1/m} \tag{4}$$

where k_s is Sips constant and m is a constant that represents deviation from ideal heterogeneity of the adsorption system and is typically considered as intensity factor with values above 1 indicative of a heterogeneous adsorption. Equation (4) reduces to its Langmuir form (Equation (1)) in cases when m equals unity. The Sips model is known to be best applied in cases where the system operates at higher-pressure conditions. The Toth isotherm [36] (Equation (5)), on the other hand, is another hybrid combination of the Langmuir and Freundlich isotherm that is reported to have better data fitting in applications at a wider variety of pressures compared to the Sips model.

$$Q = q_{sat} \left\{ \left(\frac{(k_t P)^n}{(1 + k_t)^n} \right) \right\}^{\frac{1}{n}} \tag{5}$$

where k_t and n are Toth constants specific for adsorbate-adsorbent pairs, while n indicates the affinity of the adsorption. A value of n close to 1 is an indication of the system heterogeneity. As with the Sips model, a value of 1 reduced the Toth equation back into Langmuir isotherm. The applicability of this model in a good range of system pressure application resulted in its wider application in gas solid adsorption systems.

Statistical analyses were used to evaluate the fit of the experimental data to the various models. The most reported parameter used is based on the average absolute relative deviation (AARD), coefficient of determination (R^2) (Equations (6) and (7), respectively).

$$AARD(100\%) = \frac{100}{N} \sum_{i=1}^{N} \left| \frac{Q_{exp} - Q_{pred}}{Q_{exp}} \right| \tag{6}$$

$$R^2 = 1 - \frac{\sum_{i=1}^{N} \left(Q_{pred} - Q_{exp,i} \right)^2}{\sum_{i=1}^{N} \left(Q_{pred-i} - \overline{Q}_{exp} \right)^2} \tag{7}$$

where Q_{pred} is the predicted amounts, Q_{exp} is the values of Q obtained experimentally, and N is the number of the experimental data points used in the isotherm fit.

3.2. Isoteric Heats of Adsorption, IAST Selectivity and Adsorption Kinetics

The isosteric heat of adsorption makes it possible to characterize the surface properties of adsorbents, catalysts, and other materials, and provides information on the homogeneity and heterogeneity of surface. The calculation of the isosteric heats of adsorption (Q_{st} (kJ mol^{-1})) is essential for an understanding of the strength of the interactions between the solid surface and the adsorbate in addition to any energetic heterogeneity in the solid surface. The calculations are typically based on the fundamental Clausius-Clapeyron equation:

$$Q_{st} = RT^2 \left(\frac{\partial \ln P}{\partial T} \right)_{qsat} \tag{8}$$

where P and T are the system pressure and temperatures and q_{sat} is the saturated equilibrium uptake amount (mmol g^{-1}).

To assess the selectivity and feasibility of the separation, the IAST selectivity was considered for C_3H_4/C_3H_6 gases on the two metal organic frameworks employed in this study. The IAST calculations facilitate the study of the selectivity in binary gas mixtures. A full mathematical description is available elsewhere [6,10,37,38].

Adsorption kinetics is fundamental for engineering evaluation of adsorption systems. Typically, models derived from Fick's law of micropore diffusion are employed to model adsorption uptake curves via the calculation of the intercystalline diffusivity (D_c) [10]. An analytical solution of the equation based on the assumption of approximate cylindrical geometry of the solids is given as:

$$\frac{q}{q_\infty} = 1 - \sum_{n=1}^{\infty} \frac{4\alpha(1+\alpha)}{4 + 4\alpha + \alpha^2\beta_n^2 \, \exp\left(-\frac{D_c}{r_c^2}\beta_n^2 t\right)} \tag{9}$$

where q_∞ is the equilibrium adsorbed amount (mmol g^{-1}), t is the time (s), and constants α and β are calculated as detailed in [10].

4. Results and Discussion

4.1. Adsorbent Characterization

The SIFSIX and NbOFFIVE metal organic frameworks were synthesized and characterized as described in our earlier work [34]. Figure 2 offers some of the main characteristics obtained, including SEM, FTIR, N$_2$ adsorption desorption, and TGA analysis. In addition, Table 1 presents the estimated surface area and pore volumes. Thermal stability is essential for proper application of microporous materials in separation processes. The stability of the prepared materials was tested in a harsh temperature environment, i.e., up to 750 °C, by thermogravimetric analysis in the presence of nitrogen gas. The TGA isotherms are given in Figure 2i for both sorbents and represent the change in the weight of the material at elevated temperatures. A steep change in the mass of the adsorbent indicates a less stable structure and the possibility of material deterioration at higher temperature settings. At 280–300 °C, both materials underwent between 10–15% decrease in mass. The decrease was much steeper in the case of SIFSIX compared to NbOFFIVE at temperatures above 300 °C. The results of TGA obtained in our study were in good agreement with those reported in Bhatt [39] and Kumar [40]. The initial loss of mass is usually due to the evaporation of free water and other volatile compounds. A large drop in mass at elevated temperatures, as is the case in SIFSIX, may result in the decomposition of the material and the loss of wall and structural integrity [30,35]. The interactions between the different functional groups in a material can clarify the gas uptake. FTIR spectroscopy (Figure 2ii) was recorded in the range of 400–4000 cm^{-1}. For NbOFFIVE-1-Ni, the peaks at 3221, 1615, 1402, and 465 cm^{-1} most likely correspond to the characteristic nickel O–H stretching. A SEM micrographs for the two prepared materials (Figure 2iv) showed multisized agglomerations in both materials with the pore size in the range of 100–600 nm. Voids are clearly present between the uniform particles. The porous nature of the materials was tested using N$_2$ adsorption.

A significant difference in pore size and volume (Figure 2iii) was found between the SIFSIX and NbOFFIVE; both were higher for SIFSIX. The pore size for this compound was nearly 2.6 times higher than that recorded for the Nb MOF (Table 1), indicating a more enhanced surface area and porosity, and the formation of a well-defined structure [17]. The shape of the adsorption and desorption isotherms (depicted in Figure 2iii) clearly indicated a reversible Type-I isotherm in accordance with IUPAC categorization. This type indicated the formation of uniform narrow mesopores and a wider distribution of the pores [10,36,41]. The initial steep rise in the N_2 adsorption–desorption isotherm at a low relative pressure (P/P_o less than 0.001) is typically associated with the formation of microporosity. The high-adsorbed volume values (Figure 2iii) at relative pressures higher than 0.8 have been attributed to the N_2 capillary condensation in the antiparticle pores, and may indicate expandable pores and a flexible structure which may lead to what is known as gate-opening effect [35]. The N_2 adsorption–desorption isotherms did not indicate the presence of a clear hysteresis loop at low P/P_o ratio. A slight hysteresis effect was present at higher P/P_o ratio.

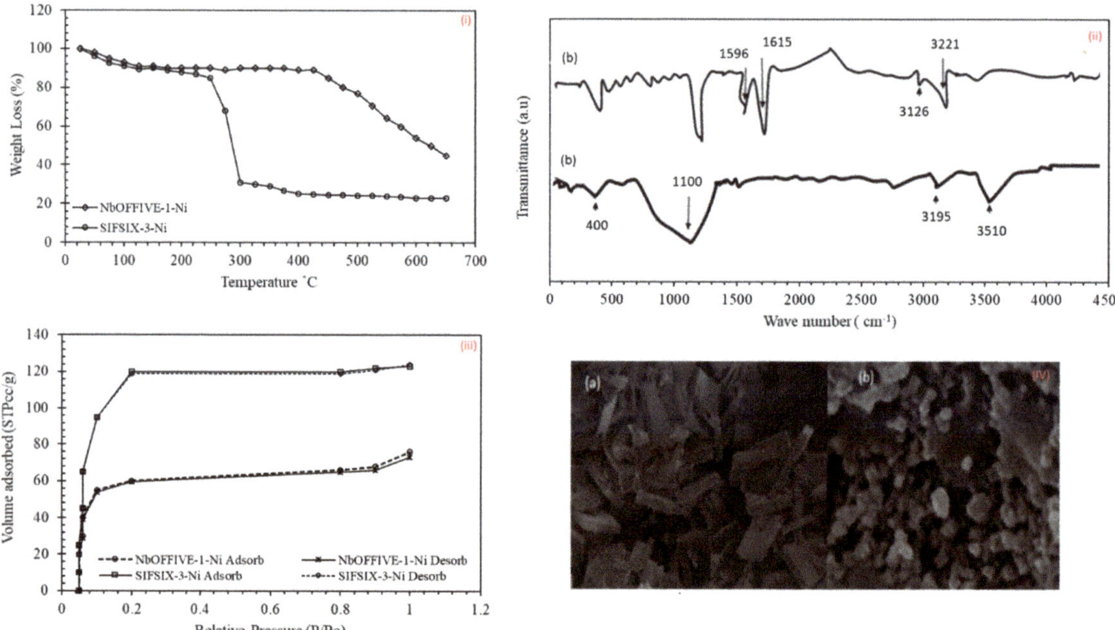

Figure 2. Characterization results for SIFSIX-3-Ni and NbOFFIVE-1-Ni; (**i**) TGA; (**ii**) FTIR; (**iii**) N_2 adsorption/desorption; (**iv**) SEM images (**a**) NbOFFIVE-1-Ni, (**b**) SIFSIX-3-Ni.

It has been reported that the separation of C_3H_6/C_3H_8, for example, is dominated by size selection separation [1]. It is hence expected that the separation of C_3H_4/C_3H_6 will follow a similar mechanism as in both cases of fluorinated metal organic frameworks (NbOFFIVE and SIFSIX were deployed under similar experimental conditions). The restrictive pore size of the metal organic frameworks due to the prevention of the pyrazine moieties (Figure 1) affects the size selection between gas species. From the N_2 adsorption data, BET areas were estimated to be 248 and 368 m^2 g^{-1} for NbOFFIVE and SIFSIX, respectively (Table 1). The experimentally obtained total pore volumes were 0.095 and 0.167 cm^3 g^{-1} for NbOFFIVE and SIFSIX, respectively.

4.2. C_3H_4/C_3H_6 Uptake and Separation

The potential of using the microporous metal organic frameworks for the selective adsorption–desorption and molecular interactions and binding nature of propyne over propylene at low pressure ranges (up to 100 kPa) and various temperatures was investigated. The results are presented in Figures 3 and 4. Figure 3a,b depicts an example of adsorption and desorption gas uptake for both adsorbents at a range of pressures up to 100 kPa. The figure shows the adsorption at 300 K as an example. Similar trends were obtained at different temperatures. The figure clearly reveals that different uptake values can be obtained in the case of propyne and propylene, indicating an effective separation due to the differences in the uptake values and steepness of the isotherms obtained. Both figures clearly show that propyne is adsorbed in a larger amount at a given pressure in comparison to propylene. A maximum uptake value of 3.28 and 2.99 mmol g^{-1} was obtained for propyne on SIFSIX-3-Ni and NbOFFIVE, respectively at 300 K. The maximum absorbed values for propyne were lower than those for propyne on both adsorbents, as can be seen from Figure 3a,b. Values of 2.9 and 2.39 mmol g^{-1} were obtained for propylene on SIFSIX and NbOFFIVE, respectively. The C_3H_6 and C_3H_4 values on the SIFSIX were higher than those recorded for the NbOFFIVE-1-Ni for the entire pressure range at a given temperature. It is also evident that the difference in the adsorption of C_3H_4 and C_3H_6 on SIFSIX was wider (Figure 3a) than that obtained on the NbOFFIVE MOF (Figure 3b). This indicated that the separation of propyne from C_3H_6 on the SIFSIX was more effective than that on the NbOFFIVE based MOF, and provides evidence of the potential for C_3H_4/C_3H_6 separation applications. A value of 2.85 mmol g^{-1} was reported for SIFSIX-3-Ni at 298 K for propylene [42]. The anion pillar $NbOF_5^{2-}$ is bulkier than SiF_6^{2-} and has longer F-Nb bond length (1.89 compared to 1.68 Å for SIFSIX), resulting in a thinner and longer pore space (4.66 × 7.88) which leads to lower propyne capacity [18]. This led us to conclude that small changes to the pore structure can have a major influence on the selectivity and absorbability of the adsorbate on the selected MOFs.

In addition, study of the adsorption–desorption behavior for propyne and propylene for both adsorbents (Figure 3) was important to establish the reversibility (regeneration ability) for the adsorbents. It is clear that both adsorbent isotherms show reversibility upon uptake of propyne and C_3H_6 gases. It is also evident that the hysteresis effect is more pronounced in propyne adsorption for both sorbents compared to that for C_3H_6. This indicated a strong affinity and molecular interactions between C_3H_4 molecules and the structure of the sorbents. A less defined hysteresis effect is realized in the case of propylene adsorption on both solids. A similar effect was reported in a recent study by Lin et al. [42]. Li et al. [4] reported that some metal organic frameworks such as SIFSX-1-Cu, SIFSIX-2-Cu-i, ELM-12 have a strong interaction with C_3H_4 at pressures less than 20 kPa. These interactions were exhibited in a steep adsorption uptake isotherm at a fixed temperature over propylene with the reported benchmark uptake values [4]. One issue related to the aforementioned materials is that their pore sizes were large, allowing the passage of both gases, and hence reducing their potential selectivity.

The effect of temperature on the adsorption of C_3H_4 and C_3H_6 on both MOFs is given in Figure 4. Temperatures in the range of 300–360 K were used for pressures up to 1 kPa. It can be seen from Figure 4a–c that the temperature has a profound impact on the removal and uptake rate for all adsorbents. It is clearly shown that the uptake of the gases is lowered by the elevation in temperature. Other adsorption systems such as C_3H_6 and C_3H_8, as reported in studies, showed similar trends with respect to uptake values vs. temperature [10]. No other studies have shown the full experimental isotherm results for propyne and C_3H_6 systems. In all cases, the adsorption of C_3H_4 is higher than that for propylene at all temperature ranges and in both adsorbents. In addition, SIFSIX outperformed NbOFFIVE for the selectivity of C_3H_4. It can also be noted that a steep increase in the adsorption capacity at the lower pressure range is more apparent in the case of C_3H_4 and SIFSIX combination at all temperature ranges in comparison with the other adsorbent/adsorbate systems (Figure 4a). A similar result was reported, but in the C_3H_6 and C_3H_8 systems on different metal organic frameworks [10]. To facilitate a good comparison, Figure 4c also shows experimental uptake isotherms at a selected temperature for both adsorbents and adsorbate systems. The superior uptake capacity of SIFSIX for propyne is evident, especially at low-pressure ranges (<20 kPa), indicating a strong interaction between the SIFSIX and C_3H_4. Comparisons between the uptake capacity and those published in the literature are presented in Table 2. This higher affinity and uptake values may also be related to the better BET surface area and pore structure of the SIFSIX compared to NbOFFIVE metal organic frameworks (Table 1). Figure 4c shows that the gap between the two isotherms for C_3H_4 and C_3H_6 are larger in SIFSIX compared to that in NbOFFIVE. A marked gap means that the separation is feasible for this system, and will be further studied via IAST analysis. The kinetic diameter of the molecule will have a major effect on the selectivity of one gas over another. Propyne is a linear molecule ($4.16 \times 4.01 \times 6.51$ Å), while C_3H_6 is a larger molecule which is typically curve-shaped with $4.64 \times 4.16 \times 6.44$ Å [4]. Irrespective of the linear shape, the existence of the methyl group in both gases makes their kinetic diameters quite close (4.2 and 4.6 Å for C_3H_4 and C_3H_6, respectively). This small difference is one of the main reasons for the difficulty and energy intensity of their separation. SIFSIX and NbOFFIVE have a range of reported dimensions, depending on methods of preparation, post-treatment and experimental conditions, (Table 1), which clearly indicates that SIFSIX is more suited for the adsorption of C_3H_4 from C_3H_6. A micropore analysis of the SIFSIX and kinetic diameter (average 4.2) will allow both C_3H_4 and C_3H_6 to enter the pores. Nonetheless, it seems that a large number of anions (SiF_6^{2-}) in the pores and channels of the framework [4] provides a better binding ability for alkynes, as compared to alkenes, creating a preferred sieving effect towards propyne. This results in the excellent selectivity and separation ability of SIFSIX compared to NbOFFIVE. The maximum equilibrium adsorption value reported here for the SIFSIX MOF was higher than that reported for the same material and C_3H_4 separation (1.87 mmol g^{-1}) [4]. The difference may be attributed to the different $v/v\%$ ratios used in the reported study. In addition, the value recorded in our work is similar to the benchmark uptake value reported for UTSA-200 ([Cu(azpy)$_2$(SiF$_6$)]$_n$; azpy = 4,4'-azopyridine) in a recent work by Li et al. (4). The main reason for the good reported selectivity is the small aperture size of the UTSA-200 (3.4 Å), meaning that size exclusion is the only dominating mechanism in this case, compared to stronger interactions between SIFSIX and the C_3H_4 gas. Yang et al. [18] reported that the precise tuning of the size of the inorganic anion hybrid ultramicroporous materials based on SiF_6^{2-} and $NbOF_5^{2-}$ serves as a single molecule trap for propyne.

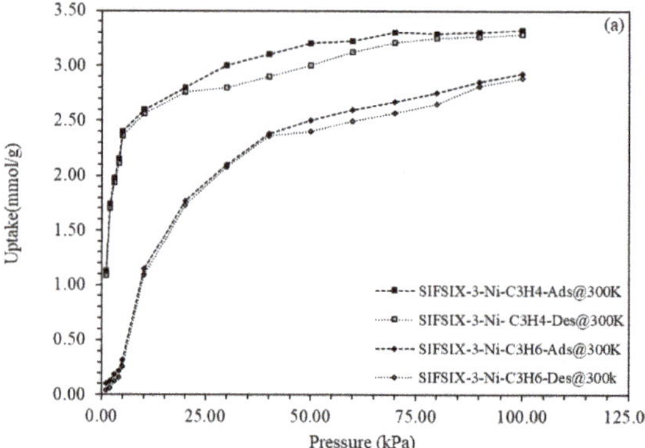

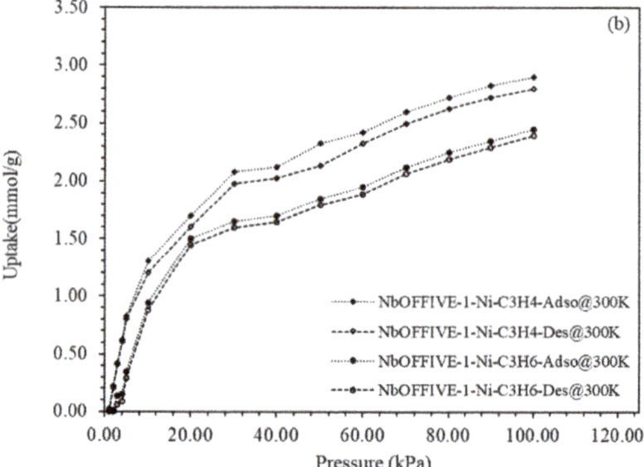

Figure 3. (**a**) adsorption/ desorption of C_3H_4 on SIFSIX-3-Ni at 300 K and (**b**) adsorption/ desorption of C_3H_6 on NbOFFIVE-1-Ni at 300 K.

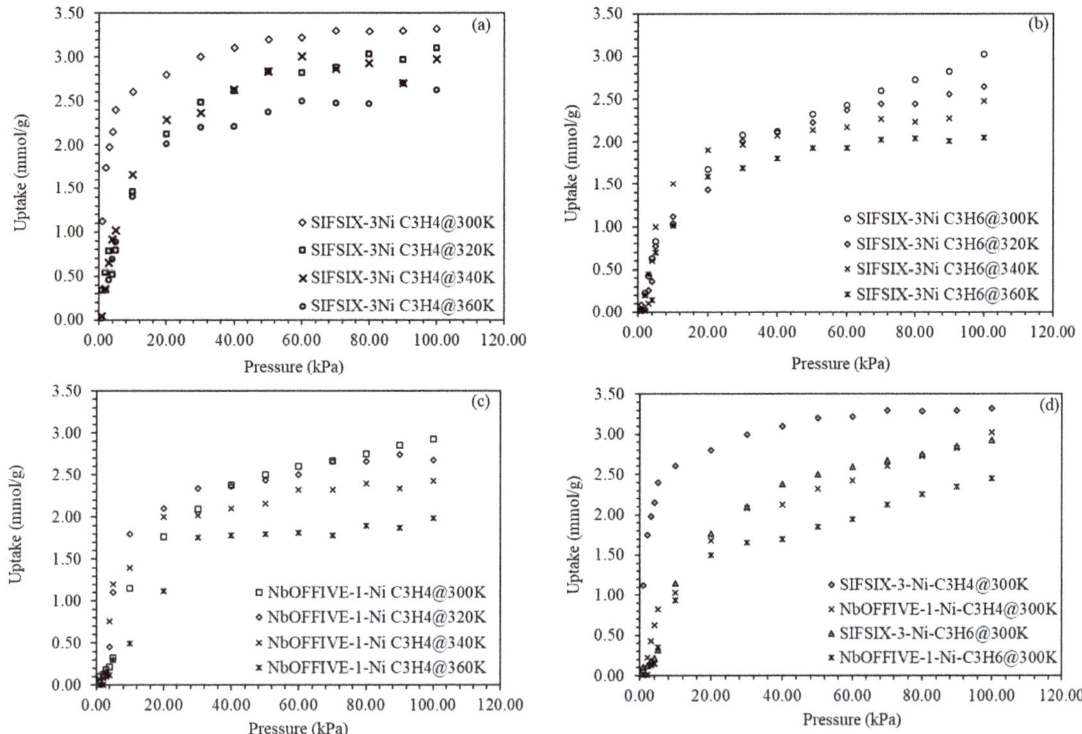

Figure 4. Full experimental isotherms at different temperatures (300–360 K) (**a**) C_3H_4 and SIFSIX-3-Ni (**b**) C_3H_6 and SIFSIX-3-Ni (**c**) C_3H_4 and NbOFFIVE-1-Ni (**d**) C_3H_6 and NbOFFIVE-1-Ni.

Table 2. Uptake adsorption values reported in the literature.

Material	Adsorption Uptake (mmol g^{-1})	Ref.
	C_3H_4 from C_3H_4/C_3H_6	
SIFSIX-3-Ni	3.2	This study
NbOFFIVE-1-Ni	2.9	This study
SIF-Six-2-Cu-i	1.73	[3]
SIFSIX-3-Ni	2.7	[3]
SIFSIX-1-Cu	0.19	[18]
SIFSIX-2-Cu-i	0.2	[18]
SIFSIX-3-Ni	2.65	[18]
[Cu(dhbc)$_2$(4,4′-bipy)]	0.25	[43]
NK-MOF-Ni	1.83	[3]
NK-MOF-Cu	1.76	[3]

4.3. IAST Selectivity and Isosteric Heats of Adsorption

Figure 5a shows the isosteric heats of sorption (Q_{st}) as a function of uptake amounts of propyne and propylene. The trends in Figure 5a clearly indicate that the Q_{st} of C_3H_4 is higher than that for C_3H_6 for both SIFSIX and NBOFFIVE metal organic frameworks. A higher Q_{st} value is indicative of strong affinity and interactions between the solid pore structure and the gas being adsorbed. At zero uptake, the maximum value of Q_{st} is attained.

For SIFSIX and C_3H_4 system, the Q_{st} zero was around 45 kJ Kmol^{-1} and for propylene system, the maximum value was around 30 kJ Kmol^{-1}. Similar trend was observed in the case of NbOFFIVE with the gas systems where isosteric heats of adsorption at zero uptakes were 38 and 30 kJ Kmol^{-1}, respectively, for propyne and propylene systems. In both cases, the values of Q_{st} at zero coverage were higher for C_3H_4 than for C_3H_6, indicating a stronger affinity and interactions between the C_3H_4 molecules and the MOF structures. In addition, the value for propyne for the SIFSIX is higher than that obtained for NbOFFIVE, also indicating the stronger affiliation and intermolecular interactions between the SIF and propyne gas. In addition, the general trend for all cases is that Q_{st} decreases gradually with the increase in uptake rate of the gases on the pore structure of the solids. The continuous decrease at a similar rate indicates the homogeneity of the pore environment. Typically, as Q_{st} increases with higher uptake rates, heterogeneity of the surface is usually present [10]. In addition, the differences between the propyne and propylene for both solids is a good indication of the separation possibilities of the two gases.

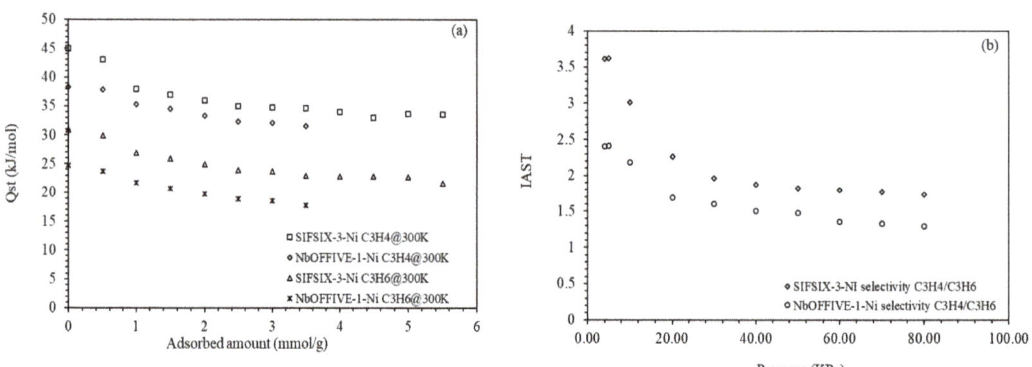

Figure 5. (**a**) Isosteric heats of sorption for SIFSIX and NbOFFIVE metal organic frameworks for both C_3H_4 and C_3H_6 as a function of adsorption uptake, (**b**) IAST selectivity for C_3H_4/C_3H_6 system on both adsorbents at 300 K.

To assess the separation ability of the two compounds, the selectivity (a key factor and application index for industry) of the ideal adsorbed solution theory (IAST) were calculated and are presented in Figure 5b for 10/90 v/v binary gas mixtures of C_3H_4 and C_3H_6 at 300 K. The IAST is typically employed, as the isotherms for binary systems of gases are difficult to measure. In the IAST estimations, it was assumed that the gas species had formed an idea mixture. This approximation has been used in many studies of metal organic frameworks and different gas species. Full details on the and mathematical treatment of Raoult's law are given elsewhere [10]. As indicated in Figure 5b, the selectivity of propyne over propylene for both metal organic frameworks was higher at lower pressure operations compared to pressures above 20 kPa. In some studies using different flexible structures, metal organic framework trends for propane/propylene adsorption, for example, showed higher selectivity at pressures above 60 kPa, indicating increased adsorption due to the "gate opening" effect of the MOF structures used [10]. The trend observed in this investigation led us to the conclusion that the structure of the used metal organic frameworks did not change with an increase of pressure in the range used in this study. In addition, the SIFSIX showed a higher selectivity between the two gases compared to those found for NbOFFIVE. As there is no published data on the selectivity of the gases and MOF combinations used, it is difficult to draw comparisons with other works. However, the trends reported here lean towards the conclusion of the size sieving effect and strong interactions between the propyne and SIFSIX at low pressure range. A Q_{st} value of 68 kJ mol^{-1} at 0.003 bar was reported by Yang et al. [18] for SIFSIX, and was attributed to the effect of single-molecule trap and the strong interaction between the MOF and the gas molecules.

4.4. Modeling of the Equilibrium and Kinetic Adsorption for C_3H_4/C_3H_6 on Metal Organic Frameworks

Experimental data were fitted to isotherms in an effort to help predict the general behavior of the adsorbent adsorbate interactions and uptake values. Isotherms were divided into one-parameter isotherms, such as Langmuir and Freundlich, or a multicontent hybrid combination of the two models such as the Sips and Toth models, as explained in an earlier section. The affinity of the system for adsorption was indicated by the value of the calculated constants. In the case of Langmuir and Freundlich, the linearized form of the Equations (1) and (3) is used to estimate the equilibrium adsorption uptake amount (q_{sat}) under a given pressure and temperature system. It is also related to the maximum adsorbed amount Q. The experimental results obtained for both metal organic frameworks and the two gases were analyzed by regression analyses, and all model fits are represented in Figure 6. In the Freundlich model, a value of $1/n$ close to zero indicates a heterogeneous surface, while a value of n greater than 1 indicates strong affinity [41]. From the trends shown in Figure 6a,b, it can be seen that the two linearized models did not represent the adsorption data well, as reflected by the lowered AARD (Table 3) for the range of pressures for the SIFSIX and propyne systems. Having said that, both the Freundlich and Langmuir models represented the data between the other adsorbent adsorbate combinations. This may have been due to stronger monolayer type interactions between the propyne and SIFSIX compared to the NbOFFIVE gas system. Even in the case of propylene and SIFSIX, both isotherms managed to represent the data in a more accurate manner. It can also be seen that both models underestimated the experimental uptake data in most cases. This was more apparent at higher pressure regions, especially in the Nb gas adsorbent-adsorbate systems. Apart from SIFSIX/propyne, the isotherm models represented the data at lower pressure ranges. The Sips (combined Langmuir and Freundlich isotherm) and the Toth model regression fitting are given in Figure 6c,d. Both models offered a better fit of the experimental data over the entire range of pressures. Again, it can be seen that in the case of the SIFSIX and propyne combination, the Sips model was less accurate compared to the Toth model.

In addition to equilibrium and selectivity studies, information related to the adsorption system kinetics is key in designing of adsorption systems. The most reliable method of addressing the transient uptake curves is to use micropores diffusion models that are based on Fick's law of diffusion (Equation (9)). The modeled data are given in Figure 7, where the fractional uptake ratio is expressed as a function of time (solid lines are model-estimated values). It can be seen that the model fit the adsorption uptake data well at all combinations of metal organic frameworks and gases, indicating a diffusion based process. It can also be seen that the model fits the propyne and SIFSIX combination best. This was expected, given the availability of passage through the SIFSIX structure due to the suitability of the kinetic size of both the adsorbent and adsorbate. From Equation (9) and upon fitting of the experimental data, the values of D_c/r_c^2 were calculated and are reported in Table 3. It can be seen from Figure 7 that the value of the time constant of micropore diffusion D_c/r_c^2 increased with increase in temperature. In addition, the propyne time constant was smaller than that of propylene for both adsorbents owing to the ease of diffusion due to the size acceptance of propyne compared to that of propylene.

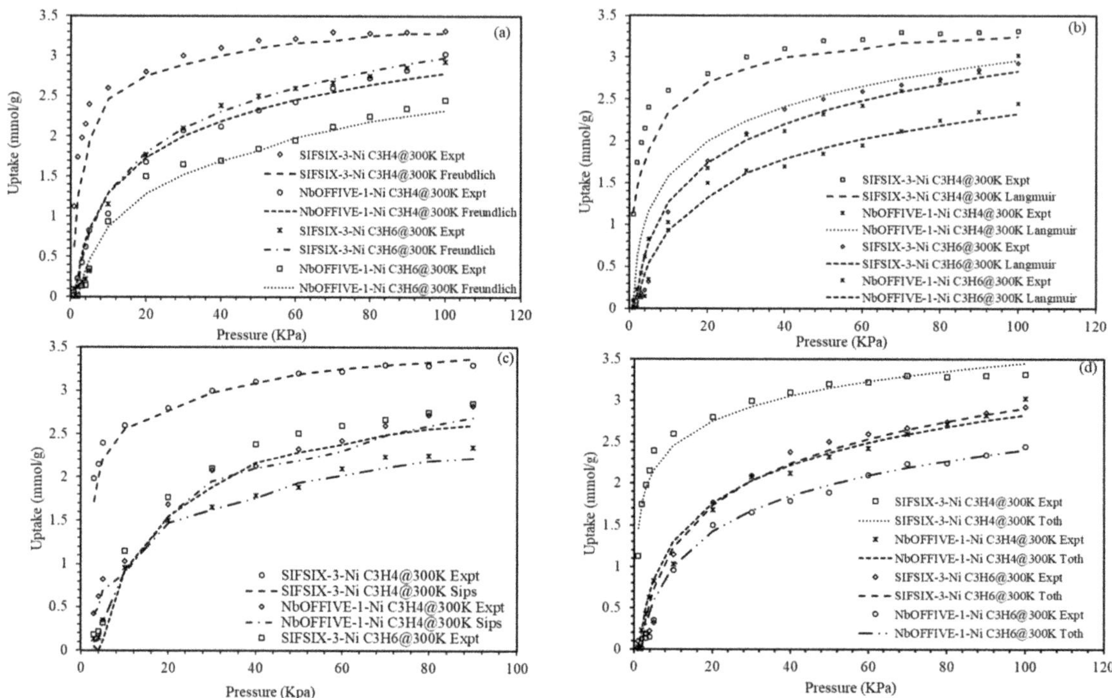

Figure 6. Fitted adsorption isotherms at temperatures of 300 K (**a**) Freundlich, (**b**) Langmuir, (**c**) Sips, (**d**) Toth.

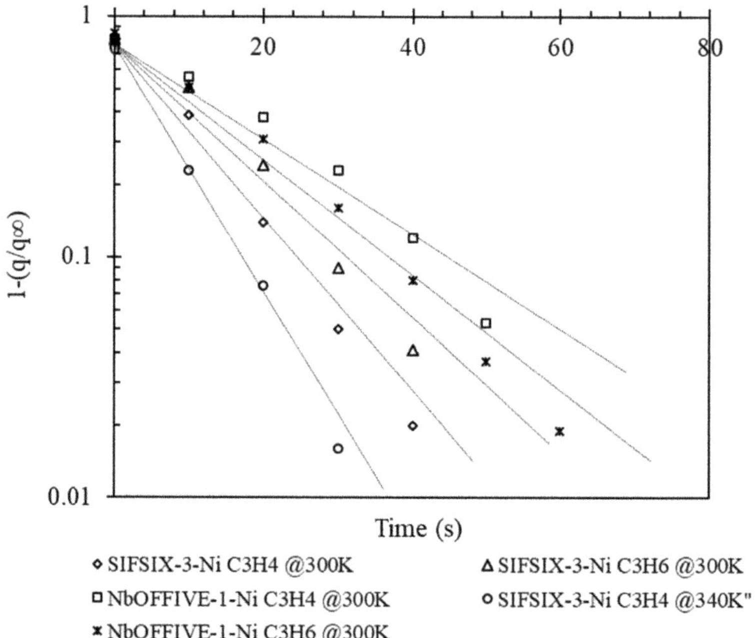

Figure 7. Fractional adsorption rate for C_3H_4/C_3H_6 on SIFSIX-3-Ni and NbOFFIVE-1-Ni.

4.5. Breakthrough and Cyclic Breakthrough Experiments

To confirm the validity of the kinetic effect and the ability and extent of the separation on the SIFSIX and NBOFFIVE metal organic frameworks, studies on binary mixtures of C_3H_4/C_3H_6 (10/90 v/v) were conducted. It can be clearly seen from the trends in Figure 8a that the heavier propylene molecules were eluted first from the column, in contrast to the lighter propyne molecules, upon adsorption on SIFSIS-3-Ni. Similar trends were found for the Nb gas combination (not shown). This also confirms earlier findings about the stronger bindings and affinities found between propyne and the SIFSIX adsorbent. The trends (Figure 8a) show that the breakthrough for propylene occurs after nearly 4–5 min and yields a 99.98% purity gas, while propyne was slowly eluted and becomes detectable in the outgas after nearly 16 min. This confirms the suitability of SIFSIX for the appropriate separation of C_3H_6/C_3H_4 systems. Accordingly, high purity propylene gas can be effectively separated and collected within the 10-min gap between the breakthrough of propyne and propylene from the column confirming the potential application of SIFSIX-3-Ni for the separation. Effective durability and recyclability are very important parameters in the selection and operation of adsorbent in gas separations. Eight cycles of adsorption and regeneration were conducted in order to examine the recyclability of the adsorbents and its effectiveness in repeated application. After each breakthrough test, the experimental breakthrough column was simply heated to around 423 K for 20 min [44] to allow the desorption of the gas adsorbed in the solid. Figure 8b shows the uptake ratios for eight consecutive cycles using the same experimental condition and adsorbent-adsorbate combinations. As can be seen, the adsorption capacity of the solids was not accepted with repeated use, confirming the suitability of solids for potential C_3H_4/C_3H_6 application.

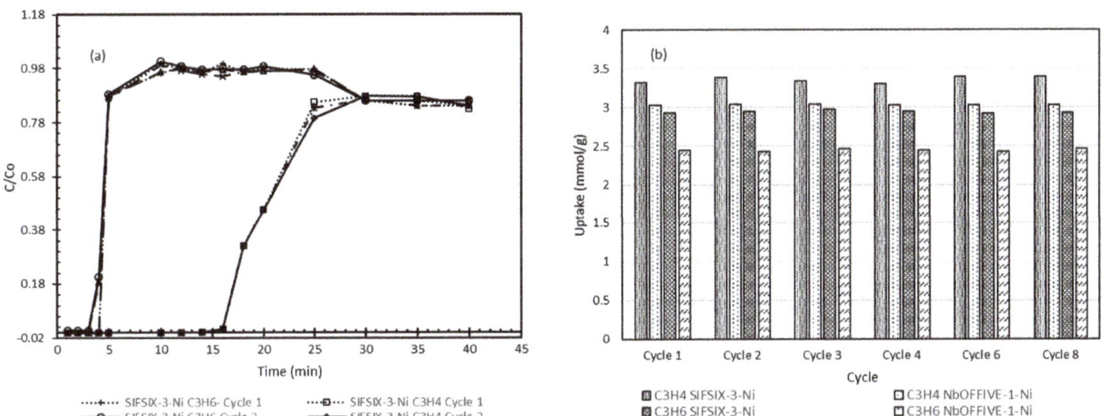

Figure 8. (**a**) Dynamic breakthrough curves for the SIFSIX and NbOFFIVE and C_3H_4/C_3H_6 systems at 300 K (4 cycles are shown); (**b**) uptake amounts of C_3H_4/C_3H_6 after 8 adsorption cycles at 300 K.

Table 3. Modeling isotherm statistical parameters, isosteric heats of adsorption and micropores diffusion time constants for MOFs and C_3H_4/C_3H_6 at 300 K.

	SIFSIX-3-Ni		NbOFFIVE-1-Ni	
	C_3H_4	C_3H_8	C_3H_4	C_3H_8
Langmuir				
q_{sat} (mmol/g)	3.32	2.93	3.03	2.45
k_l	0.23	0.21	0.034	0.0022
R_l	0.82	0.65	0.74	0.62
AARD (%)	13.7	12.2	10.3	9.7
Freundlich				
n	0.298	0.228	0.265	0.2007
k_f	0.342	0.432	0.665	0.453
AARD (%)	9.2	8.5	6.8	6.3
Sips				
q_{sat} (mmol/g)	0.047	0.837	1.179	1.231
K_s (mmol/gbar)	0.0087	0.0076	0.0061	0.0052
m	0.067	0.025	0.0289	0.088
AARD (%)	2.3	1.7	3.2	2.1
Toth				
q_{sat} (mmol/g)	4.09	3.98	3.54	2.48
k_t	0.042	0.076	0.0342	0.066
n	0.203	0.019	0.187	0.0172
AARD(%)	0.03	0.04	0.05	0.04
Q_{st} (J/mol)	45.0	38.3	30.8	24.7
D_c/r^2_c (s^{-1})	9.34×10^{-3}	5.23×10^{-3}	6.14×10^{-3}	4.12×10^{-3}

5. Conclusions

This investigation reports the effective separation of propyne (C_3H_4) and propylene (C_3H_6) using the fluorinated metal organic framework, SIFSIX-3-Ni. The SIFSIX MOF showed a better adsorption capacity compared to another pyrazine based MOF (NbOFFIVE-1-Ni) under the same experimental conditions. Characterization of the adsorbents showed developed micropores and a stable structure with a BET area of around 248 m^2 g^{-1}. The maximum uptake recorded for the SIFSIX was in excess of 3.2 mmol g^{-1} for C_3H_4 compared to 2.99 mmol g^{-1} for C_3H_6. Size sieving and thermodynamic interactions were thought to be the main separation mechanisms, as indicated by the high isosteric heat of adsorption towards propyne on SIFSIX. The selectivity of propyne over propylene on the used metal organic frameworks can be attributed to kinetic (size exclusion) and thermodynamic (pore and surface interactions) combination. Isotherm models fitted the Toth model well for all combinations of solid metal organic frameworks and gases at the full applied range of temperatures. The smaller kinetic diameter of propyne and the strong interactions make its adsorption easier on SIFSIX-3-Ni and facilitate its possible application in the separation of propyne and propylene binary mixtures. A 10-min time difference between the breakthrough of C_3H_6 and the lighter C_3H_4 was evident from the dynamic breakthrough curves, indicating great potential for the application of SISFIX-3-Ni for propyne/propylene separation.

Author Contributions: Conceptualization, M.K. and G.W.; methodology, M.K. and F.A.; software, M.K.; validation, M.K. and F.A.; formal analysis, M.K.; investigation, M.K.; data curation, M.K.;

writing—original draft preparation, M.K.; writing—review and editing, G.W. and F.A.; visualization, M.K.; supervision, M.K.; project administration, M.K. and F.A.; funding acquisition, M.K. and G.W. All authors have read and agreed to the published version of the manuscript.

Funding: Qatar National Research Fund under the National Priorities Research Program award number NPRP10-0107-170119.

Acknowledgments: The work was made possible by a grant from the Qatar National Research Fund under the National Priorities Research Program award number NPRP10-0107-170119. Its content are solely the responsibility of the authors and do not necessarily represent the official views of QNRF. The authors acknowledge the CLU unit at Qatar University for helping in sample analysis.

Conflicts of Interest: The authors declare no conflict of interest.

References

1.	Wang, Y.; Zhao, D. Beyond Equilibrium: Metal–Organic Frameworks for Molecular Sieving and Kinetic Gas Separation. *Cryst. Growth Des.* **2017**, *17*, 2291–2308. [CrossRef]
2.	Da Silva, F.A.; Rodrigues, A.E. Adsorption equilibria and kinetics for propylene and propane over 13X and 4A zeolite pellets. *Ind. Eng. Chem. Res.* **1999**, *38*, 2051–2057. [CrossRef]
3.	Li, L.; Guo, L.; Zheng, F.; Zhang, Z.; Yang, Q.; Yang, Y.; Ren, Q.; Bao, Z. Calcium-Based Metal–Organic Framework for Simultaneous Capture of Trace Propyne and Propadiene from Propylene. *ACS Appl. Mater. Interfaces* **2020**, *12*, 17147–17154. [CrossRef] [PubMed]
4.	Li, L.; Wen, H.-M.; He, C.; Lin, R.-B.; Krishna, R.; Wu, H.; Zhou, W.; Li, J.; Li, B.; Chen, B. A Metal–Organic Framework with Suitable Pore Size and Specific Functional Sites for the Removal of Trace Propyne from Propylene. *Angew. Chem.* **2018**, *130*, 15403–15408. [CrossRef]
5.	Cui, W.; Hu, T.; Bu, X. Metal–Organic Framework Materials for the Separation and Purification of Light Hydrocarbons. *Adv. Mater.* **2020**, *32*, e1806445. [CrossRef]
6.	Li, L.; Duan, Y.; Liao, S.; Ke, Q.; Qiao, Z.; Wei, Y. Adsorption and separation of propane/propylene on various ZIF-8 poly-morphs: Insights from GCMC simulations and the ideal adsorbed solution theory (IAST). *Chem. Eng. J.* **2020**, *386*, 123945. [CrossRef]
7.	Martins, V.F.D.; Ribeiro, A.M.; Plaza, M.G.; Santos, J.C.; Loureiro, J.M.; Ferreira, A.F.P. Gas-phase simulated moving bed: Pro-pane/propylene separation on 13X zeolite. *J. Chromatogr. A* **2015**, *1423*, 136–148. [CrossRef]
8.	Mundstock, A.; Wang, N.; Friebe, S.; Caro, J. Propane/propene permeation through Na-X membranes: The interplay of separa-tion performance and pre-synthetic support functionalization. *Microporous Mesoporous Mater.* **2015**, *215*, 20–28. [CrossRef]
9.	Kim, J.; Hong, S.H.; Park, D.; Chung, K.; Lee, C. Separation of propane and propylene by desorbent swing adsorption using zeolite 13X and carbon dioxide. *Chem. Eng. J.* **2021**, *410*, 128276. [CrossRef]
10.	Abedini, H.; Shariati, A.; Khosravi-Nikou, M.R. Adsorption of propane and propylene on M-MOF-74 (M = Cu, Co): Equilibrium and kinetic study. *Chem. Eng. Res. Des.* **2020**, *153*, 96–106. [CrossRef]
11.	Dobladez, J.A.D.; Maté, V.I.Á.; Torrellas, S.Á.; Larriba, M. Separation of the propane propylene mixture with high recovery by a dual PSA process. *Comput. Chem. Eng.* **2020**, *136*, 106717. [CrossRef]
12.	Grande, C.A.; Gascon, J.; Kapteijn, F.; Rodrigues, A.E. Propane/propylene separation with Li-exchanged zeolite 13X. *Chem. Eng. J.* **2010**, *160*, 207–214. [CrossRef]
13.	Jorge, M.; Lamia, N.; Rodrigues, A.E. Molecular simulation of propane/propylene separation on the metal–organic framework CuBTC. *Colloids Surf. A Physicochem. Eng. Asp.* **2010**, *357*, 27–34. [CrossRef]
14.	Lamia, N.; Wolff, L.; Leflaive, P.; Gomes, P.S.; Grande, C.A.; Rodrigues, A.E. Propane/Propylene Separation by Simulated Moving Bed I. Adsorption of Propane, Propylene and Isobutane in Pellets of 13X Zeolite. *Sep. Sci. Technol.* **2007**, *42*, 2539–2566. [CrossRef]
15.	Li, L.; Lin, R.-B.; Wang, X.; Zhou, W.; Jia, L.; Li, J. Kinetic separation of propylene over propane in a microporous metal-organic framework. *Chem. Eng. J.* **2018**, *354*, 977–982. [CrossRef]
16.	Plaza, M.; Ribeiro, A.; Ferreira, A.; Santos, J.; Lee, U.-H.; Chang, J.-S. Propylene/propane separation by vacuum swing adsorption using Cu-BTC spheres. *Sep. Purif. Technol.* **2012**, *90*, 109–119. [CrossRef]
17.	Wang, S.; Zhang, Y.; Tang, Y.; Wen, Y.; Lv, Z.; Liu, S.; Li, X.; Zhou, X. Propane-selective design of zirconium-based MOFs for propylene purification. *Chem. Eng. Sci.* **2020**, *219*. [CrossRef]
18.	Yang, L.; Cui, X.; Yang, Q.; Qian, S.; Wu, H.; Bao, Z. A Single-Molecule Propyne Trap: Highly Efficient Removal of Propyne from Propylene with Anion-Pillared Ultramicroporous Materials. *Adv. Mater.* **2018**, *30*, 1705374. [CrossRef]
19.	Yang, L.; Cui, X.; Zhang, Y.; Yang, Q.; Xing, H. A highly sensitive flexible metal–organic framework sets a new benchmark for separating propyne from propylene. *J. Mater. Chem. A* **2018**, *6*, 24452–24458. [CrossRef]
20.	Das, M.C.; Guo, Q.; He, Y.; Kim, J.; Zhao, C.-G.; Hong, K.; Xiang, S.; Zhang, Z.; Thomas, K.M.; Krishna, R.; et al. Interplay of Metalloligand and Organic Ligand to Tune Micropores within Isostructural Mixed-Metal Organic Frameworks (M′MOFs) for Their Highly Selective Separation of Chiral and Achiral Small Molecules. *J. Am. Chem. Soc.* **2012**, *134*, 8703–8710. [CrossRef]
21.	Li, L.; Lin, R.-B.; Krishna, R.; Wang, X.; Li, B.; Wu, H. Flexible–robust metal–organic framework for efficient removal of propyne from propylene. *J. Am. Chem. Soc.* **2017**, *139*, 7733–7736. [CrossRef] [PubMed]

22. Wang, X.; Krishna, R.; Li, L.; Wang, B.; He, T.; Zhang, Y.-Z.; Li, J.-R.; Li, J. Guest-dependent pressure induced gate-opening effect enables effective separation of propene and propane in a flexible MOF. *Chem. Eng. J.* **2018**, *346*, 489–496. [CrossRef]
23. Bulánek, R.; Koudelková, E.; de Oliveira Ramos, F.S.; Trachta, M.; Bludský, O.; Rubeš, M. Experimental and theoretical study of propene adsorption on alkali metal exchanged FER zeolites. *Microporous Mesoporous Mater.* **2019**, *280*, 203–210. [CrossRef]
24. Li, X.; Shen, W.; Zheng, A. The influence of acid strength and pore size effect on propene elimination reaction over zeolites: A theoretical study. *Microporous Mesoporous Mater.* **2019**, *278*, 121–129. [CrossRef]
25. Liu, G.; Labreche, Y.; Chernikova, V.; Shekhah, O.; Zhang, C.; Belmabkhout, Y.; Eddaoudi, M.; Koros, W.J. Zeolite-like MOF nanocrystals incorporated 6FDA-polyimide mixed-matrix membranes for CO_2/CH_4 separation. *J. Membr. Sci.* **2018**, *565*, 186–193. [CrossRef]
26. Yilmaz, G.; Keskin, S. Molecular modeling of MOF and ZIF-filled MMMs for CO_2/N_2 separations. *J. Membr. Sci.* **2014**, *454*, 407–417. [CrossRef]
27. Elsaidi, S.K.; Mohamed, M.H.; Banerjee, D.; Thallapally, P.K. Flexibility in Metal–Organic Frameworks: A fundamental understanding. *Coord. Chem. Rev.* **2018**, *358*, 125–152. [CrossRef]
28. Li, H.; Li, L.; Lin, R.-B.; Zhou, W.; Zhang, Z.; Xiang, S.; Chen, B. Porous metal-organic frameworks for gas storage and separation: Status and challenges. *EnergyChem* **2019**, *1*, 100006. [CrossRef]
29. Zhang, J.; Chen, Z. Metal-organic frameworks as stationary phase for application in chromatographic separation. *J. Chromatogr. A* **2017**, *1530*, 1–18. [CrossRef]
30. Khraisheh, M.; Mukherjee, S.; Kumar, A.; Al Momani, F.; Walker, G.; Zaworotko, M.J. An overview on trace CO_2 removal by advanced physisorbent materials. *J. Environ. Manag.* **2020**, *255*, 109874. [CrossRef]
31. Madden, D.; Albadarin, A.B.; O'Nolan, D.; Cronin, P.; Perry, I.V.J.J.; Solomon, S. Metal-Organic Material Polymer Coatings for Enhanced Gas Sorption Performance and Hydrolytic Stability Under Humid Conditions. *ACS Appl. Mater. Interfaces* **2020**. [CrossRef] [PubMed]
32. Madden, D.G.; O'Nolan, D.; Chen, K.-J.; Hua, C.; Kumar, A.; Pham, T.; Forrest, K.A.; Space, B.; Perry, J.J.; Khraisheh, M.; et al. Highly selective CO_2 removal for one-step liquefied natural gas processing by physisorbents. *Chem. Commun.* **2019**, *55*, 3219–3222. [CrossRef]
33. Cadiau, A.; Adil, K.; Bhatt, P.; Belmabkhout, Y.; Eddaoudi, M. A metal-organic framework–based splitter for separating propyl-ene from propane. *Science* **2016**, *353*, 137–140. [CrossRef]
34. Khraisheh, M.; Almomani, F.; Walker, G. Solid Sorbents as a Retrofit Technology for CO_2 Removal from Natural Gas Under High Pressure and Temperature Conditions. *Sci. Rep.* **2020**, *10*, 1–12. [CrossRef]
35. Ullah, S.; Bustam, M.A.; Al-Sehemi, A.G.; Assiri, M.A.; Abdul Kareem, F.A.; Mukhtar, A. Influence of post-synthetic graphene oxide (GO) functionalization on the selective CO_2/CH_4 adsorption behavior of MOF-200 at different temperatures; an experimental and adsorption isotherms study. *Microporous Mesoporous Mater.* **2020**, *296*, 110002. [CrossRef]
36. Ullah, S.; Bustam, M.A.; Assiri, M.A.; Al-Sehemi, A.G.; Gonfa, G.; Mukhtar, A.; Kareem, F.A.A.; Ayoub, M.; Saqib, S.; Mellon, N.B. Synthesis and characterization of mesoporous MOF UMCM-1 for CO_2/CH_4 adsorption; an experimental, isotherm modeling and thermodynamic study. *Microporous Mesoporous Mater.* **2020**, *294*, 109844. [CrossRef]
37. Weston, M.H.; Colón, Y.J.; Bae, Y.-S.; Garibay, S.J.; Snurr, R.Q.; Farha, O.K.; Hupp, J.T.; Nguyen, S.T. High propylene/propane adsorption selectivity in a copper(catecholate)-decorated porous organic polymer. *J. Mater. Chem. A* **2013**, *2*, 299–302. [CrossRef]
38. Kim, A.-R.; Yoon, T.-U.; Kim, E.-J.; Yoon, J.W.; Kim, S.-Y.; Yoon, J.W. Facile loading of Cu (I) in MIL-100 (Fe) through redox-active Fe (II) sites and remarkable propylene/propane separation performance. *Chem. Eng. J.* **2018**, *331*, 777–784. [CrossRef]
39. Bhatt, P.M.; Belmabkhout, Y.; Cadiau, A.; Adil, K.; Shekhah, O.; Shkurenko, A.; Barbour, L.J.; Eddaoudi, M. A Fine-Tuned Fluorinated MOF Addresses the Needs for Trace CO_2 Removal and Air Capture Using Physisorption. *J. Am. Chem. Soc.* **2016**, *138*, 9301–9307. [CrossRef]
40. Kumar, A.; Hua, C.; Madden, D.G.; O'Nolan, D.; Chen, K.-J.; Keane, L.-A.J.; Perry, J.J.; Zaworotko, M.J. Hybrid ultramicroporous materials (HUMs) with enhanced stability and trace carbon capture performance. *Chem. Commun.* **2017**, *53*, 5946–5949. [CrossRef]
41. Ullah, S.; Bustam, M.A.; Assiri, M.A.; Al-Sehemi, A.G.; Sagir, M.; Abdul Kareem, F.A. Synthesis, and characterization of metal-organic frameworks -177 for static and dynamic adsorption behavior of CO_2 and CH_4. *Microporous Mesoporous Mater.* **2019**, *288*, 109569. [CrossRef]
42. Lin, Z.-T.; Liu, Q.-Y.; Yang, L.; He, C.-T.; Li, L.; Wang, Y.-L. Fluorinated Biphenyldicarboxylate-Based Metal–Organic Framework Exhibiting Efficient Propyne/Propylene Separation. *Inorg. Chem.* **2020**, *59*, 4030–4036. [CrossRef]
43. Li, L.; Krishna, R.; Wang, Y.; Yang, J.; Wang, X.; Li, J. Exploiting the gate opening effect in a flexible MOF for selective adsorption of propyne from C1/C2/C3 hydrocarbons. *J. Mater. Chem. A* **2016**, *4*, 751–755. [CrossRef]
44. Wu, H.; Yuan, Y.; Chen, Y.; Xu, F.; Lv, D.; Wu, Y. Efficient adsorptive separation of propene over propane through a pillar-layer cobalt-based metal–organic framework. *AIChE J.* **2020**, *66*, e16858. [CrossRef]

Article

Metal(II) Coordination Polymers from Tetracarboxylate Linkers: Synthesis, Structures, and Catalytic Cyanosilylation of Benzaldehydes

Yu Li [1,†], Chumin Liang [1,†], Xunzhong Zou [1], Jinzhong Gu [2], Marina V. Kirillova [3] and Alexander M. Kirillov [3,4,*]

[1] Guangdong Research Center for Special Building Materials and Its Green Preparation Technology/Foshan Research Center for Special Functional Building Materials and Its Green Preparation Technology, Guangdong Industry Polytechnic, Guangzhou 510300, China; liyu@gdip.edu.cn (Y.L.); liangchumin@gdip.edu.cn (C.L.); zouxunzhong@gdip.edu.cn (X.Z.)

[2] College of Chemistry and Chemical Engineering, Lanzhou University, Lanzhou 730000, China; gujzh@lzu.edu.cn

[3] Centro de Química Estrutural and Departamento de Engenharia Química, Instituto Superior Técnico, Universidade de Lisboa, Av. Rovisco Pais, 1049-001 Lisbon, Portugal; kirillova@tecnico.ulisboa.pt

[4] Research Institute of Chemistry, Peoples' Friendship University of Russia (RUDN University), 6 Miklukho-Maklaya st., 117198 Moscow, Russia

* Correspondence: kirillov@tecnico.ulisboa.pt; Tel.: +351-218-419-396

† These authors contributed equally to this work.

Abstract: Three 2D coordination polymers, $[Cu_2(\mu_4\text{-dpa})(bipy)_2(H_2O)]_n \cdot 6nH_2O$ (**1**), $[Mn_2(\mu_6\text{-dpa})(bipy)_2]_n$ (**2**), and $[Zn_2(\mu_4\text{-dpa})(bipy)_2(H_2O)_2]_n \cdot 2nH_2O$ (**3**), were prepared by a hydrothermal method using metal(II) chloride salts, 3-(2′,4′-dicarboxylphenoxy)phthalic acid (H_4dpa) as a linker, as well as 2,2′-bipyridine (bipy) as a crystallization mediator. Compounds **1–3** were obtained as crystalline solids and fully characterized. The structures of **1–3** were established by single-crystal X-ray diffraction, revealing 2D metal-organic networks of **sql**, **3,6L66**, and **hcb** topological types. Thermal stability and catalytic behavior of **1–3** were also studied. In particular, zinc(II) coordination polymer **3** functions as a highly active and recoverable heterogeneous catalyst in the mild cyanosilylation of benzaldehydes with trimethylsilyl cyanide to give cyanohydrin derivatives. The influence of various parameters was investigated, including a time of reaction, a loading of catalyst and its recycling, an effect of solvent type, and a substrate scope. As a result, up to 93% product yields were attained in a catalyst recoverable and reusable system when exploring 4-nitrobenzaldehyde as a model substrate. This study contributes to widening the types of multifunctional polycarboxylic acid linkers for the design of novel coordination polymers with notable applications in heterogeneous catalysis.

Keywords: hydrothermal synthesis; coordination polymers; crystal structures; metal-organic frameworks; carboxylate ligands; heterogeneous catalysis

Citation: Li, Y.; Liang, C.; Zou, X.; Gu, J.; Kirillova, M.V.; Kirillov, A.M. Metal(II) Coordination Polymers from Tetracarboxylate Linkers: Synthesis, Structures, and Catalytic Cyanosilylation of Benzaldehydes. *Catalysts* **2021**, *11*, 204. https://doi.org/10.3390/catal11020204

Academic Editors: Jorge Bedia and Carolina Belver

Received: 26 December 2020
Accepted: 29 January 2021
Published: 3 February 2021

1. Introduction

Functional coordination polymers (CPs) and derived materials have been of a special focus in recent years owing to important structural characteristics of these compounds [1–3], intrinsic properties [4,5], and a broad diversity of applications [6–10] including in the field of catalysis [11–17]. The development of new catalytic systems incorporating coordination polymers with target structures and functionalities continues to be a challenging area, since the assembly of CPs can be affected by a diversity of factors. These include the nature of metal centers, organic linkers and supporting ligands, stoichiometry, and various reaction conditions [18–24].

Aromatic carboxylic acids with several COOH groups are the most common building blocks for constructing functional CPs [14,15,17,19]. Within a diversity of such carboxylic acids, semi-flexible linkers are especially captivating due to unique geometrical arrangements and conformational flexibility, which may lead to crystallization of unexpected metal-organic architectures [21–23,25].

On the other hand, cyanosilylation of carbonyl substrates is an interesting reaction for C-C bond formation [26,27], which is used for the preparation of cyanohydrins—key precursors for some pharmaceutical and fine chemistry products [28,29]. The use of transition metal complexes or coordination polymers in cyanosilylation reactions is gaining relevance as these compounds can behave as low-cost, efficient, and recyclable catalysts [25,30,31].

Considering our research focus on the design of CPs and catalytic systems on their basis [14,15,25], the main goal of this study consisted in the preparation of new metal-organic architectures, followed by their characterization and catalytic application in cyanosilylation of benzaldehydes. Thus, we selected an unexplored ether-linked tetracarboxylic acid, 3-(2′,4′-dicarboxylphenoxy)phthalic acid (H$_4$dpa, Scheme 1), and tested it as a principal building block for generating copper(II), manganese(II), and zinc(II) CPs. The use of H$_4$dpa as a linker can be justified by a number of relevant features of this tetracarboxylic acid. These features include (i) up to 9 possible sites for coordination (eight O-carboxylate sites and an O-ether site); (ii) two aromatic rings that can provide certain spatial flexibility and conformational adaptation owing to their separation by O-ether group; (iii) the fact that this carboxylic acid possesses good stability under hydrothermal conditions and remains little-explored in the synthesis of CPs.

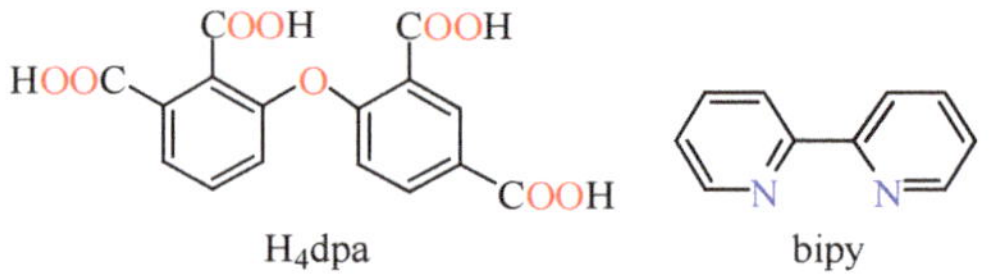

Scheme 1. Structural formulae of H$_4$dpa and bipy.

Thus, this work reports on the hydrothermal synthesis, characterization, thermal behavior, crystal structures, and catalytic application of 2D CPs derived from H$_4$dpa as a linker and bipy as a crystallization mediator. The obtained products [Cu$_2$(μ_4-dpa)(bipy)$_2$(H$_2$O)]$_n\cdot$6nH$_2$O (**1**), [Mn$_2$(μ_6-dpa)(bipy)$_2$]$_n$ (**2**), and [Zn$_2$(μ_4-dpa)(bipy)$_2$(H$_2$O)$_2$]$_n\cdot$2nH$_2$O (**3**) were also screened as potential catalysts for mild cyanosilylation of benzaldehydes into cyanohydrins. The influence of different reaction conditions and substrate scope was investigated, showing that the zinc(II) CP **3** is a particularly effective and recyclable heterogeneous catalyst.

2. Results and Discussion

2.1. Hydrothermal Synthesis

Aqueous medium mixtures composed of Cu(II), Mn(II), or Zn(II) chlorides with H$_4$dpa as a linker, NaOH as a base for deprotonation of carboxylic acid groups, and 2,2′-bipyridine as a mediator of crystallization were subjected to hydrothermal synthesis (3 days, 160 °C), resulting in the formation of three 2D coordination polymers as crystalline solids. These were formulated as [Cu$_2$(μ_4-dpa)(bipy)$_2$(H$_2$O)]$_n\cdot$6nH$_2$O (**1**), [Mn$_2$(μ_6-dpa)(bipy)$_2$]$_n$ (**2**), and [Zn$_2$(μ_4-dpa)(bipy)$_2$(H$_2$O)$_2$]$_n\cdot$2nH$_2$O (**3**) on the basis of standard solid-state characterization methods, namely infrared spectroscopy (IR), elemental analysis (EA), thermogravimetric analysis (TGA), powder (PXRD) and single-crystal X-ray diffraction. All compounds represent 2D metal-organic networks that are driven by ether-bridged tetracarboxylate nodes that show two distinct coordination modes, namely μ_4-dpa^{4-} in **1** and **3**, or μ_6-dpa^{4-} in **2** (Scheme 2).

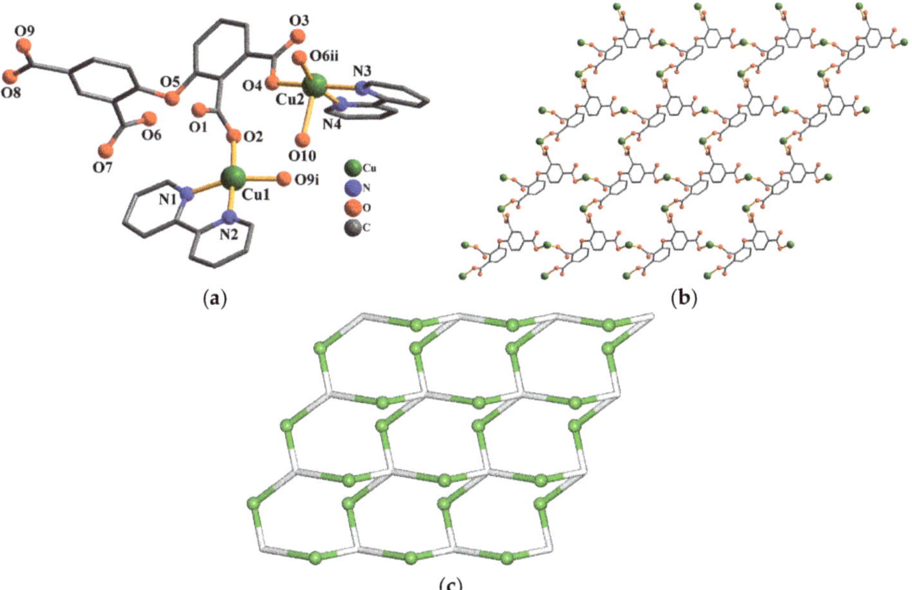

Scheme 2. Coordination modes of μ_4- or μ_6-dpa^{4-} linkers in **1–3**.

2.2. Structure of $[Cu_2(\mu_4\text{-}dpa)(bipy)_2(H_2O)]_n \cdot 6H_2O$ (**1**)

The structure of a 2D CP **1** (Figure 1) comprises two Cu(II) centers (Cu1, Cu2), one μ_4-dpa^{4-} linker, two bipy moieties, and one terminal H_2O ligand per asymmetric unit. The Cu1 center is 4-coordinated and exhibits a distorted $\{CuN_2O_2\}$ seesaw geometry. It is formed by two carboxylate O atoms from a pair of μ_4-dpa^{4-} ligands and two bipy N atoms (Figure 1a). The Cu2 atom is five-coordinated and features a distorted $\{CuN_2O_3\}$ square-pyramidal geometry. It is constructed from two oxygen atoms from two μ_4-dpa^{4-} blocks, two bipy N atoms, and a terminal H_2O ligand.

Figure 1. Structural fragments of **1**. (**a**) Coordination environment around Cu(II) atoms; H atoms are omitted for clarity. Symmetry codes: i = x + 1, y, z + 1; ii = x + 1, y, z. (**b**) 2D metal-organic layer; view along the b axis. (**c**) Topological view of 2D layer with a **sql** topology; view along the b axis; Cu centers (green balls), centroids of μ_4-dpa^{4-} nodes (gray).

The Cu–O [1.942(2)–2.345(3) Å] and Cu–N [1.992(3)–2.058(3) Å] bonds are typical for such a type of compounds [14,18,30]. The dpa^{4-} ligand behaves as a μ_4-linker with carboxylate moieties being monodentate (Scheme 2, mode I). In μ_4-dpa^{4-}, a dihedral angle between aromatic cycles and a C_{ar}–O_{ether}–C_{ar} angle are 86.34 and 115.93°, correspondingly. The μ_4-dpa^{4-} linkers connect four Cu atoms to assemble a 2D metal-organic layer (Figure 2b) which, after simplification, is described as a mononodal 4-linked net. It possesses a **sql** (Shubnikov tetragonal plane net) topology with a $(4^4.6^2)$ point symbol (Figure 1c) [32,33].

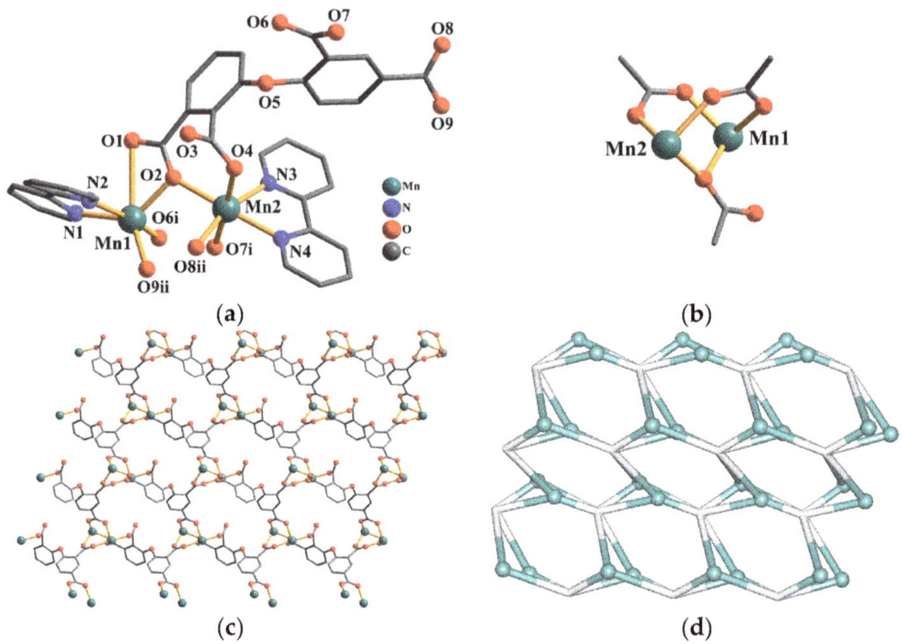

Figure 2. Structural fragments of **2**. (**a**) Coordination environment around Mn(II) atoms; H atoms are omitted for clarity. Symmetry codes: i = $x + 1, y, z$; ii = $x + 1, -y + 3/2, z + 1/2$. (**b**) Mn$_2$ subunit. (**c**) 2D metal-organic layer; view along the b axis. (**d**) Topological representation of 2D layer with a **3,6L66** topology; view along the b axis; Mn(II) nodes (turquoise balls), centroids of μ_6-dpa^{4-} nodes (gray).

2.3. Structure of $[Mn_2(\mu_6\text{-}dpa)(bipy)_2]_n$ (**2**)

The structure of **2** also shows a 2D coordination polymer network (Figure 2). Per asymmetric unit, there are two manganese(II) atoms, one μ_6-dpa^{4-} block, and two bipy ligands. The Mn1 center is five-coordinated and shows a distorted {MnN$_2$O$_3$} trigonal-bipyramidal environment. It is built from three oxygen atoms from three μ_6-dpa^{4-} linkers and a pair of bipy N donors (Figure 2a). The six-coordinate Mn2 center is bound by four oxygen atoms from three μ_6-dpa^{4-} ligands and a pair of bipy N atoms, thus creating a distorted {MnN$_2$O$_4$} octahedral geometry. The Mn–O [2.087(3)–2.217(4) Å] and Mn–N [2.226(4)–2.277(4) Å] bonds agree with those in related compounds [19,21,22]. The tetracarboxylate dpa^{4-} moiety behaves as a μ_6-linker (Scheme 2, mode II) with the carboxylate functionalities being monodentate or μ-bridging bidentate. In μ_6-dpa^{4-}, the relevant angles are 79.83° (dihedral angle among aromatic moieties) and 116.37° (C$_{ar}$–O$_{ether}$–C$_{ar}$). The manganese(II) centers are held together by three carboxylate groups from three μ_6-dpa^{4-} blocks, thus generating a dimanganese(II) subunit [Mn1·Mn2 3.4679(6) Å] (Figure 2b). Such Mn$_2$ subunits are additionally connected by carboxylate groups of μ_6-dpa^{4-} to form a 2D metal-organic layer (Figure 2c). Regarding topology, this 2D layer is built from 3-linked Mn1/Mn2 nodes as well as the 6-linked μ_6-dpa^{4-} nodes (Figure 2d). The resulting net is thus classified as a binodal 3,6-linked layers of a **3,6L66** topological type [34]. It is described by a $(4^3.6^{12})(4^3)_2$ point symbol, wherein the $(4^3.6^{12})$ and (4^3) indices correspond to the Mn(II) and μ_6-dpa^{4-} nodes, respectively.

2.4. Structure of $[Zn_2(\mu_4\text{-}dpa)(bipy)_2(H_2O)_2]_n \cdot 2H_2O$ (**3**)

In the structure of a 2D coordination polymer **3** (Figure 3), there are two zinc(II) atoms (Zn1, Zn2), one μ_4-dpa^{4-} block, two bipy moieties, two terminal water ligands and a couple of lattice H$_2$O molecules. Both Zn atoms are five-coordinated, showing distorted {ZnN$_2$O$_3$} square pyramidal geometries. These are constructed from two O donors from a pair of μ_4-dpa^{4-} linkers, a water ligand, and a pair of bipy nitrogen atoms. The Zn–O [1.973(3)–2.093(4) Å] and Zn–N [2.119(3)–2.181(3) Å] bond lengths are typical for compounds having an O–Zn–N environment [2,30,35,36]. The dpa^{4-} block adopts a μ_4-coordination fashion (Scheme 2, mode I). In μ_4-dpa^{4-}, the relevant angles are 72.59° (dihedral angle among two aromatic moieties) and 117.72° (C$_{ar}$–O$_{ether}$–C$_{ar}$). The μ_4-dpa^{4-} blocks interconnect four Zn(II) centers to furnish a 2D layer (Figure 3b). Its topological classification discloses an uninodal 3-linked layer with an **hcb** (Shubnikov hexagonal plane net/(6, 3)) topology and a (6^3) point symbol (Figure 3c) [37,38].

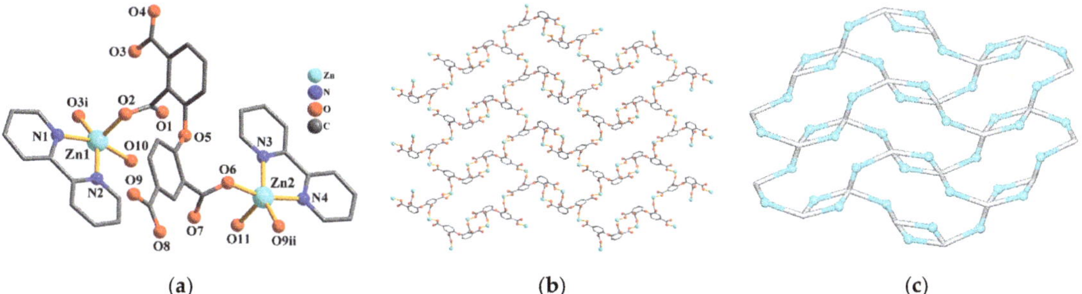

Figure 3. Structural fragments of **3**. (**a**) Coordination environment around Zn(II) atoms; H atoms are omitted for clarity. Symmetry codes: i = −x, −y + 1, −z; ii = −x + 1/2, y − 1/2, −z + 1/2. (**b**) 2D metal-organic layer; view along the *a* axis. (**c**) Topological representation of 2D layer with an **hcb** topology; view along the *a* axis; Zn(II) centers (cyan balls), centroids of μ_4-dpa^{4-} nodes (gray).

2.5. TGA & PXRD

Thermal behavior of compounds **1–3** was investigated by TGA on gradual heating from 30 to 800 °C under nitrogen flow (Figure S2). CP **1** reveals a loss of six lattice and one coordinated H$_2$O molecules in the 33–104 °C interval (calcd. 13.8%; exptl. 13.5%). After dehydration, the network of **1** maintains its stability until 220 °C. CP **2** does not encompass water as a lattice solvent or ligand and shows thermal stability up to 338 °C. For CP **3**, a mass loss in the 86–171 °C window refers to an elimination of two crystallization and two coordinated H$_2$O moieties (calcd. 8.4%; exptl. 8.6%); the obtained sample maintains stability until 218 °C.

To confirm a phase purity, the microcrystalline powders of **1–3** were analyzed by PXRD (Figure S2). The obtained experimental plots well agree with the patterns calculated from single-crystal X-ray diffraction data, what confirms a purity of the bulk solids **1–3** prepared via hydrothermal synthetic procedure.

2.6. Cyanosilylation of Benzaldehydes

Coordination polymers **1–3** were tested as catalysts in the mild, heterogeneous cyanosilylation of benzaldehyde substrates with TMSCN (trimethylsilyl cyanide) [33,35]. 4-Nitrobenzaldehyde was used as model substrate and converted into the corresponding cyanohydrin product, 2-(4-nitrophenyl)-2-[(trimethylsilyl)oxy]acetonitrile (Table 1, Scheme 3). An influence of catalyst loading, solvent type, reaction time, catalyst recycling and substrate scope was investigated.

The cyanosilylation reaction almost does not undergo when catalyst is absent or when using a carboxylic acid ligand or a metal salt precursor as potential catalysts (3−8% yields; Table 1, entries 1−3). The product yields are higher when using **1** (37%, entry 4) and **2** (27%, entry 5) as catalysts, and significantly higher in the presence of **3** (92%, entry 6). Although it is not possible to establish a clear relationship between structural features and catalytic activity of the obtained 2D compounds, we can speculate that a superior activity exhibited by the coordination polymer **3** may eventually be related to the presence of accessible and unsaturated zinc(II) centers on a surface of catalyst particles, together with a higher Lewis acidity of the zinc sites [30,39]. Given a superior activity of zinc(II) derivative **3**, the reaction catalyzed by this CP was optimized further.

Table 1. Cyanosilylation of 4-nitrobenzaldehyde with TMSCN catalyzed by coordination polymers. [a]

Entry	Catalyst	Time, h	Catalyst Loading, mol%	Solvent	Yield [b], %
1	blank	12	-	CH_2Cl_2	3
2	H_3dpna	12	3.0	CH_2Cl_2	6
3	$ZnCl_2$	12	3.0	CH_2Cl_2	8
4	**1**	12	3.0	CH_2Cl_2	37
5	**2**	12	3.0	CH_2Cl_2	27
6	**3**	12	3.0	CH_2Cl_2	92
7	**3**	1	3.0	CH_2Cl_2	40
8	**3**	2	3.0	CH_2Cl_2	62
9	**3**	4	3.0	CH_2Cl_2	73
10	**3**	6	3.0	CH_2Cl_2	80
11	**3**	8	3.0	CH_2Cl_2	84
12	**3**	10	3.0	CH_2Cl_2	88
13	**3**	12	2.0	CH_2Cl_2	73
14	**3**	12	4.0	CH_2Cl_2	93
15	**3**	12	3.0	$CHCl_3$	82
16	**3**	12	3.0	CH_3CN	76
17	**3**	12	3.0	THF	68
18	**3**	12	3.0	CH_3OH	80

[a] Conditions: 4-nitrobenzaldehyde (0.5 mmol), TMSCN (1.0 mmol), solvent (2.5 mL), room temperature (~25 °C).
[b] Yields based on ^{1}H NMR (nuclear magnetic resonance) analysis: (moles of product per mol of benzaldehyde substrate) × 100%.

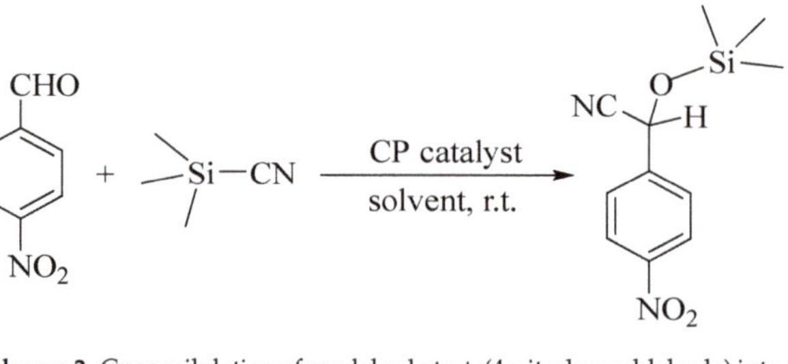

Scheme 3. Cyanosilylation of model substrate (4-nitrobenzaldehyde) into the corresponding cyanohydrin.

We found that there is an yield growth from 40 to 92% on increasing the time of reaction in the 1–12 h interval (Table 1, entries 6–12; Figure S7). The catalyst **3** loading of 2, 3, or 4 mol% has also a notable effect as attested by an yield increase from 73 to 92 and 93%, respectively (entries 6, 13, 14). Although dichloromethane (CH_2Cl_2) was used as a standard solvent to achieve the highest yield (92%), the cyanosilylation reaction also undergoes quite effectively in alternative solvents such as chloroform ($CHCl_3$, 82%), tetrahydrofuran (THF, 68%), acetonitrile (CH_3CN, 76%), and methanol (CH_3OH, 80%) (Table 1, entries 15–18).

Furthermore, in the reactions catalyzed by **3**, substrate scope was studied using different functionalized benzaldehydes (Table 2). The corresponding cyanohydrin products are obtained in yields ranging from 51 to 87%. The substrates with an electron-withdrawing group (R = NO_2, Cl) generally show a higher reactivity (Table 2, entries 2–5), which can be explained by an increased substrate electrophilicity. As expected, benzaldehyde (R = H) and substrates with an electron-donating group (R = OH, CH_3) reveal lower product yields (entries 6, 7).

Table 2. Substrate scope in cyanosilylation of various benzaldehydes with TMSCN catalyzed by **3**. [a]

Entry	Substrate (R-C$_6$H$_4$CHO)	Yield [b], %
1	R = H	61
2	R = 2-NO$_2$	82
3	R = 3-NO$_2$	87
4	R = 4-NO$_2$	92
5	R = 4-Cl	62
6	R = 4-OH	56
7	R = 4-CH$_3$	51

[a] Conditions: functionalized benzaldehyde (0.5 mmol), TMSCN (1.0 mmol), catalyst **3** (3.0 mol.%), CH$_2$Cl$_2$ (2.5 mL), room temperature (~25 °C). [b] Yields based on ^{1}H NMR analysis: (moles of product per mol of functionalized benzaldehyde substrate) × 100%.

Given a remarkable activity of CP **3** in these cyanosilylation reactions, the stability of the catalyst and its possible recycling were studied. Thus, in the end of the cyanosilylation of 4-nitrobenzaldehyde (conditions of entry 6, Table 1), the catalyst was isolated via centrifugation, washed by CH$_2$Cl$_2$, air-dried, and reused in the next reaction runs. These tests indicate that the coordination polymer **3** behaves as a stable catalyst which maintains its activity for a minimum of 4 cyanosilylation cycles, as attested by resembling product yields in the 92−87% range (Figure S8). In addition, PXRD data of the used catalyst (Figure S9) show that its metal-organic structure remains stable.

3. Experimental

3.1. Chemicals & Equipment

All chemicals were obtained from commercial sources. In particular, 3-(2′,4′-Dicarboxylphenoxy)phthalic acid (H$_4$dpa) was acquired from Jinan Henghua Sci. & Tec. Co., Ltd. (Jinan, China). Infrared (IR) spectroscopy measurements (KBr discs) were performed on a Bruker EQUINOX 55 spectrometer (Bruker Corporation, Billerica, MA, USA). Elemental analyses (C, H, N) were carried out on an Elementar Vario EL device (Elementar, Langenselbold, Germany). LINSEIS STA PT1600 thermal analyzer (Linseis Messgeräte GmbH, Selb, Germany) was used for TGA (thermogravimetric analysis) measurements under N$_2$ atmosphere at 10 °C/min heating rate. Excitation/emission spectroscopic data were obtained using an Edinburgh FLS920 fluorescence spectrometer (Edinburgh Instruments, Edinburgh, England). Rigaku-Dmax 2400 diffractometer (Cu-Kα radiation; λ = 1.54060 Å; Rigaku Corporation, Tokyo, Japan) was used to obtain powder X-ray diffraction (PXRD) patterns. Solution ^{1}H NMR (nuclear magnetic resonance) spectra were measured on a JNM ECS 400M spectrometer (JEOL Ltd., Tokyo, Japan).

3.2. Hydrothermal Synthesis & Analytical Data

In a general procedure, metal(II) chloride (0.2 mmol: CuCl$_2$·2H$_2$O (34.1 mg) for **1**, MnCl$_2$·4H$_2$O (39.6 mg) for **2**, or ZnCl$_2$ (27.3 mg) for **3**), H$_4$dpa (0.1 mmol, 34.6 mg), bipy (0.2 mmol, 31.2 mg), NaOH (0.4 mmol, 16.0 mg), and H$_2$O (10 mL) were added into a Teflon-lined stainless steel vessel (volume: 25 mL) and stirred for 15 min at ambient temperature. Then, the vessel was closed and kept in an oven at 160 °C. After 3 days at this temperature, the vessel was gradually (10 °C/h) cooled down to ambient temperature. The reaction mixture was then transferred to a glass flask and the crystals of products were decanted or filtered off, followed by washing with H_2O and drying in air to produce compounds **1–3**.

[Cu$_2$(μ_4-dpa)(bipy)$_2$(H$_2$O)]$_n$·6nH$_2$O (1). Blue block-shaped crystals, yield: 55% based on H_4dpa. Calcd for $C_{36}H_{36}Cu_2N_4O_{16}$: C 47.63, H 4.00, N 6.17%. Found: C 47.86, H 3.98, N 6.21%. IR (KBr, cm^{-1}): 3748 w, 3082 w, 1599 s, 1568 s, 1497 w, 1475 w, 1448 m, 1368 s, 1288 w, 1240 m, 1164 w, 1137 w, 1084 w, 1062 w, 1032 w, 974 w, 921w, 854 w, 823 w, 770 m, 735 w, 663 w.

[Mn$_2$(μ_6-dpa)(bipy)$_2$]$_n$ (2). Yellow block-shaped crystals, yield: 52% based on H_4dpa. Calcd for $C_{36}H_{22}Mn_2N_4O_9$: C 56.56, H 2.90, N 7.33%. Found: C 56.37, H 2.89, N 7.38%. IR (KBr, cm^{-1}): 1639 s, 1598 s, 1569 m, 1470 w, 1438 w, 1376 s, 1240 m, 1182 w, 1157 w,1128 w, 1087 w, 1063 w, 1013 w, 968 w, 923 w, 848 w, 820 w, 770 m, 736 w, 696 w, 675 w, 646 w.

[Zn$_2$(μ_4-dpa)(bipy)$_2$(H$_2$O)$_2$]$_n$·2nH$_2$O (3). Colorless block-shaped crystals, yield: 48% based on H_4dpa. Calcd for $C_{36}H_{30}Zn_2N_4O_{13}$: C 50.43, H 3.53, N 6.53%. Found: C 50.64, H 3.55, N 6.51%. IR (KBr, cm^{-1}): 3383 w, 3106 w, 1614 s, 1569 s, 1470 m, 1438 s, 1392 s, 1322 w, 1244 w, 1157 w, 1083 w, 1063 w, 1021 w, 972 w, 918 w, 877 w, 766 m, 733 w, 687 w, 650 w.

3.3. Single-Crystal X-ray Diffraction

For CPs **1–3**, data were collected on a Bruker APEX-II CCD diffractometer (Bruker, Karlsruhe, Germany) (graphite-monochromated CuK_α radiation, λ = 1.54178 Å). SADABS was used for semi-empirical absorption corrections. Crystal structures were determined by direct methods, followed by the refinement (full-matrix least-squares on F^2) with SHELXS-97 and SHELXL-97 [40]. With an exception of hydrogen atoms, all other atoms were subjected to an anisotropic refinement (full-matrix least-squares on F^2). Except hydrogen atoms in water, all H atoms were placed in calculated positions (fixed isotropic thermal parameters); these were taken into account at the final stage of full-matrix least-squares refinement in structure factor calculations. Water hydrogen atoms were found by difference maps and constrained to the respective oxygen centers. In **1**, some molecules of solvent are heavily disordered and thus were detached by applying SQUEEZE in PLATON [41]. Elemental and thermogravimetric analyses confirmed the number of crystallization H_2O molecules. For **1–3**, final crystal data are summarized in Table 3. Representative bond distances and angles (Table S1) as well as hydrogen bond parameters (Table S2) are listed in Supplementary Materials. CCDC-2043757–2043759 contain crystallographic data for **1–3**.

For topological analysis of 2D layers in **1–3**, an underlying net concept was followed [42]. Such simplified networks were constructed by reducing bridging ligands to their centroids and removing terminal ligands, preserving a ligand–metal center connectivity.

3.4. Catalytic Cyanosilylation

Typical reaction mixtures were prepared as follows: in a small vessel, solid catalyst (typically 3 mol%) was suspended in dichloromethane (2.5 mL) and then an aldehyde substrate (0.50 mmol) and a cyanosilylation agent trimethylsilyl cyanide (1.0 mmol) were added. The reaction was kept under stirring at room temperature (~25 °C) for the desired time. Then, the catalyst was separated by centrifugation and the filtrate was subjected to solvent evaporation under reduced pressure to form a crude solid. This solid was dissolved in CDCl$_3$ and analyzed by ^{1}H NMR spectroscopy for product quantification (for details, see Supplementary Materials, Figures S5 and S6). In the catalyst recycling experiments, the catalyst was centrifuged, washed with CH_2Cl_2, dried at room temperature, and reused in subsequent steps that were done as described above. Blank tests without any catalyst or using metal salt or H_4dpa as catalyst were also carried out for comparative purposes.

Table 3. Crystal data for CPs **1–3**.

Compound	1	2	3
Chemical formula	$C_{36}H_{36}Cu_2N_4O_{16}$	$C_{36}H_{22}Mn_2N_4O_9$	$C_{36}H_{30}Zn_2N_4O_{13}$
Molecular weight	907.78	764.45	857.38
Crystal system	Monoclinic	Monoclinic	Monoclinic
Space group	$P2_1/n$	$P2_1/c$	$P2_1/n$
$a/Å$	10.3607 (5)	10.5312 (3)	7.9352 (2)
$b/Å$	35.8638 (11)	21.2195 (5)	13.1264 (4)
$c/Å$	11.3823 (5)	15.1835 (4)	34.5048 (10)
$\alpha/(°)$	90	90	90
$\beta/(°)$	116.385 (6)	107.751 (3)	94.158 (3)
$\gamma/(°)$	90	90	90
$V/Å$	3788.8 (3)	3231.47 (17)	3584.58 (18)
Z	4	4	4
$F\,(000)$	1624	1552	1752
Crystal size/mm	$0.25 \times 0.23 \times 0.21$	$0.18 \times 0.16 \times 0.15$	$0.26 \times 0.24 \times 0.20$
θ range for data collection	4.508–69.993	3.699–66.992	3.604–70.060
Limiting indices	$-12 \leq h \leq 12,\ -43 \leq k \leq 30,\ -13 \leq l \leq 13$	$-12 \leq h \leq 11,\ -25 \leq k \leq 25,\ -18 \leq l \leq 17$	$-8 \leq h \leq 9,\ -1 \leq 15,\ -41 \leq l \leq 40$
Reflections collected/unique (R_{int})	19,512/7079 (0.0494)	20,432/5753 (0.0993)	13,030/6658 (0.0337)
$D_c/(Mg \cdot cm^{-3})$	1.402	1.571	1.589
μ/mm^{-1}	1.892	6.917	2.285
Data/restraints/parameters	7079/0/469	5753/0/460	6658/0/496
Goodness-of-fit on F^2	1.047	1.074	1.017
Final R indices [$(I \geq 2\sigma(I))$] R_1, wR_2	0.0537, 0.1444	0.0707, 0.1680	0.0515, 0.1169
R indices (all data) R_1, wR_2	0.0710, 0.1546	0.0901, 0.1800	0.0619, 0.1238
Largest diff. peak and hole/(e·$Å^{-3}$)	1.195 and -0.648	0.859 and -1.089	0.857 and -1.096

4. Conclusions

In this study, we explored a still poorly studied tetracaboxylic acid, 3-(2′,4′-dicarboxyl phenoxy)phthalic acid (H_4dpa), as a multifunctional linker for the hydrothermal synthesis of new CPs. Hence, a simple synthetic procedure led to the formation of a series of copper(II), manganese(II), and zinc(II) coordination polymers. These compounds feature distinct types of 2D metal-organic layers driven by the μ_4-dpa^{4-} or μ_6-dpa^{4-} linkers, as well as different topologies ranging from **sql** (**1**) and **3,6L66** (**2**) to **hcb** (**3**).

Catalytic activity of **1–3** was also explored in the mild cyanosilylation of benzaldehyde substrates with TMSCN, showing that the zinc(II) CP **3** functions as an effective and reusable heterogeneous catalyst to give cyanohydrin products in up to 93% yields. The effects of various reactions parameters as well as substrate scope were studied.

As main novelty features of this work, we can highlight (i) an application of H_4dpa as an underexplored multifunctional linker for design of new CPs, (ii) a synthesis of three structurally and topologically distinct metal-organic architectures, and (iii) notable catalytic activity of a zinc(II) derivative in the cyanosilylation reactions. Besides, the obtained results widen the applications of coordination polymers in heterogeneous catalysis [43,44]. Further research on assembling related types of CPs and exploring their potential in other catalytic transformations are currently underway.

Supplementary Materials: The following data are available online at https://www.mdpi.com/article/10.3390/catal11020204/s1. Figure S1: FT-IR spectra, Figure S2: TGA curves, Figures S3 and S9: PXRD patterns, Figure S4: emission spectra, Figures S5–S8: supplementary catalysis data, Tables S1 and S2: selected structural parameters for 1–3. CCDC-2043757–2043759.

Author Contributions: Conceptualization, J.G., X.Z. and A.M.K.; data curation, C.L., Y.L.; funding acquisition, Y.L., J.G. and A.M.K.; investigation, C.L., X.Z., Y.L., J.G. and M.V.K.; methodology, J.G.; visualization, J.G. and M.V.K.; writing—original draft, Y.L., J.G., X.Z., M.V.K. and A.M.K.; writing—review and editing, A.M.K. All authors have read and agreed to the published version of the manuscript.

Funding: This work has been supported by Science and Technology Planning Project of Guangzhou (201904010381), the Foundation for Science and Technology (FCT) and Portugal 2020 (projects CEECIND/03708/2017, LISBOA-01-0145-FEDER-029697, PTDC/QUI-QIN/3898/2020, and UID/QUI/00100/2013). This paper has been supported by the RUDN University Strategic Academic Leadership Program.

Conflicts of Interest: The authors declare no conflict of interest.

References

1. Kaskel, S. (Ed.) *The Chemistry of Metal-Organic Frameworks, 2 Volume Set: Synthesis, Characterization, and Applications*; John Wiley & Sons: Weinheim, Germany, 2016.
2. Batten, S.R.; Neville, S.M.; Turner, D.R. *Coordination Polymers: Design, Analysis and Application*; RSC: Cambridge, UK, 2009.
3. MacGillivray, L.R.; Lukehart, C.M. (Eds.) *Metal-Organic Framework Materials*; John Wiley & Sons: Chichester, UK, 2014.
4. Janiak, C. Engineering coordination polymers towards applications. *Dalton Trans.* **2003**, *2003*, 2781–2804. [CrossRef]
5. Kitagawa, S.; Kitaura, R.; Noro, S.-I. Functional Porous Coordination Polymers. *Angew. Chem. Int. Ed.* **2004**, *43*, 2334–2375. [CrossRef] [PubMed]
6. Tibbetts, I.; Kostakis, G.E. Recent Bio-Advances in Metal-Organic Frameworks. *Molecules* **2020**, *25*, 1291. [CrossRef] [PubMed]
7. Gómez-Avilés, A.; Muelas-Ramos, V.; Bedia, J.; Rodriguez, J.J.; Belver, C. Thermal post-treatments to enhance the water sta-bility of NH2-MIL-125(Ti). *Catalysts* **2020**, *10*, 603. [CrossRef]
8. Yu, L.; Dong, X.L.; Gong, Q.H.; Acharya, S.R.; Lin, Y.H.; Wang, H.; Han, Y.; Thonhauser, T.; Li, J. Splitting mono- and di-branched alkane isomers by a robust aluminum-based metal-organic framework material with optimal pore dimensions. *J. Am. Chem. Soc.* **2020**, *142*, 6925–6929. [CrossRef]
9. Wang, H.; Li, J. Microporous Metal–Organic Frameworks for Adsorptive Separation of C5–C6 Alkane Isomers. *Acc. Chem. Res.* **2019**, *52*, 1968–1978. [CrossRef] [PubMed]
10. Cui, Y.; Yue, Y.; Qian, G.; Chen, B. Luminescent Functional Metal–Organic Frameworks. *Chem. Rev.* **2011**, *112*, 1126–1162. [CrossRef]
11. Bedia, J.; Muelas-Ramos, V.; Peñas-Garzón, M.; Gómez-Avilés, A.; Rodriguez, J.J.; Belver, C. A Review on the Synthesis and Characterization of Metal Organic Frameworks for Photocatalytic Water Purification. *Catalysts* **2019**, *9*, 52. [CrossRef]
12. Karabach, Y.Y.; Guedes da Silva, M.F.C.; Kopylovich, M.N.; Gil-Hernández, B.; Sanchiz, J.; Kirillov, A.M.; Pombeiro, A.J.L. Self-assembled 3D heterometallic Cu^{II}/Fe^{II} coordination polymers with octahedral net skeletons: Structural features, molecular magnetism, thermal and oxidation catalytic properties. *Inorg. Chem.* **2010**, *49*, 11096–11105. [CrossRef]
13. Xiao, J.-D.; Jiang, H. Metal–Organic Frameworks for Photocatalysis and Photothermal Catalysis. *Acc. Chem. Res.* **2019**, *52*, 356–366. [CrossRef]
14. Gu, J.Z.; Wen, M.; Cai, Y.; Shi, Z.F.; Arol, A.S.; Kirillova, M.V.; Kirillov, A.M. Metal−organic architectures assembled from multifunctionalpolycarboxylates: Hydrothermal self-assembly, structures, andcatalytic activity in alkane oxidation. *Inorg. Chem.* **2019**, *58*, 2403–2412. [CrossRef] [PubMed]
15. Gu, J.Z.; Wen, M.; Cai, Y.; Shi, Z.F.; Nesterov, D.S.; Kirillova, M.V.; Kirillov, A.M. Cobalt(II) Coordination Polymers Assem-bled from UnexploredPyridine-Carboxylic Acids: Structural Diversity and CatalyticOxidation of Alcohols. *Inorg. Chem.* **2019**, *58*, 5875–5885. [CrossRef] [PubMed]
16. Zhang, Y.Y.; Huang, C.; Mi, L.W. Metal-organic frameworks as acid- and/or base-functionalized catalysts for tandem reac-tions. *Dalton Trans.* **2020**, *49*, 14723–14730. [CrossRef] [PubMed]
17. Agarwal, R.A.; Gupta, A.K.; De, D. Flexible Zn-MOF Exhibiting Selective CO_2 Adsorption and Efficient Lewis Acidic Catalytic Activity. *Cryst. Growth Des.* **2019**, *19*, 2010–2018. [CrossRef]
18. Liu, J.-J.; Lu, Y.; Lu, W. Metal-dependent photosensitivity of three isostructural 1D CPs based on the 1,1′-bis(3-carboxylatobenzyl)-4,4′-bipyridinium moiety. *Dalton Trans.* **2020**, *49*, 4044–4049. [CrossRef]
19. Gu, J.Z.; Cui, Y.H.; Liang, X.X.; Wu, J.; Lv, D.Y.; Kirillov, A.M. Srtructurally distinct metal−organic and H-bonded networks derivedfrom 5-(6-carboxypyridin-3-yl)isophthalic acid: Coordination and template effect of 4,4′-bipyridine. *Cryst. Grwoth Des.* **2016**, *16*, 4658–4670. [CrossRef]
20. Han, S.-D.; Chen, Y.-Q.; Zhao, J.-P.; Liu, S.-J.; Miao, X.-H.; Hu, T.-L.; Bu, X.-H. Solvent-induced structural diversities from discrete cup-shaped Co8 clusters to Co8 cluster-based chains accompanied by in situ ligand conversion. *Cryst. Eng. Comm.* **2014**, *16*, 753–756. [CrossRef]
21. Zou, X.Z.; Wu, J.; Gu, J.Z.; Zhao, N.; Feng, A.S.; Li, Y. Syntheses of two nickel(II) coordination compounds based on a rigid linear tricarboxylic acid. *Chin. J. Inorg. Chem.* **2019**, *35*, 1705–1711.
22. Gu, J.; Gao, Z.; Tang, Y. pH and Auxiliary Ligand Influence on the Structural Variations of 5(2′-Carboxylphenyl) Nicotate Coordination Polymers. *Cryst. Growth Des.* **2012**, *12*, 3312–3323. [CrossRef]
23. Li, Y.; Wu, J.; Gu, J.Z.; Qiu, W.D.; Feng, A.S. Temperature-dependent syntheses of two manganese(II) coordination com-pounds based on an ether-bridged tetracarboxylic acid. *Chin. J. Struct. Chem.* **2020**, *39*, 727–736.
24. Hou, Y.L.; Peng, Y.L.; Diao, Y.X.; Liu, J.; Chen, L.Z.; Li, D. Side chain induced self-assembly and selective catalytic oxidation activity of copper(I)-coper(II)-N4 complexes. *Cryst. Growth Des.* **2020**, *20*, 1237–1241. [CrossRef]
25. Gu, J.Z.; Wan, S.M.; Kirillova, M.V.; Kirillov, A.M. H-bonded and metal(II)-organic architectures assembled from an unex-plored aromatic tricarboxylic acid: Structural variety and functional properties. *Dalton Trans.* **2020**, *49*, 7197–7209. [CrossRef] [PubMed]

26. Gregory, R.J.H. Cyanohydrins in Nature and the Laboratory: Biology, Preparations, and Synthetic Applications. *Chem. Rev.* **1999**, *99*, 3649–3682. [CrossRef] [PubMed]

27. Brunel, J.M.; Holms, I.P. Chemically catalyzed asymmetric cyanohydrin syntheses. *Angew. Chem. Int. Ed.* **2004**, *43*, 2752–2778. [CrossRef]

28. Lacour, M.-A.; Rahier, N.J.; Taillefer, M. Mild and Efficient Trimethylsilylcyanation of Ketones Catalysed by PNP Chloride. *Chem. Eur. J.* **2011**, *17*, 12276–12279. [CrossRef]

29. Mowry, D.T. The preparation of nitriles. *Chem. Rev.* **1948**, *42*, 189–283. [CrossRef]

30. Karmakar, A.; Paul, A.; Rúbio, G.M.D.M.; Da Silva, M.F.C.G.; Pombeiro, A.J.L. Zinc(II) and Copper(II) Metal-Organic Frameworks Constructed from a Terphenyl-4,4″-dicarboxylic Acid Derivative: Synthesis, Structure, and Catalytic Application in the Cyanosilylation of Aldehydes. *Eur. J. Inorg. Chem.* **2016**, *2016*, 5557–5567. [CrossRef]

31. Xi, Y.M.; Ma, Z.Z.; Wang, L.N.; Li, M.; Li, Z.J. Three-dimensional Ni(II)-MOF containing anasymmetric pyridyl-carboxylate ligand: Catalytic cyanosilylation of aldehydes and inhibits human promyelocytic leukemia cancer cells. *J. Clust. Sci.* **2019**, *30*, 1455–1464. [CrossRef]

32. Gu, J.-Z.; Liang, X.-X.; Cui, Y.-H.; Wu, J.; Shi, Z.-F.; Kirillov, A.M. Introducing 2-(2-carboxyphenoxy)terephthalic acid as a new versatile building block for design of diverse coordination polymers: Synthesis, structural features, luminescence sensing, and magnetism. *CrystEngComm* **2017**, *19*, 2570–2588. [CrossRef]

33. Kirillov, A.M.; Karabach, Y.Y.; Kirillova, M.V.; Haukka, M.; Pombeiro, A.J.L. Topologically Unique 2D Heterometallic CuII/Mg Coordination Polymer: Synthesis, Structural Features, and Catalytic Use in Alkane Hydrocarboxylation. *Cryst. Growth Des.* **2012**, *12*, 1069–1074. [CrossRef]

34. Chen, J.; Li, Y.; Gu, J.; Kirillova, M.V.; Kirillov, A.M. Introducing a flexible tetracarboxylic acid linker into functional coordination polymers: Synthesis, structural traits, and photocatalytic dye degradation. *New J. Chem.* **2020**, *44*, 16082–16091. [CrossRef]

35. Yu, Y. Auxiliary ligands induced two new Zn(II) compounds with the structural variations from 2D layer to 3D framework: Syntheses, structures and photoluminescent properties. *J. Mol. Struct.* **2017**, *1149*, 761–765. [CrossRef]

36. Wang, L.-N.; Zhang, Y.-H.; Jiang, S.; Liu, Z.-Z. Three coordination polymers based on 3-(3′,5′-dicarboxylphenoxy)phthalic acid and auxiliary N-donor ligands: Syntheses, structures, and highly selective sensing for nitro explosives and Fe3+ ions. *CrystEngComm* **2019**, *21*, 4557–4567. [CrossRef]

37. Gu, J.-Z.; Liang, X.-X.; Cai, Y.; Wu, J.; Shi, Z.-F.; Kirillov, A.M. Hydrothermal assembly, structures, topologies, luminescence, and magnetism of a novel series of coordination polymers driven by a trifunctional nicotinic acid building block. *Dalton Trans.* **2017**, *46*, 10908–10925. [CrossRef]

38. Jaros, S.W.; Da Silva, M.F.C.G.; Florek, M.; Smoleński, P.; Pombeiro, A.J.L.; Kirillov, A.M. Silver(I) 1,3,5-Triaza-7-phosphaadamantane Coordination Polymers Driven by Substituted Glutarate and Malonate Building Blocks: Self-Assembly Synthesis, Structural Features, and Antimicrobial Properties. *Inorg. Chem.* **2016**, *55*, 5886–5894. [CrossRef] [PubMed]

39. Cui, X.; Xu, M.-C.; Zhang, L.-J.; Yao, R.-X.; Zhang, X.-M. Solvent-free heterogeneous catalysis for cyanosilylation in a dynamic cobalt-MOF. *Dalton Trans.* **2015**, *44*, 12711–12716. [CrossRef]

40. Sheldrick, G.M. Crystal structure refinement with SHELXL. *Acta Cryst.* **2015**, *71*, 3–8. [CrossRef]

41. Van de Sluis, P.; Spek, A.L. Platon Program. *Acta Cryst.* **1990**, *46*, 194–201.

42. Blatov, V.A.; Shevchenko, A.P.; Proserpio, D.M. Applied Topological Analysis of Crystal Structures with the Program Package ToposPro. *Cryst. Growth Des.* **2014**, *14*, 3576–3586. [CrossRef]

43. Llabrés i Xamena, F.X.; Gascon, J. (Eds.) *Metal-Organic Frameworks as Heterogeneous Catalysts*; Royal Society of Chemistry: Cambridge, UK, 2013.

44. García, H.; Navalón, S. (Eds.) *Metal-Organic Frameworks: Applications in Separations and Catalysis*; John Wiley & Sons: Weinheim, Germany, 2018.

Article

ZIF-67 Derived MnO$_2$ Doped Electrocatalyst for Oxygen Reduction Reaction

Usman Salahuddin [1,2], Naseem Iqbal [1,*], Tayyaba Noor [3], Saadia Hanif [1], Haider Ejaz [1], Neelam Zaman [1] and Safeer Ahmed [4]

[1] U.S.-Pakistan Center for Advanced Studies in Energy (USPCASE), National University of Sciences and Technology, Islamabad 44000, Pakistan; usmanscme@gmail.com (U.S.); saadiah_hanif@yahoo.com (S.H.); haider@uspcase.nust.edu.pk (H.E.); neelamzaman6650@gmail.com (N.Z.)

[2] Institute of Material Science, University of Connecticut, Storrs, CT 06268, USA

[3] School of Chemical and Materials Engineering (SCME), National University of Sciences and Technology, Islamabad 44000, Pakistan; tayyaba.noor@scme.nust.edu.pk

[4] Department of Physical Chemistry, Quaid-e-Azam University, Islamabad 45320, Pakistan; safeerad@qau.edu.pk

* Correspondence: naseem@uspcase.nust.edu.pk

Abstract: In this study, zeolitic imidazolate framework (ZIF-67) derived nano-porous carbon structures that were further hybridized with MnO$_2$ were tested for oxygen reduction reaction (ORR) as cathode material for fuel cells. The prepared electrocatalyst was characterized by X-ray powder diffraction (XRD), scanning electron microscopy (SEM) and Energy Dispersive X-ray Analysis (EDX). Cyclic voltammetry was performed on these materials at different scan rates under dissolved oxygen in basic media (0.1 M KOH), inert and oxygen rich conditions to obtain their I–V curves. Electrochemical impedance spectroscopy (EIS) and Chronoamperometry was also performed to observe the materials' impedance and stability. We report improved performance of hybridized catalyst for ORR based on cyclic voltammetry and EIS results, which show that it can be a potential candidate for fuel cell applications.

Keywords: fuel cell; oxygen reduction reaction (ORR); metal organic frameworks (MOFs)

Citation: Salahuddin, U.; Iqbal, N.; Noor, T.; Hanif, S.; Ejaz, H.; Zaman, N.; Ahmed, S. ZIF-67 Derived MnO$_2$ Doped Electrocatalyst for Oxygen Reduction Reaction. *Catalysts* **2021**, *11*, 92. https://doi.org/10.3390/catal11010092

Received: 9 December 2020
Accepted: 5 January 2021
Published: 12 January 2021

Publisher's Note: MDPI stays neutral with regard to jurisdictional claims in published maps and institutional affiliations.

1. Introduction

Consumption and production of energy is the sign of industrial growth and progress of any country, as energy develops everything, and around 85% of our energy commitments depend upon fossil fuels [1,2]. However, energy resources such as fossil fuels reduce speedily due to escalating life standards and growing populations. In addition, the economic growth of developed countries, industrial civilization and modern lifestyles rely on the energy withdrawal from gas and oil supplies [3,4]. For intermittent energy generation technologies to strengthen their foothold, energy storage solutions need to become better performing and economically viable [5].

Recently, fuel cells have been considered to be a promising energy resource in contrast to other substitutes that convert chemical energy into electrical energy during a catalytic reaction [6,7]. Various types of fuel cells are available among these; the polymer electrolyte membrane fuel cell (PEMFC) and alkaline fuel cell (AFC) have the advantage of having smaller size, light weight and astonishing power density [8]. Consequently, they can be utilized for stationary and portable applications. They can perform work constantly at low temperature and give high current densities [9]. Cost and stability, however, are the two main factors that delay the commercialization of fuel cells at a large scale.

Since the key cost is because of extreme and ineffective utilization of platinum based electro catalysts, Pt electrodes present the ideal catalytic activity for ORR (oxygen reduction reaction), thus serving as a standard electrode for all the catalysts prepared up until now [10]. Time is needed to prepare various new non-noble metal catalysts which have generated a lot of attraction because of their vastly effectual catalytic properties [11,12].

Multiple techniques have been tested to address the catalysis, including the use of microspheres, nanoparticles, perovskites, etc., of which metal organic frameworks are also part [13–17]. Recently, metal organic frameworks (made up of organic ligands and Inorganic metal ions) have been the subject of significant attention in the field of electrochemistry because they have a variety of structures with large surface area, large pore volume, high porosity and tunable pore size, and are being tested as an economically viable substitutes for noble metal nano-composites [18,19]. Wang et. al. designed a carbon matrix with nitrogen phosphorous doping using Cu-MOF, showing an extraordinary performance as an electro-catalyst for hydrogen evolution reaction (HER) and ORR [20].

In addition, the nano sized pores present in the metal organic frameworks, when turned into porous carbons, make the access to guest molecules much easier, thus increasing the likelihood of an active site being available [21]. Moreover, nano carbons formed from metal organic frameworks are formed as sheets, nanotubes and multiple other forms which can act as high-performance nonmetal catalysts. Besides, for improving their mechanical strength and conductivity, they are transformed conventionally into NPC (nitrogen doped nanoporous carbon), which has shown outstanding performance in the electrochemical field. Gai et al. modify an electrode by NPC prepared from ZIF-8 for the detection of uric acid, ascorbic acid and dopamine. Rizvi et.al. reported Cu-MOF Derived Cu@AC electrocatalyst for ORR in PEMFC. The composite Cu@AC (1:1) shows the peak current density of 2.11 mA cm^{-2} in 0.1 M KOH at a potential of 0.9 V with a scan rate of 50 mV s^{-1}, which shows superior activity compared to commercial grade Pt/C, having a peak current density of 1.37 mA cm^{-2} at a potential of 0.86 V [10]. Moreover, bimetallic MOFs have been utilized to boost the catalyst electrocatalytic performance [22,23]. Yoon el al. reported new bimetallic 2D MOFs (Co$_x$Ni$_y$-CATs) for electrochemical reduction of oxygen; the two metal ions, i.e., Co^{2+} and Ni^{2+}, are rationally controlled in Co$_x$Ni$_y$-CATs (a bimetallic catalyst) for efficient performance in the oxygen reduction reaction (ORR) [24].

Electro catalysts derived from metal organic frameworks, i.e., ZIFs (Zeolitic imidazole frameworks), which are rich in transition metals, i.e., Zn^{+2}, CO^{+2} and nitrogen and carbon and preparation of Zeolitic frameworks, which are single-site solid catalysts with effective and uniform catalytic activity, can be accomplished via the use of metal organic frameworks [25,26]. The metal organic framework is a nitrogen and carbon precursor with a transition metal and is heat treated at 800–1000 °C to form nitrogen doped electro catalyst [27]. ZIFs have been utilized in the production of ORR catalysts where a metal–nitrogen–carbon structure is formed when ZIF-67 is pyrolyzed in the presence of iron carrier, which showed effective electrochemical activity owing partially to the increased surface area provided to the active metals [28,29]. ZIF-67 can also provide the basis for creating tunable structures owing to the ordered arrangement of atoms in the framework. The formation of nanocrystals of carbon decorated with cobalt catalyst has been reported with the ability to catalyze ORR and to perform this function in symmetry. Cobalt containing ZIF-67 based catalysts can also perform catalysis under special preparatory conditions [30,31]. This ability to create uniform crystals can also be utilized in conjunction with the flexibility of carbon materials for ORR [32,33].

Besides cobalt and nickel, manganese has also been shown to perform catalytic activity pertaining to oxygen reduction, which has sparked interest in utilizing this capability in ZIF-67 based carbon electrodes [34,35]. Work with ZIF-67 involving the use of magnesium oxides has yielded remarkable enhancement in the catalytic capability of carbon-based electrodes. In work utilizing Mn$_3$O$_4$ and Co$_3$O$_4$, aimed at catalyzing water splitting reactions and oxygen reduction, the reversible overpotentials were reported to be better than those shown by electrodes containing noble metals like platinum and ruthenium [36,37].

In this paper, we have followed a novel approach to recommend a new material for ORR reaction in fuel cells. ZIF-67 derived nanoporous carbon was modified with MnO$_2$ particles using a simple hydrothermal process to enhance its ORR performance. Both

ZIF-67 derived nanoporous carbon and the modified sample were tested through cyclic voltammetry to analyze the difference in their individual performance.

2. Results and Discussion

2.1. Characterization of Prepared Catalyst

The morphology of synthesized catalysts such as ZIF-67, ZIF derived carbon nanotubes (ZCNT) and Manganese oxide doped ZCNT (ZCNT-M) was analyzed by scanning electron microscopy as illustrated in Figure 1a–c. The rhombic dodecahedron shaped nano crystal of the ZIF-67 is well preserved, as shown in Figure 1a.

(a)

(b)   (c)

Figure 1. SEM images of (**a**) ZIF-67, (**b**) ZCNT and (**c**) ZCNT-M.

From Figure 1b,c, it was observed that after pyrolysis of ZIF-67, carbon nanotubes (CNTs) are visible in SEM images of ZCNT and ZCNT-M. Moreover, at high magnification, the SEM image shows that the obtained ZIF-67 surface was smooth, and their dodecahedron-shaped crystals were closely affixed to the CNTs. In addition, the surface pop and shacks of ZIF-67 nanoparticles more evidently approve that nanoparticles of ZIF-67 were in situ grown on CNTs' surfaces.

The EDS analysis of prepared catalysts such as ZCNT and ZCNT-M shows the presence of manganese, cobalt, oxygen and carbon without any impurity. Table 1 shows the weight percentages of the following element. ZCNT has the maximum carbon percentage while other samples such as ZCNT-M have comparatively lower percentages of carbon, as the unstable organic groups have evaporated after heating, which also reduces the carbon percentage. Moreover, after Mn loading, the relative wt. % of carbon decreases in ZCNT-M, correspondingly.

Table 1. EDS outcome of ZCNT, ZCNT-M.

Sample Element	ZCNT	ZCNT-M
C wt %	50.02	15.98
O wt %	19.07	47.91
Co wt%	30.91	9.99
Mn wt%	-	24.80

Moreover, EDS elemental mapping images, i.e., Figure 2a,b, illustrate that the uniform loading of Mn in the sample and the elemental composition match well with the expected ratio of elemental weight and atomic %.

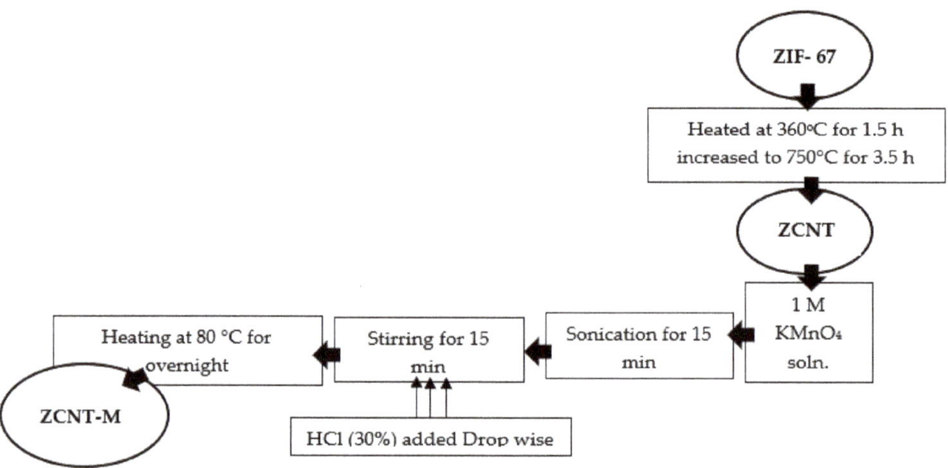

Figure 2. Elemental Mapping of ZCNT-M. (**a**) map showing C, O, Mn & Co distribution, (**b**) mapshowing Mn distribution.

Figure 3a illustrates the XRD pattern of ZIF-67. The presence of characteristic peaks indicates the successful synthesis of material, i.e., 7.2° (011), 10.4° (002), 12.7° (112), 14.7° (022), 16.4° (013), 18° (222), 22.1° (114), 26.5° (134), 29.6° (044), 31.3° (244), 32.5° (235), and 43.1° (100) [38].

Figure 3b illustrates the XRD pattern of prepared ZCNT. The peak at 26.3 (002) confirms the presence of graphitized CNTs and other peaks at 44.36°, 51.67° and 75.98° correspond to Co (111), Co (200) and Co (220) [39].

Figure 3c illustrates the XRD pattern of prepared ZCNT-M sample. The presence of characteristic peaks corresponds to cobalt carbide (JCPDS card number 43-1144) [40], cobalt oxide (JCPDS card number 43-1003) [41], manganese oxide (JCPDS card number 44-0141) [42] and cobalt manganese oxide (JCPDS card number 32-0297) [43].

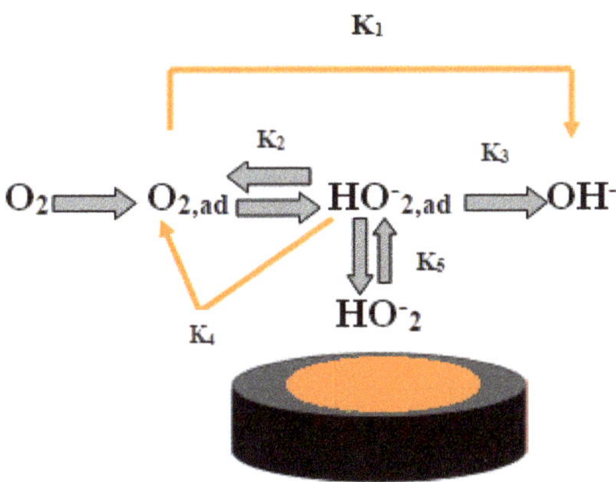

Figure 3. XRD plot for (**a**) ZIF-67 (**b**) ZCNT (**c**) ZCNT-M.

2.2. Electrochemical Analyses

Eco Chemie Autolab PGSTAT 302 potentiostat/galvanostat (Utrech, The Netherlands) was used to perform cyclic voltammetry measurements and GPES software 4.9 was used to run the experiments on the equipment. A three-electrode system was used in which Ag/AgCl was used as the reference electrode, platinum wire as the counter electrode, while glassy carbon (7.065 mm^2) was used as the working electrode.

Moreover, for the modification of working electrode (GCE) ink is deposited on its surface and ink is prepared by adding ZCNT and ZCNT-M catalyst in 100 μL ethanol with 20 μL Nafion (5 wt %) as the binding and conducting agent to form the catalyst ink, which was later deposited (20 μL) on the glassy carbon electrode and allowed to dry. All the prepared composites were tested for different techniques, i.e., cyclic voltammetry, chronoamperometry and electrochemical impedance spectroscopy (EIS) in 0.1 M KOH (an electrolyte) by using the same method of preparing ink. In addition, ZCNT and ZCNT-M samples were tested under inert, dissolved oxygen and oxygen rich conditions. A frequency range of 10 to 40 kHz with a scan rate amplitude of 50 mVs^{-1} was used for electrochemical impedance spectroscopy under potentiostatic mode. Chronoamperometry was also performed for 3600 s.

2.3. Electrochemical Evaluation of Prepared Catalsyts

At first, ZIF-67 derived CNTs (ZCNT) and ZIF-67 derived CNTs/MnO$_2$ (ZCNT-M) were compared. To ensure that dissolved oxygen is the only analyte present within the KOH solution and there are no other analyte species to react with the electrode, the solution was purged with argon gas for 2–3 min and then the response was recorded and compared with oxygen dissolved solution. Figure 4 shows that in the presence of oxygen, the prepared catalysts' reduction current was noticeably increased, which may be attributed to the presence of continuously regenerated reaction centers that might lead to current value amplification during the reduction process [44].

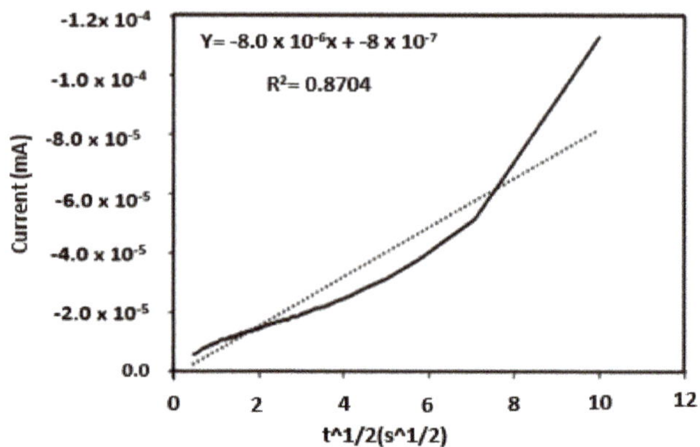

Figure 4. ZCNT-M Performance under dissolved O_2 and Ar purged environment.

With an optimal flow rate of oxygen gas and at a scan rate of 50 mV/s, cyclic voltammetry of prepared catalysts was performed; in this study, the electrochemical activity of ZCNT was compared with ZCNT-M to obtain the values of peak current density, onset potential and peak potentials for oxygen reduction reactions, as illustrated in Figure 5. From the figure, it can be observed that ORR performance of ZCNT-M is much better than ZCNT due to the addition of MnO_2, and it has markedly increased the current densities up to 6.56 mA/cm^2 for ORR with an ORR onset potential (V vs. RHE) of 1.02 V, which is comparable with that of Pt/C (1.01 V), illustrating that there is a current density of 5.02 mA/cm^2 [45]. In comparison to ZCNT-M, commercial MnO_2 shows remarkably low current density with 0.25 mA/cm^2 and with onset potential (V vs. RHE) of 0.96 V, as reported by Huang et al. and Chhetri et al. [46,47]. The increased current densities of ZCNT-M can be attributed to good catalytic ORR activity of MnO_2; as with the ZIF-67 derived CNTs (ZCNT), it enhances the surface area and conductivity of prepared catalysts to a significant level. Thus, as-prepared ZCNT-M composite was used as an efficient non-precious cathodic electrocatalyst with preferable ORR stability, enhanced electron-transport performance, and elevated antitoxic property in alkaline media for ORR.

Figure 6 illustrates the effect of scan rate on the current density of the prepared sample. All the tests were executed with a diverse range of scan rate values such as 5 mV/s, 15 mV/s, 25 mV/s and 50 mV/s in an alkaline media (0.1 M KOH). Prepared ink composition (i.e., 3 mg per catalyst) remained the same in all the experiments. The current density of ZCNT-M in 0.1M KOH for ORR increased gradually because of electroactive species' easy access to the surface of the electrode in the lesson time period [5]; also, at high scan rates, non-electrolytic species were not able to be reduced or oxidized into products. Consequently, only electroactive products were liable for high current density values, this remarkable response of catalyst is linked to the improved extent of reaction. Moreover, a slight shift in peaks was observed, which indicated a slight irreversibility during reaction.

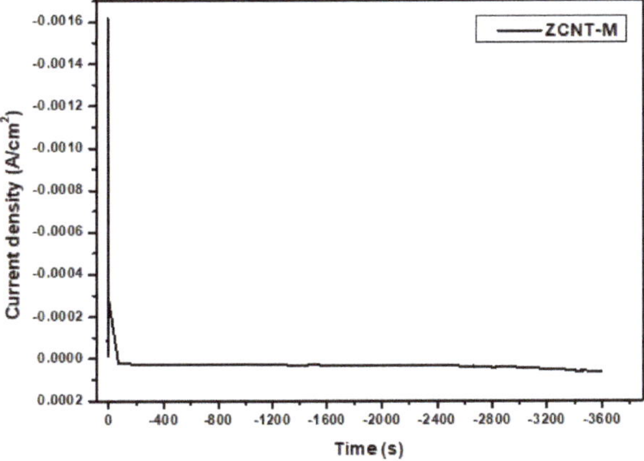

Figure 5. Cylic voltammograms of ZCNT and ZCNT-M for ORR.

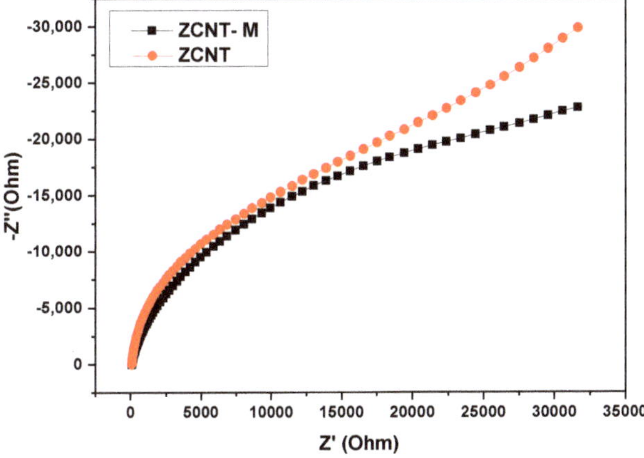

Figure 6. Cyclic voltammograms at different scan rates for ZCNT-M.

Furthermore, linear sweep voltammetry was performed with oxygen purging to study the effect of increasing analyte concentration on current density, as shown in Figure 7. Pure oxygen was purged through the electrolyte for one, two and three minutes, respectively, before performing the linear cyclic voltammetry experiment. A continuous supply of oxygen was maintained during the experimental run. Oxygen purging showed a marked increase in the current density relative to the dissolved oxygen case for ORR. The graph below shows that peak current densities increased as the amount of oxygen present in the electrolyte was increased; however, the current density decreased beyond two minutes of oxygen purging. A possible explanation for this decrease is the saturation of the electrolyte with analyte along with a decrease in oxygen diffusion to the electrode surface.

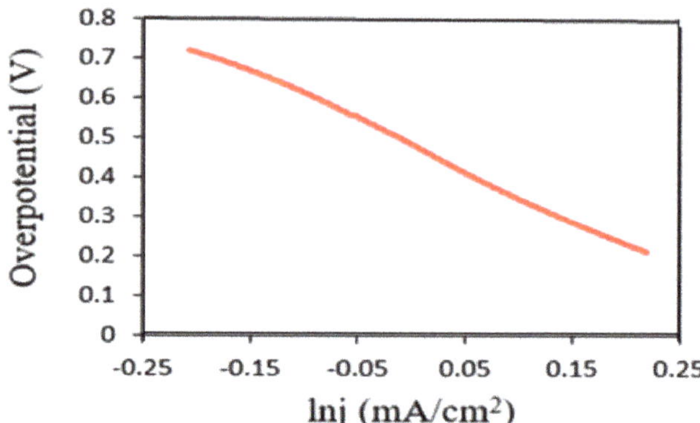

Figure 7. Current densities at different oxygen purging durations for ZCNT-M.

Furthermore, the kinetics of ORR reactions were found to be diffusion controlled. A plot between the square root of scan rate and peak current density was made for ZCNT-M, as shown in Figure 8.

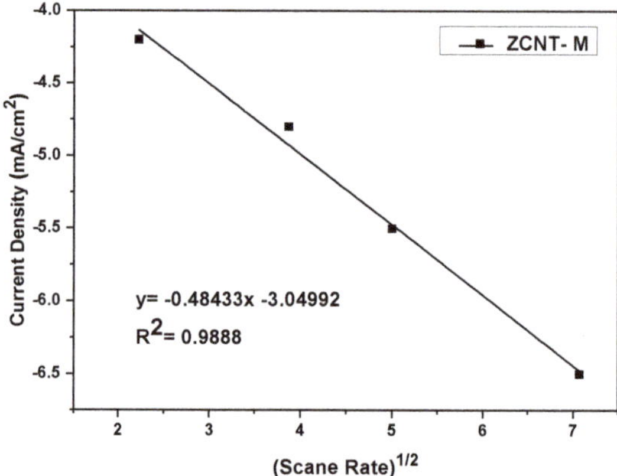

Figure 8. Scan rate vs. peak current density for ZCNT-M.

The figure illustrates that the square roots of scan rates and current densities have a linear relationship, while this linear plot is relative to $D^{1/2}$ to obtain the slope value. Moreover, diffusion coefficients were calculated using the Randles-Sevcik Equation (1) [48].

$$Ip = 0.4463nFAC\sqrt{\frac{nFvD}{RT}} \tag{1}$$

where D is the diffusion coefficient, A is the active surface area (cm^2), C is the molar bulk concentration of 0.1 M KOH, v is the scan rate (V s^{-1}), and n is the number of electrons transferred.

The diffusion coefficients for ORR of ZCNT-M are calculated as $D_{ORR} = 6.6 \times 10^{-4}$ cm^2/s. These results support that a diffusion-controlled mechanism is followed by electrocatalytic oxygen evaluation reaction and oxidation reduction reaction.

Finally, to analyze the trend of overpotential with the current density, Tafel plots (Figure 9) were made which were then used to calculate the exchange current densities value of 1.49×10^{-3} A/cm^2. Firstly, overpotential is calculated by using the formula such as $E—E_o$ [19]. In order to comprehend the reaction kinetic performance, Tafel slopes were calculated by using the subsequent Equation (2).

$$d\eta/d \ln |j| = -RT/\alpha nF \qquad (2)$$

where α was calculated by using Equation (3) [42]:

$$E_p - E_p/2 = 1.857\ RT/\alpha F \qquad (3)$$

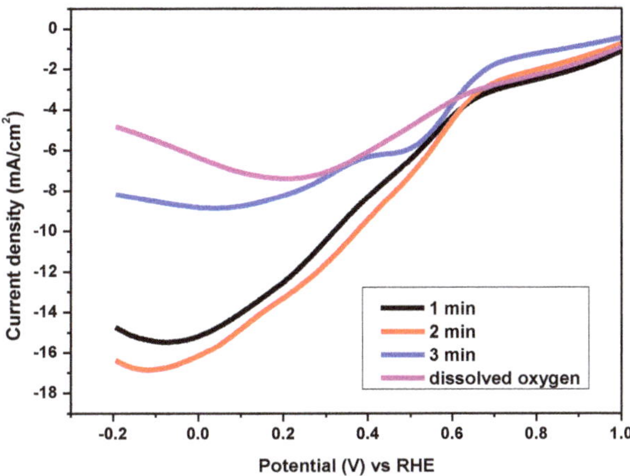

Figure 9. Tafel plots for ZCNT-M.

For the oxygen reduction reaction, n was deemed to be 4. The Tafel slope value for ZCNT-M is calculated and obtained in the range of 165–200 mV/dec, and the value of slope is determined such that if >118 mV/dec, then the rate determining steps are ascribed via (i) ongoing chemical oxidation, (ii) the resulting chemical combination and (iii) the transfer of electrons occurring via an oxide layer. In order to elude the confusion, the outcomes collected from the Tafel slopes will be referred as "cathodic quantities" and the mechanisms for ORR can be established precisely by these approaches [17]. The outcomes are in accord with the literature, where the first C–H bond breaking in ORR occurs because of the low potential region along with the rate determining step through the first electron transfer, while in the high potential region, the increase in slope values is because of poisonous intermediate species having less exposure [10].

To understand the activity of the modified electrode in a better way, electrochemical impedance spectroscopy was performed using the same three electrode systems in 0.1 M KOH solution under the potentiostatic mode. The Nyquist plot below in Figure 10 represents two regions, presenting an idea regarding solution resistance (Rs) and charge transfer resistance (Rct); the small semicircle clearly shows that the charge transfer resistance for ZCNT-M is lower in comparison to ZCNT. Moreover, corresponding low Rct and Rs values are liable for higher catalyst activity as well [5,49]. The decreased value of resistance in ZCNT-M can be attributed in good catalytic ORR activity of MnO_2, as it improves the surface area and conductivity of prepared catalysts to a significant level. Similarly, the high electronic and ionic conductivity of ZCNT-M may possibly be responsible for the straight line. This significant reduction in charge transfer resistance in ZCNT-M clearly favors ORR reactions in ZCNT-M as compared to ZCNT.

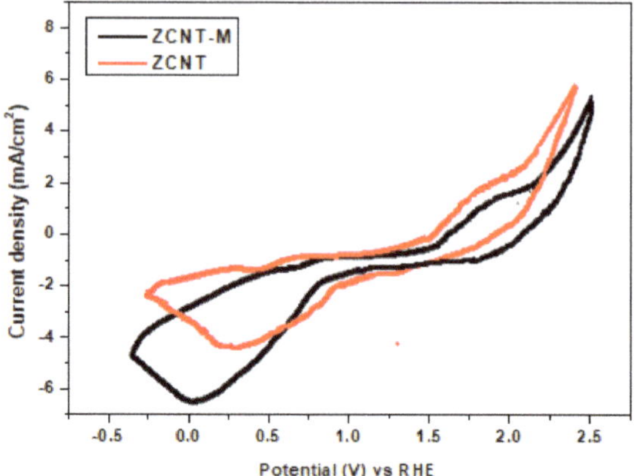

Figure 10. Electrochemical impedance spectroscopy (EIS) plot for ZCNT-M and ZCNT.

The prepared catalysts' stability was determined via the chronoamperometry technique, a key parameter to accomplish the practical application of synthesized samples. The stability test of ZCNT-M was carried out in 0.1 M KOH solution at a potential of 0.1 V for 3600 s in the similar three electrode setup and subsequent electrolyte. At the start, the current dropped substantially very quickly, which can be justified with the following reasons: (a) adsorption of reaction intermediate to the electrode surface; (b) blockage of active site due to evolved oxygen accumulation on the surface of the electrode [10] and (c) flake off material caused by extreme bubbling, but later it adopted a fairly stable trend for the remainder of the hour, as shown in the Figure 11 below. Moreover, Figure 12 shows the plot between the current and square root of current and describes a linear trend over time.

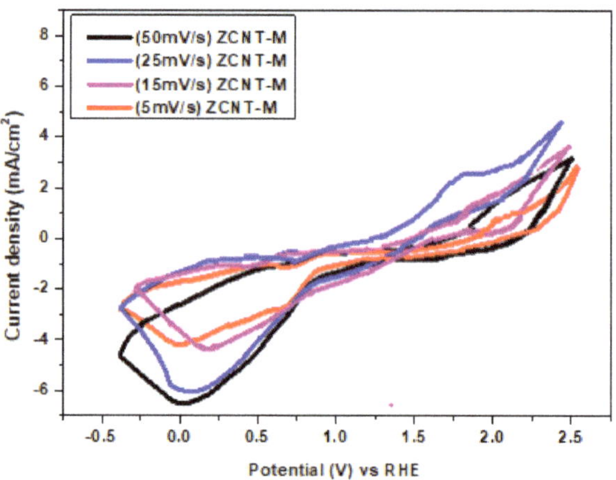

Figure 11. Chronoamperometric plot for ZCNT-M.

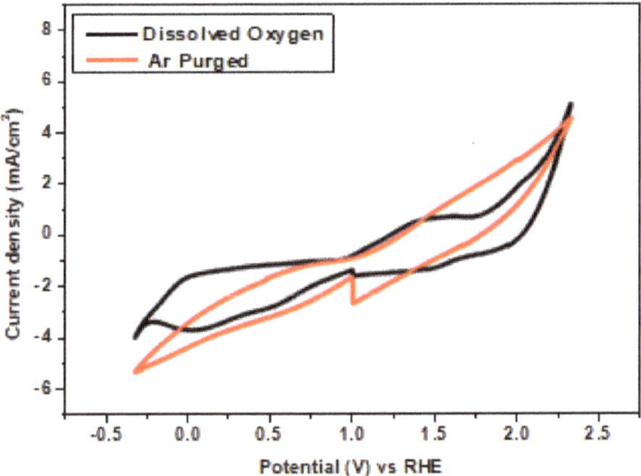

Figure 12. Chronoamperometric plot between current and square root of current for ZCNT-M.

Figure 13 shows the mechanism of ORR in basic media (KOH); it describes the complex reaction pathway by which reductive splitting of the oxygen O–O bond occurs on the catalyst adsorbed surface. Here, k_1 epitomizes the direct reduction of O_2 to OH^- ion without any intermediate formation.

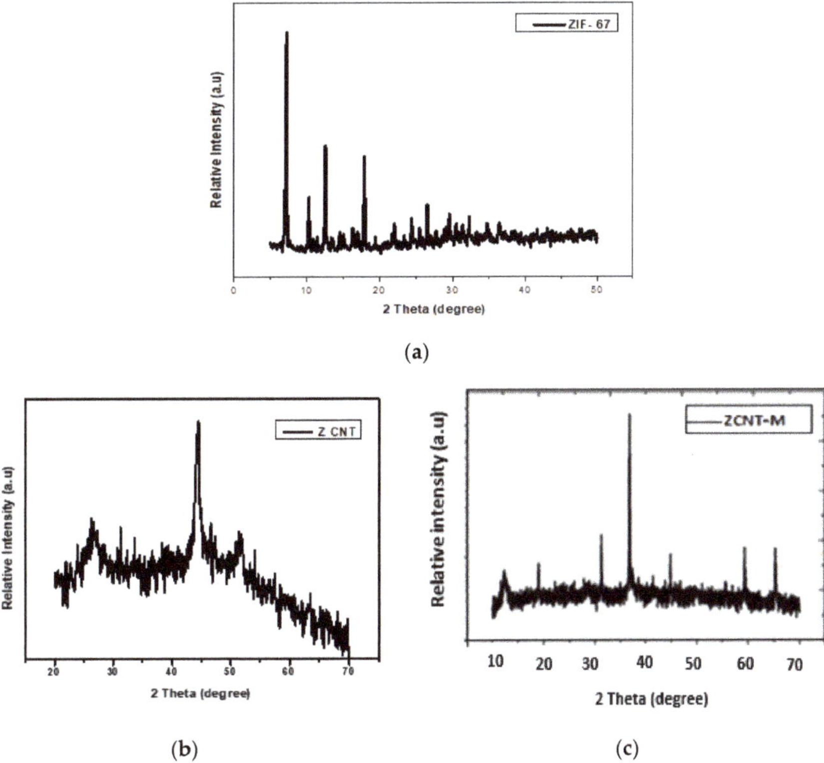

(a)

(b)   (c)

Figure 13. Oxygen reduction reactions (ORR) mechanism on the surface of electrode.

In addition, the k_2 is a comprehensive rate constant for the adsorbed peroxide formation, and might implicate other rate constants that are associated to both the disproportionation reaction and intermediate formation of the adsorbed super oxide; besides, k_3 is the rate constant for peroxide reduction, k_4 refers to the catalytic decaying of adsorbed peroxide on the electrode surface, and k_5 represents rate constants for peroxide desorption and adsorption processes [50].

3. Experimental

3.1. Characterization

To study the surface morphologies of prepared catalysts, SEM analysis was conducted with VEGA3 TESCON at the voltage of 20 kV. For elemental analysis of prepared catalysts, EDS analysis was conducted. Moreover, the crystal structure and phase purity of the prepared catalyst was established by XRD analysis (D8 Advanced Diffractometer) by using Jade 6.0 with diffraction angle (2θ), at a range of 10–$70°$, with the step size of $4°/s$.

3.2. Synthesis of ZIF-67

A quantity of 1.97 g of 2-methylimidazole was dissolved in a 40 mL of 50/50 (*v/v* %) of ethanol and methanol. Furthermore, 1.746 g of $Co(NO_3)_2 \cdot 6H_2O$ were mixed with ethanol and methanol mixture, keeping the ratios as before. The two solutions were then stirred together for 20 min and kept at room temperature for 20 h. After centrifugation, washing and drying, a purple precipitate was obtained [33].

3.3. Synthesis of Mesoporous Carbon

ZIF-67 was heated to 350 °C and was maintained at that temperature for 1.5 h using a tube furnace under reducing atmosphere (H_2/Ar). The temperature was then increased to 750 °C with a ramp rate of 2 °C/min and was sustained at that temperature for 3.5 h. The furnace was naturally allowed to cool down. The sample was then treated with H_2SO_4, centrifuged, washed, and dried [51].

3.4. Synthesis of MnO_2-Doped Mesoporous Carbon

Next, 1 M solution of $KMnO_4$ was prepared in deionized water. Nano-porous carbon particles derived from ZIF-67 were dispersed in the 100 mL of solution using a bath sonicator for 15 min. The mixture was then stirred for 30 min and HCl (30%) was added dropwise to the mixture. The mixture was then transferred into a Teflon lined autoclave and heated in a box furnace at 80 °C for 3 h. The heated suspension was then filtered and washed using ethanol/water mixture and eventually dried in a vacuum oven at 80 °C overnight. The dried sample is the desired product (ZIF-67 derived nano-porous carbon and MnO_2 hybrid). Figure 14 illustrates the synthesis route of ZCNT and ZCNT-M.

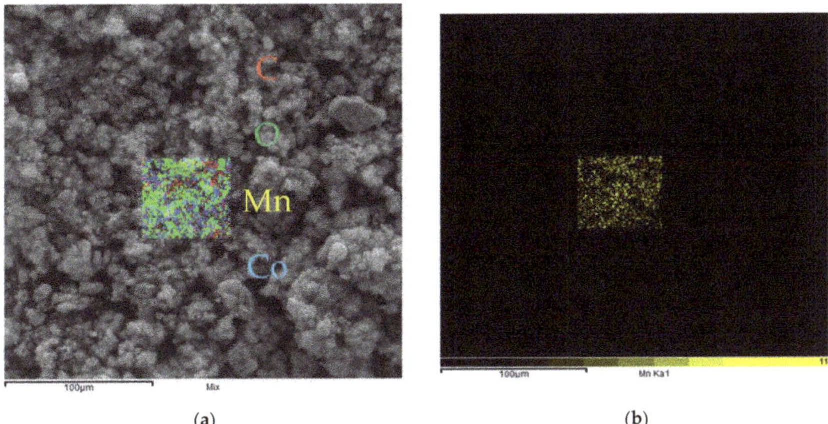

(a)   (b)

Figure 14. Synthesis route of ZCNT and ZCNT-M.

4. Conclusions

ZIF-67 derived nano-porous carbon that is further doped with MnO_2 particles offers a potential material to be used as ORR catalyst in fuel cells, and the material shows good ORR performance. Onset potentials, peak potential and peak current densities were calculated using current vs. voltage plots obtained through cyclic voltammetry. The modified material ZCNT-M showed better performance as compared to ZCNT, as observed through cyclic voltammetry and EIS. The material showed a performance enhancement up to a certain level of oxygen purging, as compared to its performance with dissolved oxygen in the electrolyte. In addition, electrochemical stability was tested by using chronoamperometry, showing a sudden decrease in current and stable performance up to 3600s. Ultimately, it can be concluded from this work that the good catalytic ORR activity of ZIF-67 derived $CNTs/MnO_2$ (ZCNT-M) is due to the incorporation of MnO_2 not only enhancing the surface area, but also the conductivity of prepared catalysts to a significant level.

Author Contributions: Conceptualization, U.S.; Data curation, U.S., T.N., S.H., H.E. and N.Z.; Formal analysis, N.I., H.E. and S.A.; Investigation, T.N.; Methodology, U.S., N.I., S.H., N.Z. and S.A.; Supervision, T.N. and S.A.; Writing–original draft, U.S., S.H., H.E. and N.Z.; Writing–review & editing, N.I. All authors have read and agreed to the published version of the manuscript.

Funding: This research received no external funding.

Acknowledgments: Research work was done at laboratories at USPCAS-E, NUST and the Department of Physical Chemistry, Quaid-e-Azam University, Islamabad, Pakistan.

Conflicts of Interest: The authors declare no conflict of interest.

References

1. Pathak, S. Energy Crisis: A Review. *Int. J. Eng. Res. Appl.* **2014**, *4*, 845–851.
2. Noor, T.; Ammad, M.; Zaman, N.; Iqbal, N.; Yaqoob, L.; Nasir, H. A highly efficient and stable copper BTC metal organic framework derived electrocatalyst for oxidation of methanol in DMFC application. *Catal. Lett.* **2019**, *149*, 3312–3327. [CrossRef]
3. Shafiee, S.; Topal, E. When will fossil fuel reserves be diminished? *Energy Policy* **2009**, *37*, 181–189. [CrossRef]
4. Qureshi, M.N. *Energy Crisis in Pakistan: A Threat to National Security*; ISSRA: Islamabad, Pakistan, 2009.
5. Yaqoob, L.; Noor, T.; Iqbal, N.; Nasir, H.; Sohail, M.; Zaman, N.; Usman, M. Nanocomposites of cobalt benzene tricarboxylic acid MOF with rGO: An efficient and robust electocatalyst for oxygen evaluation reaction (OER). *Renew. Energy* **2020**. [CrossRef]
6. Baker, B.S.; Ghezel-Ayagh, H.G. Fuel Cell System. U.S. Patents 4532192A, 30 July 1985.
7. Wahab, A.; Iqbal, N.; Noor, T.; Ashraf, S.; Raza, M.A.; Ahmad, A.; Khan, U.A. Thermally reduced mesoporous manganese MOF@ reduced graphene oxide nanocomposite as bifunctional electrocatalyst for oxygen reduction and evolution. *RSC Adv.* **2020**, *10*, 27728–27742. [CrossRef]
8. Ren, Y.; Chia, G.H.; Gao, Z. Metal–organic frameworks in fuel cell technologies. *Nano Today* **2013**, *8*, 577–597. [CrossRef]

9. Strahl, S.; Costa-Castelló, R. Temperature control of open-cathode PEM fuel cells. *IFAC-PapersOnLine* **2017**, *50*, 11088–11093. [CrossRef]

10. Rizvi, S.A.M.; Iqbal, N.; Haider, M.D.; Noor, T.; Anwar, R.; Hanif, S. Synthesis and Characterization of Cu-MOF Derived Cu@ AC Electrocatalyst for Oxygen Reduction Reaction in PEMFC. *Catal. Lett.* **2019**, *150*, 1397–1407. [CrossRef]

11. Zhang, P.; Sun, F.; Xiang, Z.; Shen, Z.; Yun, J.; Cao, D. ZIF-derived in situ nitrogen-doped porous carbons as efficient metal-free electrocatalysts for oxygen reduction reaction. *Energy Environ. Sci.* **2014**, *7*, 442–450. [CrossRef]

12. Sarwar, E.; Noor, T.; Iqbal, N.; Mehmood, Y.; Ahmed, S.; Mehek, R. Effect of Co–Ni Ratio in Graphene Based Bimetallic Electro–catalyst for Methanol Oxidation. *Fuel cells* **2018**, *18*, 189–194. [CrossRef]

13. Wang, L.; Zhao, X.; Lu, Y.; Xu, M.; Zhang, D.; Ruoff, R.S.; Stevenson, K.J.; Goodenough, J.B. CoMn$_2$O$_4$ spinel nanoparticles grown on graphene as bifunctional catalyst for lithium-air batteries. *J. Electrochem. Soc.* **2011**, *158*, A1379–A1382. [CrossRef]

14. Yang, W.; Salim, J.; Ma, C.; Ma, Z.; Sun, C.; Li, J.; Chen, L.; Kim, Y. Flowerlike Co$_3$O$_4$ microspheres loaded with copper nanoparticle as an efficient bifunctional catalyst for lithium–air batteries. *Electrochem. Commun.* **2013**, *28*, 13–16. [CrossRef]

15. Jin, C.; Yang, Z.; Cao, X.; Lu, F.; Yang, R. A novel bifunctional catalyst of Ba$_{0.9}$Co$_{0.5}$Fe$_{0.4}$Nb$_{0.1}$O$_{3-\delta}$ perovskite for lithium–air battery. *Int. J. Hydrogen Energy* **2014**, *39*, 2526–2530. [CrossRef]

16. Farrusseng, D.; Aguado, S.; Pinel, C. Metal–organic frameworks: Opportunities for catalysis. *Angew. Chem. Int. Ed.* **2009**, *48*, 7502–7513. [CrossRef] [PubMed]

17. Haider, M.D.; Iqbal, N.; Rizvi, S.A.M.; Noor, T.; Hanif, S.; Anwar, R. ZIF-67 derived Cu doped electrocatalyst for oxygen reduction reaction. *J. Electrochem. Energy Convers. Storage* **2020**, *18*, 021001. [CrossRef]

18. Xia, B.Y.; Yan, Y.; Li, N.; Wu, H.B.; Lou, X.W.D.; Wang, X. A metal–organic framework-derived bifunctional oxygen electrocatalyst. *Nat. Energy* **2016**, *1*, 15006. [CrossRef]

19. Noor, T.; Zaman, N.; Nasir, H.; Iqbal, N.; Hussain, Z. Electro catalytic study of NiO-MOF/rGO composites for methanol oxidation reaction. *Electrochim. Acta* **2019**, *307*, 1–12. [CrossRef]

20. Wang, R.; Dong, X.Y.; Du, J.; Zhao, J.Y.; Zang, S.Q. MOF-Derived bifunctional Cu3P nanoparticles coated by a N, P–codoped carbon shell for hydrogen evolution and oxygen reduction. *Adv. Mater.* **2018**, *30*, 1703711. [CrossRef]

21. Yaqoob, L.; Noor, T.; Iqbal, N.; Nasir, H.; Zaman, N. Development of nickel-BTC-MOF-derived nanocomposites with rGO towards electrocatalytic oxidation of methanol and its product analysis. *Catalysts* **2019**, *9*, 856. [CrossRef]

22. Ghoshal, S.; Zaccarine, S.; Anderson, G.C.; Martinez, M.B.; Hurst, K.E.; Pylypenko, S.; Pivovar, B.S.; Alia, S.M. ZIF 67 Based Highly Active Electrocatalysts as Oxygen Electrodes in Water Electrolyzer. *ACS Appl. Energy Mater.* **2019**, *2*, 5568–5576. [CrossRef]

23. Wang, H.; Wei, L.; Liu, J.; Shen, J. Hollow bimetal ZIFs derived Cu/Co/N co-coordinated ORR electrocatalyst for microbial fuel cells. *Int. J. Hydrogen Energy* **2020**, *45*, 4481–4489. [CrossRef]

24. Yoon, H.; Lee, S.; Oh, S.; Park, H.; Choi, S.; Oh, M. Synthesis of Bimetallic Conductive 2D Metal-Organic Framework (CoxNiy-CAT) and Its Mass Production: Enhanced Electrochemical Oxygen Reduction Activity. *Small* **2019**, *15*, 1805232. [CrossRef]

25. Wang, C.; Liu, D.; Lin, W. Metal–organic frameworks as a tunable platform for designing functional molecular materials. *J. Am. Chem. Soc.* **2013**, *135*, 13222–13234. [CrossRef] [PubMed]

26. Hanif, S.; Iqbal, N.; Shi, X.; Noor, T.; Ali, G.; Kannan, A. NiCo-N-doped carbon nanotubes based cathode catalyst for alkaline membrane fuel cell. *Renew. Energy* **2020**, *154*, 508–516. [CrossRef]

27. Wu, G.; Zelenay, P. Nanostructured nonprecious metal catalysts for oxygen reduction reaction. *Acc. Chem. Res.* **2013**, *46*, 1878–1889. [CrossRef] [PubMed]

28. Palaniselvam, T.; Biswal, B.P.; Banerjee, R.; Kurungot, S. Zeolitic Imidazolate Framework (ZIF)-Derived, Hollow-Core, Nitrogen-Doped Carbon Nanostructures for Oxygen-Reduction Reactions in PEFCs. *Chem. A Eur. J.* **2013**, *19*, 9335–9342. [CrossRef]

29. Hanif, S.; Shi, X.; Iqbal, N.; Noor, T.; Anwar, R.; Kannan, A. ZIF derived PtNiCo/NC cathode catalyst for proton exchange membrane fuel cell. *Appl. Catal. B Environ.* **2019**, *258*, 117947. [CrossRef]

30. Xia, W.; Zhu, J.; Guo, W.; An, L.; Xia, D.; Zou, R. Well-defined carbon polyhedrons prepared from nano metal–organic frameworks for oxygen reduction. *J. Mater. Chem. A* **2014**, *2*, 11606–11613. [CrossRef]

31. Aijaz, A.; Masa, J.; Rösler, C.; Xia, W.; Weide, P.; Botz, A.J.; Fischer, R.A.; Schuhmann, W.; Muhler, M. Co@Co$_3$O$_4$ encapsulated in carbon nanotube-grafted nitrogen-doped carbon polyhedra as an advanced bifunctional oxygen electrode. *Angew. Chem. Int. Ed.* **2016**, *55*, 4087–4091. [CrossRef] [PubMed]

32. Meng, F.; Zhong, H.; Bao, D.; Yan, J.; Zhang, X. In situ coupling of strung Co4N and intertwined N–C fibers toward free-standing bifunctional cathode for robust, efficient, and flexible Zn–air batteries. *J. Am. Chem. Soc.* **2016**, *138*, 10226–10231. [CrossRef]

33. Ahmad, R.; Iqbal, N.; Baig, M.; Noor, T.; Ali, G.; Gul, I. ZIF-67 derived NCNT/S@Ni(OH)$_2$ decorated Ni foam based electrode material for high-performance supercapacitors. *Electrochem. Acta* **2020**, *364*, 137147. [CrossRef]

34. Yang, J.; Xu, J.J. Nanoporous amorphous manganese oxide as electrocatalyst for oxygen reduction in alkaline solutions. *Electrochem. Commun.* **2003**, *5*, 306–311. [CrossRef]

35. Lima, F.H.; Calegaro, M.L.; Ticianelli, E.A. Investigations of the catalytic properties of manganese oxides for the oxygen reduction reaction in alkaline media. *J. Electroanal. Chem.* **2006**, *590*, 152–160. [CrossRef]

36. Masa, J.; Xia, W.; Sinev, I.; Zhao, A.; Sun, Z.; Grützke, S.; Weide, P.; Muhler, M.; Schuhmann, W. MnxOy/NC and CoxOy/NC nanoparticles embedded in a nitrogen-doped carbon matrix for high-performance bifunctional oxygen electrodes. *Angew. Chem. Int. Ed.* **2014**, *53*, 8508–8512. [CrossRef] [PubMed]

37. Liang, Y.; Wang, H.; Zhou, J.; Li, Y.; Wang, J.; Regier, T.; Dai, H. Covalent hybrid of spinel manganese–cobalt oxide and graphene as advanced oxygen reduction electrocatalysts. *J. Am. Chem. Soc.* **2012**, *134*, 3517–3523. [CrossRef] [PubMed]
38. Noor, T.; Raffi, U.; Iqbal, N.; Yaqoob, L.; Zaman, N. Kinetic evaluation and comparative study of cationic and anionic dyes adsorption on Zeolitic imidazolate frameworks based metal organic frameworks. *Mater. Res. Express* **2019**, *6*, 125088. [CrossRef]
39. Shi, X.; Iqbal, N.; Kunwar, S.; Wahab, G.; Kasat, H.; Kannan, A.M. PtCo@ NCNTs cathode catalyst using ZIF-67 for proton exchange membrane fuel cell. *Int. J. Hydrogen Energy* **2018**, *43*, 3520–3526. [CrossRef]
40. Fan, Q.; Guo, Z.; Li, Z.; Wang, Z.; Yang, L.; Chen, Q.; Liu, Z.; Wang, X. Atomic layer deposition of cobalt carbide thin films from cobalt amidinate and hydrogen plasma. *ACS Appl. Electron. Mater.* **2019**, *1*, 444–453. [CrossRef]
41. Yang, Z.; Wang, S.; Liu, Y.; Lei, X. Cobalt oxide microtubes with balsam pear-shaped outer surfaces as anode material for lithium ion batteries. *Ionics* **2015**, *21*, 2423–2430. [CrossRef]
42. Johnson, C.; Dees, D.; Mansuetto, M.; Thackeray, M.; Vissers, D.; Argyriou, D.; Loong, C.-K.; Christensen, L. Structural and electrochemical studies of α-manganese dioxide (α-MnO2). *J. Power Sources* **1997**, *68*, 570–577. [CrossRef]
43. Ahmadian, H.; Veisi, H.; Karami, C.; Sedrpoushan, A.; Nouri, M.; Jamshidi, F.; Alavioon, I. Cobalt manganese oxide nanoparticles as recyclable catalyst for efficient synthesis of 2-aryl-1-arylmethyl-1H-1, 3-benzimidazoles under solvent-free conditions. *Appl. Organomet. Chem.* **2015**, *29*, 266–269. [CrossRef]
44. Rapson, T.D.; Kusuoka, R.; Butcher, J.; Musameh, M.; Dunn, C.J.; Church, J.S.; Warden, A.C.; Blanford, C.F.; Nakamura, N.; Sutherland, T.D. Bioinspired electrocatalysts for oxygen reduction using recombinant silk films. *J. Mater. Chem. A* **2017**, *5*, 10236–10243. [CrossRef]
45. Wang, W.; Geng, J.; Kuai, L.; Li, M.; Geng, B. Porous Mn_2O_3: A Low-Cost Electrocatalyst for Oxygen Reduction Reaction in Alkaline Media with Comparable Activity to Pt/C. *Chem. Eur. J.* **2016**, *22*, 9909–9913. [CrossRef] [PubMed]
46. Huang, B.; Zhang, X.; Cai, J.; Liu, W.; Lin, S. A novel MnO_2/rGO composite prepared by electrodeposition as a non-noble metal electrocatalyst for ORR. *J. Appl. Electrochem.* **2019**, *49*, 767–777. [CrossRef]
47. Chhetri, B.P.; Parnell, C.M.; Wayland, H.; RanguMagar, A.B.; Kannarpady, G.; Watanabe, F.; Albkuri, Y.M.; Biris, A.S.; Ghosh, A. Chitosan-derived NiO-Mn_2O_3/C nanocomposites as non-precious catalysts for enhanced oxygen reduction reaction. *ChemistrySelect* **2018**, *3*, 922–932. [CrossRef]
48. Atabaki, M.M.; Kovacevic, R. Graphene composites as anode materials in lithium-ion batteries. *Electron. Mater. Lett.* **2013**, *9*, 133–153. [CrossRef]
49. Bae, S.H.; Kim, J.E.; Randriamahazaka, H.; Moon, S.Y.; Park, J.Y.; Oh, I.K. Seamlessly conductive 3D nanoarchitecture of core-shell Ni-Co nanowire network for highly efficient oxygen evolution. *Adv. Energy Mater.* **2017**, *7*, 1601492. [CrossRef]
50. Lin, X. The Kinetic and Mechanism of the Oxygen Reduction Reaction on Pt, Au, Cu, PtCu/C and CuAu/C in Alkaline Media. Master's Thesis, The Ohio State University, Columbus, OH, USA, 2016.
51. Ahmad, R.; Iqbal, N.; Noor, T. Development of ZIF-Derived Nanoporous Carbon and Cobalt Sulfide-Based Electrode Material for Supercapacitor. *Materials* **2019**, *12*, 2940. [CrossRef]

Article

Efficient Photocatalytic CO_2 Reduction with MIL-100(Fe)-CsPbBr$_3$ Composites

Ruolin Cheng [1], Elke Debroye [1], Johan Hofkens [1,*] and Maarten B. J. Roeffaers [2,*]

[1] Department of Chemistry, Faculty of Sciences, KU Leuven, Celestijnenlaan 200F, 3001 Leuven, Belgium; ruolin.cheng@kuleuven.be (R.C.); elke.debroye@kuleuven.be (E.D.)

[2] Center for Membrane Separations, Adsorption, Catalysis and Spectroscopy for Sustainable Solutions (cMACS), KU Leuven, Celestijnenlaan 200F, 3001 Leuven, Belgium

* Correspondence: johan.hofkens@kuleuven.be (J.H.); maarten.roeffaers@kuleuven.be (M.B.J.R.); Tel.: +32-16-327804 (J.H.); +32-16-327449 (M.B.J.R.)

Received: 4 November 2020; Accepted: 18 November 2020; Published: 20 November 2020

Abstract: Bromide-based metal halide perovskites (MHPs) are promising photocatalysts with strong blue-green light absorption. Composite photocatalysts of MHPs with MIL-100(Fe), as a powerful photocatalyst itself, have been investigated to extend the responsiveness towards red light. The composites, with a high specific surface area, display an enhanced solar light response, and the improved charge carrier separation in the heterojunctions is employed to maximize the photocatalytic performance. Optimization of the relative composition, with the formation of a dual-phase $CsPbBr_3$ to $CsPb_2Br_5$ perovskite composite, shows an excellent photocatalytic performance with 20.4 μmol CO produced per gram of photocatalyst during one hour of visible light irradiation.

Keywords: metal halide perovskites; metal-organic framework; photocatalysis

---

1. Introduction

The capture of CO_2 and further conversion of this greenhouse gas into chemical fuels (CO, CH_4, CH_3OH, etc.) has been a hot topic during the past decades [1]. The reduction of CO_2 is complicated by its inherent chemical stability [2,3]. Despite this issue, photocatalytic CO_2 reduction is still seen as one of the promising ways to sustainably produce chemicals [4]. To date, a wide variety of photocatalysts such as TiO_2, ZnO, CdS, perovskite oxides and their composites, and metal (complex) functionalized derivatives, have been tested. But the obtained performances under solar irradiation are still limited, mainly due to the low efficiency of using visible light.

Over the last decades, metal-organic frameworks (MOFs) have been one of the fastest developing materials, exhibiting unique properties such as structural flexibility, large specific surface area, tunable but uniform cavities, and easy ligand functionalization [5,6]. MOF-5 was the first MOF reported to have photocatalytic activity in UV-induced phenol photodegradation [7]. Fu et al. successfully employed an amine-functionalized Ti-based MOF, NH_2-MIL-125(Ti), in CO_2 photoreduction [8]. Ever since, MOFs are not only used as matrix for other semiconductors or noble metals but are also seen as promising photocatalysts on their own [9–11].

Recently, all-inorganic $CsPbBr_3$ perovskites have emerged as promising photocatalysts in various applications ranging from dye degradation [12] to selective organic reactions [13] and renewable fuel generation from CO_2 and water [14]. These materials combine the excellent optoelectronic properties of organic-inorganic hybrid perovskites, such as a bandgap largely overlapping with the visible part of the electromagnetic spectrum, a high absorption coefficient, and long-range charge transport, with good temperature stability [15,16]. However, the performances of MHPs are limited by their poor structural stability in a humid atmosphere. Heterojunction formation with materials such as graphene oxide, MOFs, TiO_2, C_3N_4, etc., has been explored to further improve charge carrier separation and protect the

MHPs from a polar environment [17–21]. The use of MOFs in combination with MHPs is motivated by the composites improved stability as compared to the parent MHP material in photocatalytic reactions in the presence of water.

Herein, we report the excellent performance of $CsPbBr_3$/MIL-100(Fe) composites for the photocatalytic reduction of CO_2. Among the reported MOF photocatalysts, Fe-based MIL-100 was chosen due to its low cost, high chemical and water stability, and the intense and complementary visible light absorption to the far red [22,23]. We reasoned that the $CsPbBr_3$/MIL-100(Fe) composite should show an extended and improved absorption and hence has the potential to be an efficient photocatalyst. The composite material was generated through the in situ synthesis of MIL-100(Fe) on perovskite particles, which tends to improve the stability of the composite and offers more efficient electron transfer [24]. MIL-100(Fe) introduction endows the composite with a high specific surface area and enhanced visible light response.

2. Results

Pure $CsPbBr_3$ was synthesized using an anti-solvent method and MIL-100(Fe) was synthesized by a modified non-hydrothermal method (see Section 3). The composite photocatalysts were obtained by an in situ growth method and were named after the amount of MIL-100(Fe) precursor ($Fe(NO_3)_3 \cdot 9H_2O$) added (see Section 3).

In the first step, X-ray diffraction (XRD) was used to investigate the crystallinity of the pure and composite materials (Figure 1a). The XRD patterns clearly show the orthorhombic $CsPbBr_3$ structure and in the composite materials, additional diffraction peaks at 3.4° and 11° from MIL-100(Fe) appear (Figure 1b) [25,26]. With an increasing amount of MIL-100(Fe) loaded in the composites, we observe a gradual transformation of $CsPbBr_3$ (PDF#18-0364) to $CsPb_2Br_5$ (PDF#25-0211). Figure 1c illustrates the perovskites' crystallographic structure transformation. This phase transformation could be ascribed to the excessive H_2O in the Fe precursor, which partially converts the $CsPbBr_3$. To evaluate the necessity of H_2O during the in-situ growth of the MOF, the Fe precursor was dried ahead of the synthesis. With this dried $Fe(NO_3)_3$ precursor, the desired MIL-100(Fe) could not be generated (Figure 1b), indicating the critical role of water during the MIL-100(Fe) formation, in line with literature reports [10].

The Fourier-transform infrared (FTIR) spectra of all composites exhibit typical MIL-100(Fe) bands (Figure 2a) [27]. The bands at 1625 and 1380 cm^{-1} are related to the stretching vibrations of carboxyl groups [28], the band at 1446 cm^{-1} can be ascribed to the O–H stretching vibration, while the ones corresponding to the bending vibration of C–H (at 759 cm^{-1}) and C=C (at 711 cm^{-1}) originate from the benzene ring. As expected, the peak intensities of the Fe-O stretching vibration (at 491 cm^{-1}) and free C=O stretching vibration from unreacted H_3BTC (at 1716 cm^{-1}) enhanced with increasing the amount of MIL-100(Fe) loaded in the composites [29]. Further, a broad peak at 3000–3500 cm^{-1} indicates the presence of a significant amount of adsorbed H_2O in the composites.

Thermogravimetric analysis (TGA) was performed in O_2 to quantify the MIL-100(Fe) amount in the composites based on the oxidation of the organic linker (Figure 2b). MIL-100(Fe) exhibits a clear two-step weight loss: (1) below 200 °C, the weight loss is associated with the removal of adsorbed H_2O, and H_2O coordinated to the iron trimers, (2) around 300 °C, H_3BTC decomposes [30]. This two-step weight loss was identified in all the composites. $CsPbBr_3$ shows a good thermal stability to about 500 °C. Hence, the weight loss due to H_3BTC decomposition was used to determine the relative amount of MIL-100(Fe) in the composites. As listed in Table 1, the MIL-100(Fe) content in the composites varies from 9 wt% in p-30Fe to 53 wt% in p-180Fe.

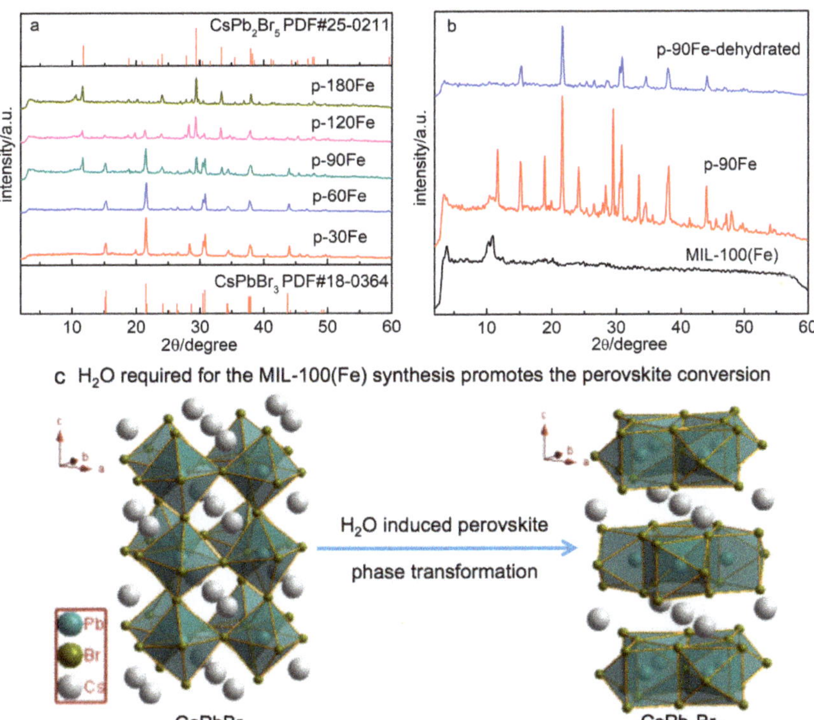

Figure 1. (**a**,**b**) XRD patterns of the as-prepared photocatalysts, and (**c**) visualization of the perovskites' crystallographic structure transformation.

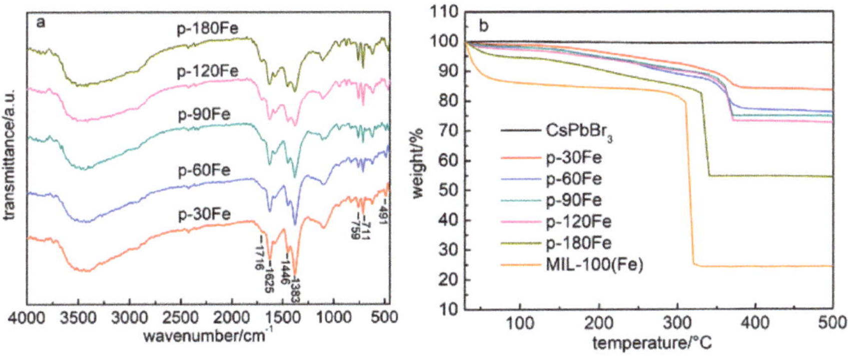

Figure 2. (**a**) FTIR spectra, and (**b**) TGA thermograms of the as-prepared photocatalysts.

Table 1. Weight ratio of MIL-100(Fe) in the composites.

Sample	p-30Fe	p-60Fe	p-90Fe	p-120Fe	p-180Fe
MIL-100(Fe) content/wt%	9	14	18	27	53

The surface chemical composition and chemical states of the as-synthesized composites were further revealed by X-ray photoelectron spectroscopy (XPS). The XPS spectrum (Figure 3a) of p-90Fe shows distinct peaks from both of MIL-100(Fe) and $CsPbBr_3$. Relevant high-resolution spectra of p-90Fe (Figure S1) display peaks of Cs 3d, Pb 4f, and Br 3d in the range of 720–745, 134–148, and 64–74 eV, respectively, which are well-matched with those of $CsPbBr_3$ [31,32]. Two dominant peaks of Fe (Figure 3b) at 724.6 and 711.7 eV are attributed to Fe $2p_{1/2}$ and Fe $2p_{3/2}$, respectively. Additional satellite peak appearing at 717.2 eV corresponds to Fe^{3+} in MIL-100(Fe) [33].

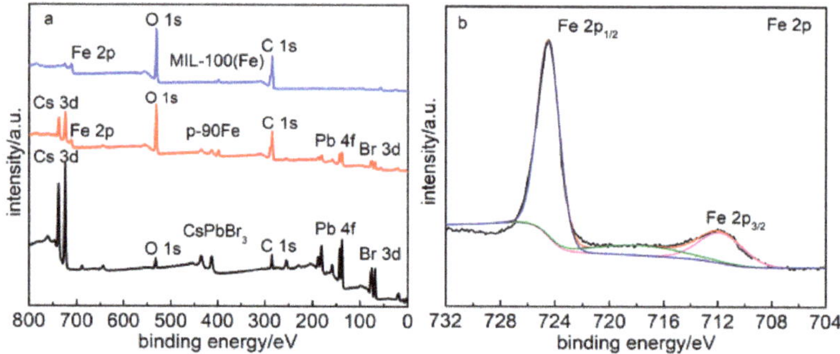

Figure 3. XPS spectra of p-90Fe: (**a**) survey and (**b**) Fe 2p.

The response of MIL-100(Fe) in the visible light region arises from the direct excitation of the Fe–O clusters. As shown in the absorption spectra in Figure 4a (as obtained by UV-vis diffuse reflectance, see Section 3), p-30Fe, p-60Fe, and p-90Fe basically maintain the pattern of $CsPbBr_3$. p-120Fe and p-180Fe exhibit a pattern similar to MIL-100(Fe), due to the larger amount of MIL-100(Fe) included. The addition of MIL-100(Fe) enhances the composites' optical response in the visible light region, especially above 550 nm. The absorption edges around 550 nm gradually blue shift, corresponding to the increased bandgaps (E_g) (calculated by the Tauc plots, Figure 4b) of the composites consisting of more MIL-100(Fe) (Table 2). This blue-shift arises from the phase transformation from $CsPbBr_3$ (E_g = 2.3 eV) to $CsPb_2Br_5$ (E_g = 3.0 eV) and the more dominant role of MIL-100(Fe) in the composite photocatalysts upon increasing the Fe content.

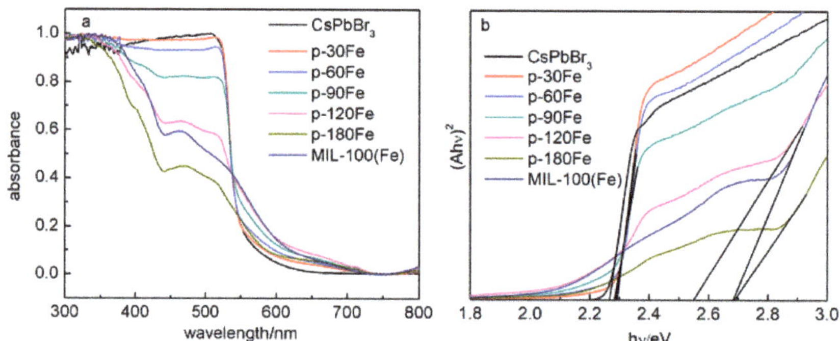

Figure 4. (**a**) UV-vis absorption spectra of parent compounds and composites, and (**b**) Tauc plots of MIL-100(Fe) and the composites. The band gap energies of the photocatalysts, listed in Table 2, were estimated following Kubelka-Munk transformation. Each bandgap is determined by the intersection point of the corresponding tangent line and horizontal axis.

Table 2. Bandgap data of the prepared samples.

Sample	MIL-100(Fe)	CsPbBr₃	p-30Fe	p-60Fe	p-90Fe	p-120Fe	p-180Fe
E_g/eV	2.68	2.27	2.29	2.29	2.30	2.55	2.69

Figure 5 and Figure S2 show the steady-state photoluminescence (PL) spectra of the photocatalysts at an excitation wavelength of 380 nm. As shown in Figure S2, $CsPbBr_3$ has significantly higher peak intensity than the composites. The broad emission peak from 400 to 450 nm can be attributed to the 1,3,5-benzene tricarboxylic acid linkers in the MIL-100(Fe) structure [34]. The PL emission spectra of p-30Fe, p-60Fe and p-90Fe show a gradual blue shift and weakening intensity of the peak at around 530 nm. Different from 3D $CsPbBr_3$, Cs^+ ions separate the layers of 2D $CsPb_2Br_5$, and the excitons are slowed down by the layered structure [35]. Thus, with more $CsPb_2Br_5$ formed, the PL peaks of p-120Fe and p-180Fe get more blue shifted [36,37]. p-90Fe possesses the lowest PL intensity among all as-synthesized $CsPbBr_3$/MIL-100(Fe) samples.

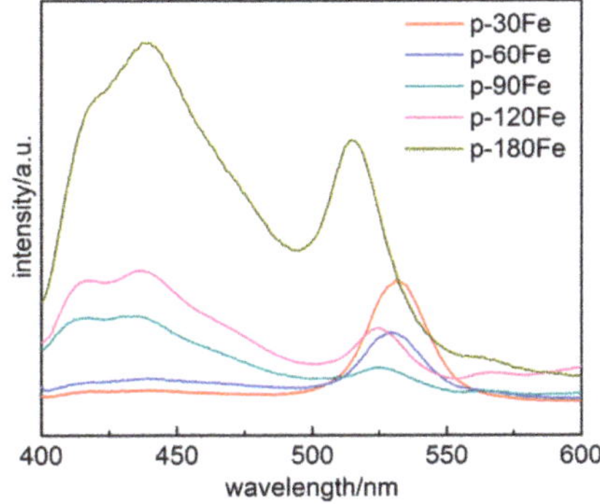

Figure 5. Steady-state PL spectra at an excitation wavelength of 380 nm of the composites. PL spectra of the parent compounds in Supplementary Materials.

As shown in Figure S3a, the pure $CsPbBr_3$ sample displays nanoparticles with a uniform cubic shape. MIL-100(Fe) synthesized in HF-free conditions crystallizes into small nanoparticles with no clearly visible shape compared to the classic octahedral shapes (Figure S3b); this is in line with previous reports [38]. In the composite materials, e.g., p-90Fe, the cubic $CsPbBr_3$ morphology is lost and only agglomerated particles are observed (Figure S3c). Selected location elemental analysis by energy-dispersive X-ray spectroscopy (EDS) confirms the co-existence of both $CsPbBr_3$ and MIL-100(Fe) (Figure S3d).

The components' dispersion and interaction were studied by confocal laser scanning fluorescence measurements on p-90Fe, using a 375 nm laser. Figure 6d showed the wide field image of p-90Fe. Based on the PL spectra mentioned above, two emission channels at 430–470 nm and 505–540 nm were chosen to visualize the MIL-100(Fe) (red colored) (Figure 6a) and $CsPbBr_3$ (green colored) (Figure 6b) distribution, respectively. Both channels were acquired with the same excitation power. As $CsPbBr_3$ has a significantly stronger signal than MIL-100(Fe) upon excitation (Figure S2), Figure 6a was adjusted to be 25% brighter than Figure 6b.

As shown in Figure 6c, yellow shaded areas appear where both signals overlap. Based on particle count, ca. 80% particles were yellowish-green. These uniformly dispersed yellow shades reflect the close contact between MOF and MHP parts in the composite.

The complementary light absorption by $CsPbBr_3$ and MIL-100(Fe) should result in an efficient photocatalytic activity under simulated solar irradiation. Here, CO_2 photoreduction under visible light irradiation was chosen to test the photoactivity of the different catalysts. The reaction was performed under 1 bar hydrated CO_2 atmosphere, at ambient temperature. CO was found as the only photoreduction product generated from CO_2, and no other carbonaceous products was detected. A series of control experiments were also conducted. No appreciable amounts of CO or other hydrocarbons were detected in the absence of light irradiation, or photocatalyst, or CO_2 (under wet He atmosphere).

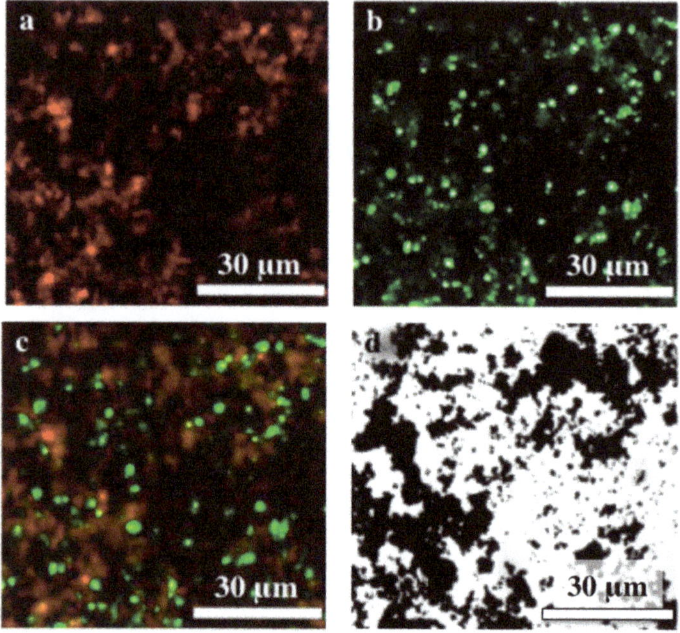

Figure 6. Confocal fluorescence scanning images of (**a**) 430–470 nm emission and (**b**) 505–540 nm emission, (**c**) overlapped image of (**a**,**b**), and (**d**) wide field image, using a 375 nm laser excitation of p-90Fe.

Figure 7a shows the time-dependent CO production on the as-synthesized samples during a 4 h experiment. The CO production rate over pure $CsPbBr_3$ and MIL-100(Fe) are similar, about 4.5 µmol g^{-1} h^{-1}. The composite materials show significantly higher activity, and p-90Fe exhibits a maximum CO production of 20.4 µmol g^{-1} h^{-1}, which is about 4.5 times higher than the pure constituents. Upon increasing the load of MIL-100(Fe), lower photocatalytic activity is obtained. The decrease in photocatalytic activity may be caused by an increasing amount of $CsPb_2Br_5$ in the composites, which is not active upon visible light irradiation. The critical role of H_2O was revealed by a test reaction on p-90Fe in high purity CO_2 gas without H_2O. The CO yield over p-90Fe dropped from 20.4 to 5.3 µmol g^{-1} h^{-1}. The composites' stability was evaluated by four consecutive runs (4 h each, in total 16 h) on p-90Fe. As shown in Figure 7b, the composite exhibits no significant deactivation, and its crystal structure is well maintained after the 16 h photocatalytic reaction (Figure S4).

Two reference samples, p-post and p-mix, with the same amount of MIL-100(Fe) loaded as p-90Fe, were constructed. p-post was synthesized by the anti-solvent deposition of $CsPbBr_3$ onto MIL-100(Fe) [39]. p-mix was prepared through ultrasonically mixing the $CsPbBr_3$ and MIL-100(Fe) powders. Under visible light irradiation, the CO production rates over p-post and p-mix are only 8.7 µmol g^{-1} h^{-1} and 6.8 µmol g^{-1} h^{-1}, nearly one-third of that obtained via the newly introduced in situ growth route. A summary of the reported photocatalytic CO_2 reduction performance on perovskite-based and traditional photocatalysts under various illumination conditions is listed in Table S1.

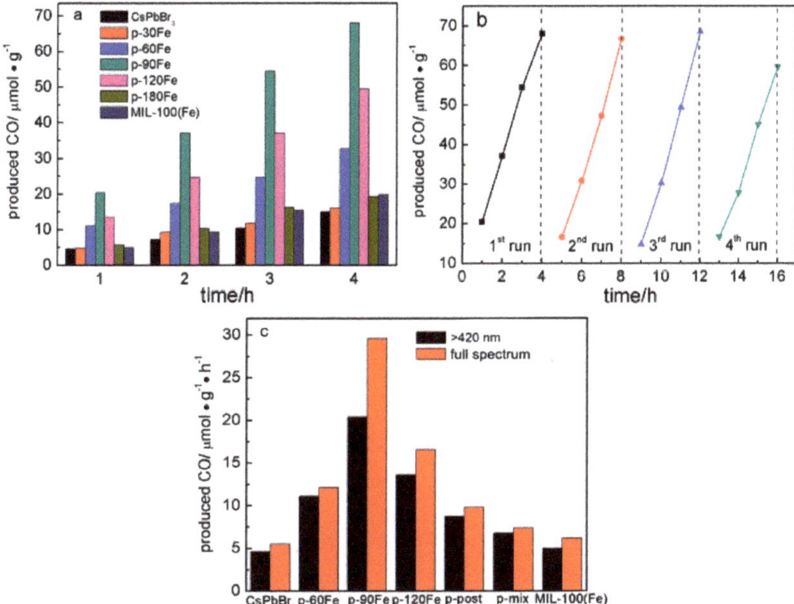

Figure 7. (**a**) Time-dependent CO generation over the synthesized MIL-100(Fe), CsPbBr$_3$, and composites, (**b**) Stability test on p-90Fe for four consecutive runs (4 h each, in total 16 h), and (**c**) CO generation over the photocatalysts under different illumination conditions.

It is observed that CsPbBr$_3$ gradually transforms to CsPb$_2$Br$_5$ in the composites, and CsPb$_2$Br$_5$ is a large bandgap material without visible light response. Therefore, full-spectrum (300–800 nm) measurements were performed on the selected samples to investigate the photocatalytic contribution of CsPb$_2$Br$_5$ in the composites. As shown in Figure 7c, only slight improvements in CO generation were found on CsPbBr$_3$, p-post, and p-mix, compared to that under visible light irradiation (420–800 nm). The CO generation significantly enhanced from 20.4 to 29.6 μmol g^{-1} h^{-1} on p-90Fe and from 13.6 to 16.6 μmol g^{-1} h^{-1} on p-120Fe. Hence, CsPb$_2$Br$_5$, used to be seen as the undesirable byproduct of CsPbBr$_3$, can contribute to the composites' photoactivity.

It is acknowledged that the photocatalysts' performance can be influenced by the specific surface area. First, gas sorption measurements were performed (Figure S5a). N$_2$ physisorption revealed a negligible surface area for the pure CsPbBr$_3$ microcrystals. The composite materials show a tremendously increased specific surface area between 130 and 400 m^2/g (Table 3); the specific surface area increases with the amount of MIL-100(Fe). This surface area enhancement favors the exposure of active sites and the adsorption of CO$_2$ molecules. For the composites, a type IV isotherm is observed related to the existence of a mesoporous structure. Further, Figure S5b shows the CO$_2$ adsorption isotherms of the as-obtained samples. The CO$_2$ uptakes in the composites are 15 to 30 times higher than that in pure CsPbBr$_3$, which benefits the photocatalytic efficiency. Figure S6 shows the N$_2$ physisorption isotherms and related pore size distribution of p-90Fe before and after the reaction. After the reaction, the surface area reduced from 201 m^2/g to 155 m^2/g, the average pore size decreased from 4.5 nm to 3.7 nm, and the pore volume dropped from 0.21 cm^3/g to 0.11 cm^3/g. As can be seen in Figure S6b, p-90Fe has dominant peaks at 1.2 and 2.0 nm, which is the typical pore size distribution of MIL-100(Fe) [40]. After the reaction, the portion of the mesopores around 10 nm decreased, which may be due to the influence of H$_2$O on the particle size and shape during the reaction.

Table 3. The specific surface area of the prepared composites.

Sample	p-60Fe	p-90Fe	p-120Fe	p-180Fe
$S_{BET}/m^2\,g^{-1}$	130	201	277	390

As shown in Figure 8, the PL decay plots of $CsPbBr_3$ and composite p-90Fe are fitted with a biexponential decay function. The short and long PL lifetimes can be assigned to two different physical origins. The short (τ_1) and long (τ_2) lifetimes are related to the trap-assisted and exciton recombination, respectively [13]. The average lifetime (τ) of $CsPbBr_3$ exhibits an obvious decrease from 4.4 to 1.6 ns after adding MOF, resulting from the suppressed exciton recombination. MIL-100(Fe) here functions as a quencher of $CsPbBr_3$, endowing the composite with a more effective electron extraction [41].

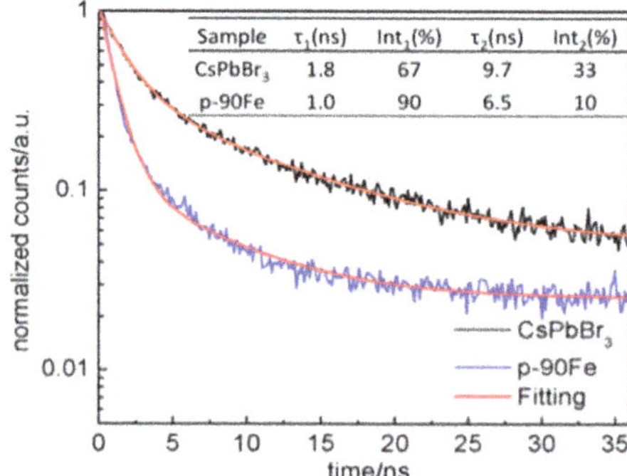

Figure 8. Time-resolved PL spectra of pure $CsPbBr_3$ and p-90Fe fitted with a biexponential decay kinetic, including the corresponding fitting parameters.

The photocatalytic activity of MIL-100(Fe) mainly originates from the direct excitation of the $Fe_3-\mu_3-oxo$ clusters inside the structure. Under light irradiation, Fe–O clusters in MIL-100(Fe) can be excited, transferring an electron from the O^{2-} to Fe^{3+} for the formation of Fe^{2+}, which is responsible for CO_2 reduction over pure MIL-100(Fe) [22,42]. The combination of the bandgap data with valence band measurements (Figure S7 and Table 2) allow to determine the VB and CB edge potentials of $CsPbBr_3$ to be 0.82 and −1.45 eV, respectively. The VB and CB edge potentials of MIL-100(Fe) are calculated to be 1.75 and −0.93 eV, respectively. Therefore, a typical type II heterojunction was formed by the perfect band structure matching of $CsPbBr_3$ and MIL-100(Fe) [42].

A possible mechanism for the visible-light driven photocatalytic CO_2 reduction over the composite is proposed, as shown in Figure 9. Photo-induced electron and hole pairs are generated on $CsPbBr_3$ and MIL-100(Fe) and tend to transfer. The electrons in the conduction band of $CsPbBr_3$ will transfer to that of MIL-100(Fe), where CO_2 would be reduced to CO. The holes on the valence band of MIL-100(Fe) would migrate to that of $CsPbBr_3$, where H_2O will be trapped to generate O_2.

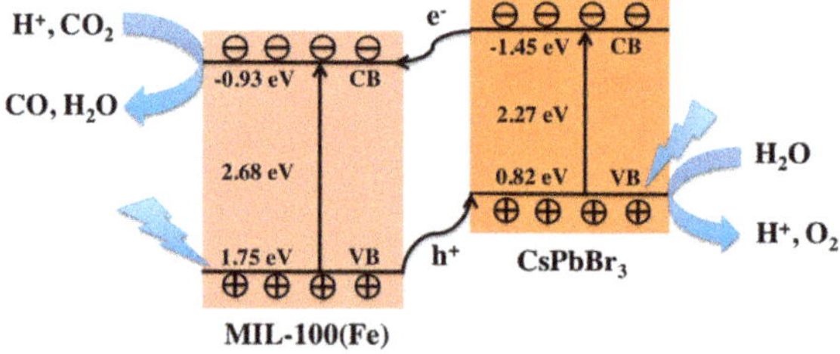

Figure 9. Schematics of the CO_2 photoreduction process on $CsPbBr_3$/MIL-100(Fe) under visible light irradiation.

With H_2O involved in the synthesis, $CsPbBr_3$ gradually converts to $CsPb_2Br_5$, which has no visible light response. Revealed by the XRD and UV-vis absorption spectra, p-90Fe has the highest amount of MIL-100(Fe) in the composite while maximally retaining $CsPbBr_3$. MIL-100(Fe) greatly increases the surface area and enhances the visible light absorption ability of $CsPbBr_3$. Upon visible light irradiation, the separation and transfer of the photogenerated charge carriers is promoted in the composites, resulting in enhanced photocatalytic performance. Further increasing the amount of MIL-100(Fe) increases the specific surface area and CO_2 uptake, but at the expense of $CsPbBr_3$.

3. Materials and Methods

3.1. Catalyst Synthesis

3.1.1. Synthesis of $CsPbBr_3$

$CsPbBr_3$ was synthesized by the anti-solvent method. 2.5 mmol cesium bromide (CsBr, 99.9%, Alfa Aesar, Kandel, Germany) and 2 mmol lead (II) bromide ($PbBr_2$, 99.999%, Alfa Aesar, Kandel, Germany) were dissolved in 15 mL dimethyl sulphoxide (DMSO, ≥99.9%, ACS reagent, Sigma-Aldrich, Overijse, Belgium) and stirred for 12 h. The solution was quickly added into 150 mL toluene under stirring. The obtained product was collected by centrifugation, washed with toluene, and dried in a vacuum oven at 80 °C.

3.1.2. Synthesis of Pure MIL-100(Fe)

In a typical procedure, 2.02 g iron (III) nitrate nonahydrate ($Fe(NO_3)_3 \cdot 9H_2O$)(99%, ACROS Organics, Geel, Belgium), and 0.7 g 1,3,5-benzenetricarboxylic acid (H_3BTC, 95%, Sigma-Aldrich, Overijse, Belgium) were added to 5 mL H_2O and stirred for 30 min at RT. This mixture was then heated to 95 °C and maintained at this temperature for 12 h. After cooling down the mixture, the obtained orange solid product was collected by centrifugation, washed with distilled water, and dried at 80 °C.

3.1.3. Synthesis of the $CsPbBr_3$/MIL-100(Fe) Composites

The $CsPbBr_3$/MIL-100(Fe) composites were obtained by in situ growth. The Fe precursor solution for MIL-100(Fe) was made first, by adding 30, 60, 90, 120, and 180 mg $Fe(NO_3)_3 \cdot 9H_2O$ into 3 mL 1-propanol (99.5%, ACS agent, Fisher Chemical, Merelbeke, Belgium). Next, 0.1 g $CsPbBr_3$ was added to this Fe precursor solution and the mixture was stirred for 12 h at RT. The H_3BTC powder was then added with a 3:1 molar ratio of H_3BTC to $Fe(NO_3)_3 \cdot 9H_2O$. The mixture was heated to 95 °C and kept at this temperature for 12 h. The resulting orange solid product was collected by centrifugation, washed by toluene, and dried at 100 °C in a vacuum oven. The obtained samples were named as p-xFe (x = the weight of $Fe(NO_3)_3 \cdot 9H_2O$ added).

For comparison, $CsPbBr_3$ was loaded onto MIL-100(Fe) by anti-solvent deposition [39]. This sample is named as p-post. Furthermore, a physical mixture of the pure $CsPbBr_3$ and MIL-100(Fe) was prepared by ultrasonically mixing the powders, named as p-mix. The weight ratios of MIL-100(Fe) in both p-post and p-mix are 18 wt%. Finally, 90 mg $Fe(NO_3)_3 \cdot 9H_2O$ was dehydrated in a vacuum oven at 80 °C for 4 h before the composite synthesis, and the obtained sample was named as p-90Fe-dehydrated.

3.2. Characterization

X-ray diffraction (XRD) patterns were recorded on a Malvern PANalytical Empyrean system (Malvern PANalytical, Hoeilaart, Belgium). Scanning electron microscopy (SEM) images and the corresponding energy dispersive X-ray spectroscopy (EDS) data were taken with a FEI-Q FEG250 system (Thermo Fisher Scientific, Zaventem, Belgium). X-ray photoelectron spectroscopy (XPS) data were recorded on an ESCALAB 250 spectrometer (Thermo Fisher Scientific, Beijing, China) with a monochromatized Al Ka X-ray as the excitation source, and the binding energies were calibrated by the C1s peak at 284.8 eV. Gas sorption measurements were performed on a 3Flex Surface Analyzer

(Micromeritics, Unterschleissheim, Germany) where nitrogen (N_2) adsorption-desorption isotherms were acquired at −196 °C and CO_2 adsorption isotherm were acquired at 0 °C. Before the measurement, all the samples were degassed at 150 °C for 8 h under flowing N_2. Thermogravimetric analysis (TGA) was performed from room temperature to 800 °C at a heating rate of 10 °C min^{-1} in O_2 on a TA-TGA Q500 (TA Instruments, Antwerp, Belgium). Fourier transformed infrared (FTIR) spectra were collected using a Bruker FTIR spectrophotometer (Bruker, Kontich, Belgium). UV-Vis diffuse reflectance spectra were recorded with a Lambda-950 UV-Vis spectrometer (PerkinElmer, Mechelen, Belgium). The steady-state photoluminescence (PL) spectra were acquired on an Edinburgh FLS980 instrument (Edinburgh Instruments Ltd, Livingston, UK). The samples were prepared by filling a thin quartz cuvette (Macro cell 110-QS, 1 mm light path, Hellma, Kruibeke, Belgium) with the same volume of powders. Time-resolved PL spectra were acquired by a home-built confocal FLIM microscope, equipped with a single-photon counting device (Picoquant, Berlin, Belgium) Powders of $CsPbBr_3$ and p-90Fe were dispersed in toluene and sonicated for 15 min. The suspension was then dropped onto a 20 × 20 mm cover slide and dried in the vacuum oven at 80 °C overnight. The excitation source was a 5 MHz pulsed 485 nm laser diode, and the emission was filtered by a 530 ± 25 nm band pass filter. Confocal fluorescence images were acquired on a Fluoview FV1000 confocal microscope (Olympus, Tokyo, Japan). An Olympus 20 × 0.75 NA air immersion objective lens was used. A 375 nm laser was used as the excitation source, and the detectors for fluorescence emission were set at 430–470 and 505–540 nm. The image size was 1024 × 768 pixels with a pixel dwell time of 4 μs.

3.3. Photocatalytic CO_2 Reduction Measurement

The photocatalytic reduction of CO_2 was performed in a homemade Pyrex reactor (volume: 150 mL). Visible light was generated by a 300 W Xe lamp with a 420 cut-off filter (Newport, Darmstadt, Germany) and positioned 5 cm away from the photocatalytic reactor. In a typical sample preparation, 20 mg photocatalyst was uniformly dispersed on a flat glass plate with an area of 4 cm^2. The as-prepared sample plate was left in the vacuum oven at 80 °C overnight to remove the residual solvent. Before the reaction, helium flowed through the reactor for about 20 min to eliminate the air inside. Then, a mixture of CO_2 and water vapor, generated by passing high purity CO_2 (99.99%) gas through a water bubbler, flowed through the reactor for another 40 min in the dark. Afterwards, the reactor was closed off and light irradiation was started. The gas sample was evaluated every 1 h by gas chromatography (GC-2014, Shimadzu, Tokyo, Japan) equipped with a ShinCarbon packed column with a flame ionization detector (FID) and a thermal conductivity detector (TCD). The carrier gas used in the GC-2014 was high purity helium. After the 16 h stability test, the sample was collected and treated at 80 °C in the vacuum oven overnight.

4. Conclusions

In summary, we have reported the in situ synthesis of $CsPbBr_3$/MIL-100(Fe) for CO_2 photoreduction under visible light irradiation. Compared to anti-solvent deposition and physical mixing methods, the composites obtained by this new route show a significantly improved CO_2 photoreduction'activity. H_2O is necessary for the in situ growth of the MOF structure, but excessive H_2O initiates the transformation of $CsPbBr_3$ to $CsPb_2Br_5$. Through optimization of the composite material composition, the highest CO production rate is 20.4 μmol g^{-1} h^{-1}, which is about 4.5 times that of the parent materials. The composite showed good stability in a 16 h photocatalytic reaction, where H_2O is involved as the reactant. The introduction of MIL-100(Fe) endowed the composites with a largely increased surface area and enhanced light-harvesting capability in the visible light region. The perfect band matching between $CsPbBr_3$ and MIL-100(Fe) attributes to better electron-hole separation and transfer. The findings here could serve as a steppingstone for further developing MHP photocatalysts, involving MOF-based heterojunctions.

Supplementary Materials: The following are available online at http://www.mdpi.com/2073-4344/10/11/1352/s1, Figure S1: XPS spectra of p-90Fe, Figure S2: Photoluminescence spectra of the as-prepared samples at an excitation wavelength of 380 nm, Figure S3: Typical SEM images of $CsPbBr_3$, MIL-100(Fe) and p-90Fe, and EDS pattern of the selected region in p-90Fe, Figure S4: XRD pattern of p-90Fe before and after a 16 h photocatalytic reaction, Figure S5: N_2 adsorption-desorption isotherms and CO_2 adsorption isotherms of the photocatalysts, Figure S6: N_2 adsorption-desorption isotherms and pore size distribution curves of the p-90Fe before and after reaction, Figure S7: Valence band XPS spectra of MIL-100(Fe), $CsPbBr_3$ and p-90Fe, Table S1: Summary of the reported photocatalytic CO_2 reduction performance of perovskite-based and traditional photocatalysts under various illumination conditions.

Author Contributions: Conceptualization, R.C. and J.H.; methodology, R.C., E.D., M.B.J.R. and J.H.; formal analysis, R.C.; writing—original draft preparation, R.C.; writing—review and editing, M.B.J.R., E.D. and J.H.; supervision, J.H. and M.B.J.R. All authors have read and agreed to the published version of the manuscript.

Funding: This research was funded by the European Union (Horizon 2020) Marie Sklodowska-Curie innovation program (Grant No. 722591, in the form of a Ph.D. fellowship to R.C.), the Research Foundation—Flanders (FWO) through a postdoctoral fellowship to E.D. (FWO Grant No. 12O3719N), and research projects to J.H. and M.B.J.R (FWO Grant Nos. G098319N and ZW15_09-GOH6316), the KU Leuven Research Fund (C14/15/053 and C14/19/079), and the Flemish government through long term structural funding Methusalem (CASAS2, Meth/15/04).

Conflicts of Interest: The authors declare no conflict of interest.

References

1. Sun, Z.X.; Wang, H.Q.; Wu, Z.B.A.; Wang, L.Z. g-C_3N_4 based composite photocatalysts for photocatalytic CO_2 reduction. *Catal. Today* **2018**, *300*, 160–172. [CrossRef]

2. Indrakanti, V.P.; Kubicki, J.D.; Schobert, H.H. Photoinduced activation of CO_2 on Ti-based heterogeneous catalysts: Current state, chemical physics-based insights and outlook. *Energy Environ. Sci.* **2009**, *2*, 745–758. [CrossRef]

3. Tran, P.D.; Wong, L.H.; Barber, J.; Loo, J.S.C. Recent advances in hybrid photocatalysts for solar fuel production. *Energy Environ. Sci.* **2012**, *5*, 5902–5918. [CrossRef]

4. Habisreutinger, S.N.; Schmidt-Mende, L.; Stolarczyk, J.K. Photocatalytic reduction of CO_2 on TiO_2 and other semiconductors. *Angew. Chem. Int. Ed.* **2013**, *52*, 7372–7408. [CrossRef] [PubMed]

5. Zhao, M.T.; Huang, Y.; Peng, Y.W.; Huang, Z.Q.; Ma, Q.L.; Zhang, H. Two-Dimensional metal-organic framework nanosheets: Synthesis and applications. *Chem. Soc. Rev.* **2018**, *47*, 6267–6295. [CrossRef] [PubMed]

6. Tu, W.G.; Xu, Y.; Yin, S.M.; Xu, R. Rational design of catalytic centers in crystalline frameworks. *Adv. Mater.* **2018**, *30*, 1707582–1707610. [CrossRef]

7. Alvaro, M.; Carbonell, E.; Ferrer, B.; Xamena, F.X.L.I.; Garcia, H. Semiconductor behavior of a metal-organic framework (MOF). *Chem. Eur. J.* **2007**, *13*, 5106–5112. [CrossRef]

8. Fu, Y.H.; Sun, D.R.; Chen, Y.J.; Huang, R.K.; Ding, Z.X.; Fu, X.Z.; Li, Z.H. An amine-functionalized titanium metal-organic framework photocatalyst with visible-light-induced activity for CO_2 reduction. *Angew. Chem. Int. Ed.* **2012**, *51*, 3364–3367. [CrossRef]

9. Liang, Z.B.; Qu, C.; Xia, D.G.; Zou, R.Q.; Xu, Q. Atomically dispersed metal sites in MOF-based materials for electrocatalytic and photocatalytic energy conversion. *Angew. Chem. Int. Ed.* **2018**, *57*, 9604–9633. [CrossRef]

10. Qiu, J.H.; Zhang, X.G.; Feng, Y.; Zhang, X.F.; Wang, H.T.; Yao, J.F. Modified metal-organic frameworks as photocatalysts. *Appl. Catal. B* **2018**, *231*, 317–342. [CrossRef]

11. Yang, Z.W.; Xu, X.Q.; Liang, X.X.; Lei, C.; Wei, Y.L.; He, P.Q.; Lv, B.L.; Ma, H.C.; Lei, Z.Q. MIL-53(Fe)-graphene nanocomposites: Efficient visible-light photocatalysts for the selective oxidation of alcohols. *Appl. Catal. B* **2016**, *198*, 112–123. [CrossRef]

12. Zhao, Y.; Wang, Y.; Liang, X.; Shi, H.; Wang, C.; Fan, J.; Hu, X.; Liu, E. Enhanced photocatalytic activity of Ag-$CsPbBr_3$/CN composite for broad spectrum photocatalytic degradation of cephalosporin antibiotics 7-ACA. *Appl. Catal. B* **2019**, *247*, 57–69. [CrossRef]

13. Huang, H.W.; Yuan, H.F.; Janssen, K.P.F.; Solis-Fernandez, G.; Wang, Y.; Tan, C.Y.X.; Jonckheere, D.; Debroye, E.; Long, J.L.; Hendrix, J.; et al. Efficient and selective photocatalytic oxidation of benzylic alcohols with hybrid organic-inorganic perovskite materials. *ACS Energy Lett.* **2018**, *3*, 755–759. [CrossRef]

14. Huang, H.W.; Pradhan, B.; Hofkens, J.; Roeffaers, M.B.J.; Steele, J.A. Solar-Driven metal halide perovskite photocatalysis: Design, stability, and performance. *ACS Energy Lett.* **2020**, *5*, 1107–1123. [CrossRef]

15. Kang, B.; Biswas, K. Exploring polaronic, excitonic structures and luminescence in Cs_4PbBr_6/$CsPbBr_3$. *J. Phys. Chem. Lett.* **2018**, *9*, 830–836. [CrossRef]

16. Dana, J.; Maity, P.; Jana, B.; Maiti, S.; Ghosh, H.N. Concurrent ultrafast electron- and hole-transfer dynamics in $CsPbBr_3$ perovskite and quantum Dots. *ACS Omega* **2018**, *3*, 2706–2714. [CrossRef] [PubMed]

17. Cheng, R.L.; Jin, H.D.; Roeffaers, M.B.J.; Hofkens, J. Incorporation of cesium lead halide perovskites into g-C_3N_4 for photocatalytic CO_2 Reduction. *ACS Omega* **2020**, *5*, 24495–24503. [CrossRef] [PubMed]

18. Xu, Y.F.; Yang, M.Z.; Chen, B.X.; Wang, X.D.; Chen, H.Y.; Kuang, D.B.; Su, C.Y. A $CsPbBr_3$ perovskite quantum dot/graphene oxide composite for photocatalytic CO_2 reduction. *J. Am. Chem. Soc.* **2017**, *139*, 5660–5663. [CrossRef]

19. Kong, Z.C.; Liao, J.F.; Dong, Y.J.; Xu, Y.F.; Chen, H.Y.; Kuang, D.B.; Su, C.Y. Core@shell $CsPbBr_3$@zeolitic imidazolate framework nanocomposite for efficient photocatalytic CO_2 reduction. *ACS Energy Lett.* **2018**, 2656–2662. [CrossRef]

20. Mollick, S.; Mandal, T.N.; Jana, A.; Fajal, S.; Desai, A.V.; Ghosh, S.K. Ultrastable luminescent hybrid bromide perovskite@MOF nanocomposites for the degradation of organic pollutants in water. *ACS Appl. Nano Mater.* **2019**, *2*, 1333–1340. [CrossRef]

21. Wan, S.P.; Ou, M.; Zhong, Q.; Wang, X.M. Perovskite-type $CsPbBr_3$ quantum dots/UiO-66(NH$_2$) nanojunction as efficient visible-light-driven photocatalyst for CO_2 reduction. *Chem. Eng. J.* **2019**, *358*, 1287–1295. [CrossRef]

22. Wang, D.K.; Wang, M.T.; Li, Z.H. Fe-Based metal-organic frameworks for highly selective photocatalytic benzene hydroxylation to phenol. *ACS Catal.* **2015**, *5*, 6852–6857. [CrossRef]

23. Liang, R.W.; Luo, S.G.; Jing, F.F.; Shen, L.J.; Qin, N.; Wu, L. A simple strategy for fabrication of Pd@MIL-100(Fe) nanocomposite as a visible-light-driven photocatalyst for the treatment of pharmaceuticals and personal care products (PPCPs). *Appl. Catal. B* **2015**, *176*, 240–248. [CrossRef]

24. Hong, J.D.; Chen, C.P.; Bedoya, F.E.; Kelsall, G.H.; O'Hare, D.; Petit, C. Carbon nitride nanosheet/metal-organic framework nanocomposites with synergistic photocatalytic activities. *Catal. Sci. Technol.* **2016**, *6*, 5042–5051. [CrossRef]

25. Zheng, J.J.; Jiao, Z.B. Modified Bi_2WO_6 with metal-organic frameworks for enhanced photocatalytic activity under visible light. *J. Colloid Interface Sci.* **2017**, *488*, 234–239. [CrossRef]

26. Zhang, C.F.; Qiu, L.G.; Ke, F.; Zhu, Y.J.; Yuan, Y.P.; Xu, G.S.; Jiang, X. A novel magnetic recyclable photocatalyst based on a core-shell metal-organic framework Fe_3O_4@MIL-100(Fe) for the decolorization of methylene blue dye. *J. Mater. Chem. A* **2013**, *1*, 14329–14334. [CrossRef]

27. Parvinizadeh, F.; Daneshfar, A. Fabrication of a magnetic metal-organic framework molecularly imprinted polymer for extraction of anti-malaria agent hydroxychloroquine. *New J. Chem.* **2019**, *43*, 8508–8516. [CrossRef]

28. Chaturvedi, G.; Kaur, A.; Umar, A.; Khan, M.A.; Algarni, H.; Kansal, S.K. Removal of fluoroquinolone drug, levofloxacin, from aqueous phase over iron based MOFs, MIL-100(Fe). *J. Solid State Chem.* **2020**, *281*, 121029–121038. [CrossRef]

29. Mahmoodi, N.M.; Abdi, J.; Oveisi, M.; Asli, M.A.; Vossoughi, M. Metal-Organic framework (MIL-100 (Fe)): Synthesis, detailed photocatalytic dye degradation ability in colored textile wastewater and recycling. *Mater. Res. Bull.* **2018**, *100*, 357–366. [CrossRef]

30. Huang, J.; Zhang, X.B.; Song, H.Y.; Chen, C.X.; Han, F.Q.; Wen, C.C. Protonated graphitic carbon nitride coated metal-organic frameworks with enhanced visible-light photocatalytic activity for contaminants degradation. *Appl. Surf. Sci.* **2018**, *441*, 85–98. [CrossRef]

31. Li, Z.J.; Hofman, E.; Li, J.; Davis, A.H.; Tung, C.H.; Wu, L.Z.; Zheng, W.W. Photoelectrochemically active and environmentally stable $CsPbBr_3$/TiO_2 core/shell nanocrystals. *Adv. Funct. Mater.* **2018**, *28*, 1704288–1704294. [CrossRef]

32. Pal, P.; Saha, S.; Banik, A.; Sarkar, A.; Biswas, K. All-solid-state mechanochemical synthesis and post-synthetic transformation of inorganic perovskite-type halides. *Chem. Eur. J.* **2018**, *24*, 1811–1815. [CrossRef] [PubMed]

33. He, X.; Nguyen, V.; Jiang, Z.; Wang, D.W.; Zhu, Z.; Wang, W.N. Highly-Oriented one-dimensional MOF-semiconductor nanoarrays for efficient photodegradation of antibiotics. *Catal. Sci. Technol.* **2018**, *8*, 2117–2123. [CrossRef]

34. Araya, T.; Chen, C.C.; Jia, M.K.; Johnson, D.; Li, R.; Huang, Y.P. Selective degradation of organic dyes by a resin modified Fe-based metal-organic framework under visible light irradiation. *Opt. Mater.* **2017**, *64*, 512–523. [CrossRef]

35. Tang, X.S.; Hu, Z.P.; Yuan, W.; Hu, W.; Shao, H.B.; Han, D.J.; Zheng, J.F.; Hao, J.Y.; Zang, Z.G.; Du, J.; et al. Perovskite $CsPb_2Br_5$ microplate laser with enhanced stability and tunable properties. *Adv. Opt. Mater.* **2017**, *5*, 1600788–1600795. [CrossRef]

36. Tang, X.S.; Han, S.; Zu, Z.Q.; Hu, W.; Zhou, D.; Du, J.; Hu, Z.P.; Li, S.Q.; Zang, Z.G. All-Inorganic perovskite $CsPb_2Br_5$ microsheets for photodetector application. *Front. Phys.* **2018**, *5*. [CrossRef]

37. Song, P.J.; Qiao, B.; Song, D.D.; Liang, Z.Q.; Gao, D.; Cao, J.Y.; Shen, Z.H.; Xu, Z.; Zhao, S.L. Colour- and structure-stable $CsPbBr_3$-$CsPb_2Br_5$ compounded quantum dots with tuneable blue and green light emission. *J. Alloys Compd.* **2018**, *767*, 98–105. [CrossRef]

38. Gargiulo, V.; Alfe, M.; Raganati, F.; Lisi, L.; Chirone, R.; Ammendola, P. BTC-Based metal-organic frameworks: Correlation between relevant structural features and CO_2 adsorption performances. *Fuel* **2018**, *222*, 319–326. [CrossRef]

39. Pu, Y.C.; Fan, H.C.; Liu, T.W.; Chen, J.W. Methylamine lead bromide perovskite/protonated graphitic carbon nitride nanocomposites: Interfacial charge carrier dynamics and photocatalysis. *J. Mater. Chem. A* **2017**, *5*, 25438–25449. [CrossRef]

40. Tan, F.C.; Liu, M.; Li, K.Y.; Wang, Y.R.; Wang, J.H.; Guo, X.W.; Zhang, G.L.; Song, C.S. Facile synthesis of size-controlled MIL-100(Fe) with excellent adsorption capacity for methylene blue. *Chem. Eng. J.* **2015**, *281*, 360–367. [CrossRef]

41. Zhan, X.S.; Jin, Z.W.; Zhan, J.R.; Bai, D.L.; Bian, H.; Wang, K.; Sun, J.; Wang, Q.; Liu, S.F. All-Ambient processed binary $CsPbBr_3$-$CsPb_2Br_5$ perovskites with synergistic enhancement for high-efficiency Cs-Pb-Br-based solar cells. *ACS Appl. Mater. Interfaces* **2018**, *10*, 7145–7154. [CrossRef] [PubMed]

42. Wang, D.K.; Song, Y.J.; Cai, J.Y.; Wu, L.; Li, Z.H. Effective photo-reduction to deposit Pt nanoparticles on MIL-100(Fe) for visible-light-induced hydrogen evolution. *New J. Chem.* **2016**, *40*, 9170–9175. [CrossRef]

Publisher's Note: MDPI stays neutral with regard to jurisdictional claims in published maps and institutional affiliations.

 *catalysts*

Article

Testing Metal–Organic Framework Catalysts in a Microreactor for Ethyl Paraoxon Hydrolysis

Palani Elumalai [1], Nagat Elrefaei [2], Wenmiao Chen [1], Ma'moun Al-Rawashdeh [2,*] and Sherzod T. Madrahimov [1,*]

[1] Chemistry Department, Texas A&M University at Qatar, Doha, Qatar; palani.elumalai@qatar.tamu.edu (P.E.); cwm-tamu@tamu.edu (W.C.)

[2] Chemical Engineering Department, Texas A&M University at Qatar, Doha, Qatar; nagat.elrefaei@qatar.tamu.edu

* Correspondence: mamoun.al-rawashdeh@qatar.tamu.edu (M.A.-R.); sherzod.madrahimov@qatar.tamu.edu (S.T.M.)

Received: 17 September 2020; Accepted: 1 October 2020; Published: 9 October 2020

Abstract: We explored the practical advantages and limitations of applying a **UiO-66**-based metal–organic framework (MOF) catalyst in a flow microreactor demonstrated by the catalytic hydrolysis of ethyl paraoxon, an organophosphorus chemical agent. The influences of the following factors on the reaction yield were investigated: a) catalyst properties such as crystal size (14, 200, and 540 nm), functionality (NH_2 group), and particle size, and b) process conditions: temperature (20, 40, and 60 °C), space times, and concentration of the substrate. In addition, long-term catalyst stability was tested with an 18 h continuous run. We found that tableting and sieving is a viable method to obtain MOF particles of a suitable size to be successfully screened under flow conditions in a microreactor. This method was used successfully to study the effects of crystal size, functionality, temperature, reagent concentration, and residence time. Catalyst particles with a sieved fraction between 125 and 250 µm were found to be optimal. A smaller sieved fraction size showed a major limitation due to the very high pressure drop. The low apparent activation energy indicated that internal mass transfer may exist. A dedicated separate study is required to assess the impact of pore diffusion and site accessibility.

Keywords: MOF; catalyst; microreactor; kinetic studies

---

1. Introduction

In the early 1990s, microreactors and flow chemistry emerged and established themselves as powerful tools for exploratory research, piloting, and industrial production. Due to the characteristically small diameter of the reaction channel in the micron-to-millimeter range, a very high surface-to-volume ratio is realized and mixing and heating are significantly accelerated, which enable demanding chemical processes to operate safely and under closely controlled conditions [1,2]. The very low holdup of the reaction volume ensures an entirely new level of inherent safety in the process, as evidenced in the application of this technology to toxic or explosive reactions [3–6]. The narrow residence time distribution approaching that of an ideal plug flow reactor [7–9] makes it beneficial for maximizing the selectivity of parallel and sequential reactions and in the synthesis of monodispersed nanoparticles [10–12]. Running reactions under flow microconditions enables much better quality control compared to batch reactions, as processes can be run under continuous monitoring using online sensors. As such, flow microreactors are ideal for fully automated systems to synthesize, screen, and optimize chemical research [13–15].

Metal–organic frameworks (MOFs) are an emerging class of porous materials constructed from metal-containing secondary building units (SBUs) and organic linkers [16]. The highly porous cavities decorated with functional ligands and active metal SBUs produce MOFs with exceptional potential for the fabrication of a variety of heterogeneous catalysts. For example, the hydrolysis of chemical warfare agents (CWAs) and simulants with highly stable Zr(IV)-based MOFs such as **UiO-66, NU-901, NU-1000,** and **MOF-808** has been explored because of their ultrahigh stabilities in aqueous solvents. Using Lewis acidic zirconium clusters as active sites has also been studied [17–19]. Although MOF or MOF-based composite catalysts have been proven to be effective for the degradation of CWAs, most examples were tested in batch conditions in basic solutions. In general, there are few examples of catalytic testing of MOF-based catalysts in flow reactors as shown in Scheme 1 [20,21].

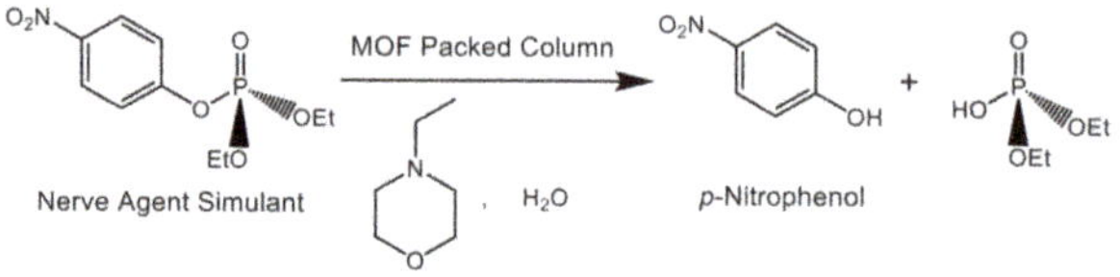

Scheme 1. Schematic illustration of the catalytic hydrolysis of CWAs under microflow conditions.

Catalyst testing in flow microreactors has many advantages over traditional solid catalyst testing in batch reactors [20]. First, many operating parameters such as temperature, pressure, and feed concentrations can be easily and quickly varied in flow microreactors, which provide an insight into the reaction mechanism and kinetics; second, consumption of chemicals and waste production is significantly reduced; and third, it enables easy testing of catalyst reusability, and catalyst stability under reaction conditions via long-term testing on a stream and changing the feed purity. Additionally, flow microreactors enable the study of the direct leaching of the active metal or organic components by chemical analysis of the filtrate and effluent. Considering the potential of MOFs as catalytic materials and flow microreactors as a powerful tool for catalyst testing, studying how to use MOF catalysts in a flow microreactor is valuable for future research, especially during the exploration and screening phase.

A major limitation to loading the as-synthesized MOF catalyst powder in the flow microreactor is the small particle size, ranging from nanometers up to a few micrometers, which results in a very high pressure drop. To eliminate this drop, the catalyst particles must be enlarged to a sufficient size. Several compounding methods have been proposed in the literature to enlarge MOF catalyst particle size. These include coating the catalyst on the wall of the reaction channel, using a monolith, producing catalyst microfibers, or using 3D printing to construct smart designs [22,23]. Each of these concepts has its own advantages and specific challenges. Although these are promising solutions for final commercial applications, testing catalysts in powder form is sometimes inevitable in the research stage. This is especially true at the initial exploratory phase of research, in which determining intrinsic reaction kinetics is essential, the quantities of available catalysts are small, and supply is limited. The intrinsic kinetics are better tested without the presence of binders, as they could affect activity and catalyst particle size needs to remain small to avoid mass and heat transfer limitations.

The most common method used to increase catalyst particle size is tableting the as-synthesized powder via mechanical compression, followed by crushing and sieving the tablet to obtain particles of the desired size. This is advantageous and easy as no binding material or additional complex treatment steps that could affect the catalyst properties are needed. The only concern is that applying mechanical compression to MOFs could cause a decrease in surface area due to destruction of the crystalline structure [24–26]. Some MOFs collapse when submitted to mechanical compression beyond a certain pressure [27]. The zirconium metal–organic framework **UiO-66** chosen for this study possesses exceptional mechanical stability in a highly porous system and can be processed through the pelleting, crushing, and sieving procedure without significant mechanical damage [28–30].

The aim of this paper is to demonstrate the capability of MOFs as a catalytic material in a flow microreactor. The emphasis is mainly on demonstrating if increasing particle size by tableting and crushing is a viable option for efficient catalyst screening and testing. To this end, the detoxification of ethyl-paraoxon (pesticide) through a hydrolysis reaction was studied in the loaded capillary flow reactor. Several parameters were tested including different MOF crystal sizes and functionality. The operating temperature, concentration, and residence time were also tested. Finally, the catalyst was tested for a long period of time under continuous flow to check its stability.

2. Results

2.1. MOF Synthesis, Loading Sieved Catalysts in the Capillary Flow Reactor and Analysis of the Catalyst Bed

All reagents were purchased from commercial sources and used without further purification. **UiO-66-14nm (1)**, **UiO-66-200nm (2)**, **UiO-66-540nm (3)**, **UiO-66-NH$_2$-14nm (4)**, and ethyl-paraoxon were synthesized according to literature procedures (see the Supplementary Materials) [31,32].

To prepare the particles suitable for microreactor loading, the as-synthesized MOF powders were tableted for 5 min using a bench-top tablet press at 10 tons for a tablet diameter of 13 mm. The tablet was then crushed using a hand mortar and sieved. Particle sizes in the sieved fraction ranged from 45 to 125 μm and from 125 to 250 μm were collected for reactor loading. The process was repeated several times until a sufficient amount of powder in the targeted ranges was obtained. The sieved catalyst was then loaded into the 15 cm capillary tube with an internal diameter of 1.55 mm. The capillary loaded with the catalyst was secured from both sides by inserting glass wool. A quantity of catalyst ranging from 35 to 60 mg was loaded into each capillary, resulting in loaded catalyst lengths of 28 to 37 mm. The six capillary tubes prepared according to this method are summarized in Table 1. For example, reactor **1a** represents the capillary tube loaded with a 30 mm catalyst bed of 35 mg of particles, with particle size fractions between 125 and 250 μm and prepared with **UiO66_14nm and UiO66** MOF with an average crystal size of 14 nm.

Table 1. Catalyst-loaded capillary reactors used in this study.

Reactor	Loaded UiO66	Amount (mg)	Length (mm)	Sieved Fraction (μm)
1a	UiO66_14nm	35	30	125–250
1b	UiO66_14nm	45	28	45–125
2a	UiO66_200nm	39	29	125–250
3a	UiO66_540nm	60	37	125–250
4a	UiO66-NH$_2$_14nm	28	36	125–250
4b	UiO66-NH$_2$_14nm	35	30	45–125

Table 2 presents the Langmuir surface area of as-synthesized MOF samples after tableting and sieving and a sample analyzed after the reaction. The surface area measurements for as-synthesized MOF was similar to what is reported in the literature [33,34]. The surface area of tableted MOF **1a** significantly decreased compared to as-synthesized **UiO66**, **1**, although no damage to the crystal structure was incurred as observed by the powder X-ray diffraction (PXRD) patterns, which are shown in the supplementary material. A similar decrease in the surface area after the tableting and sieving process was also observed for catalyst **4a**. A significant decrease in surface area for catalysts **2a** and **3a**, with larger crystal sizes, has also been previously reported [35]. We also observed that the surface area of catalyst **1a^{AR}** recovered after the reaction decreased compared to the surface area measured before loading, which may be due to the substrate or product being trapped in the pores of the catalyst during the flow reaction. The as-synthesized, tableted, and sieved catalysts were also analyzed by transmission electron microscopy (TEM), scanning electron microscopy (SEM), and energy dispersive X-ray analysis EDX-elemental mapping and PXRD techniques. Despite the observed decrease in the surface area, these measurements showed that the structure and integrity of the particles were preserved during the flow reactions, in agreement with previous observations on the mechanical and chemical stability of **UiO66** MOFs. Detailed TEM, SEM, and EDX-elemental mapping and Brunauer-Emmett-Teller (BET) and PXRD analysis data are provided in the Supplementary Materials, Figures S1–S17.

Table 2. Langmuir surface areas of prepared **UiO66**-based MOFs.

MOF	Langmuir Surface Area (m^2 g^{-1})
1	1280
1a	1102
1a ^{AR}	1218
2a	628
3a	230
4	1317
4a	1246

^{AR} After catalytic run.

2.2. Catalyst Testing for Ethyl Paraoxon Hydrolysis and Analysis

The experimental setup for the flow microreactor is shown in Figure 1. It consisted of a syringe pump that enabled the microreactor flow rates to be controlled. The 12 mL syringe was filled with the buffer solution of the 3.4 g/L (0.012 mol/L) organophosphorus agent and connected to the reactor via capillary tubes. The reactions were conducted at different temperatures by placing the microreactor in a heated water bath. The products were collected in small batches of less than 0.2 mL and the yields were measured offline via UV spectroscopy by comparing the absorbance of p-nitrophenoxide at 405 nm in the product mixture to the calibration curve. The calibration curve and further details about the UV analysis are provided in the Supplementary Materials, Figure S18.

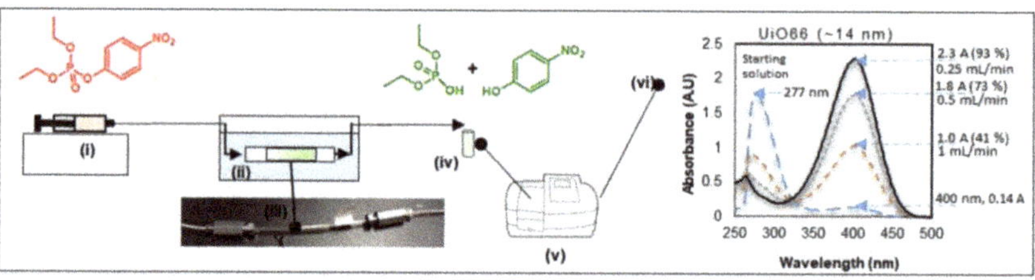

Figure 1. Experimental testing setup and analysis. (**i**) Syringe pump, (**ii**) water bath, (**iii**) loaded MOF catalyst in a capillary flow reactor, (**iv**) collected sample in a vial, (**v**) UV–Vis spectroscopy, (**vi**) example of one of the results, including reference starting solution.

After the catalyst was loaded, the loaded microreactors were tested for a pressure drop at three flow rates of 0.5, 1, and 1.5 mL/min using water. A very high pressure drop was observed in reactors **1b** and **4b** (Table 1) with sieved particle sizes between 45 and 125 µm that resulted in catalyst particle movement and the formation of a segregated channeled catalyst bed. For reactors loaded with larger sieved particle sizes of 125 to 250 µm, the catalyst bed was stable and demonstrated reproducible performance. MOF catalyst particles larger than 125 µm were therefore tested in catalytic runs in the capillary flow reactor.

Figure 2 summarizes all of the flow reactions conducted in this study. Results are presented in terms of yield versus liquid hourly space velocity (LHSV, flow rate divided by amount of loaded catalyst). For reactor **1a** tested in an ambient temperature, yields in excess of 93% in less than 15 s were obtained, corresponding to an LHSV of 7.1 L/min/kg (a liquid flow rate of 0.25 mL/min). The yield decreased linearly as the flow rate increased. A good level of reproducibility was observed, with a yield fluctuation of ±2%. Reactor **1b**, loaded with the same MOF catalyst but a smaller particle size (with a sieved fraction between 45 and 125 µm), demonstrated a significantly lower yield (see Figure S19 in Supplementary Materials). As mentioned earlier, the catalyst bed was not stable for sieved fractions smaller than 125 µm, resulting in segregation and channeling because of the very high pressure drop.

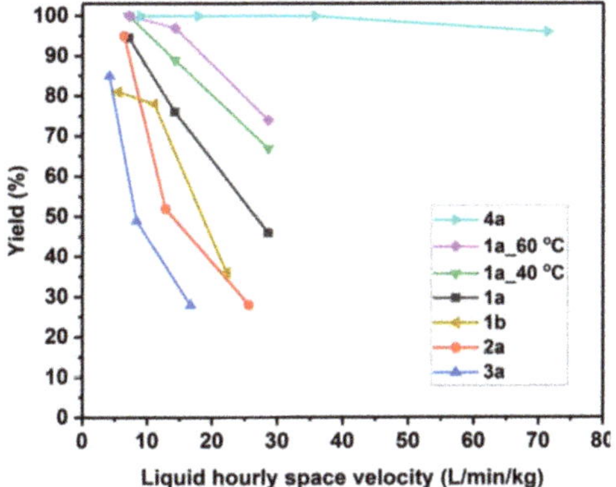

Figure 2. The plot of yield vs. liquid hourly space velocity (LHSV) for reactors 1–4.

UiO66 crystal sizes of 14, 200, and 540 nm, **1a**, **2a**, and **3a**, respectively, were tested, as shown in Figure 2. These MOFs were all prepared according to the procedure described by Morris et al., in which the average MOF particle size was controlled by the amount of acid used in the synthesis, whereby using a larger amount of acid in the MOF synthesis leads to the formation of MOFs with larger crystal size [31]. Increasing the crystal sizes decreased the yield at a given LHSV. This is consistent with the results reported in the literature. Since the reaction is considered to occur at the surface of the crystal, catalysts with a smaller crystal size will have more surface area accessible and more activity [36]. The yield trend for larger crystal sizes of 200 nm (**2a**) and 540 nm (**3a**) was different than that of **1a**, indicating that other factors influence the activity. It could be that when crystal size increases, the interparticle pore structure in the crystal aggregate is altered to reduce the accessibility of the catalytically active sites. The effect of the –NH$_2$ group in the MOF linker on catalyst activity was also tested. As expected, the presence of the –NH$_2$ group in the MOF resulted in a significant increase in catalyst activity, in agreement with prior reports [37].

The long-term stability of the catalyst under the flow conditions was tested by running the reference catalyst in reactor **1a** for 18 h at a liquid flow rate of 0.5 mL/min at 20 °C with periodic sampling, as shown in Figure 3. As can be seen in the figure, catalyst activity remained level throughout the testing period. This excellent stability over time suggests good potential for future practical applications of MOF catalysts in flow reactors.

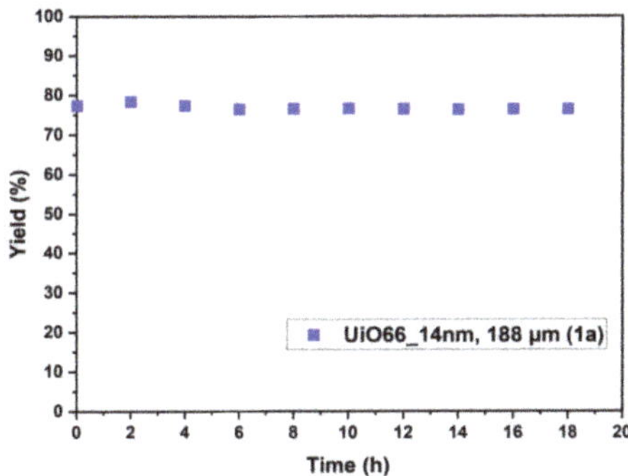

Figure 3. Stability of reaction yield for reference case of reactor **1a** tested for 18 h at liquid flow rate of 0.5 mL/min and 20 °C.

2.3. Catalyst Kinetic Evaluation Based on Initial Rates

The effect of temperature on catalyst performance is shown in Figure 2 for reactor **1a**. The flow microreactor enabled multiple space times and temperatures to be easily tested. This is a significant advantage over batch reactors where each condition requires a separate reaction setup. The kinetic measurements of catalyst **1a** were found to nicely fit first-order reaction kinetics, as shown in Figure 4. Using the Arrhenius plot, the apparent activation energy was estimated as 8.8 kJ/mol. Although the response to temperature is evident in Figure 4, the process may be limited by internal mass transfer limitation, as the apparent activation energy (Ea) value could be considered low [38]. To provide conclusive answers on mass transfer limitations, a dedicated study in which various particle size ranges and loading lengths are tested at different temperatures is required.

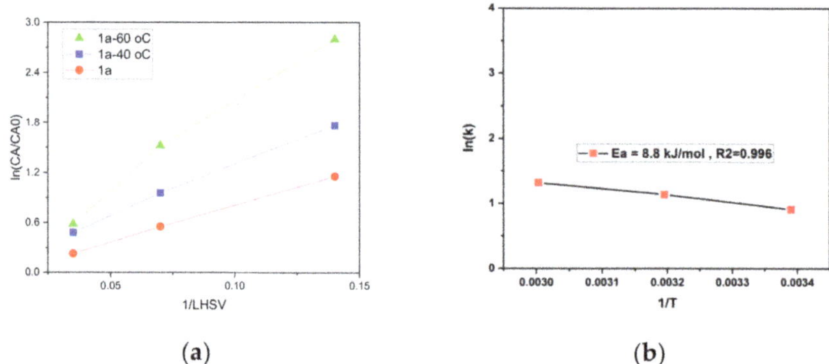

Figure 4. Arrhenius plot of catalyst **1a** loaded in reactor **1**, which followed first-order reaction kinetics.

3. Materials and Methods

3.1. Materials

All air- or water-sensitive reactions were carried out in a nitrogen atmosphere using oven-dried glassware. All syntheses of oven-based **UiO66** and the derivative MOFs were carried out in explosion-proof HERAtherm OMS-100 (Thermo Fisher Scientific, Waltham, MA, USA) ovens that had been pre-heated to a specific temperature. All sonication was carried out with a Fisher Scientific Ultrasonic Cleaner FS60 (Thermo Fisher Scientific, Waltham, MA, USA). Anhydrous solvents in Sure/Seal™ bottles were purchased from the Aldrich Chemical Company (Milwaukee, WI, USA) and used as received inside a nitrogen-filled KIYON Glovebox (Korea Kiyon Glovebox System, Seoul, South Korea). All other reagents were purchased from the Aldrich Chemical Company (Milwaukee, WI, USA) and used without further purification unless otherwise noted.

Powder XRD spectra were collected using a Rigaku Ultima IV multipurpose X-ray diffractometer equipped with Cu-Kα radiation and a fixed monochromator. The XRD was operated at 40 KV and 40 mA and a fixed time-scan mode with a 0.02-degree step width and 1 s/step count time used for data collected from 5 to 90 degrees.

For scanning electron microscopy/energy dispersive X-ray spectroscopy (SEM/EDS), an FEI Quanta 400 environmental scanning electron microscope was used to collect high-resolution SEM images. The SEM was operated at an acceleration voltage of 30 KV and the working distance was 6.5 mm. An EDAX Apollo EDS system was used for EDS signal collection and analysis.

Transmission electron microscopy images were collected using a Thermo Fisher company TalosF200x model with a super X EDS system. Powder 1 or 2 was dispersed in isopropyl alcohol and sonicated for about 5 min. An amount of 20 μL of the dispersed solution was dropped over a 300 mesh copper/lacey carbon grid and dried at room temperature for TEM analysis.

For N_2 gas sorption analysis, the MOF samples for analysis were heated under vacuum overnight at 150 °C to remove the solvents trapped within the pores prior to analysis. Following this, an ~50 mg sample was transferred to pre-weighed sample tubes and degassed at 150 °C for 3 h by a Micromeritics Flow prep 060 sample degassing system. After degassing, the MOF sample tubes were re-weighed to obtain the mass for the samples. Sorption data with the Brunauer–Emmett–Teller method (BET) and the Langmuir surface area (m^2/g) measurements were collected at 77 K with N_2 on a Micromeritics TriStar II 3020 surface area and porosity system adsorption analyzer. The BET surface area and pore volume of prepared MOFs are shown in Supplementary Materials Table S1.

For the continuous-flow reaction, the MOF catalyst was loaded in transparent 1/8″ Radel® tubing with an inner diameter of 1.55 mm. Glass wool was used to hold the catalyst inside the reactor. A syringe pump was used to pump the reaction solution through 1/16″ peek tubing connected to the packed reactor capillary and placed in a water bath to control the temperature, as shown in Figure 1. Before starting all flow runs, the packed microreactor was run under the flow of pure solvent to remove any components from the catalyst bed and perform the pressure drop test. Then a syringe was loaded with the chemical solution of ethyl paraoxon at a concentration of 0.061 M in a buffer solution of 0.1 M N-methylmorpholine. Multiple samples were collected from each condition and analyzed using UV–Vis measurements calibrated beforehand based on the product concentration.

UV–Vis spectra were recorded using quartz cells with a path length of 10 mm at room temperature with a Perkin Elmer UV/Vis/NIR spectrophotometer model Lambda 950; analysis was carried out using the UVwinLab software. The spectral baseline was corrected with Cary Win UV software. Progress of the reaction was monitored by following the p-nitrophenoxide absorbance at 405 nm to avoid overlapping absorptions with other species. Yields were calculated based on the calibration run of para-nitrophenol solutions in known concentrations (Supplementary Materials Figure S16).

3.2. Synthesis

3.2.1. **UiO66** MOF Nano-Catalyst Synthesis

Synthesis of UiO-66_14 nm (1) and UiO-66-NH$_2$_14 nm (4)

All **UiO-66** MOF materials with 14 (**1**), 200 (**2**), and 540 (**3**) nm and **UiO-66-NH$_2$** MOF materials in 14 nm (**4**) were prepared and activated according to previously reported procedure by Morris et al. [31] with slight modifications. The 1,4-benzenedicarboxylic acid (500 mg, 3.0 mmol) was dissolved in 10 mL of N,N-dimethylformamide (DMF). In a separate vial, zirconyl chloride octahydrate (270 mg, 0.83 mmol) was dissolved in 30 mL of DMF. The two solutions were mixed together in a 100 mL Erlenmeyer flask and acetic acid (3 mL) was added to the reaction mixture. The solution was mixed well to obtain a homogeneous solution. The homogeneous solution was separated into seven 15 mL glass vials, with approximately 6 mL in each vial. The solution vials were heated at 90 °C for 18 h to yield **UiO-66** with an average size of 14 nm (**UiO-66_14nm** (**1**). The MOF nanoparticles were purified by centrifugation (10,000 rpm, 30 min) followed by a solvent exchange (3× DMF and 3× H_2O) over a 24 h period. Similar procedure was followed for **UiO-66-NH$_2$_14nm** (**4**) for MOF synthesis and purification. 2-amino-1-4-benzenedicarboxylic acid was used in this reaction instead of 1,4-benzenedicarboxylic acid, but the remaining materials were the same as for **UiO-66_14nm** (**1**). To confirm the formation of product, the crystallinity and particle size of all the synthesized MOF materials were analyzed by powder X-ray diffraction (PXRD) as well as TEM and SEM EDX-elemental mapping. The analysis data obtained were verified against the reported results.

Synthesis of UiO-66_200 nm (2) and UiO-66_540 nm (3)

The same synthetic conditions mentioned earlier were used, but the volume of acetic acid was changed from 3 to 21 and 35 mL for **UiO-66_200** (**2**) and **UiO-66_540 nm** (**3**), respectively [31]. Based on the methodology used in this paper, increasing the amount of acetic acid in the MOF synthesis procedure results in larger crystals.

3.2.2. Synthesis of Paraoxon-Ethyl

Paraoxon-ethyl was prepared by following the synthesis method reported by Tamilselvi et al. [32]. Diethyl chlorophosphate (0.860 mL, 5 mmol), p-nitrophenol (0.696 g, 5 mmol), and triethylamine (0.7 mL) were mixed in diethyl ether (20 mL). After stirring the reaction mixture for 12 h at room temperature, the reaction mixture was poured into water and the compound was extracted from the aqueous layer with diethyl ether. The combined organic fractions were evaporated to dryness to produce a yellow oil, which was subjected to reverse-phase flash chromatography to obtain pure paraoxon. The formation of paraoxon-ethyl and purity were verified by analyzing with ^{1}H and ^{31}P NMR spectroscopy and comparing the data to reported results.

4. Conclusions

This study demonstrated the application of flow microreactors packed with catalysts derived from **UiO-66** MOF for the hydrolysis of ethyl paraoxone. Through the tableting and sieving procedure, catalysts with a particle size between 125 and 250 µm were found to be most suitable for kinetic testing of this application. Catalysts with a smaller particle size demonstrated a significant pressure drop that resulted in the segregation and channeling of the packed catalyst bed. Catalysts derived from MOFs with smaller crystal sizes proved more effective than catalysts derived from MOFs with larger crystal sizes. The catalyst displayed an excellent long-term stability of more than 18 h of continuous operation. The presence of –NH$_2$ functionality greatly enhanced catalyst activity in agreement with previous reports. The flow microreactor setup also allowed us to conduct easy kinetic investigations to deduce the first-order reaction kinetics and the apparent activation energy for the reaction.

Supplementary Materials: The following are available online at http://www.mdpi.com/2073-4344/10/10/1159/s1: detailed synthesis methods of 1–4, ethyl-paraoxon, and complete analysis data from Figures S1–S19, Table S1.

Author Contributions: Conceptualization, S.T.M., and M.A.-R.; funding acquisition, S.T.M., and M.A.-R.; methodology, P.E, S.T.M., and M.A.-R.; project execution, P.E., N.E., and W.C.; Supervision, S.T.M. and M.A.-R.; writing—original draft, P.E., S.T.M., and M.A.-R. writing—review and editing, P.E., S.T.M., and M.A.-R. All authors have read and agreed to the published version of the manuscript.

Funding: This research was supported by TAMUQ seed and startup research funding and the Qatar National Research Fund, National Priority Research Program grant NPRP9-377-1-080.

Conflicts of Interest: The authors declare no conflict of interest.

References

1. Inoue, T.; Schmidt, M.A.; Jensen, K.F. Microfabricated multiphase reactors for the direct synthesis of hydrogen peroxide from hydrogen and oxygen. *Ind. Eng. Chem. Res.* **2007**, *46*, 1153–1160. [CrossRef]
2. Plouffe, P.; Macchi, A.; Roberge, D.M. From batch to continuous chemical synthesis-a toolbox approach. *Org. Process Res. Dev.* **2014**, *18*, 1286–1294. [CrossRef]
3. Wada, Y.; Schmidt, M.A.; Jensen, K.F. Flow distribution and ozonolysis in gas-liquid multichannel microreactors. *Ind. Eng. Chem. Res.* **2006**, *45*, 8036–8042. [CrossRef]
4. Gutmann, B.; Cantillo, D.; Kappe, C.O. Continuous-flow technology-A tool for the safe manufacturing of active pharmaceutical ingredients. *Angew. Chem. Int. Ed.* **2015**, *54*, 6688–6728. [CrossRef]
5. Chambers, R.D.; Fox, M.A.; Holling, D.; Nakano, T.; Okazoe, T.; Sandford, G. Elemental fluorine Part 16. Versatile thin-film gas-liquid multi-channel microreactors for effective scale-out. *Lab Chip* **2005**, *5*, 191–198. [CrossRef]
6. Leclerc, A.; Alame, M.; Schweich, D.; Pouteau, P.; Delattre, C.; de Bellefon, C. Gas-liquid selective oxidations with oxygen under explosive conditions in a micro-structured reactor. *Lab Chip* **2008**, *8*, 814–817. [CrossRef] [PubMed]

7. Schwolow, S.; Hollmann, J.; Schenkel, B.; Röder, T. Application-oriented analysis of mixing performance in microreactors. *Org. Process Res. Dev.* **2012**, *16*, 1513–1522. [CrossRef]

8. Gobert, S.R.L.; Kuhn, S.; Braeken, L.; Thomassen, L.C.J. Characterization of Milli- and Microflow Reactors: Mixing Efficiency and Residence Time Distribution. *Org. Process Res. Dev.* **2017**, *21*, 531–542. [CrossRef]

9. Reckamp, J.M.; Bindels, A.; Duffield, S.; Liu, Y.C.; Bradford, E.; Ricci, E.; Susanne, F.; Rutter, A. Mixing Performance Evaluation for Commercially Available Micromixers Using Villermaux-Dushman Reaction Scheme with the Interaction by Exchange with the Mean Model. *Org. Process Res. Dev.* **2017**, *21*, 816–820. [CrossRef]

10. Hartman, R.L. Managing solids in microreactors for the upstream continuous processing of fine chemicals. *Org. Process Res. Dev.* **2012**, *16*, 870–887. [CrossRef]

11. Nakamura, H. Nanoparticle Synthesis in Microreactors. In *Encyclopedia of Microfluidics and Nanofluidics*; Springer US: Berlin/Heidelberg, Germany, 2008; pp. 1437–1446.

12. Wagner, J.; Köhler, J.M. Continuous synthesis of gold nanoparticles in a microreactor. *Nano Lett.* **2005**, *5*, 685–691. [CrossRef] [PubMed]

13. Reizman, B.J.; Jensen, K.F. Feedback in Flow for Accelerated Reaction Development. *Acc. Chem. Res.* **2016**, *49*, 1786–1796. [CrossRef] [PubMed]

14. McMullen, J.P.; Jensen, K.F. Rapid determination of reaction kinetics with an automated microfluidic system. *Org. Process Res. Dev.* **2011**, *15*, 398–407. [CrossRef]

15. May, S.A. Flow Chemistry, Continuous Processing, and Continuous Manufacturing: A Pharmaceutical Perspective. *J. Flow Chem.* **2017**, *7*, 137–145. [CrossRef]

16. Li, J.R.; Kuppler, R.J.; Zhou, H.C. Selective gas adsorption and separation in metal-organic frameworks. *Chem. Soc. Rev.* **2009**, *38*, 1477–1504. [CrossRef]

17. Kalaj, M.; Denny, M.S.; Bentz, K.C.; Palomba, J.M.; Cohen, S.M. Nylon–MOF Composites through Postsynthetic Polymerization. *Angew. Chem. Int. Ed.* **2019**, *58*, 2336–2340. [CrossRef]

18. Chen, Z.; Hanna, S.L.; Redfern, L.R.; Alezi, D.; Islamoglu, T.; Farha, O.K. Reticular chemistry in the rational synthesis of functional zirconium cluster-based MOFs. *Coord. Chem. Rev.* **2019**, *386*, 32–49. [CrossRef]

19. Chen, Z.; Ma, K.; Mahle, J.J.; Wang, H.; Syed, Z.H.; Atilgan, A.; Chen, Y.; Xin, J.H.; Islamoglu, T.; Peterson, G.W.; et al. Integration of Metal-Organic Frameworks on Protective Layers for Destruction of Nerve Agents under Relevant Conditions. *J. Am. Chem. Soc.* **2019**, *141*, 20016–20021. [CrossRef]

20. Ji, P.; Feng, X.; Oliveres, P.; Li, Z.; Murakami, A.; Wang, C.; Lin, W. Strongly Lewis Acidic Metal-Organic Frameworks for Continuous Flow Catalysis. *J. Am. Chem. Soc.* **2019**, *141*, 14878–14888. [CrossRef]

21. Dhakshinamoorthy, A.; Navalon, S.; Asiri, A.M.; Garcia, H. Metal organic frameworks as solid catalysts for liquid-phase continuous flow reactions. *Chem. Commun.* **2019**, *56*, 26–45. [CrossRef]

22. Chen, Y.; Huang, X.; Zhang, S.; Li, S.; Cao, S.; Pei, X.; Zhou, J.; Feng, X.; Wang, B. Shaping of Metal–Organic Frameworks: From Fluid to Shaped Bodies and Robust Foams. *J. Am. Chem. Soc.* **2016**, *138*, 10810–10813. [CrossRef]

23. Furukawa, S.; Reboul, J.; Phane Diring, S.; Sumida, K.; Kitagawa, S. Structuring of metal-organic frameworks at the mesoscopic/macroscopic scale. *Chem. Soc. Rev.* **2014**, *5700*, 5700. [CrossRef] [PubMed]

24. Gascon, J.; Corma, A.; Kapteijn, F.; Llabrés i Xamena, F.X. Metal Organic Framework Catalysis: Quo vadis? *ACS Catal.* **2014**, *4*, 361–378. [CrossRef]

25. Tan, J.C.; Cheetham, A.K. Mechanical properties of hybrid inorganic-organic framework materials: Establishing fundamental structure-property relationships. *Chem. Soc. Rev.* **2011**, *40*, 1059–1080. [CrossRef] [PubMed]

26. Serra-Crespo, P.; Stavitski, E.; Kapteijn, F.; Gascon, J. High compressibility of a flexible metal-organic framework. *Rsc Adv.* **2012**, *2*, 5051–5053. [CrossRef]

27. Montoro, C.; Linares, F.; Quartapelle Procopio, E.; Senkovska, I.; Kaskel, S.; Galli, S.; Masciocchi, N.; Barea, E.; Navarro, J.A.R. Capture of nerve agents and mustard gas analogues by hydrophobic robust MOF-5 type metal-organic frameworks. *J. Am. Chem. Soc.* **2011**, *133*, 11888–11891. [CrossRef]

28. Yot, P.G.; Yang, K.; Ragon, F.; Dmitriev, V.; Devic, T.; Horcajada, P.; Serre, C.; Maurin, G. Exploration of the mechanical behavior of metal organic frameworks UiO-66(Zr) and MIL-125(Ti) and their NH2 functionalized versions. *Dalt. Trans.* **2016**, *45*, 4283–4288. [CrossRef]

29. Wu, H.; Yildirim, T.; Zhou, W. Exceptional mechanical stability of highly porous zirconium metal-organic framework UiO-66 and its important implications. *J. Phys. Chem. Lett.* **2013**, *4*, 925–930. [CrossRef]

30. Van De Voorde, B.; Stassen, I.; Bueken, B.; Vermoortele, F.; De Vos, D.; Ameloot, R.; Tan, J.C.; Bennett, T.D. Improving the mechanical stability of zirconium-based metal-organic frameworks by incorporation of acidic modulators. *J. Mater. Chem. A* **2015**, *3*, 1737–1742. [CrossRef]
31. Morris, W.; Briley, W.E.; Auyeung, E.; Cabezas, M.D.; Mirkin, C.A. Nucleic acid-metal organic framework (MOF) nanoparticle conjugates. *J. Am. Chem. Soc.* **2014**, *136*, 7261–7264. [CrossRef]
32. Tamilselvi, A.; Mugesh, G. Hydrolysis of Organophosphate Esters: Phosphotriesterase Activity of Metallo-β-lactamase and Its Functional Mimics. *Chem. A Eur. J.* **2010**, *16*, 8878–8886. [CrossRef] [PubMed]
33. Cavka, J.H.; Jakobsen, S.; Olsbye, U.; Guillou, N.; Lamberti, C.; Bordiga, S.; Lillerud, K.P. A New Zirconium Inorganic Building Brick Forming Metal Organic Frameworks with Exceptional Stability. *J. Am. Chem. Soc.* **2008**, *6*, 13850–13851. [CrossRef] [PubMed]
34. Azarifar, D.; Daliran, S.; Ghorbani-Vaghei, R.; Oveisi, R.A. A Multifunctional Zirconium-Based Metal–Organic Framework for the One-Pot Tandem Photooxidative Passerini Three-Component Reaction of Alcohols. *ChemCatChem* **2017**, *9*, 1992–2000. [CrossRef]
35. Xuechuan, G.; Ruixue, C.; Jia, G.; Liu, Z. Size and surface controllable metal–organic frameworks (MOFs) for fluorescence imaging and cancer therapy. *Nanoscale* **2018**, *10*, 6205–6211.
36. Užarević, K.; Wang, T.C.; Moon, S.Y.; Fidelli, A.M.; Hupp, J.T.; Farha, O.K.; Friščić, T. Mechanochemical and solvent-free assembly of zirconium-based metal-organic frameworks. *Chem. Commun.* **2016**, *52*, 2133–2136. [CrossRef] [PubMed]
37. Katz, M.J.; Moon, S.Y.; Mondloch, J.E.; Beyzavi, M.H.; Stephenson, C.J.; Hupp, J.T.; Farha, O.K. Exploiting parameter space in MOFs: A 20-fold enhancement of phosphate-ester hydrolysis with UiO-66-NH2. *Chem. Sci.* **2015**, *6*, 2286–2291. [CrossRef]
38. Marin, G.B.; Yablonsky, G.S. *Kinetics of Chemical Reactions: Decoding Complexity*; John Wiley & Sons: Hoboken, NJ, USA, 2011; Volume 428.

 *catalysts*

Article

High Catalytic Efficiency of a Layered Coordination Polymer to Remove Simultaneous Sulfur and Nitrogen Compounds from Fuels

Fátima Mirante [1], Ricardo F. Mendes [2], Filipe A. Almeida Paz [2,*] and Salete S. Balula [1,*]

[1] REQUIMTE/LAQV & Department of Chemistry and Biochemistry, Faculty of Sciences, University of Porto, 4169-007 Porto, Portugal; fatimaisabelmirante@gmail.com

[2] CICECO-Aveiro Institute of Materials, Department of Chemistry, University of Aveiro, Campus Universitário de Santiago, 3810-193 Aveiro, Portugal; rfmendes@ua.pt

* Correspondence: filipe.paz@ua.pt (F.A.A.P.); sbalula@fc.up.pt (S.S.B.)

Received: 1 June 2020; Accepted: 24 June 2020; Published: 2 July 2020

Abstract: An ionic lamellar coordination polymer based on a flexible triphosphonic acid linker, $[Gd(H_4nmp)(H_2O)_2]Cl_2 \cdot H_2O$ (1) (H_6nmp stands for nitrilo(trimethylphosphonic) acid), presents high efficiency to remove sulfur and nitrogen pollutant compounds from model diesel. Its oxidative catalytic performance was investigated using single sulfur (1-BT, DBT, 4-MDBT and 4,6-DMDBT, 2350 ppm of S) and nitrogen (indole and quinolone, 400 ppm of N) model diesels and further, using multicomponent S/N model diesel. Different methodologies of preparation followed (microwave, one-pot, hydrothermal) originated small morphological differences that did not influenced the catalytic performance of catalyst. Complete desulfurization and denitrogenation were achieved after 2 h using single model diesels, an ionic liquid as extraction solvent ($[BMIM]PF_6$) and H_2O_2 as oxidant. Simultaneous desulfurization and denitrogenation processes revealed that the nitrogen compounds are more easily removed from the diesel phase to the $[BMIM]PF_6$ phase and consequently, faster oxidized than the sulfur compounds. The lamellar catalyst showed a high recycle capacity for desulfurization. The reusability of the diesel/H_2O_2/$[BMIM]PF_6$ system catalyzed by lamellar catalyst was more efficient for denitrogenation than for desulfurization process using a multicomponent model diesel. This behavior is not associated with the catalyst performance but it is mainly due to the saturation of S/N compounds in the extraction phase.

Keywords: layered coordination polymer; oxidative desulfurization; denitrogenation extraction; hydrogen peroxide; lanthanides

---

1. Introduction

One of the main aims of Green chemistry is to minimize the negative impact of the petroleum and chemical industries on the environment and human health. The major sources of air pollution in urban areas are the road fuels [1,2], releasing to the atmosphere different pollutants such as carbon monoxide, ammonia, sulfur dioxide (SO_2), nitrogen oxides (NO_x) and particulate matter. SO_2 and NO_x emissions can have adverse effects for the environment and for human health [3], with harsh regulations being implemented for sulfur content in road fuels to decrease SO_2 emissions [4–6]. In the European Union, strict policies implemented the limit to sulfur level in diesel from 2000 ppm in 1993 to 10 ppm presently. On the other hand, the nitrogen containing levels in fuels are not regulated. NO_x emissions results from either the oxidation of nitrogen present compounds in fuels or the oxidation of atmospheric nitrogen at high temperatures [7]. These emissions have already been restrained: the limit for diesel powered light duty vehicles decreased from 0.18 g km^{-1} for the Euro V standard to 0.08 g km^{-1} for Euro VI [8,9].

The industrial process is able to efficiently remove the heterocyclic sulfur and nitrogen compounds (such as benzothiophenes, dibenzothiophenes, thiophenes, quinolines, indoles, carbazoles, acridines, and pyridines, Figure S1 in Electronic Supporting Information) from fuels by hydrotreating processes which require severe conditions (high temperatures >350 °C and high pressure 20–130 atm H_2) [4,10]. The presence of nitrogen compounds in fuels even in low concentrations (<100 ppm) affects the efficacy of hydrodesulfurization (HDS) reactions since these compounds are strong inhibitors and promoting catalytic deactivation, caused by their competition with sulfur compounds for the active sites of the hydrotreating catalysts [10–13]. Novel technologies capable of complementing or replacing the industrial HDS and hydrodenitrogenation (HDN) processes are needed in order to obtain ultra-low sulfur and nitrogen levels in fuels by more attractive cost-effective methods.

In the literature, various examples dedicated to the elimination of sulfur compounds from the fuels are described. The most promising processes are the oxidative desulfurization, combining with the liquid-liquid extraction [4–6]. Most of these processes use hydrogen peroxide as the oxidant, because of the high active oxygen content and with water being the sole by-product.

Solid catalysts that allow an easy separation from the reaction media and easy recyclability in consecutive reaction cycles are required. Metal-Organic Frameworks (MOFs) and Coordination Polymers (CPs) comprising organic bridging ligands and metallic centers emerged as promising materials because of their considerable structural diversity and unique properties, such as high porosity, large surface areas and in certain cases high thermal, chemical and hydrolytic stabilities. The investigation and use of these materials as heterogeneous catalysts for desulfurization processes is, however, relatively scarce. Composite materials based on MOFs/CPs have been more regularly applied, employed as host materials to support active homogenous catalysts in their pores or matrices [14–17]. Taking advantage of the possibility of some synthetic modification strategies, the catalytic activity of these materials can be greatly improved [18]. In the last decade, we have been focusing on the design of networks based on polyphosphonic acid ligands and rare-earth metal cations which typically induce the formation of highly robust dense networks [19]. The crystalline hybrid materials designed and developed by our groups typically have a high concentration of acid protons and solvent (typically water) molecules [20]. We are looking to take advantage of this particular structural feature to design better-performing compounds that take advantage of proton mobility and/or exchange. These acid properties can be an important advantage for oxidative desulfurization.

In the present work, a positively charged lamellar coordination polymer based on a flexible triphosphonic acid linker, $[Gd(H_4nmp)(H_2O)_2]Cl_2$ H_2O (**1**) [H_6nmp stands for nitrilo(trimethylphosphonic) acid], was used as an acid heterogeneous catalyst in oxidative desulfurization and denitrogenation. Environmentally-friendly conditions (low temperature and H_2O_2/S molar ratio, and an ionic liquid as solvent [BMIM]PF_6) were employed. The stability and the recycle capacity of the catalyst was also investigated.

2. Results and Discussion

2.1. Preparation of the Catalyst

Our research group have shown the potential of MOFs and CPs as heterogeneous catalysts in various reactions with great industrial interest. The materials tested were based on polyphosphonic acid, self-assembled with various lanthanide cations. The high number of phosphonate groups proved to be important for the catalytic process: the local large concentrations of acidic protons and solvent molecules allowed high conversions rates with low reaction times for reactions such as sulfoxidation of thioanisole [21], conversion of styrene oxide [22] and conversion of benzaldehyde into (dimethoxymethyl)benzene [23]. We further explore the catalytic activity of a positively charged layered CP obtained by combination of nitrilo(trimethylphosphonic) acid (H_6nmp) and Gd^{3+} ions (Figure 1), $[Gd(H_4nmp)(H_2O)_2]Cl·2H_2O$ (**1**). When prepared using a simple one-pot method, this material showed a high catalytic activity in four different organic reactions: alcoholysis of styrene oxide, acetalization of benzaldehyde and cyclohexanaldehyde and ketalization of cyclohexanone, with conversion above 95% after 1–4 h of reaction [24]. In this work we further explore the catalytic activity of this charged 2D material (charge balanced by chloride ions), in the desulfurization and denitrogenation of a multicomponent model diesel. Compound **1** was prepared using three different methods: one-pot (as reported previously), microwave-assisted and hydrothermal syntheses (Figure 1). All materials present the same crystalline phase with slight differences in crystal morphologies and sizes (Figure S2 in ESI). As previously reported, the use of hydrochloric acid is crucial for the preparation of **1**: the acid allows the protonation of the organic linker, retarding the coordination process, leading to the formation of a more crystalline material. Not only that, the acid is the source of chloride anions that are present in the interlayer spaces, being responsible for the CP charge balancing (we note that the use of $GdCl_3$ as the metallic source does not originate compound **1**). For additional structural details on this compound we refer the reader to our past publication [24].

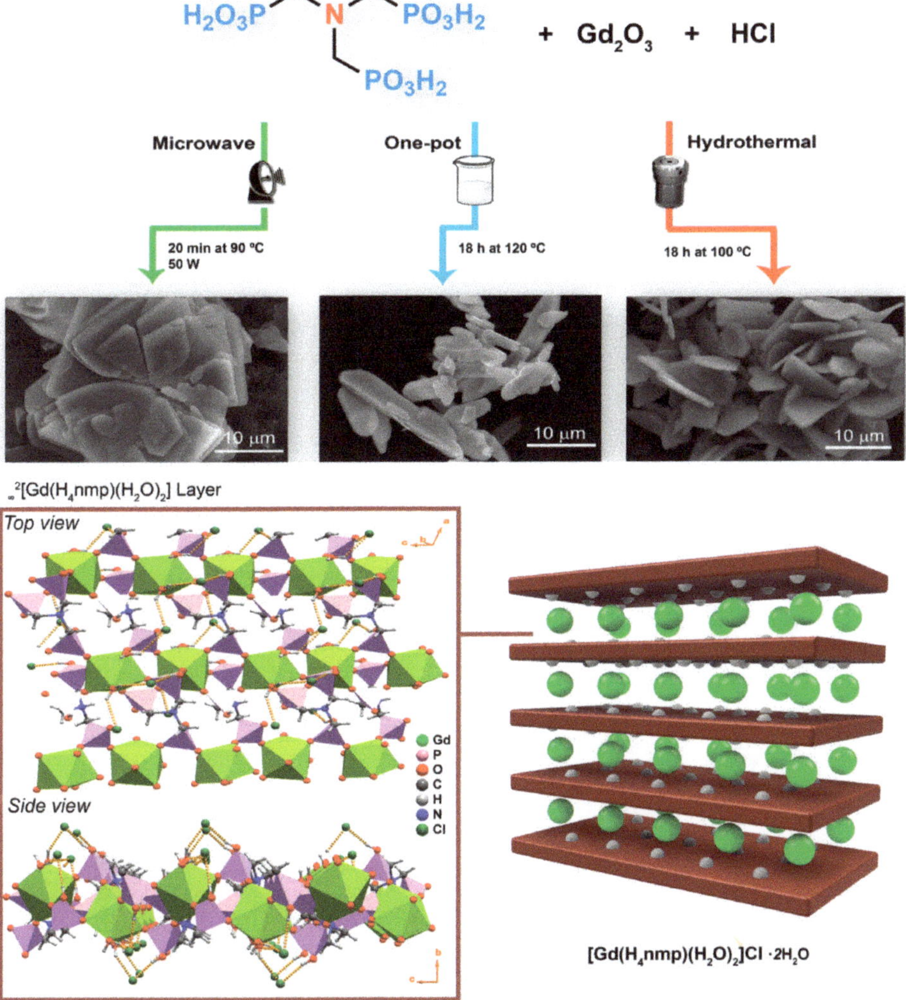

Figure 1. Schematic representation of the synthesis and structural features of [Gd(H$_4$nmp)(H$_2$O)$_2$]Cl·2H$_2$O (**1**), with emphasis in the chloride anions present in the interlayer spaces.

2.2. Optimization of Desulfurization Process

[Gd(H$_4$nmp)(H$_2$O)$_2$]Cl·2H$_2$O (**1**) was used as catalyst in the oxidative desulfurization (ODS) of a model diesel composed by the most refractory sulfur compounds toward HDS process in liquid fuels, namely 1-benzothiophene (1-BT), dibenzothiophene (DBT), 4-methyldibenzothiophene (4-MDBT) and 4,6-dimethyldibenzothiophene (4,6-DMDBT) in *n*-octane with a total sulfur concentration of approximately 500 ppm (0.0156 mol dm^{-3}) for each compound (for more detailed information of the experimental conditions the reader is directed to the ESI).

Initially catalyst **1**, prepared by the three different methods, was tested and their ECODS profiles are presented in Figure 2. The ionic liquid 1-butyl-3- methylimidazolium hexafluorophosphate ([BMIM]PF$_6$) was used as extraction solvent. Similar desulfurization efficiency was obtained following different preparation procedures. An initial extraction desulfurization of approximately 40% was achieved after 10 min at 70 °C. After the addition of the oxidant the desulfurization increased more appreciably after 40 min and complete desulfurization was found near 2 h. Therefore, the method of catalyst preparation does not seem to have a significant effect in the catalytic efficiency. A small difference is observed between the 40 and 70 min mark and it can be attributed to the overall average particle size. While all methods originated in crystals with plate-like morphology, small differences in size were indeed observed. Contrary to that observed for other materials, the hydrothermal synthesis allows the formation of regular plates with the average size of *ca.* 10 m. On the other hand, microwave synthesis originated plates as agglomerates ranging from 15 to 30 m. Crystals obtained by the one-pot method presented a more irregular crystal morphology, with sizes varying between 5 and 15 m. Nonetheless, the overall small difference of particle size does not seem to have a direct influence in the catalyst activity itself, achieving the same desulfurization efficiency after 2 h of reaction. Their similar crystallinity may indicate that the active catalytic centers in the various catalytic samples are identical.

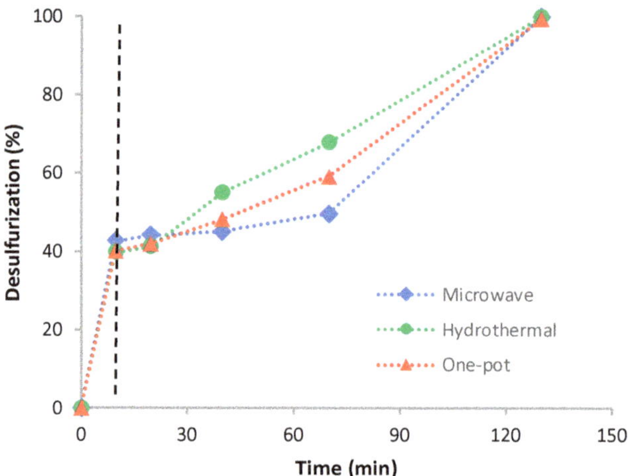

Figure 2. Desulfurization of a multicomponent model diesel (2350 ppm S) catalyzed by [Gd(H$_4$nmp)(H$_2$O)$_2$]Cl·2H$_2$O (**1**) (20 mg) prepared by different methods (microwave, hydrothermal and one-pot), using equal volumes of model diesel and [BMIM]PF$_6$ as extraction solvent and H$_2$O$_2$ (0.64 mmol, 30% aq.) at 70 °C. The vertical dashed line indicates the instant that oxidative catalytic reaction was started by addition of oxidant.

Because the catalytic activity was virtually the same for **1** obtained by the various synthetic methods, the following optimization of the ECODS system was performed using [Gd(H$_4$nmp)(H$_2$O)$_2$]Cl·2H$_2$O prepared by the one-pot approach (**1op**). Various parameters were investigated, such as extraction solvent, temperature, and oxidant amount, in order to improve the catalytic performance, the sustainability and cost-efficiency of the process. Three different extraction solvents were investigated: two different ionic liquids (ILs), [BMIM]PF$_6$ and [BMIM]BF$_4$, and the polar organic acetonitrile (MeCN). Figure 3 displays the desulfurization profiles using the various extraction solvents (1:1 model diesel/extraction solvent) and also when no solvent is used. The ECODS system model diesel/[BMIM]PF$_6$ was the most efficient, achieving complete desulfurization after 2 h of reaction. In the absence of extraction solvent, practically no oxidative desulfurization occurred, what indicates an inefficiency of this layered catalyst in the model diesel phase. Using acetonitrile, an initial extraction of 48% of desulfurization was achieved after 10 min. This desulfurization was not, however, increased during the oxidative desulfurization stage, indicating that the catalyst is not active using this solvent extraction. When the solvent extraction used was the IL [BMIM]BF$_4$, a lower initial extraction was achieved (22%) and the desulfurization increased after the addition of the oxidant during the first 30 min to 43%, stabilizing this result after this time. Therefore, the combination of the catalyst **1op** with the [BMIM]PF$_6$ solvent promoted the highest catalytic efficiency, achieving complete desulfurization after only 2 h. ECODS process performed with the 1:1 model diesel/[BMIM]PF$_6$ system without catalyst did not resulted any oxidative desulfurization, i.e, after the initial extraction the desulfurization did not increased after oxidant addition. This result also indicates that the absorptive capacity of the material is negligible. The outstanding catalytic result achieved with the solvent [BMIM]PF$_6$ can be explained by its immiscibility with the oxidant, creating a three phases system (diesel/H$_2$O$_2$/[BMIM]PF$_6$). This prevents direct contact between the catalyst and the oxidant, avoiding a possible catalyst deactivation.

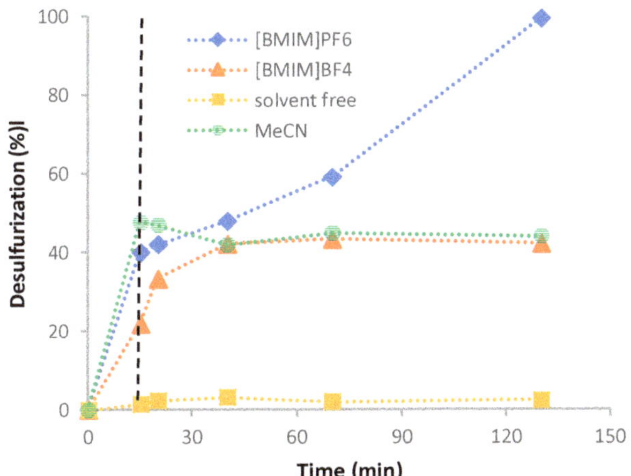

Figure 3. Desulfurization of a multicomponent model diesel (2350 ppm S) catalyzed by [Gd(H$_4$nmp)(H$_2$O)$_2$]Cl·2H$_2$O (**1op**) (20 mg), using different extraction solvents (MeCN, [BMIM]PF$_6$/[BMIM]BF$_4$) or absence of this solvent, H$_2$O$_2$ (0.64 mmol, 30% aq.) as oxidant at 70 °C. The vertical dashed line indicates the instant that oxidative catalytic reaction was started by addition of oxidant.

Comparing the oxidative desulfurization of each sulfur-based compound, it was possible to notice that after the initial extraction, 1-BT and DBT are more easily transferred from model diesel to the extraction solvent. As 1-BT possesses the smallest molecular size, this transfer is easier, while the presence of methyl substituents at the sterically hindered positions in 4-MDBT and 4,6-DMDBT makes the extraction more challenging [25–27]. During the oxidative catalytic stage, the 1-BT is the most difficult to be oxidized because after 1 h its desulfurization was 67% in contrast to the 92%, 88% and 80% of total desulfurization for DBT, 4-MDBT and 4,6-DMDBT, respectively (Figure 4). The lower electron density of 1-BT compared with the other studied compounds may explain its lower reactivity [27–29]. The studied dibenzothiophene derivative exhibit similar electron densities on the sulfur atom and their distinct desulfurization performance is probably caused the steric hindrance promoted by the methyl groups.

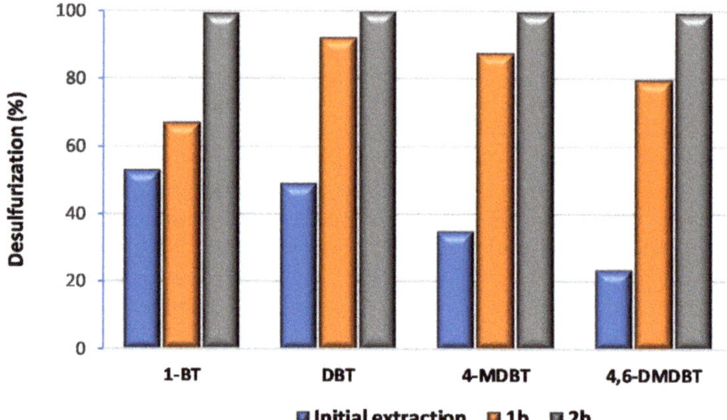

Figure 4. Desulfurization results obtained for each sulfur compounds in the model diesel (2350 ppm S) catalyzed by [Gd(H$_4$nmp)(H$_2$O)$_2$]Cl·2H$_2$O (**1op**) (20 mg), using [BMIM]PF$_6$, as extraction solvent, H$_2$O$_2$ (0.64 mmol, 30% aq.) as oxidant at 70 °C.

The effect of temperature (50, 70 and 80 °C) on the oxidative desulfurization process was analyzed and results are summarized in Figure 5. Higher temperatures were not investigated because decomposition of H$_2$O$_2$ could become significant above 80 °C. The increase in the reaction temperature from 50 to 70 °C led to an improvement in the desulfurization rate and resulted in a sulfur-free model diesel product after 2 h of reaction. The desulfurization profile of ECODS at 80 °C demonstrated a rapid increase of desulfurization after the first 30 min of reaction, and after 1 h of reaction 95% of total desulfurization was achieved. Complete desulfurization was, however, only found after 2 h. Therefore, the optimized temperature is 70 °C because complete desulfurization is achieved after 2 h.

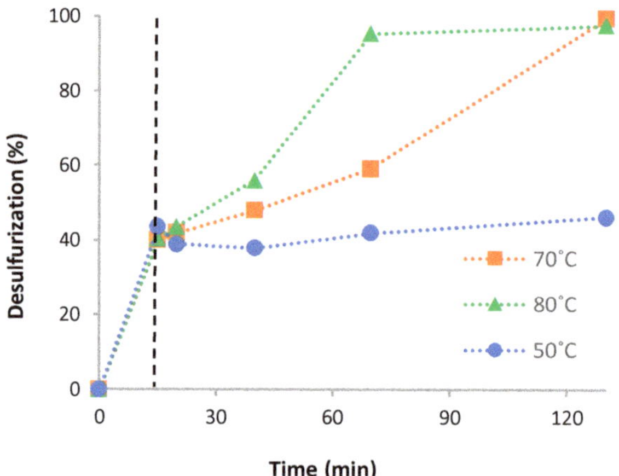

Figure 5. Desulfurization of a multicomponent model diesel (2350 ppm S) catalyzed by [Gd(H$_4$nmp)(H$_2$O)$_2$]Cl·2H$_2$O (**1op**) (20 mg), using 0.64 mmol of H$_2$O$_2$ oxidant, [BMIM]PF$_6$ as extraction solvent, at different temperatures (50, 70, 80 °C). The vertical dashed line indicates the instant that oxidative catalytic reaction was started by addition of oxidant.

The oxidant amount is another important factor because it controls, together with the catalyst, the oxidative step through oxygen donation. Three different oxidant amounts were used: 75, 50 and 25 μL, corresponding to 0.64, 0.43 and 0.21 mmol, respectively. Results obtained for the ECODS using [BMIM]PF$_6$ as extraction solvent at 70 °C, are displayed in Figure 6. Using 0.64 and 0.43 mmol of H$_2$O$_2$ a complete desulfurization of the model diesel was obtained after the 2 h. In fact, the desulfurization profile after 1 h of reaction is similar using 0.64 and 0.43 mmol of oxidant. On the other hand, using 0.21 mmol of H$_2$O$_2$ an induction period can be observed until 1 h of oxidation catalytic reaction. This is probably due to the lower interaction of the catalytic active centers with the oxidant caused by its low amount. Therefore, the optimized oxidant amount is 0.43 mmol.

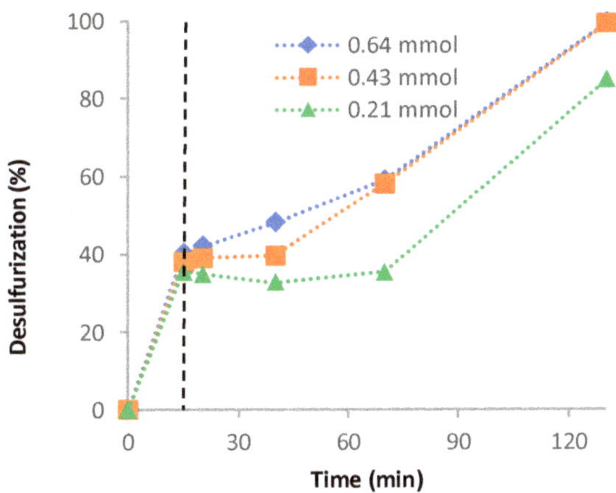

Figure 6. Desulfurization of a multicomponent model diesel (2350 ppm S) catalyzed by [Gd(H$_4$nmp)(H$_2$O)$_2$]Cl·2H$_2$O (**1op**) (20 mg), using different amounts of H$_2$O$_2$ oxidant (0.64, 0.43 and 0.21 mmol) and [BMIM]PF$_6$ as extraction solvent, at 70 °C. The vertical dashed line indicates the instant that oxidative catalytic reaction was started by addition of oxidant.

In the next step, the influence of the ratio model diesel/[BMIM]PF$_6$ solvent was studied at 70 °C using 0.43 mmol of oxidant and 20 mg of catalyst. This optimization was performed in order to improve efficiency and sustainability of the ECODS process. Two different ratios were performed: 1:1 and 1:0.5 (Figure S3 in ESI in the Supporting Information). The volume of the extraction solvent phase has an important influence in the initial extraction step because the sulfur compounds are transferred from diesel to this solvent until an equilibrium point is achieved that depends on the nature and amount of this solvent. A higher volume of [BMIM]PF$_6$ may promote a higher initial extraction of non-oxidized sulfur compounds from the model diesel to the polar phase. Figure S3 in ESI shows that the initial extraction using 1:0.5 of model diesel/[BMIM]PF$_6$ was much lower (8%) than when higher volume of extraction solvent was used (1:1 model diesel/[BMIM]PF$_6$, 31%). After the addition of the oxidant (0.43 mmol H$_2$O$_2$), the desulfurization increased much faster when a lower volume of extraction phase was employed. Using the ECODS 1:0.5 model diesel/[BMIM]PF$_6$, 99.5% of desulfurization was found after 1 h, instead of 58% which was achieved using a 1:1 model diesel/[BMIM]PF$_6$ system. The higher catalytic activity and consequent higher desulfurization efficiency obtained in the presence of a lower volume of extraction solvent, must be related to the lower dispersion of the solid catalyst in the ECODS system, which can promote a higher contact with the aqueous oxidant.

2.3. Reusability Versus Recyclability

The stability of the layered coordination polymer [Gd(H$_4$nmp)(H$_2$O)$_2$]Cl·2H$_2$O was investigated by its reusability and the recyclability in consecutive ECODS cycles. During all cycles the optimized ECODS conditions were maintained (1:0.5 model diesel/[BMIM]PF$_6$, 0.43 mmol oxidant, at 70 °C).

Two different methods were employed to recover the catalyst after each ECODS cycle. In the "reused" method, the model diesel phase was removed and fresh portions of model diesel and H_2O_2 were added without performing any other treatments. In the recycled method the solid catalyst was separated from reaction medium and washed with MeCN and dried. Figure 7 presents the results obtained for three consecutive "reused" and recycled ECODS cycles. The desulfurization efficiency decreases along the consecutive reusing cycles. This behavior has been attributed to the saturation of the extraction phase during the reused process with the oxidized sulfur species, which prevents further transfer of sulfur compounds from the non-polar phase to the IL phase, decreasing the system efficiency [30]. On the other hand, from the recycling method, complete desulfurization was achieved for the three consecutive cycles (Figure 7).

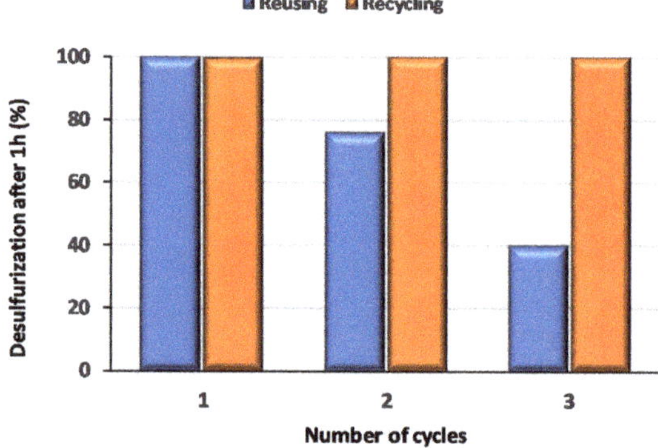

Figure 7. Desulfurization results obtained for three consecutive "reused" and recycled ECODS cycles (1 h reaction time), using [Gd(H$_4$nmp)(H$_2$O)$_2$]Cl·2H$_2$O (**1op**) (20 mg), 0.43 mmol of H_2O_2 oxidant and 1:0.5 model diesel/[BMIM]PF$_6$, at 70 °C.

2.4. Denitrogenation Process

Nitrogen compounds present in crude oils are predominantly heterocyclic aromatic compounds. and in smaller amounts as non-heterocyclic nitrogen compounds such as aliphatic amines and nitriles. While the last are easily removed by hydrotreating, heterocyclic nitrogen compounds more difficult to remove. These are classified as basic (mainly pyridine derivatives) and neutral (mainly pyrrole derivatives). Quinoline and indole (Figure S1 in ESI) are representative basic and neutral compounds that are commonly employed as the components of model fuel oils [10]. Using the optimized ECODS conditions, the oxidative denitrogenation efficiency of the layered [Gd(H$_4$nmp)(H$_2$O)$_2$]Cl·2H$_2$O (**1op**) catalyst was studied using a model diesel containing indole (200 ppm of N) and quinoline (200 ppm of N). Figure 8 presents the results obtained and it is possible to observe that most of the indole is removed from the model diesel during the first 10 min of initial extraction, instead of 80% achieved by quinoline during this first step of denitrogenation process. After the addition of H_2O_2 oxidant, i.e., during the oxidative denitrogenation stage, the removal of quinoline by its oxidation was rapidly increased and after 1 h, with both nitrogen compounds being completely extracted from model diesel. The different behaviors observed for the two nitrogen compounds can be explained by the interaction between the proton/donor molecule indole and the PF$_6^-$ anion, which may promote a selective extraction from the model diesel [31]. Quinoline, containing a six-membered pyridine ring, the electron lone pair on the N atom is not part of the aromatic system and extends in the plane of the ring, being responsible for a negative charge on the N atom, preventing a lower interaction with Lewis acidic IL [10,32–34].

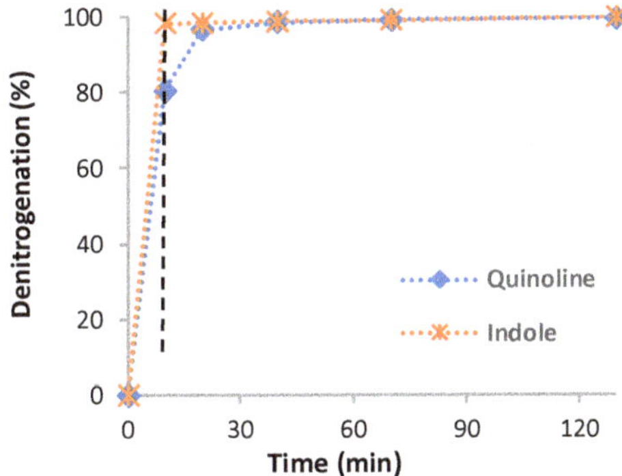

Figure 8. Denitrogenation profile for the two refractory nitrogen compounds present in diesel (400 ppm N), catalyzed by [Gd(H$_4$nmp)(H$_2$O)$_2$]Cl·2H$_2$O (**1op**, 20 mg), 0.43 mmol of H$_2$O$_2$ oxidant and 1:0.5 model diesel/[BMIM]PF$_6$, at 70 °C. The vertical dashed line indicates the instant that oxidative catalytic reaction was started by addition of oxidant.

2.5. Simultaneous Desulfurization and Denitrogenation Processes

In this work, simultaneous desulfurization and denitrogenation were also performed with a multicomponent model diesel containing the previous studied sulfur compounds (1-BT, DBT, 4-MDBT and 4,6-DMDBT, 2350 ppm of S) and indole (200 ppm of N) and quinoline (200 ppm of N). When the desulfurization and denitrogenation were performed separately, all compounds (nitrogen and sulfur) were extracted under the optimized conditions (20 mg of catalyst, 0.43 mmol of H$_2$O$_2$ and 1:0.5 model diesel/[BMIM]PF$_6$, at 70 °C). However, when denitrogenation and desulfurization were performed simultaneously, practically only nitrogen compounds were extracted from the model diesel to the [BMIM]PF$_6$ phase after the addition of the oxidant (Figure S4 in ESI). Complete removal of N compounds was also achieved after 2 h, instead of 1 h obtained with the single denitrogenation process (Figure 8). These results may be related to the use of an insufficient amount of H$_2$O$_2$, since a competitive oxidation between N and S.

One other parameter that can contribute to the absence of oxidative desulfurization is the low volume of extraction solvent to accommodate both S and N compounds. Therefore, the simultaneous desulfurization and denitrogenation was performed using 0.64 mmol of H$_2$O$_2$ and equal 1:1 of model diesel/[BMIM]PF$_6$ system. An appreciable improvement of the desulfurization profile was found while denitrogenation profile was not strongly affected, using this excess of oxidant and the equal volume of extraction solvent as the diesel phase (Figure 9). Complete denitrogenation was achieved after 2 h (99% after 1 h) and for the same time the desulfurization attained 88% (92% after 3 h). These results indicate that the removal of sulfur is affected by the presence of nitrogen, because complete desulfurization was achieved after 2 h under these experimental conditions using a single sulfur model diesel (Figure 3). This effect is mainly noticed during the oxidative catalytic stage because the initial extraction of each compound was not significantly affected by the presence of S and N in the same model diesel. Figure 10 depicts the extraction for each S and N compound. It is possible to notice that the sulfur compounds more difficult to desulfurize during the oxidative catalytic stage are 1-BT and the 4-DMDBT. Many authors reported that nitrogen compounds are significantly better extracted than sulfur compounds when ILs are used as extraction solvents [35]. Because no oxidized products were detected in the model diesel phase, the oxidative catalytic stage occurs in majority in the [BMIM]PF$_6$ phase and, therefore, the efficiency of desulfurization and denitrogenation processes are strongly dependent of the capacity of S/N extraction.

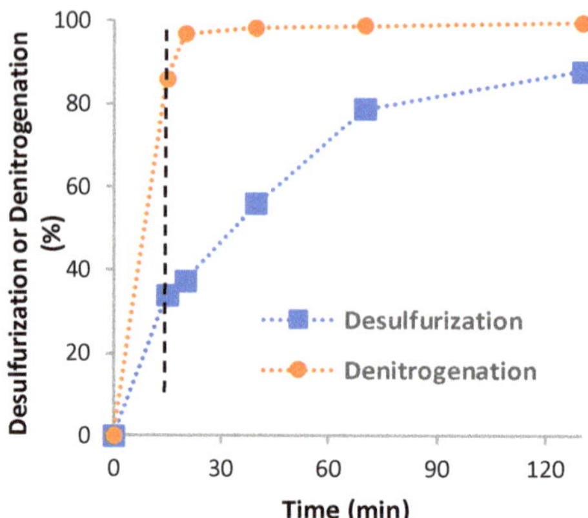

Figure 9. Denitrogenation and desulfurization profile of a model diesel containing approximately 400 ppm N and 2200 ppm of S, catalyzed by [Gd(H$_4$nmp)(H$_2$O)$_2$]Cl·2H$_2$O (**1op**) (20 mg), 0.64 mmol of H$_2$O$_2$ oxidant and 1:1 model diesel/[BMIM]PF$_6$, at 70 °C. The vertical dashed line indicates the instant that oxidative catalytic reaction was started by addition of oxidant.

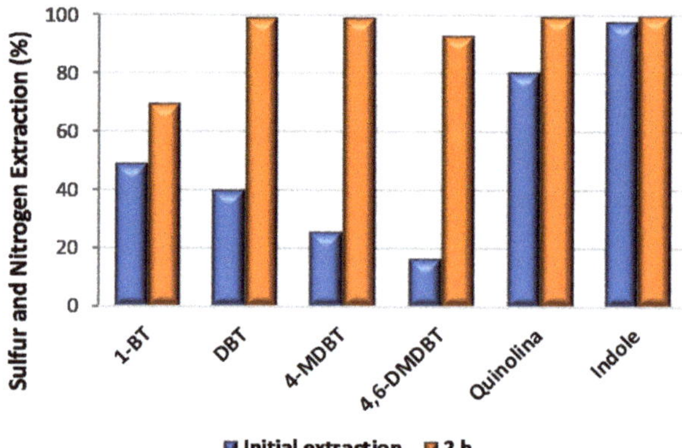

Figure 10. Desulfurization results obtained for each S and N compounds in the model diesel catalyzed by [Gd(H$_4$nmp)(H$_2$O)$_2$]Cl·2H$_2$O (**1op**) (20 mg), using [BMIM]PF$_6$, as extraction solvent, H$_2$O$_2$ (0.64 mmol, 30% aq.) as oxidant, at 70 °C.

The catalytic efficiency of [Gd(H$_4$nmp)(H$_2$O)$_2$]Cl·2H$_2$O/[BMIM]PF$_6$ system was studied for various cycles of simultaneous desulfurization and denitrogenation of multicomponent S/N model diesel. This method was chosen instead of the recycling process to face sustainability and cost effectivity that are important topics to consider for industrial application.

2.6. Reusability Studies

Reusability experiments were performed using an excess of 0.64 mmol of oxidant and a ratio of 1:1 of model diesel/[BMIM]PF$_6$. Figure 11 displays the results obtained for the three consecutive cycles presenting the data obtained after the initial extraction (before oxidant addition) and after 2 h of oxidative catalytic stage. The initial extraction is not affected by the three reusing cycles. For the denitrogenation process the transfer of N compounds to the extraction phase still increased along the cycles (86, 93, 97% for the 1st, 2nd and 3rd cycles, respectively). The oxidative denitrogenation efficiency is maintained throughout the various cycles and complete denitrogenation was achieved after 2 h. The same is not, however, observed for the oxidative desulfurization, where a decrease of efficiency was found from the 1st to the 2nd and the 3rd cycles: 88, 62 and 54%, respectively. The efficiency of the desulfurization process is, therefore, diminished by the presence of N compounds during the various consecutive cycles. The higher difficulty of desulfurization process in the presence of N compounds must be even more pronounced during the reusing cycles because the extraction phase became more saturated with N and S compounds from the 1st to the 2nd and to the 3rd cycles. Desulfurization is more sensitive to deactivation during reusing because nitrogen compounds are significantly better extracted than sulfur ones using IL as extraction solvents [35].

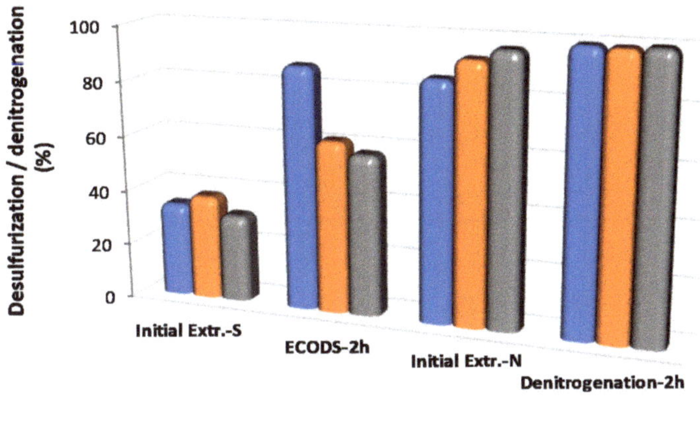

Figure 11. Desulfurization and denitrogenation results obtained for three consecutive reused cycles (initial extraction and after 2 h of catalytic oxidation), using [Gd(H$_4$nmp)(H$_2$O)$_2$]Cl·2H$_2$O (**1op**) (20 mg), 0.64 mmol of H$_2$O$_2$ oxidant and 1:1 model diesel/[BMIM]PF$_6$ system, at 70 °C.

2.7. Structural Stability of [Gd(H$_4$nmp)(H$_2$O)$_2$]Cl·2H$_2$O

The chemical robustness and structural stability of the catalyst was evaluated by performing several characterizations techniques of the solid after the ODS and ODN processes. As depicted in Figure 12, [Gd(H4nmp)(H$_2$O)$_2$]Cl·2H$_2$O (1) suffers a single-crystal-to-single-crystal transformation under these catalytic conditions. Remarkably, 1 withstands the presence of H$_2$O$_2$. Nevertheless, high temperatures led to the complete transformation of 1, even after the first cycle. [Gd(H4nmp)(H$_2$O)$_2$]Cl·2H$_2$O (1) low stability under harsh conditions was previously reported by us. [36] The catalytic structure **1** suffers a single-crystal-to-single-crystal transformation at high temperatures and relative humidity, resulting in a different 2D layered material. In this case, the combination of high temperature and the presence of the IL led to another of such structural transformation which resulted in yet another distinct, unknown phase.

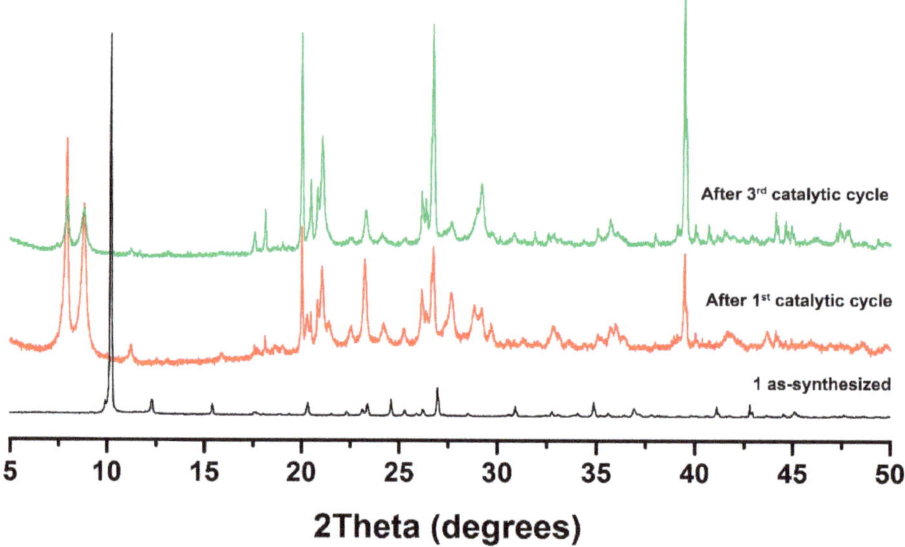

Figure 12. Powder X-ray diffraction patterns of the layered [Gd(H$_4$nmp)(H$_2$O)$_2$]Cl·2H$_2$O, before and after the 1st and 3rd ECODS cycles.

Although the structural elucidation of the catalytic material using conventional powder X-ray analysis was not possible, the remaining solid state characterization allowed us to draw some conclusions (see Section 3 in ESI document for further details). The transformation is accompanied with the release of the chloride anion (possibly in the form of hydrochloric acid, very much as recently observed by us). This release is evident by the elemental distribution analysis performed by energy dispersive X-ray spectroscopy (EDS) in the transformed material (Figure S5 in ESI), and supported by previous transformation of this material under different conditions [36]. After the catalytic process, though the particles appear contaminated with sulfur and fluoride species (which can be attributed to residual IL), the quantification of the Gd and P elements could still be performed. This led to a P:Gd ratio of 3:1 after the catalytic use (maintaining the one Gd^{3+} center per each H$_6$nmp linker), which is the same as determined for the initial layered structure. Therefore, we deduce that the observed structural transformation was not accompanied with the loss of either of the basic building units of the compound. In addition, FT-IR analysis (Figure S6 in the Supporting Information) reveals the presence of virtually the same vibrational bands as for **1**, characteristic of phosphonate-based materials, suggesting that no significant changes in coordination modes were observed.

While the crystalline structure could not unveiled, a Pawley refinement was performed (data not shown). The calculated new unit cell (cell parameters: a = 7.91 Å, b = 10.33 Å, c = 11.59 Å; α = 103.95°, β = 95.46° and γ = 90.64°) suggests a change to a less symmetric cell (from *Ia* to possibly *P*-1), with a decrease more accentuated in one of the cell axis (from 17.48 to 10.33 Å). Because 1 is a 2D layered material, the release of the chlorine anions (that are present in the interstitial space acting as charge-balancing anions) might lead to a significant decrease in the interlayer distance, resulting in registered variations of the unit cell parameters. Once again, this structural behavior agrees well with the previously reported transformation of **1** [36,37].

While the powder X-ray diffraction data confirm that the crystalline structure of layered [Gd(H$_4$nmp)(H$_2$O)$_2$]Cl·2H$_2$O is modified after one catalytic cycle, we emphasize that the new structure remained active and without structural change after three consecutive ECODS cycles (Figure 12). The new compound retains the same plate-like crystal morphology.

3. Conclusions

The ionic lamellar coordination polymer based on a flexible triphosphonic acid linker, [Gd(H$_4$nmp)(H$_2$O)$_2$]Cl$_2$ H$_2$O (H$_6$nmp stands for fornitrilo(trimethylphosphonic) acid) presented a high catalytic efficiency to oxidize the most refractory sulfur and nitrogen compounds present in real diesel (mainly dibenzothiophene derivative, indole, and quinolone). The different methods followed for its preparation (microwave, one-pot, hydrothermal) originated some morphological differences, as the size and shape of obtained particles; however, this did not influence its catalytic performance. Using the model diesel/H$_2$O$_2$/[BMIM]PF$_6$ system, complete desulfurization and denitrogenation were achieved after 2 h of reaction using sulfur or nitrogen model diesel, respectively. The ionic liquid [BMIM]PF$_6$ was the extraction solvent selected since its efficiency was higher than MeCN and other more hydrophilic ionic liquids. When the single model diesel was replaced by the multicomponent S/N model diesel, the desulfurization efficiency decreases from 100% to 88% after 2 h, while the denitrogenation effectivity was maintained. The initial extraction (before oxidant addition) for sulfur and nitrogen was maintained when single model diesels were replaced by the multicomponent diesel. However, the extraction of nitrogen compounds is higher (86%) than the sulfur compounds (36%), what contribute largely for the higher efficiency of denitrogenation process. The recycle capacity of the lamellar catalyst was studied for consecutive desulfurization processes and the catalytic efficiency was maintained between cycles. This result indicates that [Gd(H$_4$nmp)(H$_2$O)$_2$]Cl$_2$ H$_2$O is a stable catalyst, although some structural adjustment occurred to form the active heterogeneous catalyst. An improvement in the reuse capacity of the diesel/H$_2$O$_2$/[BMIM]PF$_6$ system need to be performed in the near future, since the desulfurization process catalyzed by the lamellar material loss efficiency in consecutive ECODS cycles, probably caused by the saturation of the extraction [BMIM]PF$_6$ phase with sulfur and nitrogen compounds.

Supplementary Materials: The following are available online at http://www.mdpi.com/2073-4344/10/7/731/s1, Figure S1: The representative sulfur and nitrogen compounds used in this work to prepare model diesels. 1-benzothiophene (1-BT), dibenzothiophene (DBT), 4-methyldibenzothiophene (4-MDBT) and 4,6-dimethyldibenzothiophene (4,6-DMDBT), Figure S2: Powder X-ray diffraction and SEM images of [Gd(H$_4$nmp)(H$_2$O)$_2$]Cl·2H$_2$O (1) obtained using different experimental methods (op—one-pot; mw—Microwave Assisted; ht—Hydrothermal), Figure S3: Desulfurization of a multicomponent model diesel (2350 ppm S) catalyzed by layered MOF [Gd(H$_4$nmp)(H$_2$O)$_2$]Cl·2H$_2$O (20 mg), using 0.43 mmol of H$_2$O$_2$ oxidant and different volume of [BMIM]PF6 extraction solvent (1:1 and 1:0.5 model diesel/[BMIM]PF$_6$), at 70 °C. The vertical dashed line indicates the instant that oxidative catalytic reaction was started by addition of oxidant, Figure S4: Denitrogenation and desulfurization profile of a model diesel containing approximately 400 ppm N and 2200 ppm of S, catalyzed by layered MOF [Gd(H$_4$nmp)(H$_2$O)$_2$]Cl·2H$_2$O (20 mg), 0.43 mmol of H$_2$O$_2$ oxidant and 1:0.5 model diesel/[BMIM]PF6, at 70 °C. The vertical dashed line indicates the instant that oxidative catalytic reaction was started by addition of oxidant, Figure S5: SEM, mapping and EDS spectra of compound [Gd(H$_4$nmp)(H$_2$O)$_2$]Cl·2H$_2$O (1) after catalytic use for one cycle of ECODS, with a P:Gd ratio of 3:1, Figure S6: FT-IR spectra of layered [Gd(H$_4$nmp)(H$_2$O)$_2$]Cl·2H$_2$O before and after catalytic use for one ECODS cycle.

Author Contributions: Conceptualization, F.A.A.P and S.S.B.; Data curation, F.M. and R.F.M.; Formal analysis, F.M. and R.F.M.; Funding acquisition, S.S.B.; Investigation, F.M.; Project administration, S.S.B.; Supervision, S.S.B. and F.A.A.P.; Validation, F.A.A.P. and R.F.M.; Visualization, F.M.; Writing—original draft, S.S.B.; Writing—review & editing, F.A.A.P and S.S.B. All authors have read and agreed to the published version of the manuscript.

Funding: This work was partly funded through the project REQUIMTE-LAQV (Ref. UID/QUI/50006/2019) and the project GlyGold, PTDC/CTM-CTM/31983/2017, and partially developed within the scope of the project CICECO-Aveiro Institute of Materials, UIDB/50011/2020 & UIDP/50011/2020, LAQV-REQUIMTE (UIDB/50006/2020) and CQE (UIDB/00100/2020) research units, financed by national funds through the FCT/MCTES (Fundação para a Ciência e a Tecnologia / Ministério da Ciência, Tecnologia e Ensino Superior) and when appropriate co-financed by FEDER (Fundo Europeu de Desenvolvimento Regional) under the PT2020 Partnership Agreement. FCT is also gratefully acknowledged for the Junior Research Position CEECIND/00553/2017 (to RFM).

Conflicts of Interest: The authors declare no conflict of interest.

References

1. Chossière, G.P.; Malina, R.; Allroggen, F.; Eastham, S.D.; Speth, R.L.; Barrett, S.R.H. Country- and manufacturer-level attribution of air quality impacts due to excess NOx emissions from diesel passenger vehicles in Europe. *Atmos. Environ.* **2018**, *189*, 89–97. [CrossRef]
2. Frey, H.C. Trends in onroad transportation energy and emissions. *J. Air Waste Manag. Assoc.* **2018**, *68*, 514–563. [CrossRef] [PubMed]
3. Koolen, C.D.; Rothenberg, G. Air Pollution in Europe. *ChemSusChem* **2019**, *12*, 164–172. [CrossRef] [PubMed]
4. Stanislaus, A.; Marafi, A.; Rana, M.S. Recent advances in the science and technology of ultra low sulfur diesel (ULSD) production. *Catal. Today* **2010**, *153*, 1–68. [CrossRef]
5. Chandra Srivastava, V. An evaluation of desulfurization technologies for sulfur removal from liquid fuels. *RSC Adv.* **2012**, *2*, 759–783. [CrossRef]
6. Houda, S.; Lancelot, C.; Blanchard, P.; Poinel, L.; Lamonier, C. Oxidative Desulfurization of Heavy Oils with High Sulfur Content: A Review. *Catalysts* **2018**, *8*, 344. [CrossRef]
7. Reşitoğlu, İ.A.; Altinişik, K.; Keskin, A. The pollutant emissions from diesel-engine vehicles and exhaust aftertreatment systems. *Clean Technol. Environ. Policy Vol.* **2015**, *17*, 15–27. [CrossRef]
8. Roy, S.; Baiker, A. NOx Storage–Reduction Catalysis: From Mechanism and Materials Properties to Storage–Reduction Performance. *Chem. Rev.* **2009**, *109*, 4054–4091. [CrossRef]
9. Yang, L.; Franco, V.; Mock, P.; Kolke, R.; Zhang, S.; Wu, Y.; German, J. Experimental Assessment of NOx Emissions from 73 Euro 6 Diesel Passenger Cars. *Environ. Sci. Technol.* **2015**, *49*, 14409–14415. [CrossRef]
10. Prado, G.H.C.; Rao, Y.; de Klerk, A. Nitrogen Removal from Oil: A Review. *Energy Fuels* **2017**, *31*, 14–36. [CrossRef]
11. Laredo, G.C.; Vega-Merino, P.M.; Trejo-Zarraga, F.; Castillo, J. Denitrogenation of middle distillates using adsorbent materials towards ULSD production. *Fuel Process. Echnol.* **2013**, *106*, 21–32. [CrossRef]
12. García-Gutiérrez, J.L.; Laredo, G.C.; García-Gutierrez, P.; Jiménez-Cruz, F. Oxidative desulfurization of diesel using promising heterogeneous tungsten catalysts and hydrogen peroxide. *Fuel* **2014**, *138*, 118–125. [CrossRef]
13. Rana, M.S.; Al-Barood, A.; Brouresli, R.; Al-Hendi, A.W.; Mustafa, N. Effect of organic nitrogen compounds on deep hydrodesulfurization of middle distillate. *Fuel Process. Technol.* **2018**, *177*, 170–178. [CrossRef]
14. Liu, Y.; Liu, S.; Liu, S.; Liang, D.; Li, S.; Tang, Q.; Wang, X.; Miao, J.; Shi, Z.; Zheng, Z. Facile Synthesis of a Nanocrystalline Metal–Organic Framework Impregnated with a Phosphovanadomolybdate and Its Remarkable Catalytic Performance in Ultradeep Oxidative Desulfurization. *ChemCatChem* **2013**, *5*, 3086–3091. [CrossRef]
15. Julião, D.; Gomes, A.C.; Pillinger, M.; Cunha-Silva, L.; de Castro, B.; Gonçalves, I.S.; Balula, S.S. Desulfurization of model diesel by extraction/oxidation using a zinc-substituted polyoxometalate as catalyst under homogeneous and heterogeneous (MIL-101(Cr) encapsulated) conditions. *Fuel Process. Technol.* **2015**, *131*, 78–86. [CrossRef]
16. Julião, D.; Gomes, A.C.; Pillinger, M.; Valença, R.; Ribeiro, J.C.; de Castro, B.; Gonçalves, I.S.; Cunha Silva, L.; Balula, S.S. Zinc-Substituted Polyoxotungstate@amino-MIL-101(Al)—An Efficient Catalyst for the Sustainable Desulfurization of Model and Real Diesels. *Eur. J. Inorg. Chem.* **2016**, *2016*, 5114–5122. [CrossRef]
17. Granadeiro, C.M.; Nogueira, L.S.; Julião, D.; Mirante, F.; Ananias, D.; Balula, S.S.; Cunha-Silva, L. Influence of a porous MOF support on the catalytic performance of Eu-polyoxometalate based materials: Desulfurization of a model diesel. *Catal. Sci. Technol.* **2016**, *6*, 1515–1522. [CrossRef]
18. Xu, C.; Fang, R.; Luque, R.; Chen, L.; Li, Y. Functional metal–organic frameworks for catalytic applications. *Coord. Chem. Rev.* **2019**, *388*, 268–292. [CrossRef]
19. Shimizu, G.K.H.; Vaidhyanathan, R.; Taylor, J.M. Phosphonate and sulfonate metal organic frameworks. *Chem. Soc. Rev.* **2009**, *38*, 1430–1449. [CrossRef]
20. Firmino, A.D.G.; Figueira, F.; Tome, J.P.C.; Almeida Paz, F.A.; Rocha, J. Robust Metal-Organic Frameworks assembled from tetraphosphonic ligands and lanthanides. *Coord. Chem. Rev.* **2018**, *355*, 133–149. [CrossRef]
21. Pereira, C.F.; Figueira, F.; Mendes, R.F.; Rocha, J.; Hupp, J.T.; Farha, O.K.; Simoes, M.M.Q.; Tome, J.P.C.; Paz, F.A.A. Bifunctional Porphyrin-Based Nano-Metal-Organic Frameworks: Catalytic and Chemosensing Studies. *Inorg. Chem.* **2018**, *57*, 3855–3864. [CrossRef] [PubMed]

22. Mendes, R.F.; Silva, P.; Antunes, M.M.; Valente, A.A.; Paz, F.A.A. Sustainable synthesis of a catalytic active one-dimensional lanthanide-organic coordination polymer. *Chem. Commun.* **2015**, *51*, 10807–10810. [CrossRef] [PubMed]

23. Firmino, A.D.G.; Mendes, R.F.; Antunes, M.M.; Barbosa, P.C.; Vilela, S.M.F.; Valente, A.A.; Figueiredo, F.M.L.; Tome, J.P.C.; Paz, F.A.A. Robust Multifunctional Yttrium-Based Metal Organic Frameworks with Breathing Effect. *Inorg. Chem.* **2017**, *56*, 1193–1208. [CrossRef] [PubMed]

24. Mendes, R.F.; Antunes, M.M.; Silva, P.; Barbosa, P.; Figueiredo, F.; Linden, A.; Rocha, J.; Valente, A.A.; Almeida Paz, F.A. A Lamellar Coordination Polymer with Remarkable Catalytic Activity. *Chem. A Eur. J.* **2016**, *22*, 13136–13146. [CrossRef]

25. Otsuki, S.; Nonaka, T.; Takashima, N.; Qian, W.; Ishihara, A.; Imai, T.; Kabe, T. Oxidative Desulfurization of Light Gas Oil and Vacuum Gas Oil by Oxidation and Solvent Extraction. *Energy Fuels* **2000**, *14*, 1232–1239. [CrossRef]

26. Ribeiro, S.; Barbosa, A.D.S.; Gomes, A.C.; Pillinger, M.; Gonçalves, I.S.; Cunha-Silva, L.; Balula, S.S. Catalytic oxidative desulfurization systems based on Keggin phosphotungstate and metal-organic framework MIL-101. *Fuel Process. Technol.* **2013**, *116*, 350–357. [CrossRef]

27. Xu, J.; Zhao, S.; Chen, W.; Wang, M.; Song, Y.-F. Highly Efficient Extraction and Oxidative Desulfurization System Using Na7H2LaW10O36·32 H2O in [bmim]BF4 at Room Temperature. *Chem. A Eur. J.* **2012**, *18*, 4775–4781. [CrossRef]

28. Zhu, W.; Huang, W.; Li, H.; Zhang, M.; Jiang, W.; Chen, G.; Han, C. Polyoxometalate-based ionic liquids as catalysts for deep desulfurization of fuels. *Fuel Process. Technol.* **2011**, *92*, 1842–1848. [CrossRef]

29. Ribeiro, S.O.; Nogueira, L.S.; Gago, S.; Almeida, P.L.; Corvo, M.C.; Castro, B.D.; Granadeiro, C.M.; Balula, S.S. Desulfurization process conciliating heterogeneous oxidation and liquid extraction: Organic solvent or centrifugation/water? *Appl. Catal. A Gen.* **2017**, *542*, 359–367. [CrossRef]

30. Julião, D.; Gomes, A.C.; Pillinger, M.; Valença, R.; Ribeiro, J.C.; Gonçalves, I.S.; Balula, S.S. Desulfurization of liquid fuels by extraction and sulfoxidation using H2O2 and [CpMo(CO)3R] as catalysts. *Appl. Catal. B Environ.* **2018**, *230*, 177–183. [CrossRef]

31. Jiao, T.; Zhuang, X.; He, H.; Zhao, L.; Li, C.; Chen, H.; Zhang, S. An ionic liquid extraction process for the separation of indole from wash oil. *Green Chem.* **2015**, *17*, 3783–3790. [CrossRef]

32. Liu, D.; Gui, J.; Sun, Z. Adsorption structures of heterocyclic nitrogen compounds over Cu(I)Y zeolite: A first principle study on mechanism of the denitrogenation and the effect of nitrogen compounds on adsorptive desulfurization. *J. Mol. Catal. A Chem.* **2008**, *291*, 17–21. [CrossRef]

33. Zhou, Z.Q.; Li, W.S.; Liu, J. Removal of nitrogen compounds from fuel oils using imidazolium-based ionic liquids. *Pet. Sci. Technol.* **2017**, *35*, 45–50. [CrossRef]

34. Xie, L.-L.; Favre-Reguillon, A.; Wang, X.-X.; Fu, X.; Pellet-Rostaing, S.; Toussaint, G.; Geantet, C.; Vrinat, M.; Lemaire, M. Selective extraction of neutral nitrogen compounds found in diesel feed by 1-butyl-3-methyl-imidazolium chloride. *Green Chem.* **2008**, *10*, 524–531. [CrossRef]

35. Hansmeier, A.R.; Meindersma, G.W.; de Haan, A.B. Desulfurization and denitrogenation of gasoline and diesel fuels by means of ionic liquids. *Green Chem.* **2011**, *13*, 1907–1913. [CrossRef]

36. Mendes, R.F.; Barbosa, P.; Domingues, E.M.; Silva, P.; Figueiredo, F.; Paz, F.A.A. Enhanced proton conductivity in a layered coordination polymer. *Chem. Sci.* **2020**. In Press. [CrossRef]

37. Silva, P.; Mendes, R.F.; Fernandes, C.; Gomes, A.C.; Anànias, D.; Remião, F.; Borges, F.; Valente, A.A.; Paz, F.A.A. Multifunctionality and cytotoxicity of a layered coordination polymer. *Dalton Trans.* **2020**, *49*, 3989–3998. [CrossRef]

Article

Thermal Post-Treatments to Enhance the Water Stability of NH$_2$-MIL-125(Ti)

Almudena Gómez-Avilés, Virginia Muelas-Ramos, Jorge Bedia *, Juan Jose Rodriguez and Carolina Belver *

Chemical Engineering Department, Universidad Autónoma de Madrid, Campus Cantoblanco, E-28049 Madrid, Spain; almudena.gomeza@uam.es (A.G.-A.); virginia.muelas@uam.es (V.M.-R.); juanjo.rodriguez@uam.es (J.J.R.)
* Correspondence: jorge.bedia@uam.es (J.B.); carolina.belver@uam.es (C.B.); Tel.: +34-91-497-29-11 (J.B.); +34-91-497-84-73 (C.B.)

Received: 20 April 2020; Accepted: 27 May 2020; Published: 29 May 2020

Abstract: NH$_2$-MIL-125(Ti) is a metal organic framework (MOF) based on Ti-oxo-clusters widely investigated in water-related applications. Such applications require MOFs with an excellent stability in the aqueous phase, but, despite this, the extent of MOFs' degradation in water is still not yet fully understood. In this study, we report a quantitative study of the water stability of NH$_2$-MIL-125(Ti), analyzing the ligand release along the contact time in water. This study demonstrates that NH$_2$-MIL-125(Ti) easily leached out over time while maintaining its structure. The effect of different thermal treatments applied to NH$_2$-MIL-125(Ti) was investigated to enhance its water stability. The structural and textural properties of those modified MOFs were studied in detail and those maintaining the NH$_2$-MIL-125(Ti) properties were exposed to aqueous medium. The analysis of the released ligand concentration in the filtrate can provide information on the water stability of this material.

Keywords: metal organic frameworks; NH$_2$-MIL-125(Ti); water stability; purification

1. Introduction

Metal organic frameworks (MOFs) constitute a group of materials consisting of the repetitive assembling of organic linkers and cluster metal ions. The resulting crystalline structures have a large specific surface area and a well-defined porosity. Their physicochemical properties can be tailored by selecting the nature and combination of both components yielding numerous MOFs [1,2]. MOFs have gained growing importance in many applications, such as gas storage [3,4], sensing [5,6], membrane processes [7], drug delivery [8,9], and environmental remediation [10–13]. In this last field, MOFs have been investigated in water purification, especially as: (i) adsorbents for the removal of hazardous pollutants [14–16], (ii) catalysts for the degradation of contaminants by Advanced Oxidation Processes (AOPs), such as heterogeneous Fenton and photo-assisted treatments [17–19], and (iii) membrane materials for the separation of toxic substances from water streams [20]. The growing interest in these applications is leading researchers to develop MOFs with high stability in aqueous medium.

The literature describes examples of unstable MOFs that lose their structure and pore network when exposed to water, such as MOF-5, while others showed stability in water, even after months, or have not been fully tested in this respect [21,22]. The stability of MOFs can be related to an assembly–disassembly equilibrium where both the electronic and steric effects between the linker and the metal cluster play an important role. Several studies have shown that the instability under wet conditions depends on the metal–linker bond strength [23,24]. Using highly basic ligands or acid metals results in much stronger bonds and greater stability in water, as in the cases of ZIF-8 or Al-MIL-53 [25]. In addition, the linker structure, the nature of the metal and the coordination of the clusters also play an important role. Thus, hydrophobic linkers, inert metal cations and high-coordination clusters usually improve water stability. Moreover, high-charge metal ions lead to a stronger oxo-philic tendency that significantly strengthens the metal–linker bonds and improves the chemical stability of MOF [21]. Group-IV-metal- (Ti, Zr, Hf) based MOFs are examples of high-charge metal oxo-clusters, yielding abundant MOF structures with different coordination numbers and structural diversity [26–29]. Ti-based MOFs have received great attention because of their redox activity (Ti^{3+}/Ti^{4+} transition capacity), photochemical property and biocompatibility [30]. Those based on Ti-carboxylate complexes have high stability and their functional properties can be improved by organic functionalization, mainly by amine groups [31,32]. The degradation of MOFs exposed to water can occur by two different mechanisms, ligand displacement and hydrolysis [25]. The ligand displacement involves the exchange of a ligand by a water molecule, leading to the hydration of the metal and ligand lixiviation; while, upon hydrolysis, water dissociation occurs at the metal centers, the dissociated OH groups coordinate to the metal clusters and the metal–ligand bond is broken [33].

Nevertheless, predicting the water stability of MOFs remains qualitative and the literature is controversial, mainly due to the methodology used to assess the stability. The simplest process consists of exposing the MOF to water and comparing its properties prior and after a given contact time, mainly by weight loss, X-ray diffraction and N_2 adsorption–desorption at −196 °C [29]. Water adsorption–desorption isotherms are also useful since they provide information about the hydrophilic–hydrophobic character of MOFs [34]. Additionally, however, if the MOF is being considered for application in the aqueous phase, it should also be investigated whether the sample is partially dissolved [29,35]. Recently, Taheri et al. [36] performed a quantitative study of the ligand and Zn^{2+} released when ZIF-8 was immersed in water. They observed that the lixiviation of both species occurred during the first hour and after 24 h an equilibrium was achieved. No other studies deal with the analysis of the filtrate instead of the remaining solid in the literature. Lixiviation from the MOF must be considered for the sake of application in water remediation since the released metals and/or ligands can represent some hazards for the environment.

The literature contains examples concerning the post-modifications of MOFs to improve their water stability. Wen et al. [37] coated NH_2-MIL-125(Ti) with a siloxane that increased its dye adsorption capacity while maintaining its crystal structure and morphology. Composites based on UiO-66-NH_2 and a nitrile butadiene rubber sponge are more stable in water than the bare MOF because the sponge provides a hydrophobic character that prevents hydrolysis. These composites were shown to be efficient adsorbents for 2,4-dichlorophenoxyacetic acid removal from water and were easily recoverable thanks to its sponge conformation [38]. An interesting study has recently been reported, describing the improved water stability of ZIF-8 by growing it along with UiO-66-NH_2, giving rise to a core–shell UiO-66-NH_2@ZIF-8 hybrid. This heterostructure showed excellent water stability, high adsorption capacity and selectivity for the separation of rare earths from water compared to other adsorbents [39]. Non-amine UiO-66 has been also incorporated in a composite membrane based on graphene oxide (GO) with a high water stability and highly efficient for dyes and antibiotics separation [40]. Although interesting, many of these modifications are not so easy to implement in technical applications.

Here, we focus the attention on NH_2-MIL-125(Ti), a Ti-oxo-cluster MOF, due to its notable interest in water treatment by photocatalysis, a well-known AOP for the removal of different pollutants [1,41–43]. In a previous study, we demonstrated its high photocatalytic efficiency for the removal of some emerging pollutants and its stability during the reaction, comparing its structural, textural and morphologic properties before and after use [44]. However, in a preliminary test, additional organic matter was detected in the aqueous medium after reaction, thus suggesting that certain partial dissolution of the solid occurred. Thus, in this work, we initially studied the water stability of NH_2-MIL-125(Ti) through a systematic analysis of the aqueous phase. Then, in order to improve its stability, NH_2-MIL-125(Ti) was subjected to various thermal treatments. In particular, the study highlights that the stability of NH_2-MIL-125(Ti) in water can be successfully improved, providing useful information for water-related applications.

2. Results and Discussion

2.1. Stability of the NH₂-MIL-125(Ti)

2.1. Stability of the NH_2-MIL-125(Ti)

NH_2-MIL-125(Ti) is an amine-functionalized isostructure of the MIL-125(Ti) formed by both octahedral and tetrahedral cages [45,46]. It is composed by titanium oxo-clusters and amino-terephtalate, both providing a high density of hydrophilic sites where water molecules can be adsorbed [47]. We had previously synthesized NH_2-MIL-125(Ti), doped and successfully used as solar-light driven photocatalyst [44]. Although its structural stability under the reaction conditions was confirmed by X-ray diffraction, scanning electron microscopy and N_2 adsorption–desorption at $-196\,°C$, some ligand release was detected in the aqueous medium. To the best of our knowledge, there are no reports about the amount of linker leached when NH_2-MIL-125(Ti) is put in contact with water. Thus, to determine the amount of released linker, NH_2-MIL-125(Ti) was suspended in aqueous medium and the filtrates were analyzed, the leachate percentage being determined by:

$$\text{Leachate } (\%) = 100 \times \frac{C_{\text{linker}}}{C_{NH_2\text{-MIL-125(Ti)}}} \cdot \frac{M_{NH_2\text{-MIL-125(Ti)}}}{6 \times M_{\text{linker}}} \tag{1}$$

where C_{linker} is the linker concentration dissolved in the liquid phase (mg·L^{-1}), $C_{NH_2\text{-MIL-125(Ti)}}$ is the concentration of the MOF suspended in water, $M_{NH_2\text{-MIL-125(Ti)}}$ and M_{linker} are the molecular weight values of the MOF ($1653.74\ \text{g·mol}^{-1}$) and the linker (2-amino benzene dicarboxylate, NH_2-BDC, $179.12\ \text{g·mol}^{-1}$), respectively. Figure 1a shows the evolution of dissolved linker and the corresponding leachate percentage upon contact time. NH_2-MIL-125(Ti) undergoes relatively high linker leaching in water, which increases significantly over time to reach $40\ \text{mg·L}^{-1}$ after 24 h, about 25% of the initial linker content of bare NH_2-MIL-125(Ti). This leaching process occurs continuously over time, without reaching equilibrium. Similar behavior has recently been reported for ZIF-8, whose lixiviation started during the first hour and required one day to achieve equilibrium [36]. This leaching is detrimental regarding the potential water-related applications of NH_2-MIL-125(Ti). For the sake of improving the stability of NH_2-MIL-125(Ti) by reducing ligand leaching as much as possible, this MOF was subjected to different thermal treatments at different temperatures and holding times and under different atmospheres. The temperature was selected after studying the thermal behavior of this MOF by thermogravimetric analysis in air (Figure 1b). NH_2-MIL-125(Ti) is thermally robust and suffers a strong weight loss of 54% within the 300 to 350 °C range, due to the oxidation of the organic linkers [36,48]. After increasing the temperature up to 500 °C, a small weight loss (9%) occurs, corresponding to the removal of the hydroxo-groups in the metal oxo-clusters, giving rise to TiO_2, in a similar way to the analogous non-amine MIL-125(Ti) [46,49]. Based on this analysis, the 150–300 °C range was selected for the thermal treatment addressed to stabilize the NH_2-MIL-125(Ti) without disturbing its structure and composition.

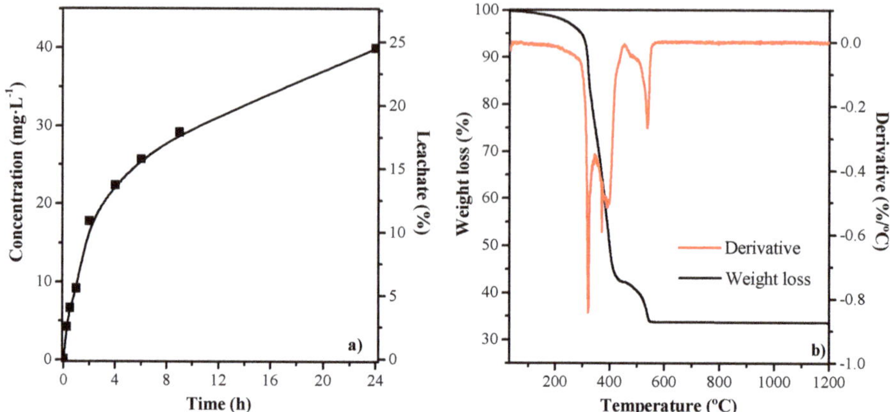

Figure 1. (**a**) Time course of linker concentration and leachate percentage in water from NH_2-MIL-125(Ti) upon contact time; (**b**) thermogravimetric analysis in air of NH_2-MIL-125(Ti) and its corresponding derivative.

2.2. Vacuum Treatment

During the synthesis of MOFs, excess ligand and solvent molecules can remain trapped in the pores of the framework, which can be detrimental for their future applications. In some cases, vacuum drying may be sufficient to purify the MOF, although it can lead to lower surface areas than expected due to the partial collapse of the structure [35]. NH_2-MIL-125(Ti) was thus subjected to vacuum treatment in a temperature range of 100 to 300 °C for 16 h to remove the excess linker and solvent molecules. Figure 2 shows the X-ray diffraction (XRD) patterns of the modified solids compared with that of the bare NH_2-MIL-125(Ti). In the notation, "V" refers to vacuum treatment, and the numbers represent the temperature (°C) and time in hours of the vacuum treatment. As can be seen, this treatment does not modify the crystalline structure of the MOF until reaching 300 °C. At this temperature, the NH_2-MIL-125(Ti) structure collapses, and the resulting material does not describe any crystalline XRD profile, appearing as an amorphous material. Before that, a reduction in the peaks' intensity as the temperature increases is evidenced, indicating a gradual loss of crystallinity. Thus, the degree of crystallinity and the crystal size, as determined by the methodologies described in Section 3.2, are collected in Table 1. The greatest reduction on crystallinity was observed when reaching 250 °C. However, since NH_2-MIL-125(Ti) is an amine-functionalized isostructure of the MIL-125(Ti), both materials would show the same XRD profile, so it would be necessary to corroborate that the amine group in the ligand is maintained. The NH_2-BDC presence was confirmed by UV-vis spectroscopy (F1 in the supplementary information). Vacuum-treated samples showed two absorption bands, at 280 and 370 nm, due to the absorption of Ti-oxo-clusters and NH_2-BDC linker, respectively. Both bands characterize the NH_2-MIL-125(Ti) and differ from the spectrum of MIL-125(Ti), which only has one band due to the absorption on Ti-oxo-clusters [49,50]. Scanning electron microscopy (SEM) and transmission electron microscopy (TEM) images of V-250-16 were taken and compared with those from the original sample (Figure 3). The MOF particles show a thin and disk-like shape with an average size close to 500 nm, very similar to those of the NH_2-MIL-125(Ti), which confirms that the vacuum treatment does not significantly modify the morphology of the MOF.

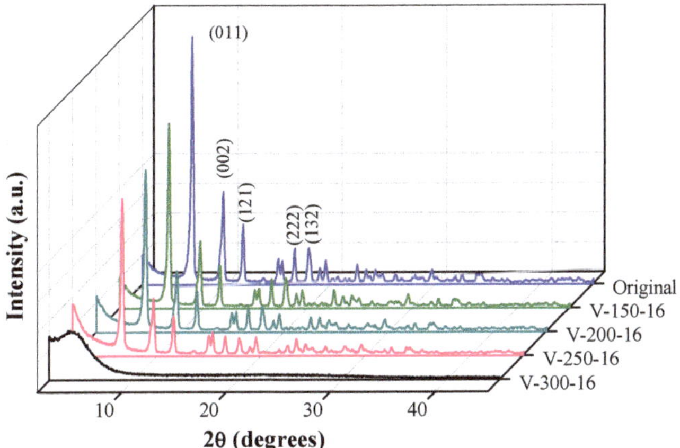

Figure 2. X-ray diffraction (XRD) patterns of the NH_2-MIL-125(Ti) treated under vacuum at different temperatures for 16 h. The original NH_2-MIL-125(Ti) pattern is included as reference.

Table 1. Crystallinity percentages and crystal size values of NH_2-MIL-125(Ti) before and after vacuum treatment at different temperatures for 16 h.

Sample	Crystallinity (%)	Relative Crystallinity [1] (%)	Crystal Size (nm)
Original	80.3	100.0	33.2
V-150-16	75.0	93.4	30.9
V-200-16	75.6	94.1	30.2
V-250-16	63.6	79.2	21.2

[1] Relative crystallinity to that of the original MOF.

Figure 3. Scanning electron microscopy (SEM) and transmission electron microscopy (TEM) images of the original NH_2-MIL-125(Ti) and V-250-16 sample.

Figure 4 shows the −196 °C N_2 adsorption–desorption isotherms of the NH_2-MIL-125(Ti) subjected to the vacuum treatments and Table 2 summarizes the porous textural characteristics derived from those isotherms. Most of the solids treated under vacuum showed isotherms similar to that of the original NH_2-MIL-125(Ti), indicative of predominantly microporous texture with some relative contribution of mesoporosity and fairly high surface area values [39,44,48]. Only the V-300-16 shows remarkable differences, with very low N_2 adsorption and S_{BET} due to the linker oxidation and structure collapse. It is worth noting that an increase in the temperature, below 300 °C, is associated with some slight increase in the microporous surface, which can be related to the removal of adsorbed species, most probably excess linker and/or solvent molecules, not removed during the washing process that block the pore network [35]. According to these results, 250 °C appears to be the best temperature to carry out the vacuum-stabilization of the NH_2-MIL-125(Ti), maintaining its structure and porosity.

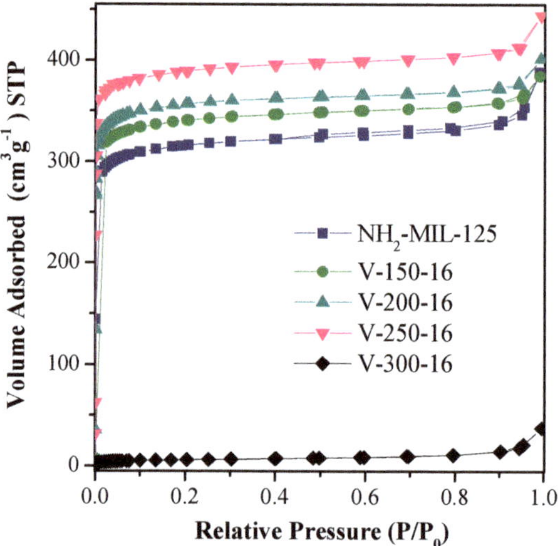

Figure 4. N_2 adsorption–desorption isotherms at −196 °C of the original NH_2-MIL-125(Ti) and after treatment under vacuum at different temperatures for 16 h.

Table 2. Porous texture characterization of the NH_2-MIL-125(Ti) and after treatment under vacuum at different temperatures for 16 h.

Sample	S_{BET} [1] $(m^2 \cdot g^{-1})$	S_{MP} [2] $(m^2 \cdot g^{-1})$	S_{EXT} [3] $(m^2 \cdot g^{-1})$	V_T [4] $(cm^3 \cdot g^{-1})$	V_{MP} [5] $(cm^3 \cdot g^{-1})$
Original	1025	940	85	0.60	0.45
V-150-16	1091	996	95	0.61	0.48
V-200-16	1103	1016	87	0.61	0.49
V-250-16	1154	1050	104	0.68	0.54
V-300-16	24	8	16	0.05	0.002

[1] S_{BET}, specific surface area; [2] S_{MP} and [3] S_{EXT}, microporous and non-microporous surface area; [4] V_T and [5] V_{MP}, total and micropore volume, respectively.

Figure 5 depicts the evolution of the linker concentration analyzed in the liquid phase and the corresponding leaching percentage, upon contact time in water. The vacuum thermal treatment allows for significantly reducing the amount of linker leaching, that effect being more pronounced at a 200–250 °C treatment temperature. However, despite this reduction, there is still a considerable amount of linker in the liquid phase, corresponding to a 15% leaching, and it is also noteworthy that the lixiviation does not reach an equilibrium, even after 24 h. Therefore V-200-16 or V-250-16 samples cannot still be considered as stable materials regarding potential applications in the aqueous phase.

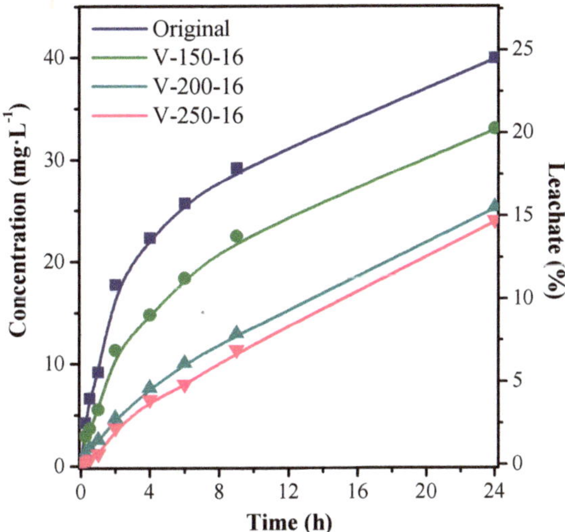

Figure 5. Linker leaching in water from NH_2-MIL-125(Ti) before and after vacuum heat treated upon contact time.

2.3. Thermal Treatment in Air Atmosphere

Figure 6 depicts the XRD diffractograms of the NH_2-MIL-125(Ti) subjected to calcination during 16 h at different temperatures in air-circulating muffle furnace. As previously observed in the vacuum treatment, the NH_2-MIL-125(Ti) structure was maintained up to 250 °C, while it collapses at 300 °C, disappearing all the characteristic peaks of NH_2-MIL-125(Ti). The degree of crystallinity and the crystal size of each sample are collected in Table 3. Increasing the temperature decreases the crystallinity of the samples, accompanied by a reduction in crystal size. These modifications can be related to the loss of solvent and excess linker molecules that are usually trapped in the MOF cages [35]. To learn more about the effect caused by air heating, additional treatments were conducted at 250 °C, but extending the treatment time. Figure 7 compares the diffraction patterns of the samples heated for 16, 48 and 72 h, where it can be seen that the characteristic peaks of the original structure are still remaining, but showing decreasing intensities as the calcination time increased, thus indicating a considerable gradual loss of crystallinity, reaching about 60% after 72 h, which means a certain amorphization of the material.

Table 3. Crystallinity percentages and crystal size values of NH_2-MIL-125(Ti) before and after heat treatment in air under different conditions. The values for M-250-48 after 24 h in contact with water are included.

Sample	Crystallinity (%)	Relative Crystallinity [1] (%)	Crystal Size (nm)
Original	80.3	100.0	33.2
M-150-16	79.2	98.5	33.5
M-200-16	75.3	93.7	32.9
M-250-16	63.7	79.2	31.1
M-250-48	50.0	62.2	31.4
M-250-72	34.1	42.4	32.9
M-250-48-Exposed	49.7	61.8	31.1

[1] Relative crystallinity to that of the original MOF.

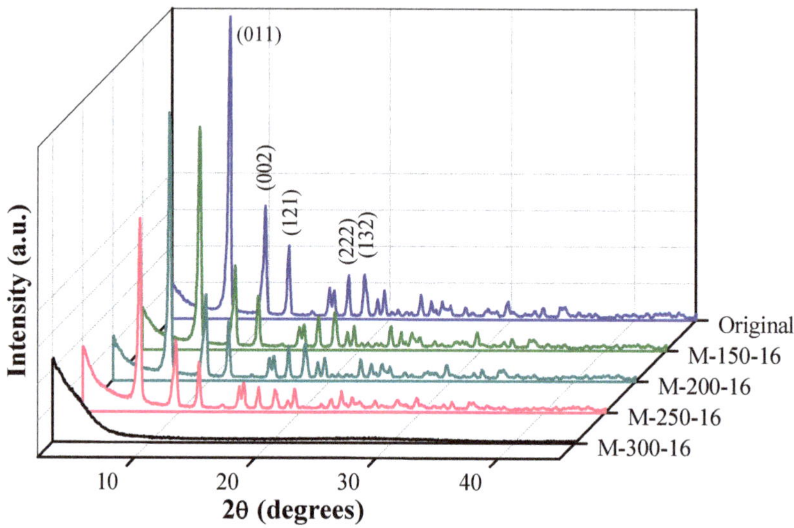

Figure 6. XRD diffraction patterns of NH_2-MIL-125(Ti) heated in air at different temperatures for 16 h. The original NH_2-MIL-125(Ti) pattern is included as reference.

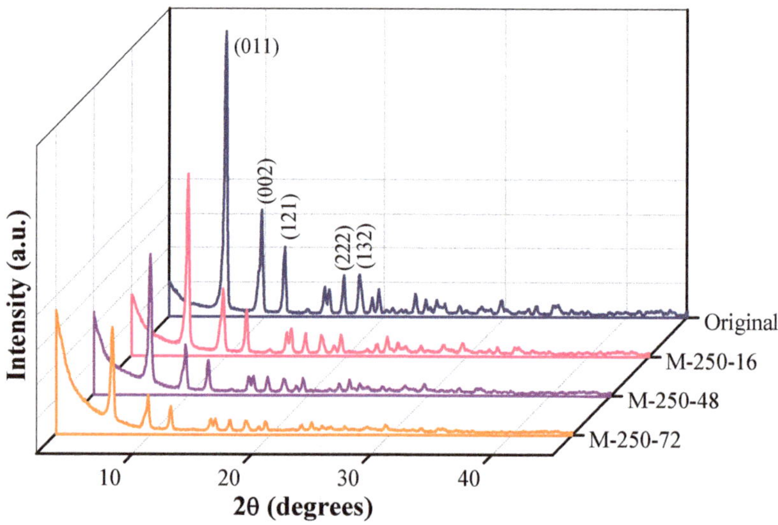

Figure 7. XRD diffraction patterns of NH_2-MIL-125(Ti) heated in air at 250 °C for 16, 48 and 72 h. The original NH_2-MIL-125(Ti) pattern is included as reference.

Once again, the presence of NH_2-BDC in all heat-treated samples was confirmed by UV-vis spectroscopy (Figure S2 in the supplementary information). The UV-vis spectra showed the two absorption bands characteristic of the NH_2-MIL-125(Ti), unlike the single band that characterizes the MIL-125(Ti) [49,51,52]. Figure 8 collects SEM and TEM images of some calcined samples that can be compared with those of the original NH_2-MIL-125(Ti) shown in Figure 3. As can be clearly observed, the morphology of M-300-16 is very different. The well-defined disk particles of the NH_2-MIL-125(Ti) disappeared and became an amorphous morphology, corroborating the structural collapse described above. However, the morphology and disk-shape of the original MOF remained after thermal treatment at 250 °C for 48 h, as well as the particle size ($\approx$ 500 nm).

Figure 8. SEM image of M-300-16. SEM and TEM images of M-250-48.

Figure 9a depicts the −196 °C N_2 adsorption–desorption isotherms of NH_2-MIL-125(Ti) subjected to thermal treatments in air and the derived porous textural characteristics are summarized in Table 4. As in the case of heat treatment under vacuum, these samples showed an isotherm typical of microporous solids [39,44,48]. Again, the calcination in air at 300 °C caused the loss of structure and the collapse of the pore network, resulting in a N_2 adsorption–desorption isotherm of a non-porous material. Again, in the range below 300 °C, increasing the calcination temperature produces some slight increase in the microporous surface. However, in the sample treated at 250 °C, the observed increase in surface area corresponds to the non-microporous one. This effect can be related with the decrease in crystallinity and increased disorder observed by XRD. When comparing the vacuum and the air thermal treatments, no obvious differences were detected, so it seems evident that temperature is the main factor governing the purification process. Regarding the effect of the treatment time, the N_2 adsorption–desorption isotherms of their corresponding samples remain typical of microporous solids (Figure 9b). Increasing the treatment time led to some moderate reduction in the BET surface area, affecting to the microporous range, while the non-microporous area was increased, probably due to the loss of crystallinity seen by XRD. Therefore, thermal treatment at moderate temperature ($\approx$250 °C) allows purifying and improving the stability of the tested MOF, while maintaining its porous structure and surface area close to 1000 $m^2 \cdot g^{-1}$, analogous to other untreated NH_2-MIL-125(Ti) [41–44].

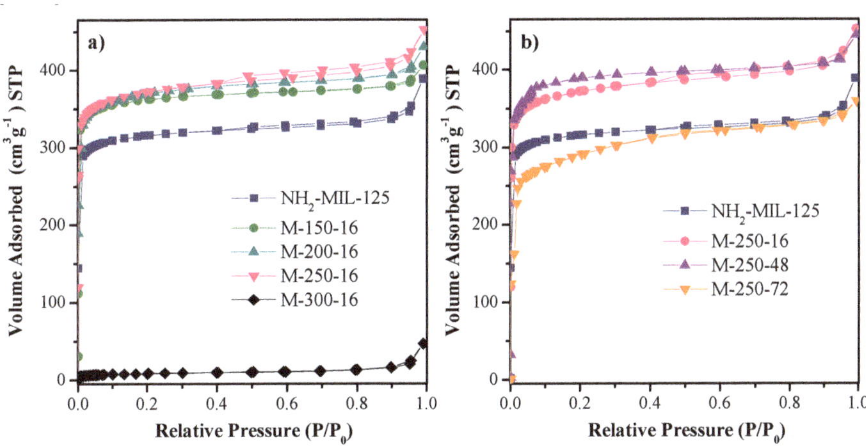

Figure 9. N_2 adsorption–desorption isotherms at −196 °C of the original NH_2-MIL-125(Ti) and after heat treatment in air at different temperatures for 16 h (**a**) and at 250 °C for different times (**b**).

Table 4. Porous texture characterization of NH$_2$-MIL-125(Ti) after heat treatment in air under different conditions.

Sample	S$_{BET}$ [1] (m^2·g^{-1})	S$_{MP}$ [2] (m^2·g^{-1})	S$_{EXT}$ [3] (m^2·g^{-1})	V$_T$ [4] (cm^3·g^{-1})	V$_{MP}$ [5] (cm^3·g^{-1})
Original	1025	940	85	0.60	0.45
M-150-16	1085	989	96	0.63	0.53
M-200-16	1173	1075	98	0.67	0.54
M-250-16	1207	1058	149	0.70	0.53
M-300-16	28	12	16	0.07	0.006
M-250-48	1046	865	181	0.56	0.35
M-250-72	962	684	278	0.56	0.33
M-250-48-Exposed	970	850	120	0.53	0.34

[1] S$_{BET}$, specific surface area; [2] S$_{MP}$ and [3] S$_{EXT}$, microporous and non-microporous surface area; [4] V$_T$ and [5] V$_{MP}$, total and micropore volume, respectively.

Figure 10 shows the results of the stability in water of the samples subjected to calcination in air at different temperatures and times. Thermal treatment in air for 16 h stabilizes the NH$_2$-MIL-125(Ti) and the leachate percentage diminishes at increasing temperature. At 250 °C, this improvement of stability in water is significantly higher than the observed with the thermal vacuum treatment (see Figure 5) and now a quasi-equilibrium state is achieved. Extending the air thermal treatment up to 48 h reduces the leaching percentage to less than 7% and maintains the quasi-equilibrium state, although no further improvement was observed at higher treatment times. Characterization of the M-250-48 sample after water exposition for 24 h was carried out to check possible structural and textural changes (Figure 11). There were no significant differences in X-ray diffractograms or N$_2$ adsorption–desorption isotherms before and after contact with water during the above-mentioned time. The S$_{BET}$ slightly decreased somewhat, from 1046 to 970 m^2·g^{-1}, but the crystallinity remained unchanged (see values in Table 3; Table 4). The NH$_2$-BDC also remained in the structure, including the presence of the amine group, as confirmed by its UV-vis spectrum (Figure S3 in the supplementary information) that presents the characteristic absorption bands of this linker. These results demonstrate that it is possible to purify and stabilize the NH$_2$-MIL-125(Ti) MOF by a fairly simple thermal treatment in air at 250 °C for 48 h, without affecting its structure and porous texture.

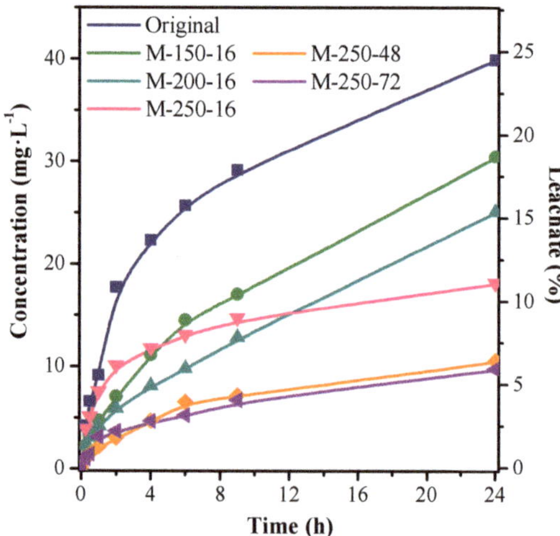

Figure 10. Linker leaching in water from the original NH$_2$-MIL-125(Ti) and after heat treatment in air for 16 h at different temperatures and at 250 °C during different times.

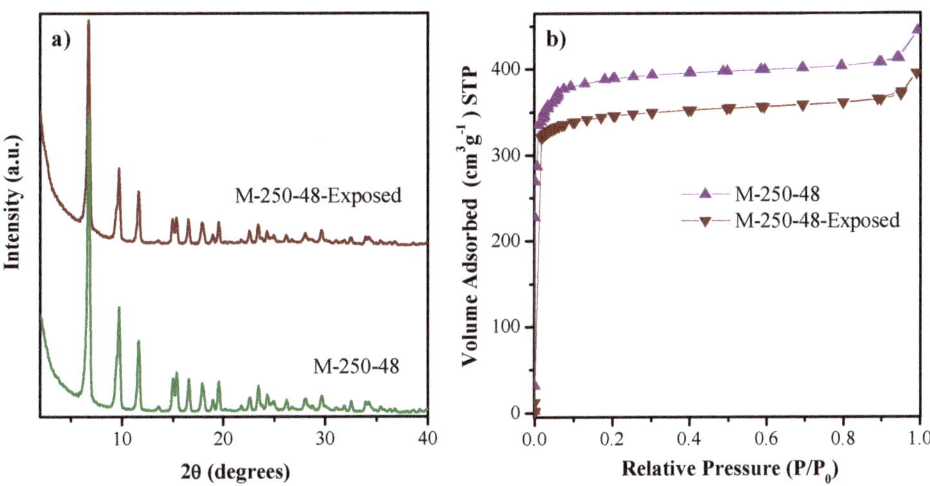

Figure 11. (a) XRD patterns and (b) N_2 adsorption–desorption isotherms at $-196\ ^\circ$C of M-250-48 before and after contact with water for 24 h.

3. Materials and Methods

3.1. NH_2-MIL-125(Ti) Synthesis

NH_2-MIL-125(Ti) was prepared following the methodology reported in our previous work (chemical information included in Figure S4 in the supplementary information) [44]. Typically, 2-amino benzene dicarboxylic acid (6 mmol, NH_2-BDC, Sigma Aldrich, Madrid, Spain, 99%) was dissolved in dimethylformamide (25 mL, DMF, Sigma Aldrich, Madrid, Spain, $\geq$ 99.8%) under stirring. Titanium isopropoxide (3 mmol, Sigma Aldrich, Madrid, Spain, $\geq$ 98%) was dropped onto the mixture and stirred until homogenized, after which methanol (25 mL, CH_3OH, Sigma Aldrich, Madrid, Spain, anhydrous 99.8%) was added. The whole mixture was kept in agitation for 30 min, drawn off to a 65 mL Teflon autoclave and placed in an oven at 150 $^\circ$C for 16 h. The resulting solid was recovered by centrifugation (5 min, 5000 rpm) and washed three times with DMF (100 mL, 30 min) and three times with methanol (100 mL, 30 min), recovering the solid in all cases by centrifugation. The final drying of the solid was performed overnight at 60 $^\circ$C.

NH_2-MIL-125(Ti) was subjected to different thermal treatments. The first series was obtained by heating the NH_2-MIL-125(Ti) at 150, 200, 250 and 300 $^\circ$C for 16 h under vacuum using a degassing station (Micromeritics VacPrep 061, Norcross, GA, USA) attached to the TriStar 123 equipment. For the second series, the NH_2-MIL-125(Ti) was heated in a muffle furnace (Fuji Electric, MOD 12 PR/400, Barcelona, Spain) in air at 150, 200, 250 and 300 $^\circ$C for 16 h, and also at 250 $^\circ$C for 48 and 72 h. The resulting samples were denoted as X-T-t, where X indicated the type of treatment, V for vacuum and M for muffle furnace, T was the treatment temperature and t the treatment time.

3.2. Characterization

Thermogravimetric analysis was carried out using a Q600 TA instruments (New Castle, DE, USA) in an alumina crucible. All measurements were carried out within the temperature range 25–1200 $^\circ$C at a heating rate of 10 $^\circ$C·min^{-1} under a continuous air flow of 20 mL·min^{-1}. X-ray diffractograms (XRD) were registered in the 2–45° 2θ range with a rate of 1.5 $^\circ$·min^{-1} and using a Cu Kα radiation. A Bruker D8 diffractometer (Bremen, Germany) was used, equipped with a Sol-X energy dispersive detector. Crystallinity percentage was calculated after diffractogram deconvolution with OriginPro by the following expression [53,54]:

$$\text{Crystallinity } (\%) = 100 \times \frac{\text{Area of crystalline peaks}}{\text{Area of all XRD pattern}} \tag{2}$$

The average crystal size was determined by using the Scherrer's equation from the reflection peak (121) of NH_2-MIL-125(Ti), not overlapped with any other peak. UV-vis diffuse reflectance spectroscopy was carried out on Shimadzu UV2600i (Kyoto, Japan) equipment, recording the spectra in the 250–700 nm range. Scanning electron microscopy images (SEM) were taken with a Hitachi S4800 microscope (Krefeld, Germany) and transmission electron microscopy images (TEM) with a TECNAI G2 20 Twin equipment (Hillsborough, NC, USA). N_2 adsorption–desorption isotherms at -196 °C were carried out with a TriStar 123 equipment (Micromeritics II 3020, Norcross, GA, USA). Solids were outgassed before testing under vacuum at 120 °C for 18 h. Specific surface area (S_{BET}) was calculated by the Brunauer–Emmet–Teller method [55], while micropore surface area (S_{MP}) was obtained using the *t-plot* method [56] and the external or non-microporous surface area (S_{EXT}) by difference between S_{BET} and S_{MP}. The total pore volume (V_T) corresponded to the amount of nitrogen adsorbed at a relative pressure of 0.99, while the micropore volume (V_{MP}) was obtained from the *t-plot* method.

3.3. Stability Testing in Water

The water stability of the NH_2-MIL-125(Ti) derived solids was tested through the leaching of the NH_2-BDC in deionized water (Type II). The solid was suspended in water (250 mg·L^{-1}) and stirred for 24 h at 25 °C and 170 rpm in a water bath orbital shaker (JULABO SW22, Seelbach, Germany). At different interval times, aliquots of 2 mL were collected and filtered with Polytetrafluoroethylene (PTFE) syringe filters (Whatman 0.45 μm). The filtrates were analyzed to determine the concentration of the ligand by high performance liquid chromatography (HPLC) (Shimadzu Prominence-I LC-2030C, Kyoto, Japan), equipped with a diode array detector (SPDM30 A) and a C18 column (Eclipse Plus, 5 μm). A gradient elution method was used for the analyses, changing the mobile phase (aqueous acetic acid solution 0.1%, Sigma Aldrich, Madrid, Spain, ≥99%) by elution with acetonitrile (HPLC grade, Scharlab, Barcelona, Spain) and using a constant flow of 0.7 mL·min^{-1}. The wavelength of the detector was fixed at 358 nm for the detection of the NH_2-BDC that appeared at a 6.3 min retention time. Solid samples were recovered from the aqueous medium by centrifugation after stability test and stored for later characterization.

4. Conclusions

Water stability of NH_2-MIL-125(Ti) was studied by analyzing quantitatively the amount of ligand released to the liquid phase along the exposition time in water. In many studies, the water stability of MOFs has been mainly investigated using the X-ray diffraction, N_2 adsorption–desorption and electron microscopy of filtered powders, comparing the properties before and after exposition. This study shows that these characterization techniques are not enough to assess that stability. We demonstrated, for the first time, the quantitative lixiviation of this MOF, which achieved values up to 25% after 24 h of contact, with water still being far from equilibrium after that time. The present study also shows that the post-modification of NH_2-MIL-125(Ti) by thermal treatments stabilizes its structure, making it less susceptible to partial dissolution. Temperature and time showed to be the determining variables, while the atmosphere (air or vacuum) showed no significant effect. Thermal treatment under these two atmospheres within 150–250 °C did not alter the structure and porous texture of the MOF, except for some slight increase in the specific surface area accompanied by a small reduction in crystallinity. Increasing the treatment temperature up to 300 °C caused the collapse of the crystalline MOF structure accompanied by the loss of the pore network. Certain reduction on the surface area were observed by extending the thermal treatment but maintaining its structure.

The results obtained provide a new understanding of the effects of thermal treatments on the water stability of NH_2-MIL-125(Ti), which was improved as the temperature of the treatment increased up to 250 °C. Some significant differences were found on the effects of the atmosphere used in the thermal treatment. Under air, greater stabilization of the MOF was achieved due to the removal of excess linker molecules, with lower leaching percentages and the attainment of a quasi-equilibrium state in a relatively low contact time in water. The best results were obtained upon thermal treatment in air at 250 °C for 48 h. After that treatment, less than 7% ligand leaching occurred upon 24 h of contact with water. These findings have significant implications regarding the potential use of MOFs in water-related applications.

Supplementary Materials: The following are available online at http://www.mdpi.com/2073-4344/10/6/603/s1, Figure S1: UV–vis diffuse absorbance spectra of NH_2-MIL-125(Ti) treated under vacuum at different temperatures for 16 h. Original NH_2-MIL-125(Ti) and MIL-125(Ti) spectra are included as reference, Figure S2: UV–vis diffuse absorbance spectra of NH_2-MIL-125(Ti) heated in air at different temperatures for 16 h (**a**) and at 250 °C for 48 and 72 h (**b**). Original NH_2-MIL-125(Ti) and MIL-125(Ti) spectra are included as reference, Figure S3: UV–vis diffuse absorbance spectra of M-250-48 before and after contact with water for 24 h. Original NH_2-MIL-125(Ti) and MIL-125(Ti) spectra are included as reference, Figure S4: (**a**) Chemical structure of Ti-oxo clusters linked to NH_2-BDC ligands, from the Cambridge Structural Database (CSD). Chemical structure of NH_2-MIL-125(Ti) viewed from a-axis (**b**) and c-axis (**c**), from cif data of Crystallography Open Database.

Author Contributions: Conceptualization, J.B., C.B. and J.J.R.; methodology, A.G.-A. and V.M.-R.; writing— original draft preparation, A.G.-A. and C.B.; writing—review and editing, A.G.-A., V.M.-R., J.B., C.B. and J.J.R.; supervision, J.B., C.B., J.J.R.; funding acquisition, J.B., C.B. and J.J.R. All authors have read and agree to the published version of the manuscript.

Funding: This research was funded by Spanish MINECO (project CTQ2016-78576-R).

Acknowledgments: V. Muelas-Ramos thanks to Spanish MCIU for BES-2017-082613 grant.

Conflicts of Interest: The authors declare no conflict of interest.

References

1. Qiu, J.; Zhang, X.; Feng, Y.; Zhang, X.; Wang, H.; Yao, J. Modified metal-organic frameworks as photocatalysts. *Appl. Catal. B Environ.* **2018**, *231*, 317–342. [CrossRef]

2. Furukawa, H.; Cordova, K.E.; O'Keeffe, M.; Yaghi, O.M. The chemistry and applications of metal-organic frameworks. *Science* **2013**, *341*, 1230444. [CrossRef] [PubMed]

3. Szczęśniak, B.; Choma, J.; Jaroniec, M. Gas adsorption properties of hybrid graphene-MOF materials. *J. Colloid Interface Sci.* **2018**, *514*, 801–813. [CrossRef]

4. Millward, A.R.; Yaghi, O.M. Metal–organic frameworks with exceptionally high capacity for storage of carbon dioxide at room temperature. *J. Am. Chem. Soc.* **2005**, *127*, 17998–17999. [CrossRef] [PubMed]

5. Rasheed, T.; Nabeel, F. Luminescent metal-organic frameworks as potential sensory materials for various environmental toxic agents. *Coord. Chem. Rev.* **2019**, *401*, 213065. [CrossRef]

6. Kreno, L.E.; Leong, K.; Farha, O.K.; Allendorf, M.; Van Duyne, R.P.; Hupp, J.T. Metal–organic framework materials as chemical sensors. *Chem. Rev.* **2012**, *112*, 1105–1125. [CrossRef] [PubMed]

7. Qiu, S.; Xue, M.; Zhu, G. Metal–organic framework membranes: From synthesis to separation application. *Chem. Soc. Rev.* **2014**, *43*, 6116–6140. [CrossRef]

8. Abánades Lázaro, I.; Forgan, R.S. Application of zirconium MOFs in drug delivery and biomedicine. *Coord. Chem. Rev.* **2019**, *380*, 230–259. [CrossRef]

9. Sun, C.-Y.; Qin, C.; Wang, X.-L.; Su, Z.-M. Metal-organic frameworks as potential drug delivery systems. *Expert Opin. Drug Deliv.* **2013**, *10*, 89–101. [CrossRef]

10. Joseph, L.; Jun, B.-M.; Jang, M.; Park, C.M.; Muñoz-Senmache, J.C.; Hernández-Maldonado, A.J.; Heyden, A.; Yu, M.; Yoon, Y. Removal of contaminants of emerging concern by metal-organic framework nanoadsorbents: A review. *Chem. Eng. J.* **2019**, *369*, 928–946. [CrossRef]

11. Li, X.; Wang, B.; Cao, Y.; Zhao, S.; Wang, H.; Feng, X.; Zhou, J.; Ma, X. Water contaminant elimination based on metal-organic frameworks and perspective on their industrial applications. *ACS Sustain. Chem. Eng.* **2019**, *7*, 4548–4563. [CrossRef]

12. Hasan, Z.; Jhung, S.H. Removal of hazardous organics from water using metal-organic frameworks (MOFs): Plausible mechanisms for selective adsorptions. *J. Hazard. Mater.* **2015**, *283*, 329–339. [CrossRef] [PubMed]

13. Bedia, J.; Muelas-Ramos, V.; Peñas-Garzón, M.; Gómez-Avilés, A.; Rodríguez, J.J.; Belver, C. A review on the synthesis and characterization of metal organic frameworks for photocatalytic water purification. *Catalysts* **2019**, *9*, 52. [CrossRef]

14. Khan, N.A.; Hasan, Z.; Jhung, S.H. Adsorptive removal of hazardous materials using metal-organic frameworks (MOFs): A review. *J. Hazard. Mater.* **2013**, *244–245*, 444–456. [CrossRef] [PubMed]

15. Pi, Y.; Li, X.; Xia, Q.; Wu, J.; Li, Y.; Xiao, J.; Li, Z. Adsorptive and photocatalytic removal of Persistent Organic Pollutants (POPs) in water by metal-organic frameworks (MOFs). *Chem. Eng. J.* **2018**, *337*, 351–371. [CrossRef]

16. Feng, M.; Zhang, P.; Zhou, H.C.; Sharma, V.K. Water-stable metal-organic frameworks for aqueous removal of heavy metals and radionuclides: A review. *Chemosphere* **2018**, *209*, 783–800. [CrossRef]

17. Sharma, V.K.; Feng, M. Water depollution using metal-organic frameworks-catalyzed advanced oxidation processes: A review. *J. Hazard. Mater.* **2019**, *372*, 3–16. [CrossRef]

18. Li, Y.; Xu, H.; Ouyang, S.; Ye, J. Metal-organic frameworks for photocatalysis. *Phys. Chem. Chem. Phys.* **2016**, *18*, 7563–7572. [CrossRef]

19. Zeng, L.; Guo, X.; He, C.; Duan, C. Metal-organic frameworks: Versatile materials for heterogeneous photocatalysis. *ACS Catal.* **2016**, *6*, 7935–7947. [CrossRef]

20. Li, W.; Zhang, Y.; Li, Q.; Zhang, G. Metal–organic framework composite membranes: Synthesis and separation applications. *Chem. Eng. Sci.* **2015**, *135*, 232–257. [CrossRef]

21. Canivet, J.; Fateeva, A.; Guo, Y.; Coasne, B.; Farrusseng, D. Water adsorption in MOFs: Fundamentals and applications. *Chem. Soc. Rev.* **2014**, *43*, 5594–5617. [CrossRef] [PubMed]

22. Furukawa, H.; Gándara, F.; Zhang, Y.-B.; Jiang, J.; Queen, W.L.; Hudson, M.R.; Yaghi, O.M. Water adsorption in porous metal–organic frameworks and related materials. *J. Am. Chem. Soc.* **2014**, *136*, 4369–4381. [CrossRef]

23. De Toni, M.; Jonchiere, R.; Pullumbi, P.; Coudert, F.X.; Fuchs, A.H. How can a hydrophobic mof be water-unstable? Insight into the hydration mechanism of IRMOFs. *Chem. Phys. Chem* **2012**, *13*, 3497–3503. [CrossRef] [PubMed]

24. Qian, X.; Yadian, B.; Wu, R.; Long, Y.; Zhou, K.; Zhu, B.; Huang, Y. Structure stability of metal-organic framework MIL-53 (Al) in aqueous solutions. *Int. J. Hydrogen Energy* **2013**, *38*, 16710–16715. [CrossRef]

25. Low, J.J.; Benin, A.I.; Jakubczak, P.; Abrahamian, J.F.; Faheem, S.A.; Willis, R.R. Virtual high throughput screening confirmed experimentally: Porous coordination polymer hydration. *J. Am. Chem. Soc.* **2009**, *131*, 15834–15842. [CrossRef]

26. Bon, V.; Senkovska, I.; Baburin, I.A.; Kaskel, S. Zr- and Hf-based metal–organic frameworks: Tracking down the polymorphism. *Cryst. Growth Des.* **2013**, *13*, 1231–1237. [CrossRef]

27. Jeremias, F.; Lozan, V.; Henninger, S.K.; Janiak, C. Programming MOFs for water sorption: Amino-functionalized MIL-125 and UiO-66 for heat transformation and heat storage applications. *Dalt. Trans.* **2013**, *42*, 15967–15973. [CrossRef]

28. Bai, Y.; Dou, Y.; Xie, L.-H.; Rutledge, W.; Li, J.-R.; Zhou, H.-C. Zr-based metal–organic frameworks: Design, synthesis, structure, and applications. *Chem. Soc. Rev.* **2016**, *45*, 2327–2367. [CrossRef]

29. Burtch, N.C.; Jasuja, H.; Walton, K.S. Water stability and adsorption in metal–organic frameworks. *Chem. Rev.* **2014**, *114*, 10575–10612. [CrossRef]

30. Zhu, J.; Li, P.Z.; Guo, W.; Zhao, Y.; Zou, R. Titanium-based metal–organic frameworks for photocatalytic applications. *Coord. Chem. Rev.* **2018**, *359*, 80–101. [CrossRef]

31. Zou, D.-H.; Cui, L.-N.; Liu, P.-Y.; Yang, S.; Zhu, Q.-Y.; Dai, J. Triphenylamine derived titanium oxo clusters: An approach to effective organic–inorganic hybrid dyes for photoactive electrodes. *Chem. Commun.* **2018**, *54*, 9933–9936. [CrossRef] [PubMed]

32. Kim, S.; Sarkar, D.; Kim, Y.; Park, M.H.; Yoon, M.; Kim, Y.; Kim, M. Synthesis of functionalized titanium-carboxylate molecular clusters and their catalytic activity. *J. Ind. Eng. Chem.* **2017**, *53*, 171–176. [CrossRef]

33. Zuluaga, S.; Fuentes-Fernandez, E.M.A.; Tan, K.; Xu, F.; Li, J.; Chabal, Y.J.; Thonhauser, T. Understanding and controlling water stability of MOF-74. *J. Mater. Chem. A* **2016**, *4*, 5176–5183. [CrossRef]

34. Canivet, J.; Bonnefoy, J.; Daniel, C.; Legrand, A.; Coasne, B.; Farrusseng, D. Structure-property relationships of water adsorption in metal-organic frameworks. *New J. Chem.* **2014**, *38*, 3102–3111. [CrossRef]

35. Howarth, A.J.; Peters, A.W.; Vermeulen, N.A.; Wang, T.C.; Hupp, J.T.; Farha, O.K. Best practices for the synthesis, activation, and characterization of metal–organic frameworks. *Chem. Mater.* **2017**, *29*, 26–39. [CrossRef]

36. Taheri, M.; Enge, T.G.; Tsuzuki, T. Water stability of cobalt doped ZIF-8: A quantitative study using optical analyses. *Mater. Today Chem.* **2020**, *16*, 100231. [CrossRef]

37. Wen, G.; Guo, Z. Facile modification of NH_2-MIL-125(Ti) to enhance water stability for efficient adsorptive removal of crystal violet from aqueous solution. *Colloids Surfaces A Physicochem. Eng. Asp.* **2018**, *541*, 58–67. [CrossRef]

38. Li, S.; Feng, F.; Chen, S.; Zhang, X.; Liang, Y.; Shan, S. Preparation of UiO-66-NH_2 and UiO-66-NH_2/sponge for adsorption of 2,4-dichlorophenoxyacetic acid in water. *Ecotoxicol. Environ. Saf.* **2020**, *194*, 110440. [CrossRef]

39. Zhang, M.; Yang, K.; Cui, J.; Yu, H.; Wang, Y.; Shan, W.; Lou, Z.; Xiong, Y. 3D-agaric like core-shell architecture UiO-66-NH_2@ZIF-8 with robust stability for highly efficient REEs recovery. *Chem. Eng. J.* **2020**, *386*, 124023. [CrossRef]

40. Fang, S.-Y.; Zhang, P.; Gong, J.-L.; Tang, L.; Zeng, G.-M.; Song, B.; Cao, W.-C.; Li, J.; Ye, J. Construction of highly water-stable metal-organic framework UiO-66 thin-film composite membrane for dyes and antibiotics separation. *Chem. Eng. J.* **2020**, *385*, 123400. [CrossRef]

41. Sun, D.; Ye, L.; Li, Z. Visible-light-assisted aerobic photocatalytic oxidation of amines to imines over NH_2-MIL-125(Ti). *Appl. Catal. B Environ.* **2015**. [CrossRef]

42. Wang, H.; Yuan, X.; Wu, Y.; Zeng, G.; Dong, H.; Chen, X.; Leng, L.; Wu, Z.; Peng, L. In situ synthesis of In_2S_3@MIL-125(Ti) core–shell microparticle for the removal of tetracycline from wastewater by integrated adsorption and visible-light-driven photocatalysis. *Appl. Catal. B Environ.* **2016**, *186*, 19–29. [CrossRef]

43. Zhu, S.-R.; Liu, P.-F.; Wu, M.-K.; Zhao, W.-N.; Li, G.-C.; Tao, K.; Yi, F.-Y.; Han, L. Enhanced photocatalytic performance of BiOBr/NH_2-MIL-125(Ti) composite for dye degradation under visible light. *Dalt. Trans.* **2016**, *45*, 17521–17529. [CrossRef] [PubMed]

44. Gómez-Avilés, A.; Peñas-Garzón, M.; Bedia, J.; Dionysiou, D.D.; Rodríguez, J.J.; Belver, C. Mixed Ti-Zr metal-organic-frameworks for the photodegradation of acetaminophen under solar irradiation. *Appl. Catal. B Environ.* **2019**, *253*, 253–262. [CrossRef]

45. Fu, Y.; Sun, D.; Chen, Y.; Huang, R.; Ding, Z.; Fu, X.; Li, Z. An amine-functionalized titanium metal–organic framework photocatalyst with visible-light-induced activity for CO_2 reduction. *Angew. Chemie Int. Ed.* **2012**, *51*, 3364–3367. [CrossRef] [PubMed]

46. Dan-Hardi, M.; Serre, C.; Frot, T.; Rozes, L.; Maurin, G.; Sanchez, C.; Férey, G. A new photoactive crystalline highly porous titanium (IV) dicarboxylate. *J. Am. Chem. Soc.* **2009**, *131*, 10857–10859. [CrossRef]

47. Zhang, Y.; Chen, Y.; Zhang, Y.; Cong, H.; Fu, B.; Wen, S.; Ruan, S. A novel humidity sensor based on NH_2-MIL-125(Ti) metal organic framework with high responsiveness. *J. Nanoparticle Res.* **2013**, *15*, 2014. [CrossRef]

48. Wang, H.; Yuan, X.; Wu, Y.; Zeng, G.; Chen, X.; Leng, L.; Wu, Z.; Jiang, L.; Li, H. Facile synthesis of amino-functionalized titanium metal-organic frameworks and their superior visible-light photocatalytic activity for Cr(VI) reduction. *J. Hazard. Mater.* **2015**, *286*, 187–194. [CrossRef]

49. Kim, S.-N.; Kim, J.; Kim, H.-Y.; Cho, H.-Y.; Ahn, W.-S. Adsorption/catalytic properties of MIL-125 and NH_2-MIL-125. *Catal. Today* **2013**, *204*, 85–93. [CrossRef]

50. Wang, F.; Li, F.-L.; Xu, M.-M.; Yu, H.; Zhang, J.-G.; Xia, H.-T.; Lang, J.-P. Facile synthesis of a Ag(i)-doped coordination polymer with enhanced catalytic performance in the photodegradation of azo dyes in water. *J. Mater. Chem. A* **2015**, *3*, 5908–5916. [CrossRef]

51. Fan, Y.-H.; Zhang, S.-W.; Qin, S.-B.; Li, X.-S.; Qi, S.-H. An enhanced adsorption of organic dyes onto NH_2 functionalization titanium-based metal-organic frameworks and the mechanism investigation. *Microporous Mesoporous Mater.* **2018**, *263*, 120–127. [CrossRef]

52. Wu, Z.; Huang, X.; Zheng, H.; Wang, P.; Hai, G.; Dong, W.; Wang, G. Aromatic heterocycle-grafted NH_2-MIL-125(Ti) via conjugated linker with enhanced photocatalytic activity for selective oxidation of alcohols under visible light. *Appl. Catal. B Environ.* **2018**, *224*, 479–487. [CrossRef]

53. Klug, H.P.; Alexander, L.E. *X-Ray Diffraction Procedures: For Polycrystalline and Amorphous Materials*, 2nd ed.; Wiley: Hoboken, NJ, USA, 1974; ISBN 978-0-471-49369-3.

54. Arbain, R.; Othman, M.; Palaniandy, S. Preparation of iron oxide nanoparticles by mechanical milling. *Miner. Eng.* **2011**, *24*, 1–9. [CrossRef]

55. Brunauer, S.; Emmett, P.H.; Teller, E. Adsorption of gases in multimolecular layers. *J. Am. Chem. Soc.* **1938**, *60*, 309–319. [CrossRef]

56. Lippens, B.C.; De Boer, J.H. Studies on pore systems in catalysts V. The t method. *J. Catal.* **1965**, *4*, 319–323. [CrossRef]

 *catalysts* 

Article

Cu^{II}- and Co^{II}-Based MOFs: $\{[La_2Cu_3(\mu\text{-}H_2O)(ODA)_6(H_2O)_3]\cdot3H_2O\}_n$ and $\{[La_2Co_3(ODA)_6(H_2O)_6]\cdot12H_2O\}_n$. The Relevance of Physicochemical Properties on the Catalytic Aerobic Oxidation of Cyclohexene

Luis Santibáñez [1,2], Néstor Escalona [3,4,5], Julia Torres [6], Carlos Kremer [6], Patricio Cancino [1,*] and Evgenia Spodine [1,2,*]

[1] Faculty of Science Chemistry and Pharmaceuticals, University of Chile, 8380544 Santiago, Chile; santibanezcamposl@gmail.com
[2] Center for the Development of Nanoscience and Nanotechnology, CEDENNA, 9160000 Santiago, Chile
[3] Faculty of Chemistry and Pharmacy, Pontifical Catholic University of Chile, 7810000 Santiago, Chile; neescalona@ing.puc.cl
[4] Department of Chemical Engineering and Bioprocesses, Ingeniery school, Pontifical Catholic University of Chile, 7810000 Santiago, Chile
[5] Millenium Nuclei on Catalytic Processes towards Sustainable Chemistry (CSC), 7810000 Santiago, Chile
[6] Departament Estrella Campos, Facultay of Chemistry, University of the Republic, 11800 Montevideo, Uruguay; jtorres@fq.edu.uy (J.T.); ckremer@fq.edu.uy (C.K.)
* Correspondence: pcancino@ciq.uchile.cl (P.C.); espodine@uchile.cl (E.S.); Tel.: +56-2-2978-2862 (P.C. & E.S.)

Received: 29 April 2020; Accepted: 14 May 2020; Published: 25 May 2020

Abstract: The aerobic oxidation of cyclohexene was done using the heterometallic metal organic frameworks (MOFs) $\{[La_2Cu_3(\mu\text{-}H_2O)(ODA)_6(H_2O)_3]\cdot3H_2O\}_n$ (LaCuODA)) (**1**) and $\{[La_2Co_3(ODA)_6(H_2O)_6]\cdot12H_2O\}_n$ (LaCoODA) (**2**) as catalysts, in solvent free conditions (ODA, oxydiacetic acid). After 24 h of reaction, the catalytic system showed that LaCoODA had a better catalytic performance than that of LaCuODA (conversion 85% and 67%). The structures of both catalysts were very similar, showing channels running along the c axis. The physicochemical properties of both MOFs were determined to understand the catalytic performance. The Langmuir surface area of LaCoODA was shown to be greater than that of LaCuODA, while the acid strength and acid sites were greater for LaCuODA. On the other hand, the redox potential of the active sites was related to Co^{II}/Co^{III} in LaCoODA and Cu^{II}/Cu^{I} in LaCuODA. Therefore, it is concluded that the Langmuir surface area and the redox potentials were more important than the acid strength and acid sites of the studied MOFs, in terms of the referred catalytic performance. Finally, the reaction conditions were also shown to play an important role in the catalytic performance of the studied systems. Especially, the type of oxidant and the way to supply it to the reaction medium influenced the catalytic results.

Keywords: metal organic framework; heterogeneous catalyst; aerobic oxidation; cyclohexene

---

1. Introduction

The oxidation of cyclic olefins is a reaction that has become of interest for many industries, such as the agrochemical, pharmaceutical, and perfumery industries, as well as the manufacture of adhesives [1–4], because the products of this reaction such as epoxides, alcohol, ketones, and aldehydes are used in commodities. However, it has become necessary to find alternative routes of production for these compounds, in order to decrease the carbon footprint and the generation of environmentally harmful sub-products [5].

Catalysis is an important tool in many industrial processes when it comes to reducing the energy cost, as well as the reaction time and the sub-product generation, enhancing the selectivity. In this sense, homogeneous and supported catalysts are classically used in the oxidation of cycloalkenes [6,7]. However, they present well known problems such as the inability to separate and recover the homogeneous catalyst, and on the other hand, the leaching of the catalyst from the support, preventing its reusability. Heterogeneous catalysts, particularly inorganic polymers, present a good alternative, as these are insoluble and stable under the usual cycloalkene oxidation conditions. The corresponding catalytic activity of inorganic polymers can be found in the literature in several reviews [8–14]. These inorganic polymers are based on metallic cores coordinated with organic linkers, with the metal centres being the catalytically active species. Transition metal ions are a good choice when designing these systems as their cost is low and their abundance is high. In this context, cobalt (II) containing molecular sieves show activity in the epoxidation with molecular oxygen of styrene at 100 °C, presenting a conversion of 45% with a selectivity of 62% to the desired epoxide [15]. The same research group reported a family of cobalt (II) exchanged zeolite X catalysts, which permitted, for the studied catalytic systems, an almost complete conversion of styrene [16].

Copper (I) species have been used as a catalyst in the aerobic oxidation of different amines to imines, with a conversion ranging in most cases from 85% to 95% using CuCl [17]. A different biomimetic Cu (I) species, $[Cu(CH_3CN)_4]PF_6$, was reported for the aerobic oxidation of secondary alcohols, yielding over a 90% conversion in most experiments [18]. A Cu(II) compound, used as a catalyst for an aerobic oxidation was $Cu_2(OH)PO_4$, and this catalyst produced a 30% conversion for styrene and 47% for cyclohexene [19]. Two catalysts, based on Cu(II) using N-benzylethanolamine or triisopropanolamine as ligands, were also used in the oxidation of cycloalkanes assisted by H_2O_2; a conversion of 23% was achieved for cycloheptane [20].

However, owing to the lower catalytic activity of some of these systems, as compared with that of the homogeneous ones, it is sometimes necessary to add co-oxidants such as TEMPO [21], isobutyraldehyde [22], TBHP [23,24], or H_2O_2 [25]. For example, $Cu_3(BTC)_2$ (BTC, 1,3,5-benzenetricarboxylate) is a Cu(II) complex assisted by TEMPO, used as a catalyst for the oxidation of benzylic alcohols, yielding 89% of conversion [5]. However, these co-catalysts sometimes produce harmful by-products, such as tert-butanol, in the case of TBHP. Therefore, aerobic oxidation using only molecular oxygen as the oxidizing agent becomes an ideal goal to achieve.

Metal organic frameworks (MOFs) can be modified by changing either the metal ions or the organic linkers, thus modifying the catalytic properties or the capability of adsorbing gases. Monometallic MOFs based on copper (II) [26,27], cobalt (II) [28,29], vanadium (IV) [30,31], or iron (III) ions [32–34] have been used in heterogeneous catalytic systems. The use of MOFs as catalysts in the aerobic oxidation of olefins has been reported by Fu et al. [35]. These researchers used catalysts based on copper (II) and cobalt (II) with 2,5-dihydroxyterephthalic acid (DOBDC), $[M_2(DOBDC)(H_2O)_2]·8H_2O$ (M= Cu^{II} or Co^{II}), and molecular oxygen to oxidize cyclohexene at 80 °C. Under these conditions and after 15 h, the reaction only produced 14.6% conversion for the copper catalyst and 10.5% for the cobalt catalyst. Tuci et al. found better results using a different cobalt (II) catalyst, $[Co(L-RR)(H_2O)]·H_2O$ (L-RR = (R,R)-thiazolidine-2,4-dicarboxylate), at 70 °C. However, molecular oxygen pressure was increased from 1 to 5 bar. A conversion of 37% with a selectivity of 49% to 2-cyclohexen-1-one was achieved [36].

Heterometallic MOFs also present activity for the oxidation of different substrates. For example, copper (II)-based MOFs with adsorbed palladium or gold nanoparticles have been used in the oxidation of benzylic alcohols [37,38], and a bimetallic catalyst of Pd–Au nanoparticles supported on an aluminium (III) MOF was used in the aerobic oxidative reaction of carbonylation of amines [39]. However, as mentioned, these systems use supported catalysts, which may not be optimal.

Our group has worked with heterogeneous Cu^{II}- 4f MOF catalysts, tuning the organic linkers [40], the *4f* lanthanide ion [41], and using different substrates for the aerobic oxidation reaction [42]. We now report the study of the effect of changing the *3d* transition metal ion of two MOFs used as catalysts in a solvent-free system, with molecular oxygen as an oxidant in the oxidation of cyclohexene, without the use of a co-catalyst. Cobalt (II) and copper (II) were chosen as the *3d* redox transition metal ions, while the *4f* ion was lanthanum (III). The studied catalysts were $\{[La_2Cu_3(\mu\text{-}H_2O)(ODA)_6(H_2O)_3]\cdot 3H_2O\}_n$ (LaCuODA) (**1**) and $\{[La_2Co_3(ODA)_6(H_2O)_6]\cdot 12H_2O\}_n$ (LaCoODA) (**2**) (H_2ODA = oxydiacetic acid).

2. Results

2.1. Characterization

2.1.1. Structural Description

Crystal structures were obtained as described in [43,44], and will be briefly described in order to understand the catalytic results. X-ray structural analyses of LaCuODA (**1**) and LaCoODA (**2**) reveal that both belong to the hexagonal crystal system, with space group *P*-62c for **1** and *P*6/mcc for **2**. The skeleton of the structures is similar in both porous MOFs. The La^{III} is coordinated by three tridentate ODA ligands, thus nine oxygen atoms define the coordination sphere of this metal centre. Each $[La(ODA)_3]^{3-}$ unit is connected to six M^{II} (M = Cu^{II}, Co^{II}) ions via single *syn-anti* μ-carboxylato-O-O' bridges. The M^{II} ions are surrounded by four oxygen atoms from $[La(ODA)_3]^{3-}$ units in (**1**) and (**2**). The difference in the coordination pattern of (**1**) and (**2**) is that Co^{II} ion in (**2**) is bound to two water molecules in the axial positions, forming a slightly distorted octahedron. On the other hand, Cu^{II} in (**1**) is bound to just one molecule of water, being penta-coordinated. In spite of this difference, the main structural feature of these MOFs is the formation of hexagonal channels along the crystallographic *c* axis (Figure 1). The maximum size of the channel can be estimated in 11.39 Å for (**1**) and 11.09 Å for (**2**). Crystallization water molecules are hosted in the channels. In the special case of (**1**), one of the crystallization water molecules per unit formula is in a fixed position forming a μ_2-H_2O bridge between La^{III} ions.

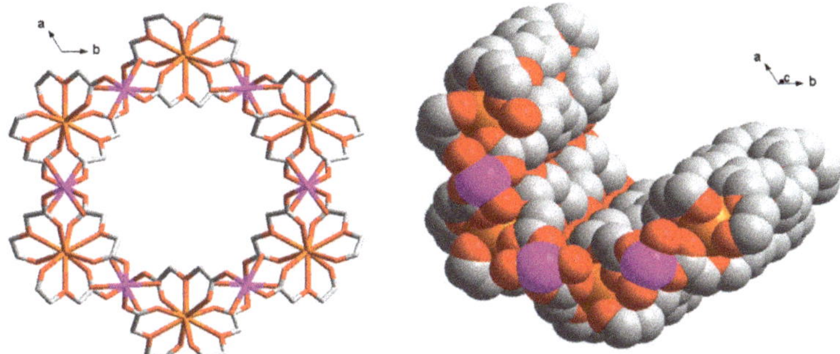

Figure 1. Left: partial view of complex (**2**) along the crystallographic *c* axis, showing the generated hexagonal channels. H atoms and crystallization water molecules are omitted for clarity. **Right**: cross-section of the channel showing the inner exposed surface. Color code: La, orange; Co, pink; O, red; C, light grey.

The determination of the acid strength for LaCoODA showed an initial potential of 24.5 mV after 4 h of stabilization with the first addition of N-butylamine, while LaCuODA presented a higher value of 93.7 mV. The ranges of initial potential defined for the strength of acid sites are as follows: potential > 100 mV, very strong acid site; 100 mV > potential > 0 mV, strong acid site; 0 mV > potential > −100 mV weak acid site; −100 mV > potential, very weak acid site. Thus, LaCuODA presents acid sites that can be defined as nearly strong, while LaCoODA presents weak acid sites. As for the number of acid sites, LaCoODA has 0.295 miliequivalent of acid sites per gram and LaCuODA has 0.783 miliequivalent of acid sites per gram (Table 1).

Table 1. Determination of acid strength and number of acid sites for LaM(ODA).

Catalyst	Initial Potential (mV)	Number of Acid Sites (meq/g)
LaCuODA (1)	93.7	0.783
LaCoODA (2)	24.5	0.295

The Thermogravimetric (TGA) profile of both complexes was reported and analyzed previously [43,44]. Compound (**1**) presents two mass losses. The first one between 56 and 94 °C corresponds to all water molecules per formula unit. The second one appears around 240 °C and can be associated with the beginning of the decomposition of the structure. Thermal analyses of (**2**) show that the loss of water involved a well-defined two-step process. The first step corresponds to twelve molecules of crystallization water (below 50 °C). The second step (between 110 and 135 °C) includes the six water molecules coordinated to Co^{II}. The decomposition of the structure is evident above 260 °C.

Table 2 shows the pore diameter and estimated surface area of the catalysts obtained by the BET and Langmuir models from the CO_2 adsorption isotherm at 273 K. The LaCoODA has a BET surface area of 762 m^2/g, which is greater than that of LaCuODA (514 m^2/g); a similar trend was obtained from the Langmuir area values. On the other hand, the pore size is the same for both catalysts, and it is consistent with the pore size described by the crystallographic data (11.39 (**1**) and 11.09 (**2**) Å). These results permit to classify the catalysts in the range of microporous materials.

Table 2. Textural properties from CO_2 adsorption measures.

Catalyst	BET Area (m^2/g)	Langmuir Area (m^2/g)	Pore Size (nm)
LaCuODA (1)	514	555	1.031
LaCoODA (2)	762	783	1.031

2.1.2. Catalytic Results

LaCuODA and LaCoODA were used as catalysts for the aerobic oxidation of cyclohexene, in the absence of additional organic solvent or co-oxidant. The obtained results are summarized in Table 3, while the obtained products are shown in Scheme 1.

Table 3. Catalytic results after 24 h of the aerobic oxidation of cyclohexene at 75 °C using LaMODA (M=Co^{II}, Cu^{II}).

Catalyst	Conversion (%)	Selectivity (%)		
		Cyclohexene Oxide	2-Cyclohexen-1-ol	2-Cyclohexen-1-one
LaCuODA (1)	67	5	40	55
LaCoODA (2)	85	0	25	75

Reactions conditions: cyclohexene (50 mmol) and (0.01 mmol) of LaM(ODA) (M = Co^{II}, Cu^{II}), 1 bar of continuous oxygen flow. The mixture is stirred (960 rpm) at 75 °C for 24 h.

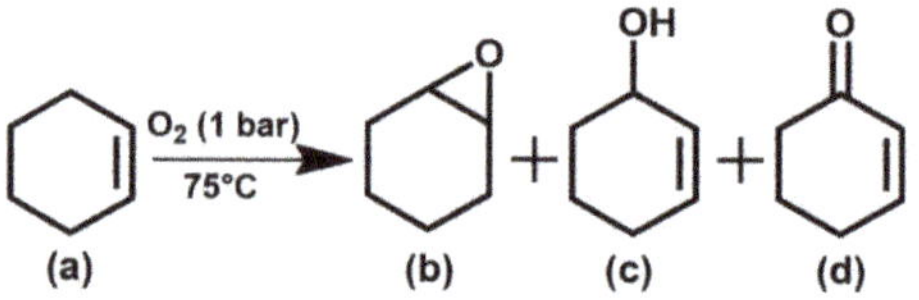

Scheme 1. Products derived from the catalytic oxidation of cyclohexene (**a**). Cyclohexene oxide (**b**); 2-cyclohexen-1-ol (**c**); 2-cyclohexen-1-one (**d**).

The conversion after 24 h of reaction shows that the MOF based on cobalt (II) has a better catalytic performance than that of the catalyst based on copper (II) (Table 2). The main product for both catalysts was 2-cyclohexen-1-one, with the selectivity for this product being 75% for LaCoODA and 55% for

LaCuODA. Thus, the influence on both the conversion and selectivity of the reaction of the nature of the *3d* transition metal ion becomes evident.

3. Discussion

To understand these results, it is important to remark that the catalysts present a few differences, mentioned in the structural description of the environment and geometry of the *3d* metal ion, being **1** square pyramidal and **2** octahedral. The difference between them is the number of water molecules in the axial positions (one and two, respectively). Taking into account the data obtained in the TGA measurements, it was assumed that, under the reaction conditions, all water molecules were removed, and thus both compounds could be considered isostructural. The crystal structure shows channels along the *c axis* for both catalysts, with water molecules inside. The TGA analyses permit to infer that, for the used reaction conditions, these water molecules could leave the channels, making the channels partially free for the entry of the substrate and the oxidant. Besides, the CO_2 absorption measurements complement the crystallographic results, showing that the BET and Langmuir surface areas are bigger for LaCoODA as compared with LaCuODA. Thus, it is possible to suppose that the catalyst based on Co^{II} has a greater amount of active sites, favoring in this way the interaction between the catalyst and the substrate/oxidant. Moreover, even though both catalysts permit to obtain the same major product (2-cyclohexen-1-one), the distribution of the products is completely different. Table 3 shows that the cobalt (II) MOF with better catalyst performance also permits to obtain a higher selectivity for 2-cyclohexen-1-one.

However, the catalysts present additional differences that could permit to explain the obtained results. Both transition metal ions have different chemical properties, such as the redox potential or the Lewis acidity. These properties can modulate the reaction mechanism and the activation of the oxidant [45]. The implied redox properties during the oxidation reaction are different because, for LaCoODA, the Co^{II}/Co^{III} couple must be active, while for LaCuODA, the redox couple is Cu^{II}/Cu^{I}. For LaCoODA, the activation will occur with the oxidation of the cobalt(II) centres through a single electron transfer to molecular oxygen and the formation of the superoxide anion [35]. In the case of LaCuODA, we recently reported the generation of a $Cu-O_2$ adduct at the beginning of the catalytic cycle of the oxidation reaction [41]. In the formation of this adduct, the Lewis acidity plays an important role, with the interaction between the catalyst and the oxidant being an acid–base interaction. [45] When we compare the acid sites and the amount of these sites in the catalyst, it is possible to conclude that the LaCuODA has stronger acid sites than LaCoODA. Apparently, the greater surface area of LaCoODA and the redox properties of the cobalt (II) ion predominate over the acid properties of the copper (II) ion, thus LaCoODA has a better catalytic performance for the oxidation reaction of cyclohexene. It is possible to conclude that the combination of the structural and physicochemical properties of the studied catalytic systems is determinant for the catalytic behavior.

As LaCoODA showed the best performance, this catalyst was used for the optimization of some catalytic parameters. The first parameter to be studied was the thermal dependence of the conversion. Figure 2 (Table S1) shows a linear dependence between the conversion and the reaction temperature. As mentioned above, the increase of the conversion with temperature can be explained by assuming that the channels present in the catalyst start to release the encapsulated water molecules, and the presence of free space in the channels facilities the interaction between the metal centres and the oxidant/substrate.

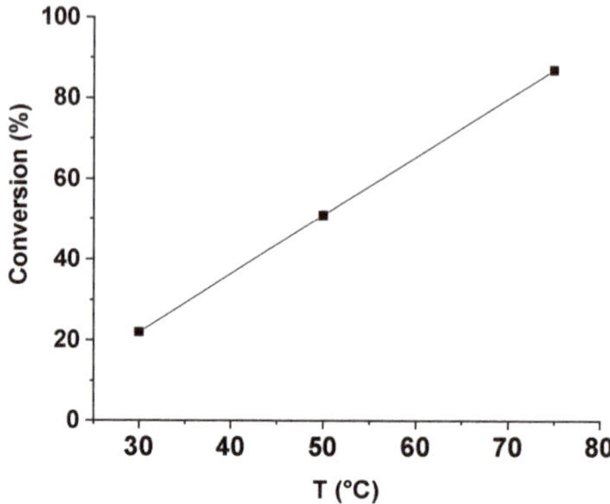

Figure 2. Thermal dependence of conversion, using LaCoODA as catalyst.

To test this assumption, we compared the catalytic performance of non-activated LaCoODA and thermally activated LaCoODA. Figure 3 (Table S2) summarizes the results, which shows that both the as prepared catalyst and the activated catalyst display a similar time dependence in their activity. As expected at short times of reaction, the oxidation process increases significantly as the reaction evolves, but it is possible to observe that, after 12 h, the activity starts to approach an asymptotic value, increasing only slightly for a sample of 24 h of reaction. The overlap of the curves indicates that the catalytic activity is similar within the experimental error for both the activated and non-activated catalyst. Apparently, the rehydration process of the catalyst is so fast that, practically, it is not possible to enhance the catalytic performance by thermal activation.

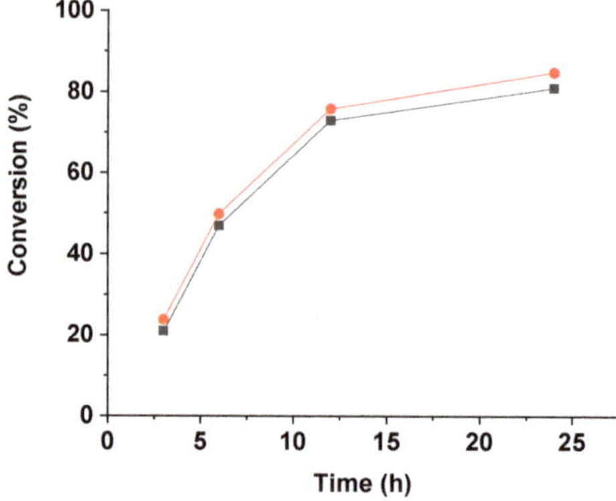

Figure 3. Time dependence of conversion using LaCoODA as catalyst. Non-activated (■); activated (●).

If we compare the results obtained with other previously reported works with similar catalysts, it is possible to find interesting aspects to discuss. For example, our group reported in 2017 [46] the catalytic performance of LaCuODA in the oxidation reaction of cyclohexene, using *tert*-butylhydroperoxide (TBHP) as an oxidant in DCE/water as a reaction medium. The achieved catalytic performance under the studied conditions in this work, that is, solvent free and O_2 as oxidant, was better in conversion and selectivity than using TBHP and the biphasic medium (Table 4). We propose that, considering that, in the DCE/water medium, the channels are fully occupied with water molecules, there must be many water molecules obstructing the interaction of the active metal centres of the catalyst and the substrate/oxidant.

Table 4. Catalytic results using LaCuODA after 24 h of oxidation of cyclohexene at 75 °C.

| Oxidant | Conversion (%) | Selectivity (%) | | | Reference |
		Cyclohexene Oxide	2-Cyclohexen-1-ol	2-Cyclohexen-1-one	
O_2	67	5	40	55	This work
TBHP *	48	5	31	56	[46]

* When using *tert*-butylhydroperoxide (TBHP) as oxidant, 1,2-cyclohexenediol was also detected as a minor reaction product (8%).

Table 5 shows the comparison between different catalytic results for Co^{II}-based MOF catalysts. All four catalysts have a distorted octahedral geometry around the cobalt(II) ion. However, one position of the octahedron is occupied by a water molecule for (**III**) and by two water molecules for **2**, **I**, and **II**. [35,36,43,44,47] Besides, all the compounds have channels with water molecules inside them, but the crystallographic diameter of the pores varies among them. Compounds **2**, **I** and **II** have a diameter *ca.* 11 Å, while compound **III** has only 5.7 Å. Despite the structural similarities of the pores of **2**, **I**, and **II**, the catalytic results reveal significant differences. Moreover, from the comparison using two sets of data (one set corresponds to 24 h of reaction, that is, catalysts **2** and **III**, and the second one to 10 and 12 hours of reaction corresponds to data for **2**, **I**, and **II**, shown in Table 5), it becomes evident that **2** has by far the best catalytic performance.

Table 5. Catalytic results of different Co^{II}-based metal organic frameworks (MOFs) using solvent free conditions.

| Catalyst | Conversion (%) | Reaction Conditions | | | Reference |
| | | Temperature | Time | O_2 Pressure | |
		(°C)	(h)	(bar)	
{[La$_2$Co$_3$(ODA)$_6$(H$_2$O)$_6$]·12H$_2$O}$_n$ (**2**)	73	75	12	1 (flow)	This work
{[La$_2$Co$_3$(ODA)$_6$(H$_2$O)$_6$]·12H$_2$O}$_n$ (**2**)	84	75	24	1 (flow)	This work
[Co$_2$(DOBDC)(H$_2$O)$_2$]·8H$_2$O (**I**)	8.1	80	10	1 (balloon)	[35]
[Co$_2$(DOBDC)(H$_2$O)$_2$]·8H$_2$O (**II**)	50	80	10	1 (balloon)	[47]
[Co(L-RR)(H$_2$O)·H$_2$O] (**III**)	28	70	24	1 (charged)	[36]

Flow: continuous flow of oxygen at 1 bar of pressure. **Balloon**: oxygen atmosphere, using a balloon fully filled with oxygen. **Charged**: the reactor is pressurized with oxygen at 1 bar of pressure.

However, the catalytic systems present some differences that could permit to explain the obtained results. As the structures do not clarify these differences, maybe the reaction conditions can give some light on the obtained data. As the temperature is quite similar for all the reported systems, the oxygen pressure is an interesting parameter to analyze. Even though the oxygen pressure used is the same for all the catalytic systems, the way of supplying the oxidant to the reacting substrate is not the same. Thus, the amount of oxygen in the reactor varies depending on the source used. That is, the pressurized oxygen in the reaction vessel has an initial finite concentration [36], while a continuous oxygen flow maintains the concentration of the oxidant in the reaction vessel. On the other hand, the oxygen concentration provided by a balloon is variable [35,47].

Therefore, it is possible to conclude that oxygen concentration is determinant in the reaction mechanism. Depending on the amount of oxygen in the reaction medium, the chance to obtain the interaction between the active site and oxidant/substrate will be modified [42], and thus the onset of the chain reactions that will form the products.

4. Materials and Methods

4.1. Synthesis

Compounds (**1**) {[La$_2$Cu$_3$(μ-H$_2$O)(ODA)$_6$(H$_2$O)$_3$]·3H$_2$O}$_n$ (LaCuODA) and (**2**) {[La$_2$Co$_3$(ODA)$_6$(H$_2$O)$_6$]·12H$_2$O}$_n$ (LaCoODA)) have been previously reported [43,44] and were prepared accordingly, with slight modifications of the synthetic route. Complex (**1**) was synthesized by direct reaction of stoichiometric amounts of LaCl$_3$·7H$_2$O (0.37 g, 1.0 mmol), CuCl$_2$·2H$_2$O (0.20 g, 1.5 mmol), and oxydiacetic acid 0.40 g, 3.0 mmol). Reagents were dissolved in water (*ca.* 35 mL); pH was adjusted with diluted ammonia to 5–6. Light blue crystals of the complex appeared after 2–3 weeks. These were separated by filtration and washed twice with water. Complex (**2**) was prepared starting from La_2O_3 (98 mg, 0.3 mmol), $Co(COOCH_3)_2 \cdot 4H_2O$ (224 mg, 0.9 mmol), and oxydiacetic acid (322 mg, 2.4 mmol). Then, 30 mL water was added, and the mixture was placed in a Teflon-lined 45 mL stainless steel acid digestion vessel and heated at 120 °C for 45 h. The resulting solution was filtered, and the filtrate was allowed to stand at room temperature. After five days, a pale red polycrystalline solid appeared. It was filtered and washed twice with water.

The purity of the solids was checked by elemental analysis (C, H) and IR spectroscopy. Calculated for $La_2Cu_3C_{24}H_{38}O_{37}$ (**1**), C, 20.8; H, 2.8. Found, C, 21.2; H, 2.9%. Calculated for $La_2Co_3C_{24}H_{60}O_{48}$ (**2**): C, 18.3; H, 4.0. Found: C, 18.5; H, 4.2%.

IR spectra were almost identical for both complexes, and can be used as a quick control of the purity of the compounds. The shift of ν_{COO} (1724, 1419 cm^{-1}) of the free ligand to *ca.* 1600 and 1430 cm^{-1} is noticeable. Besides, upon complexation, ν_{COC} (1149 cm^{-1}) for the free ligand shifts to *ca.* 1120 cm^{-1} (see Supplementary Information Figure S1a,b).

4.2. Surface Area Determination

The specific surface area (S_{BET}) was measured by CO_2 sorption measurement at 273 K, using Micrometrics 3 Flex equipment. Prior to the measurements, the samples were outgassed at 250 °C for 2 h. The BET surface area was estimated from the adsorption branch of the isotherms in the range $0.05 \leq P/P \leq 0.15$. Additionally, the surface area was also estimated using the Langmuir equation. The average pore diameter was calculated using the DFT method, applied to the CO_2 adsorption isotherm (Figures S2 and S3).

4.3. Catalytic Reactions

In a 50 mL round bottom flask provided with a refrigerant, 50 mmol of cyclohexene and 0.01 mmol of LaMODA (M = Co^{II} or Cu^{II}) were added, and the substrate to be oxidized was heated to the desired temperature. The system was connected to a 1 bar continuous oxygen flow, and the mixture was stirred at a constant speed (960 rpm). Studies of the thermal and reaction time dependence of the conversion were done with LaCo(ODA), measuring results at 30, 50, and 75 °C for 24 h and measuring results at 3, 6, 8, 12, and 24 h at 75 °C. Experiments were also done using the previously activated LaCoODA catalyst. The activation of the catalyst was performed by placing the solid in a vacuum oven at 80 °C for 2 h before its use in the catalytic reactions.

All the products were analyzed by gas chromatography, using a 5890 model SERIES II Hewlett Packard gas chromatograph equipped with a capillary non-polar Equity-1 column and an FID detector.

4.4. Determination of Acid Strength and Number of Acid Sites

The acid strength and the number of acid sites in the catalysts were determined with a potentiometric titration, as reported by Cid and Pecchi [48]. A suspension of 50.8 mg of LaCoODA (49.8 mg of LaCuODA) in 100 ml of acetonitrile was titrated with a 0.01 N solution of N-butylamine in acetonitrile, using an Ag/AgCl electrode immersed in the suspension. The first addition of 50 µL of titrant to the suspension was left stirring for 4 h, in order to stabilize the system. The first potentiometric measurement indicates the strength of the acid sites. After this first measurement was recorded, 50 µL/min of titrant was added until the potential of the system remained constant. The spent volume indicates the miliequivalents of acid sites per gram of solid. For the number of acid sites, a constant potential was achieved by the LaCoODA suspension after 0.015 miliequivalents of N-butylamine was added. In the case of the LaCuODA suspension, 0.039 miliequivalents of N-butylamine was added before a constant potential was achieved.

5. Conclusions

For the studied catalysts, the physicochemical properties controlled the catalytic performance of the reaction, with the surface area and redox properties being determinant to explain the better results of LaCoODA over LaCuODA. Therefore, it is concluded that the Langmuir surface area and the redox potentials were more important than the acid strength and acid sites of the studied MOFs, in terms of the referred catalytic performance.

The channels play an important role in the catalytic processes; the removal of the water molecules from the channels is fundamental to favor the interaction between the active sites and the oxidant/catalyst.

The type of oxidant and the way to supply it are relevant to understand the difference in the results that exists among similar compounds. The amount of oxidant available within the reaction is key to obtain better results.

Supplementary Materials: The following are available online at http://www.mdpi.com/2073-4344/10/5/589/s1, Figure S1: IR spectra of LaCuODA (a) and LaCoODA (b). Table S1: Thermal dependence of aerobic cyclohexene oxidation catalyzed by LaCoODA. Table S2: Conversions for the non-activated and thermally activated catalytic system. Selectivities for the activated catalyst are also presented.

Author Contributions: Conceptualization, P.C. and E.S.; methodology, P.C., E.S., C.K., N.E. and J.T.; formal analysis, L.S.; investigation, L.S., N.E., J.T., C.K. and E.S.; resources, E.S., N.E. and J.T. and C.K.; writing—original draft preparation, L.S., C.K., P.C. and E.S.; writing—review and editing, N.E., C.K., P.C. and E.S.; supervision, P.C. and E.S.; project administration, N.E., C.K. and E.S.; funding acquisition, N.E., C.K. and E.S. All authors have read and agreed to the published version of the manuscript.

Funding: L.S. and E.S. thank Proyecto Basal AFB180001 (CEDENNA). C.K. and J.T. thank CSIC (Uruguay) for financial support through Programa de Apoyo a Grupos de Investigación. N.E. thanks FONDEQUIP EQM 160070 and the Chilean Ministry of Economy, Development, and Tourism within the framework of the Millennium Science Initiative program for the grant "Nuclei on Catalytic Processes towards Sustainable Chemistry (CSC)".

Conflicts of Interest: The authors declare no conflict of interest.

References

1. Sheldon, A.R.; Kochi, J.K. *Metal Complex Catalyzed Oxidation of Organic Compounds*; Academic Press: New York, NY, USA, 1981.
2. Cainelli, G.; Cardillo, G. *Chromium Oxidation in Organic Chemistry*; Springer: Berlin, Germany; New York, NY, USA, 1984.
3. Satterfield, C.N. Encyclopedia of Chemical Technology. *J. Am. Chem. Soc.* **1958**, *80*, 3487. [CrossRef]
4. Chadwick, S.S. Ullmann's Encyclopedia of Industrial Chemistry. *Ref. Serv. Rev.* **1988**, *16*, 31–34. [CrossRef]
5. Dhakshinamoorthy, A.; Alvaro, M.; Garcia, H. Aerobic oxidation of benzylic alcohols catalyzed by metal-organic frameworks assisted by TEMPO. *ACS Catal.* **2011**, *1*, 48–53. [CrossRef]
6. Liniger, M.; Liu, Y.; Stoltz, B.M. Sequential ruthenium catalysis for olefin isomerization and oxidation: Application to the synthesis of unusual amino acids. *J. Am. Chem. Soc.* **2017**, *139*, 13944–13949. [CrossRef]
7. Arai, H.; Uehara, K.; Kinoshita, S.I.; Kunugi, T. Olefin Oxidation-Mercuric Salt-Active Charcoal Catalysis. *Ind. Eng. Chem. Prod. Res. Dev.* **1972**, *11*, 308–312. [CrossRef]
8. Santiago-Portillo, A.; Navalón, S.; Cirujano, F.G.; Xamena, F.X.L.I.; Alvaro, M.; Garcia, H. MIL-101 as reusable solid catalyst for autoxidation of benzylic hydrocarbons in the absence of additional oxidizing reagents. *ACS Catal.* **2015**, *5*, 3216–3224. [CrossRef]
9. Dang, D.; Wu, P.; He, C.; Xie, Z.; Duan, C. Homochiral Metal–Organic Frameworks for Heterogeneous Asymmetric Catalysis. *J. Am. Chem. Soc.* **2010**, *132*, 14321–14323. [CrossRef]
10. Goswami, S.; Jena, H.S.; Konar, S. Study of heterogeneous catalysis by iron-squarate based 3d metal organic framework for the transformation of tetrazines to oxadiazole derivatives. *Inorg. Chem.* **2014**, *53*, 7071–7073. [CrossRef]
11. Long, J.; Liu, H.; Wu, S.; Liao, S.; Li, Y. Selective oxidation of saturated hydrocarbons using Au-Pd alloy nanoparticles supported on metal-organic frameworks. *ACS Catal.* **2013**, *3*, 647–654. [CrossRef]
12. Gascon, J.; Aktay, U.; Hernandez-Alonso, M.D.; Van Klink, G.P.M.; Kapteijn, F. Amino-based metal-organic frameworks as stable, highly active basic catalysts. *J. Catal.* **2009**, *261*, 75–87. [CrossRef]

13. Jiang, H.L.; Akita, T.; Ishida, T.; Haruta, M.; Xu, Q. Synergistic catalysis of Au@Ag core-shell nanoparticles stabilized on metal-organic framework. *J. Am. Chem. Soc.* **2011**, *133*, 1304–1306. [CrossRef] [PubMed]

14. Dhakshinamoorthy, A.; Asiri, A.M.; Garcia, H. Metal-Organic Frameworks as Catalysts for Oxidation Reactions. *Chemistry* **2016**, *22*, 8012–8024. [CrossRef]

15. Tang, Q.; Zhang, Q.; Wu, H.; Wang, Y. Epoxidation of styrene with molecular oxygen catalyzed by cobalt(II)-containing molecular sieves. *J. Catal.* **2005**, *230*, 384–397. [CrossRef]

16. Sebastian, J.; Jinka, K.M.; Jasra, R.V. Effect of alkali and alkaline earth metal ions on the catalytic epoxidation of styrene with molecular oxygen using cobalt(II)-exchanged zeolite X. *J. Catal.* **2006**, *244*, 208–218. [CrossRef]

17. Patil, R.D.; Adimurthy, S. Copper-catalyzed aerobic oxidation of amines to imines under neat conditions with low catalyst loading. *Adv. Synth. Catal.* **2011**, *353*, 1695–1700. [CrossRef]

18. Xu, B.; Lumb, J.P.; Arndtsen, B.A. A TEMPO-free copper-catalyzed aerobic oxidation of alcohols. *Angew. Chem.* **2015**, *54*, 4208–4211. [CrossRef] [PubMed]

19. Meng, X.; Lin, K.; Yang, X.; Sun, Z.; Jiang, D.; Xiao, F.S. Catalytic oxidation of olefins and alcohols by molecular oxygen under air pressure over Cu2(OH)PO4 and Cu4O(PO 4)2 catalysts. *J. Catal.* **2003**, *218*, 460–464. [CrossRef]

20. Fernandes, T.A.; André, V.; Kirillov, A.M.; Kirillova, M.V. Mild homogeneous oxidation and hydrocarboxylation of cycloalkanes catalyzed by novel dicopper(II) aminoalcohol-driven cores. *J. Mol. Catal. A* **2017**, *426*, 357–367. [CrossRef]

21. Chen, S.C.; Lu, S.N.; Tian, F.; Li, N.; Qian, H.Y.; Cui, A.J.; He, M.Y.; Chen, Q. Highly selective aerobic oxidation of alcohols to aldehydes over a new Cu(II)-based metal-organic framework with mixed linkers. *Catal. Commun.* **2017**, *95*, 6–11. [CrossRef]

22. Zhang, A.; Li, L.; Li, J.; Zhang, Y.; Gao, S. Epoxidation of olefins with O2 and isobutyraldehyde catalyzed by cobalt (II)-containing zeolitic imidazolate framework material. *Catal. Commun.* **2011**, *12*, 1183–1187. [CrossRef]

23. Skobelev, I.Y.; Sorokin, A.B.; Kovalenko, K.A.; Fedin, V.P.; Kholdeeva, O.A. Solvent-free allylic oxidation of alkenes with O2 mediated by Fe- and Cr-MIL-101. *J. Catal.* **2013**, *298*, 61–69. [CrossRef]

24. Ha, Y.; Mu, M.; Liu, Q.; Ji, N.; Song, C.; Ma, D. Mn-MIL-100 heterogeneous catalyst for the selective oxidative cleavage of alkenes to aldehydes. *Catal. Commun.* **2018**, *103*, 51–55. [CrossRef]

25. Jones, C.W. *Application of Hydrogen Peroxide for the Synthesis of Fine Chemicals*; RSC Clean Technology Monographs: York, UK, 2007. [CrossRef]

26. Luz, I.; Llabrés i Xamena, F.X.; Corma, A. Bridging homogeneous and heterogeneous catalysis with MOFs: "Click" reactions with Cu-MOF catalysts. *J. Catal.* **2010**, *276*, 134–140. [CrossRef]

27. Hoang, T.T.; To, T.A.; Cao, V.T.T.; Nguyen, A.T.; Nguyen, T.T.; Phan, N.T.S. Direct oxidative C–H amination of quinoxalinones under copper-organic framework catalysis. *Catal. Commun.* **2017**, *101*, 20–25. [CrossRef]

28. Wang, Z.; Laddha, G.; Kanitkar, S.; Spivey, J.J. Metal organic framework-mediated synthesis of potassium-promoted cobalt-based catalysts for higher oxygenates synthesis. *Catal. Today* **2017**, *298*, 209–215. [CrossRef]

29. Zhang, G.; Shi, Y.; Wei, Y.; Zhang, Q.; Cai, K. Two kinds of cobalt–based coordination polymers with excellent catalytic ability for the selective oxidation of cis-cyclooctene. *Inorg. Chem. Commun.* **2017**, *86*, 112–117. [CrossRef]

30. Leus, K.; Vandichel, M.; Liu, Y.Y.; Muylaert, I.; Musschoot, J.; Pyl, S.; Vrielinck, H.; Callens, F.; Marin, G.B.; Detavernier, C.; et al. The coordinatively saturated vanadium MIL-47 as a low leaching heterogeneous catalyst in the oxidation of cyclohexene. *J. Catal.* **2012**, *285*, 196–207. [CrossRef]

31. Brown, J.W.; Nguyen, Q.T.; Otto, T.; Jarenwattananon, N.N.; Glöggler, S.; Bouchard, L.S. Epoxidation of alkenes with molecular oxygen catalyzed by a manganese porphyrin-based metal-organic framework. *Catal. Commun.* **2015**, *59*, 50–54. [CrossRef]

32. Raupp, Y.S.; Yildiz, C.; Kleist, W.; Meier, M.A.R. Aerobic oxidation of α-pinene catalyzed by homogeneous and MOF-based Mn catalysts. *Appl. Catal. A* **2017**, *546*, 1–6. [CrossRef]

33. Ruano, D.; Díaz-García, M.; Alfayate, A.; Sánchez-Sánchez, M. Nanocrystalline M-MOF-74 as heterogeneous catalysts in the oxidation of cyclohexene: Correlation of the activity and redox potential. *ChemCatChem* **2015**, *7*, 674–681. [CrossRef]

34. Dhakshinamoorthy, A.; Alvaro, M.; Garcia, H. Aerobic oxidation of styrenes catalyzed by an iron metal organic framework. *ACS Catal.* **2011**, *1*, 836–840. [CrossRef]

35. Fu, Y.; Sun, D.; Qin, M.; Huang, R.; Li, Z. Cu(ii)-and Co(ii)-containing metal-organic frameworks (MOFs) as catalysts for cyclohexene oxidation with oxygen under solvent-free conditions. *RSC Adv.* **2012**, *2*, 3309. [CrossRef]

36. Tuci, G.; Giambastiani, G.; Kwon, S.; Stair, P.C.P.C.; Snurr, R.Q.R.Q.; Rossin, A. Chiral Co(II) metal-organic framework in the heterogeneous catalytic oxidation of alkenes under aerobic and anaerobic conditions. *ACS Catal.* **2014**, *4*, 1032–1039. [CrossRef]

37. Wang, J.S.; Jin, F.Z.; Ma, H.C.; Li, X.B.; Liu, M.Y.; Kan, J.L.; Chen, G.J.; Dong, Y.B. Au@Cu(II)-MOF: Highly Efficient Bifunctional Heterogeneous Catalyst for Successive Oxidation-Condensation Reactions. *Inorg. Chem.* **2016**, *55*, 6685–6691. [CrossRef]

38. Chen, G.J.; Wang, J.S.; Jin, F.Z.; Liu, M.Y.; Zhao, C.W.; Li, Y.A.; Dong, Y.B. Pd@Cu(II)-MOF-Catalyzed Aerobic Oxidation of Benzylic Alcohols in Air with High Conversion and Selectivity. *Inorg. Chem.* **2016**, *55*, 3058–3064. [CrossRef]

39. Duan, H.; Zeng, Y.; Yao, X.; Xing, P.; Liu, J.; Zhao, Y. Tuning Synergistic Effect of Au-Pd Bimetallic Nanocatalyst for Aerobic Oxidative Carbonylation of Amines. *Chem. Mater.* **2017**, *29*, 3671–3677. [CrossRef]

40. Cancino, P.; Vega, A.; Santiago-Portillo, A.; Navalon, S.; Alvaro, M.; Aguirre, P.; Spodine, E.; García, H. A novel copper(II)-lanthanum(Iii) metal organic framework as a selective catalyst for the aerobic oxidation of benzylic hydrocarbons and cycloalkenes. *Catal. Sci. Technol.* **2016**, *6*, 3727–3736. [CrossRef]

41. Cancino, P.; Santibáñez, L.; Fuentealba, P.; Olea, C.; Vega, A.; Spodine, E. Heterometallic CuII/LnIII polymers active in the catalytic aerobic oxidation of cycloalkenes under solvent-free conditions. *Dalt. Trans.* **2018**, *47*, 13360–13367. [CrossRef]

42. Cancino, P.; Santibañez, L.; Stevens, C.; Fuentealba, P.; Audebrand, N.; Aravena, D.; Torres, J.; Martinez, S.; Kremer, C.; Spodine, E. Influence of the channel size of isostructural 3d-4f MOFs on the catalytic aerobic oxidation of cycloalkenes. *New J. Chem.* **2019**, *43*, 11057–11064. [CrossRef]

43. Liu, Q.; De Li, J.R.; Gao, S.; Ma, B.Q.; Liao, F.H.; Zhou, Q.Z.; Yu, K.B. Highly stable open frameworks of Ln-Cu (Ln = La, Ce) 3D coordination polymers containing spacious hexagon channels and unusual μ2-H$_2$O bridges. *Inorg. Chem. Commun.* **2001**, *4*, 301–304. [CrossRef]

44. Domínguez, S.; Torres, J.; Peluffo, F.; Mederos, A.; González-Platas, J.; Castiglioni, J.; Kremer, C. Mixed 3d/4f polynuclear complexes with 2,2'-oxydiacetate as bridging ligand: Synthesis, structure and chemical speciation of La-M compounds (M = bivalent cation). *J. Mol. Struct.* **2007**, *829*, 57–64. [CrossRef]

45. Corma, A.; García, H. Lewis acids as catalysts in oxidation reactions: From homogeneous to heterogeneous systems. *Chem. Rev.* **2002**, *102*, 3837–3892. [CrossRef] [PubMed]

46. Cancino, P.; Paredes-García, V.; Aliaga, C.; Aguirre, P.; Aravena, D.; Spodine, E. Influence of the lanthanide(III) ion in {[Cu3Ln2(oda)6(H2O)6]•nH2O}n (LnIII: La, Gd, Yb) catalysts on the heterogeneous oxidation of olefins. *Catal. Sci. Technol.* **2017**, *7*, 231–242. [CrossRef]

47. Sun, D.; Sun, F.; Deng, X.; Li, Z. Mixed-Metal Strategy on Metal-Organic Frameworks (MOFs) for Functionalities Expansion: Co Substitution Induces Aerobic Oxidation of Cyclohexene over Inactive Ni-MOF-74. *Inorg. Chem.* **2015**, *54*, 8639–8643. [CrossRef] [PubMed]

48. Cid, R.; Pecchi, G. Potentiometric method for determing the number and relative strength of acid sites in colored catalysts. *Appl. Catal.* **1985**, *14*, 15–21. [CrossRef]

Article

Fast Immobilization of Human Carbonic Anhydrase II on Ni-Based Metal-Organic Framework Nanorods with High Catalytic Performance

Mengzhao Jiao [1], Jie He [1], Shanshan Sun [1], Frank Vriesekoop [2], Qipeng Yuan [1], Yanhui Liu [1,3,*] and Hao Liang [1,*]

[1]   State Key Laboratory of Chemical Resource Engineering, Beijing University of Chemical Technology, Beijing 100029, China; mzjiao@163.com (M.J.); jackhe0616@163.com (J.H.); shanssun@126.com (S.S.); yuanqp@mail.buct.edu.cn (Q.Y.)

[2]   Department of Food Technology and Innovation, Harper Adams University, Newport TF10 8NB, UK; FVriesekoop@harper-adams.ac.uk

[3]   Beijing Key Laboratory of Bioprocess, Beijing University of Chemical Technology, Beijing 100029, China

*   Correspondence: liuyh@mail.buct.edu.cn (Y.L.); lianghao@mail.buct.edu.cn (H.L.); Tel.: +8610-6443-7610 (H.L.)

Received: 2 March 2020; Accepted: 15 March 2020; Published: 6 April 2020

Abstract: Carbonic anhydrase (CA) has received considerable attention for its ability to capture carbon dioxide efficiently. This study reports a simple strategy for immobilizing recombinant carbonic anhydrase II from human (hCA II) on Ni-based MOFs (Ni-BTC) nanorods, which was readily achieved in a one-pot immobilization of His-tagged hCA II (His-hCA II). Consequently, His-hCA II from cell lysate could obtain an activity recovery of 99% under optimal conditions. After storing for 10 days, the immobilized His-hCA II maintained 40% activity while the free enzyme lost 91% activity. Furthermore, during the hydrolysis of p-nitrophenyl acetic acid, immobilized His-hCA II exhibited excellent reusability and still retained more than 65% of the original activity after eight cycles. In addition, we also found that Ni-BTC had no fixation effect on proteins without histidine-tag. These results show that the Ni-BTC MOFs have a great potential with high efficiency for and specific binding of immobilized enzymes.

Keywords: one-pot hydrothermal; immobilizing recombinant; His-hCA II; Ni-BTC nanorods

---

1. Introduction

Carbonic anhydrase (CA, E.C. 4.2.1.1), a zinc metalloenzyme, can efficiently catalyze the reversible hydration of CO_2 to bicarbonate ion [1,2]. In nature, chemical fixation of CO_2 usually requires high temperatures, high pressure and long reaction times, while carbonic anhydrase can fix carbon dioxide and have a high reaction rate under low pressure conditions [3]. With the intensifying greenhouse effect, the use of carbonic anhydrase is considered as a green and efficient strategy for bio-mineralization of CO_2 [4–7]. However, the poor stability and reusability of carbonic anhydrase limits its application in practice [6,8]. Enzyme immobilization on solid supports has been developed to overcome these shortcomings [9–13]. Previous immobilization studies mainly focused on commercial carbonic anhydrase on a wide range of solid supports such as: polyurethane foam [14], SBA-15 [15], mesoporous silica [16], MIL-160 [9] and ZIF-8 [17]. However, high cost and protracted immobilization processes greatly limit the application of immobilized CA [4,10,12].

Fusion protein expression is a common strategy to obtain enzyme proteins. Due to the convenience of purification using Ni-chelating affinity chromatography, histidine tags are widely used for enzyme protein expression [18,19]. Nickel ions modified materials for enzyme purification and immobilization were developed which are based on the affinity of nickel ions and histidine labels [20–22]. However, until now, there are few piece of research on the one-pot immobilization of CA via Ni^{2+} modified materials. The introduction of histidine-tag in CA is more conducive to reducing costs and shortening immobilization time.

Metal-organic frameworks (MOFs) are a new kind of organic-inorganic hybrid porous material [23,24]. Since MOF materials are considered to be potent supporting matrices for enzyme immobilization, many kinds of commodity enzymes were immobilized successfully [25–28]. However, other researchers have spent their energy on the design of Zn^{2+}, Cu^{2+} or Fe^{3+}-based MOFs for enzyme immobilization [29–31]. Inspired by the specific affinity between Ni^{2+} and histidine-tags [32], we focus on utilizing Ni-based MOFs to achieve one-pot immobilization of histidine-tagged enzymes. To date, a number of Ni-based MOFs have been reported, however, studies on immobilized enzymes by Ni-based MOFs are rare [33]. Therefore, it is worthwhile to focus on the adsorption characteristics of Ni-based complexes itself.

Herein, the recombinant carbonic anhydrase II from human (hCA II) with His-tag and without His-tag were successfully designed and overexpressed in *Escherichia coli* (*E. coli*) (DE3). Meanwhile, Ni-based MOF (Ni-BTC) nanorods were synthesized by employing a one-pot hydrothermal process and were used as immobilization support for CA. Compared to our previous work [34] in this work we were able to carry out the protein immobilization process at room temperature, while achieving very high binding efficiencies. Our approach to realizing His-hCA II one-pot immobilization by Ni-BTC is shown in Figure 1. As a result, the His-tagged hCA II could be efficiently immobilized on Ni-BTC under optimal conditions by a simple mixing step. However, when hCA II (without His-tag) was mixed with Ni-BTC under the same conditions, the hCA II (without His-tag) could not be efficiently immobilized on Ni-BTC. This means that Ni-BTC has specific binding abilities with regards to His-hCA II. Furthermore, a specific protein purification step could be omitted in our application, which could save on production costs in practical applications. In addition, the stability of the free and immobilized His-hCA II under different conditions was also investigated.

2. Results and Discussion

2.1. Synthesis and Characterization of Ni-BTC

Because of the affinity of Ni^{2+} and histidine tags, the nickel-modified materials could easily realize the His-tagged enzyme's separation and immobilization in one step. However, most of the previously reported Ni^{2+} modified materials required complex and lengthy reaction conditions [20,21]. In this work, the light green Ni-based MOF materials were easily obtained and used as support for His-tagged enzyme immobilization.

TEM and SEM were used to characterize the morphology of Ni-BTC. As shown in Figure 1A,B respectively, the as-prepared Ni-BTC were nanorods with an average length of 3 µm and an average width was 0.5 µm. Furthermore, the Ni-BTC had a uniformly smooth long, stick-like structure and it was highly aggregated. Next, the distribution of elements in nanorods was studied by energy dispersive X-ray spectral element mapping. The elemental distribution of elements in Figure 1C was known, and we found that carbon, nitrogen, oxygen and nickel element were distributed homogeneously throughout the Ni-BTC structure. Our results with regard to the structure of synthesized Ni-BTC MOF materials (Figure 1D) appear to be in agreement with those reported elsewhere [35]. These results were also in agreement with our X-ray spectroscopy (XPS) spectrum (Figure S1). Signal peaks of Ni 2p, O 1s, N 1s, and C 1s were detected, indicating that Ni-BTC was successfully synthesized with nickel acetate and H_3BTC.

The chemical structure of Ni-BTC nanorods had been characterized previously which provided us with knowledge about the linker's integration and the oriented growth of the Ni-BTC nanorods and its cubic geometry [35]. As shown in Figure S2, Ni-BTC nanorods kept stable at temperatures below 100 °C and showed weight loss (26.1 wt %) between 100 °C and 300 °C, the reason might be the vaporization of water and solvent. Significant weight loss (43.1 wt %) occurred between 300 °C and 600 °C, which could be assigned to the decomposition of H_3BTC [34].

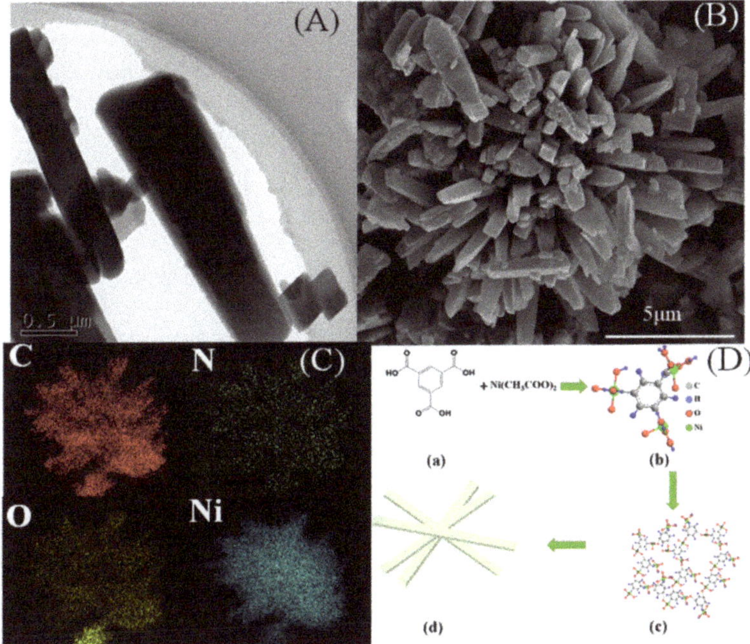

Figure 1. TEM images (**A**) and SEM images (**B**) of Ni-BTC and (**C**) corresponding carbon; nitrogen; oxygen and nickel elemental mapping of Ni-BTC; (**D**) The synthesis process and structure of the 2D Ni-MOF.

2.2. Optimal Conditions for His-hCA II Immobilization

We first prepared the His-hCA II with a histidine-tag in *E. coli* BL21 (DE3). The plasmid map of pETDuet-1-His-hCA II is shown in Figure S3. Following overexpression in *E. coli* BL21 (DE3), His-hCA II was extracted in a phosphate buffer (50 mM, pH 7.4). The purified His-hCA II appeared as a single band on the silver-stained gels (Figure S4). In order to get the best fixed conditions, 100 μL of His-hCA II crude cell lysate was incubated with Ni-BTC under different conditions. The influence of mass ratios on Ni-BTC to crude protein on protein loading and activity recovery of His-hCA II @Ni-BTC were investigated. As shown in Figure 2A, activity recovery of His-hCA II @Ni-BTC increased somewhat from a ratio of 2.5:1 to 5:1 and declined as the ratio increased further. The decline in activity recovery can possibly be attributed to the greater inclusion of heteroproteins from within the crude extract at the higher concentrations adsorbed on the Ni-BTC. Protein loading of His-hCA II @Ni-BTC as the ratio increased from 2.5:1 to 12.5:1 (Figure 2A) increased linearly. The elevated protein loading at a ratio of 12.5:1 also implies the increased adsorption of Ni-BTC on hybrid protein. On the basis of the maximum activity recovery of His-hCA II @Ni-BTC at the ratio of 5:1, we selected this protein loading ratio for further experiments. In addition, we also calculated the amount of immobilized His-hCA II (Table S1), the quality of bounding protein was calculated by the difference between the amount of protein added to the reaction mixture and that in the leachate and washing solutions after immobilization. As a result, the catalytic activity of an immobilized enzyme was 3.1 times that of a free enzyme and the protein yield of crude cell lysate was 31.8%.

The effects of immobilization temperature (Figure 2B) and time (Figure 2C) of Ni-BTC on the loading and activity recovery of His-hCA II @Ni-BTC and crude protein were also investigated. Similar to previous reports, [4,10] although the amount of protein binding increased, the activity recovery reached its maximum value at 30 °C and then decreased with an increase of the temperature from 30 °C to 50 °C. Human carbonic anhydrase has an optimum temperature close to human body temperature to retain its activity. An increase in the immobilized time showed a near linear increase in the activity recovery of immobilized His-hCA II, increasing from 61.5% to 88.2% from 15 to 60 min, and a further increase in activity recovery from 60 to 120 min to 95% (Figure 2C). Consequently, an immobilization time of 120 min was employed for further experiments. Above all, 3 mg Ni-BTC and 100 µL cell lysate were mixed with 900 µL of phosphates buffer (50 mM, pH 8.0) for half an hour at 30 °C, then collected by centrifuge for 10 minutes at 12,000 rpm.

We also investigated whether the different monomers in His-hCA II @Ni-BTC could accelerate the relative catalytic activity at equal protein concentrations of His-hCA II @Ni-BTC and His-hCA II. We found that His-hCA II @Ni-BTC and His-hCA II had similar catalytic activity ($p > 0.5$) with an activity recovery of 74.5% (Figure 2D). Whereas Ni-BTC and Ni^{2+} had very low activity recovery (<10%) and H_3BTC had almost no catalytic activity of its own. The reason for the low recovery activity observed for Ni-BTC and Ni^{2+} might be that the constituent Ni^{2+} absorbed some of the substrate and presented a minor false positive result. Hence, we conclude that the activity recovery of His-hCA II @Ni-BTC was mainly derived from the His-hCA II enzyme. In addition, we studied the specific binding ability of His-hCA II to Ni-BTC compared to hCA II (not His-tagged). In this experiment, equal quantities of hCA II and His-hCA II were incubated with different amounts of Ni-BTC at 25 °C for 30 min. The results indicate that the Ni-BTC has a much greater specific binding ability to His-hCA II compared to the non-His-tagged hCA II (Figure 2E).

The activity recovery of immobilized His-hCA II was 98.99% in optimal conditions, which proved that Ni-BTC could achieve high activity recovery of His-hCA II by a simple one-pot mixing procedure.

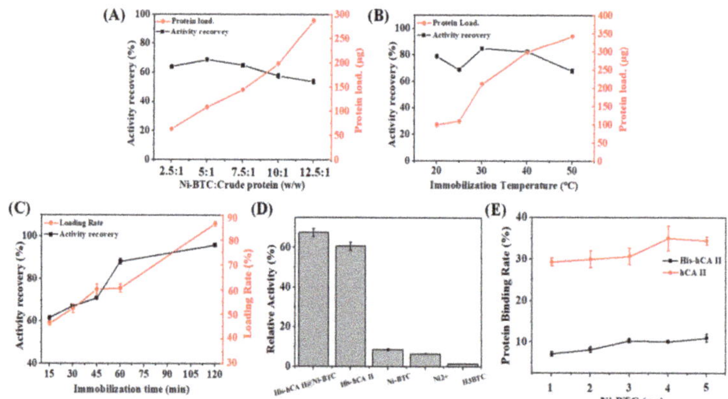

Figure 2. Optimizing immobilization conditions: (**A**) ratio of the carrier to enzyme; (**B**) immobilization temperature; (**C**) immobilization time; (**D**) the relative catalytic activity of His-hCA II @Ni-BTC, purified His-hCA II, Ni-BTC, Ni^{2+}, H3BTC; (**E**) the protein binding rate of Ni-BTC to His-hCA II and hCA II.

2.3. Enzyme (His-hCA II) Stability

The pH and thermal stability of the environmental conditions are important for practical application. The stability of free His-hCA II and His-hCA II @Ni-BTC was tested at different pH conditions (phosphate buffer, 50mM, at 30 °C, from pH 6 to 10). After 4 h of incubation at pH 6, the reaction was performed for 5 min at room temperature. Free His-hCA II maintained 30.8% of the original activity (pH 10 and 30 °C), whereas the His-hCA II @Ni-BTC retained 73.7% of its original activity (pH 10 and 30 °C) (Figure 3A). Furthermore, the free and His-hCA II @Ni-BTC maintained 59.5% and 80.7% of its original activity after incubation at pH 7 for 4 h. These results indicated that His-hCA II @Ni-BTC exhibits better stability than free hCA II under acidic condition, which is in agreement with results shown elsewhere [36]. We also tested the effects of different pH on the shedding of His-hCA II proteins from the Ni-BTC support. The results (Figure S5A) indicate that His-hCA II @Ni-BTC had different degrees of dissociation at different pH solutions. However, the degree of dissociation (protein shedding) was relatively low and had no effect on the overall catalytic activity, further demonstrating the overall pH stability of His-hCA II @Ni-BTC.

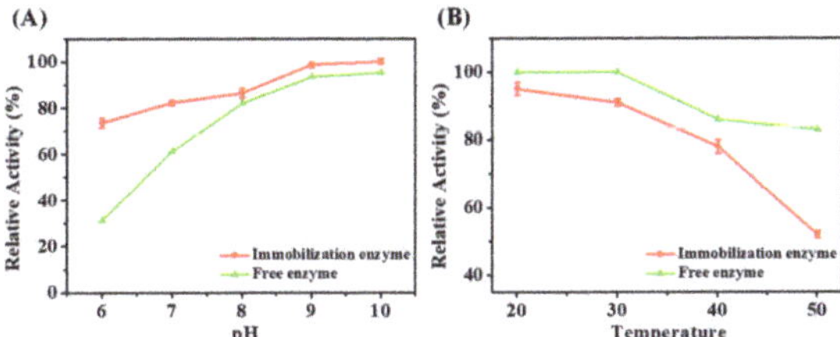

Figure 3. Stability of the immobilized and free His-hCA II (**A**) at different pHs and (**B**) temperatures.

The thermal stability of His-hCA II @Ni-BTC compared to with free His-hCA II was investigated, too. The various temperature exposures were carried out at the temperatures indicated, while the actual enzyme assays were performed at 25 °C in a phosphate buffer (50 mM, pH 8.0). The results showed that the relative activities of free enzyme and immobilized enzyme decreased significantly at temperatures over 30 °C (Figure 3B), with the immobilized enzymes performing worse than the free enzymes. Previous studies also reported a marked loss of immobilized CA activity at these temperature [4,10]. Additional experiments were conducted to detect the protein content in His-hCA II @Ni-BTC supernatant following exposure to elevated temperatures (40 and 50 °C), which showed a 23.7% and 44.8% loss of proteins, respectively (Figure S5B). According to the protein shedding rate of His-hCA II @Ni-BTC at different temperatures, Figure 3B shows that the relative activity of His-hCA II @Ni-BTC could reach 82.3% at 50 °C while the protein concentration of His-hCA II @Ni-BTC was consistent with free enzyme. It was proved that Ni-BTC has certain supporting and protective effects on enzymes. Allowing for these protein losses as lost immobilized His-hCA II @Ni-BTC, the adjusted residual His-hCA II activity outperformed the free His-hCA II. While this might not explain the exact underpinning reasons as to why the His-hAC II proteins dissociate from the Ni-BTC support, it does explain the lower activity of His-hCA II @Ni-BTC at elevated temperatures might be because the high temperature promoted His-hCA II shedding from His-hCA II @Ni-BTC. However, at the same protein concentration, as shown in Figure S6, the catalytic activity of His-hCA II @Ni-BTC was 2.4 times that of free enzymes after being maintained at 50 °C for 4 h. Thus, the protective and supporting effect of Ni-BTC on the protein under thermal conditions significantly improved the stability of the enzyme.

The storage stability of free and immobilized His-hCA II was also investigated (Figure 4A). At the end of two days of storage in a phosphate buffer (50 mM, pH 8.0) at 25 °C, the free enzyme kept 41.8% of its original activity (i.e., a 58.2% loss), whereas the immobilized enzyme retained 66.4% (i.e., a 33.6% loss) of its original activity. These losses extended further and by the end of 10 days of storage the free enzyme kept 6.4% of its original activity (i.e., a 93.6% loss), whereas the immobilized enzyme retained 41.8% (i.e., a 58.2% loss) of its original activity after 10 days of storage. During prolonged storage, denaturation and degradation of His-hCA II proteins could significantly reduce enzyme activity. However, the immobilization of His-hCA II on Ni-BTC provided physicochemical stability and mechanical protection, which enabled the His-hCA II to maintain greater stability for long-term reactions compared to free His-hCA II. Finally, the immobilized His-hCA II was tested for recycling. The immobilized His-hCA II was separated by centrifugation after running an experimental standard reaction, after which it was washed three times in water and redispersed in the buffer by vortex mixing before repeating another experimental standard reaction. There was a gradual decrease in the remaining relative enzyme activity at each recycling, however, more than 65% of original activity remained after eight cycles (Figure 4B). This might be due to the gradual shedding of some enzymes from Ni-BTC, and some denaturation during the recovery process. The notion of a gradual release of the enzyme from the Ni-BTC support was confirmed in a subsequent experiment (Figure S7). By determining the protein content in the supernatant from each cycle, we found that each cycle's process was accompanied with protein shedding.

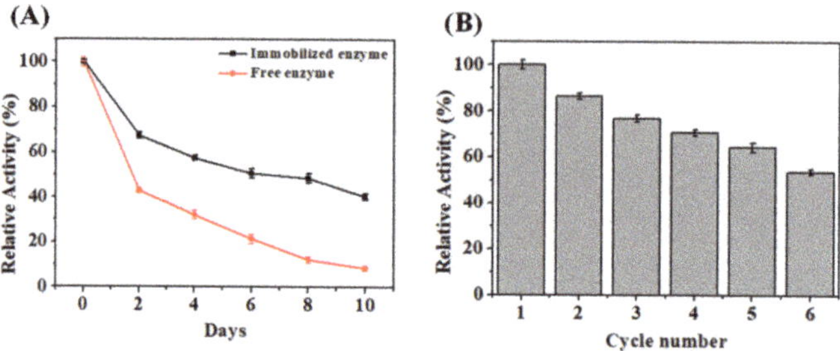

Figure 4. (**A**) Stability of the immobilized and free His-hCA II of storage stability at 25 °C; (**B**) reusability of His-hCA II @Ni-BTC.

2.4. Kinetic Parameters

To further evaluate the catalytic capacity of immobilized enzymes, we further compared the kinetics of free enzymes and His-hCA II @Ni-BTC. As presented in Table S2, the K_m and V_{max} values for free and His-hCA II @Ni-BTC were determined to be 1.82 mmol/L, 1.96 mmol/L and 0.037 mmol/min, 0.035 mmol/min for p-NPA, respectively. The substrates and possible conformational changes of the mass transfer limiting structure play an important role in increasing the K_m value of the His-hCA II @Ni-BTC after immobilization.

3. Materials and Methods

3.1. Materials

Nickel (II) acetate tetrahydrate was purchased from Yuanye Biotechnology Company. (Shanghai, China). Benzene-1, 3, 5-tricarboxylic acid (H_3BTC) was purchased from J&K China Chemical Ltd. p-nitrophenyl acetate (p-NPA) and N, N-dimethylformamide (DMF) were purchased from Aladdin (Shanghai, China). Ampicillin and isopropyl β-D-1-thiogalactopyranoside (IPTG) were obtained from Beijing QXTD-Biotechnology Co., Ltd. (Beijing, China). Milli-Q water was used to prepare all the buffers and solutions. Bovine serum albumin (BSA) and methionine were purchased from Beijing Chemical Works (Beijing, China). All other reagents and solvents were of analytical grade.

3.2. Expression and Purification of His-hCA II

The human carbonic anhydrase (hCA II) gene was cloned to the pETDuet-1 vector with restriction sites BamH I and Kpn I. The recombinant plasmids were transformed into *E. coli* BL21 (DE3) for protein synthesis. The culture medium was incubated at 37 °C for 4 h, then isopropyl-β-D-thiogalactopyranoside (IPTG) was added for induction [37], and the culture continued at 18 °C for 20 h. The cells were then harvested by high-speed centrifugation (10,000 rpm, 30 min, 4 °C) and lysed by acoustic degradation. Subsequent cell lysate centrifugation (10,000 rpm, 30 min, 4 °C) removed the crude precipitate to obtain the crude protein solution and the protein concentration of the lysate was measured by the Bradford assay using BSA as a standard [38]. The protein fraction was purified with affinity chromatography using a Ni-NTA column (Qiagen) pre-equilibrated with the lysis buffer and eluted with the lysis buffer supplemented with 300 mM of imidazole. The purified protein was analyzed by sodium salt -Polyacrylamide gel electrophoresis (SDS-PAGE).

3.3. Preparation of Ni-BTC

H_3BTC (0.5 g, 2.38 mmol) was dissolved in a DMF/ethanol (V/V = 1:1, 40 mL) mixture and moved to a metal reactor. Then 20 mL of nickel acetate solution (1.0725 g, 4.31 mmol) was added to the reaction mixture and stirred constantly. Then the reaction was finished after continuous stirring at 70 °C for 8 h. Finally, the product was separated by centrifugation and washed three times with Milli-Q water and ethanol solution (95%), successively.

3.4. Assay of His-hCA II Immobilization

3.4.1. The Immobilization of His-hCA II and Determination of Activity

The activity of free and His-hCA II @Ni-BTC activity were based on p-NPA with some modifications, as described previously [10]. The 1.5 mL total reaction volume contained 600 μL of phosphate buffer (50 mM, pH 8.0), 300 μL Milli-Q water, 500 μL freshly prepared p-NPA solution (3 mM) and 100 μL crude cell lysate or immobilized enzyme which contained the same protein content of His-hCA II. The mixture was shaken at room temperature for 5 min and then centrifuged for 10 min at 12,000 rpm. The increase of absorbance was measured by employing a spectrophotometer (Metash, UV-5100, Shanghai, China) at 348 nm. The same procedure was performed without His-hCA II solution as a control.

To optimize the immobilization conditions, His-hCA II was incubated with Ni-BTC under various conditions. In detail, 100 μL cell lysate of His-hCA II (which contained 0.4 mg protein) and 900 mL Milli-Q water were incubated with Ni-BTC. For optimizing the bonding ratio between enzyme and carrier, 100 μL His-hCA II crude cell lysate (0.4 mg total protein) was incubated with different amounts of Ni-BTC in a 1 mL system. For optimization in regard to immobilization temperature, 3 mg Ni-BTC was oscillated with 100 μL His-hCA II crude cell lysate at different temperatures ranging from 20–50 °C for 30 min. For optimization in regard to the immobilization time, 3 mg Ni-BTC was incubated with 100 μL His-hCA II crude cell lysate at 25 °C over different lengths of time (15–120 min) with shaking. After separated by centrifugation, the standard curve of protein was determined using Bradford assay with BSA, then analyzed the supernatant and the original protein concentration. The activity of enzyme was detected by the above method. In all experiments, all of the experimental results were measured by three independent experimental groups.

3.4.2. Enzyme Stability Detection

For the pH stability, the activity of free His-hCA II and His-hCA II @Ni-BTC was carried out in different phosphate buffer solutions (50 mM, pH 6–10) at 25 °C and the enzymatic activities were determined by the above method. Thermostability was also studied, free enzyme and His-hCA II @Ni-BTC were added to the phosphate buffer (50 mM, pH 8.0) over a range of temperatures, 20–50 °C, and incubated for 30 min. For the storage stability, the free enzyme and His-hCA II @Ni-BTC were kept in a phosphate buffer (50 mM, pH 8.0) at 25 °C. For the reusability of His-hCA II @Ni-BTC after the reaction, for 5 min, the immobilized enzyme was collected by centrifugation in the next cycle. Then the same substrate and buffer solution were added to the reaction system for the next cycle and the enzyme activity of each cycle was measured by the above method.

3.4.3. Assay of Kinetics of the His-hCA II

For the kinetic analysis, the reaction system was detected by the same method as enzyme activity the concentrations of p-NPA were varied from 1.0 to 5.0 mM. The K_m and V_{max} were acquired by using a Lineweaver-Burk plot.

$$\frac{1}{v} = \left(\frac{Km}{Vmax} \frac{1}{[S]} \right) + \frac{1}{Vmax}$$

3.5. Characterization

Transmission electron microscopy (TEM) analysis was performed on a JEM 1200EX transmission electron microscope (Hitachi, Tokyo, Japan). Scanning electron microscopy (SEM) analysis was performed on a Tecnai G2 F20 scanning electron microscope (FEI, USA). The X-ray diffraction (D8 Advance X-ray diffractometer, Bruker, Karlsruhe, Germany) with a Cu Kα anode ($\lambda = 0.15406$ nm) at 40 kV and 40 mA was used to measure the crystal structures of the samples. FTIR spectra were obtained by a FTIR spectrometer (8700/Continuum XL Imaging Microscope, Nicolet, Waltham, MA, USA). The spectra were collected between 400 and 4000 cm^{-1}. X-ray photoelectron spectroscopy (XPS) spectra were recorded on a Thermo Scientific K-Alpha X-ray photoelectron spectrometer. Thermogravimetric analysis (TGA) was performed using a thermogravimetric analyzer (DTG-60A, Shimadzu, Japan) in the range of 30–600 °C under a nitrogen flow (heating rate of 10 °C min^{-1}).

4. Conclusions

In conclusion, a simple and effective single pot immobilization strategy was developed to isolate histidine marker enzymes expressed in *E. coli* BL21 (DE3). Ni-BTC could be easily compounded by a hydrothermal process and showed the ability to efficiently immobilize the His-hCA II from cell lysate at high load and high activity recovery under optimal conditions. The His-hCA II @Ni-BTC exhibited excellent biocatalytic activity and reusability in the hydrolysis of p-nitrophenyl acetate. All of these results revealed the good potential of His-hCA II @Ni-BTC for biomineralization of CO_2.

Supplementary Materials: The following are available online at http://www.mdpi.com/2073-4344/10/4/401/s1, Figure S1: XPS spectra of Ni-BTC nanorods: (a) full scan, (b) Ni 2p, (c) N 1s, Figure S2: TG curves of Ni-BTC, Figure S3: The plasmid map of pETDuet-1-His-hCA II, Figure S4: SDS-PAGE of purified the recombinant hCA II from the culture supernatant. (Lane 1: His6-tagged hCA II purified using Ni-NTA column, Lane 2: supernatant of hCA II cell lysate, Lane 3: molecular weight marker), Figure S5: The protein shedding rate of immobilized enzyme at different pH and (B) the protein shedding rate of immobilized enzyme at different temperatures, Figure S6: The catalytic activity of His-hCA II@Ni-BTC and free enzyme at the same protein concentration was studied at 40 °C and 50 °C, Figure S7: The protein shedding rate of cycle process. Table S1: The amount of immobilized His-hCA II, Table S2: Michaelis-Menten kinetics parameters of immobilized and free enzymes.

Author Contributions: Conceptualization, methodology, formal analysis, validation, writing manuscript, and supervision, project management, H.L. Review, F.V. H.L. Experimental work and draft writing, M.J. Other auxiliary work, J.H., S.S., Q.Y., Y.L. All authors have read and agreed to the published version of the manuscript.

Funding: This research received no external funding.

Acknowledgments: This work was supported by the National Natural Science Foundation of China (No. 21606014 and 21878014) and the Double First-rate Program (No. ylkxj03).

Conflicts of Interest: The authors declare no competing financial interests.

References

1. Kupriyanova, E.; Pronina, N.; Los, D. Carbonic anhydrase—A universal enzyme of the carbon-based life. *Photosynthetica* **2017**, *55*, 3–19. [CrossRef]

2. Ren, S.; Feng, Y.; Wen, H.; Li, C.; Sun, B.; Cui, J.; Jia, S. Immobilized carbonic anhydrase on mesoporous cruciate flower-like metal organic framework for promoting CO_2 sequestration. *Int. J. Biol. Macromol.* **2018**, *117*, 189–198. [CrossRef] [PubMed]

3. Yadav, R.; Joshi, M.; Wanjari, S.; Prabhu, C.; Kotwal, S.; Satyanarayanan, T.; Rayalu, S. Immobilization of carbonic anhydrase on chitosan stabilized Iron nanoparticles for the carbonation reaction. *Water Air Soil Pollut.* **2012**, *223*, 5345–5356. [CrossRef]

4. Yadav, R.R.; Mudliar, S.N.; Shekh, A.Y.; Fulke, A.B.; Devi, S.S.; Krishnamurthi, K.; Juwarkar, A.; Chakrabarti, T. Immobilization of carbonic anhydrase in alginate and its influence on transformation of CO_2 to calcite. *Process Biochem.* **2012**, *47*, 585–590. [CrossRef]

5. Zhang, S.; Du, M.; Shao, P.; Wang, L.; Ye, J.; Chen, J.; Chen, J. Carbonic anhydrase enzyme-MOFs composite with a superior catalytic performance to promote CO_2 absorption into tertiary amine solution. *Environ. Sci. Technol.* **2018**, *52*, 12708–12716. [CrossRef]

6. Lee, C.H.; Jang, E.K.; Yeon, Y.J.; Pack, S.P. Stabilization of Bovine Carbonic anhydrase II through rational site-specific immobilization. *Biochem. Eng. J.* **2018**, *138*, 29–36. [CrossRef]

7. Wang, S.; Lu, G.Q.M. Effects of promoters on catalytic activity and carbon deposition of Ni/γ-Al$_2$O$_3$ catalysts in CO$_2$ reforming of CH$_4$. *J. Chem. Technol. Biotechnol.* **2015**, *75*, 589–595. [CrossRef]

8. Jin, C.; Zhang, S.; Zhang, Z.; Chen, Y. Mimic carbonic anhydrase using metal-organic frameworks for CO$_2$ capture and conversion. *Inorg. Chem.* **2018**, *57*, 2169–2174. [CrossRef]

9. Liu, Q.; Chapman, J.; Huang, A.; Williams, K.C.; Wagner, A.; Garapati, N.; Sierros, K.A.; Dinu, C.Z. User-tailored metal-organic frameworks as supports for carbonic anhydrase. *ACS Appl. Mater. Interfaces* **2018**, *10*, 41326–41337. [CrossRef]

10. Al-Dhrub, A.H.A.; Sahin, S.; Ozmen, I.; Tunca, E.; Bulbul, M. Immobilization and characterization of human carbonic anhydrase I on amine functionalized magnetic nanoparticles. *Process Biochem.* **2017**, *57*, 95–104. [CrossRef]

11. Wang, D.; Jiang, W. Preparation of chitosan-based nanoparticles for enzyme immobilization. *Int. J. Biol. Macromol.* **2019**, *126*, 1125–1132. [CrossRef] [PubMed]

12. Khodarahmi, R.; Yazdanparast, R. Refolding of chemically denatured alpha-amylase in dilution additive mode. *Biochim. Biophys. Acta* **2004**, *1674*, 175–181. [CrossRef] [PubMed]

13. Zhang, Y.H.; Chen, K.N.; Zhang, J.N.; Ren, H.; Zhang, X.Y.; Huang, J.H.; Wang, S.Z.; Fang, B.S. Preparation and evaluation of a polymer–metal–enzyme hybrid nanowire forthe immobilization of multiple oxidoreductases. *J. Chem. Technol. Biotechnol.* **2019**, *94*. [CrossRef]

14. Ozdemir, E. Biomimetic CO$_2$ sequestration: 1. immobilization of carbonic anhydrase within polyurethane foam. *Energy Fuels* **2009**, *23*, 5725–5730. [CrossRef]

15. Vinoba, M.; Kim, D.H.; Lim, K.S.; Jeong, S.K.; Lee, S.W.; Alagar, M. Biomimetic sequestration of CO$_2$ and reformation to CaCO$_3$ using bovine carbonic anhydrase immobilized on SBA-15. *Energy Fuels* **2011**, *25*, 438–445. [CrossRef]

16. Fei, X.; Chen, S.; Huang, C.; Liu, D.; Zhang, Y. Immobilization of bovine carbonic anhydrase on glycidoxypropyl-functionalized nanostructured mesoporous silicas for carbonation reaction. *J. Mol. Catal. B Enzym.* **2015**, *116*, 134–139. [CrossRef]

17. Liang, H.; Lin, F.F.; Zhang, Z.J.; Liu, B.W.; Jiang, S.H.; Yuan, Q.P.; Liu, J.W. Multicopper laccase mimicking nanozymes with nucleotides as ligands. *ACS Appl. Mater. Interfaces* **2017**, *9*, 1352–1360. [CrossRef]

18. Li, Y.; Sun, X.; Feng, Y.; Yuan, Q. Cloning, expression and activity optimization of trehalose synthase from Thermus thermophilus HB27. *Chem. Eng. Sci.* **2015**, *135*, 323–329. [CrossRef]

19. Li, Y.; Wang, Z.; Feng, Y.; Yuan, Q. Improving trehalose synthase activity by adding the C-terminal domain of trehalose synthase from Thermus thermophilus. *Bioresour. Technol.* **2017**, *245*, 1749–1756. [CrossRef]

20. Vahidi, A.K.; Yang, Y.; Ngo, T.P.N.; Li, Z. Simple and efficient immobilization of extracellular his-tagged enzyme directly from cell culture supernatant as active and recyclable nanobiocatalyst: High-Performance production of biodiesel from waste grease. *ACS Catal.* **2015**, *5*, 3157–3161. [CrossRef]

21. Cao, G.; Gao, J.; Zhou, L.; Huang, Z.; He, Y.; Zhu, M.; Jiang, Y. Fabrication of Ni 2+ -nitrilotriacetic acid functionalized magnetic mesoporous silica nanoflowers for one pot purification and immobilization of His-tagged ω-transaminase. *Biochem. Eng. J.* **2017**, *128*, 116–125. [CrossRef]

22. Wang, W.; Wang, D.I.; Li, Z. Facile fabrication of recyclable and active nanobiocatalyst: Purification and immobilization of enzyme in one pot with Ni-NTA functionalized magnetic nanoparticle. *Chem. Commun. (Camb.)* **2011**, *47*, 8115–8117. [CrossRef] [PubMed]

23. Nadar, S.S.; Rathod, V.K. Facile synthesis of glucoamylase embedded metal-organic frameworks (glucoamylase-MOF) with enhanced stability. *Int. J. Biol. Macromol.* **2017**, *95*, 511–519. [CrossRef]

24. Nadar, S.S.; Rathod, V.K. Magnetic-metal organic framework (magnetic-MOF): A novel platform for enzyme immobilization and nanozyme applications. *Int. J. Biol. Macromol.* **2018**, *120*, 2293–2302. [CrossRef] [PubMed]

25. Ling, P.; Qian, C.; Gao, F.; Lei, J. Enzyme-immobilized metal-organic framework nanosheets as tandem catalysts for the generation of nitric oxide. *Chem. Commun. (Camb.)* **2018**, *54*, 11176–11179. [CrossRef] [PubMed]

26. Chen, W.H.; Luo, G.F.; Vazquez-Gonzalez, M.; Cazelles, R.; Sohn, Y.S.; Nechushtai, R.; Mandel, Y.; Willner, I. Glucose-Responsive metal-organic-framework nanoparticles act as "smart" sense-and-treat carriers. *ACS Nano* **2018**, *12*, 7538–7545. [CrossRef]

27. Shieh, F.K.; Wang, S.C.; Yen, C.I.; Wu, C.C.; Dutta, S.; Chou, L.Y.; Morabito, J.V.; Hu, P.; Hsu, M.H.; Wu, K.C.; et al. Imparting functionality to biocatalysts via embedding enzymes into nanoporous materials by a de novo approach: Size-selective sheltering of catalase in metal-organic framework microcrystals. *J. Am. Chem. Soc.* **2015**, *137*, 4276–4279. [CrossRef]

28. Cui, J.; Feng, Y.; Jia, S. Silica encapsulated catalase@metal-organic framework composite: A highly stable and recyclable biocatalyst. *Chem. Eng. J.* **2018**, *351*, 506–514. [CrossRef]

29. Samui, A.; Chowdhuri, A.R.; Mahto, T.K.; Sahu, S.K. Fabrication of a magnetic nanoparticle embedded NH2-MIL-88B MOF hybrid for highly efficient covalent immobilization of lipase. *RSC Adv.* **2016**, *6*, 66385–66393. [CrossRef]

30. Zhao, M.; Zhang, X.; Deng, C. Rational synthesis of novel recyclable Fe3O4@MOF nanocomposites for enzymatic digestion. *Chem. Commun.* **2015**, *51*, 8116–8119. [CrossRef]

31. Govan, J.; Gun'ko, Y.K. Recent advances in the application of magnetic nanoparticles as a support for homogeneous catalysts. *Nanomaterials (Basel)* **2014**, *4*, 222–241. [CrossRef] [PubMed]

32. Yu, M.; Liu, D.; Sun, L.; Li, J.; Chen, Q.; Pan, L.; Shang, J.; Zhang, S.; Li, W. Facile fabrication of 3D porous hybrid sphere by co-immobilization of multi-enzyme directly from cell lysates as an efficient and recyclable biocatalyst for asymmetric reduction with coenzyme regeneration in situ. *Int. J. Biol. Macromol.* **2017**, *103*, 424–434. [CrossRef] [PubMed]

33. Maniam, P.; Stock, N. Investigation of porous Ni-based metal-organic frameworks containing paddle-wheel type inorganic building units via high-throughput methods. *Inorg. Chem.* **2011**, *50*, 5085–5097. [CrossRef] [PubMed]

34. He, J.; Sun, S.; Zhou, Z.; Yuan, Q.; Liu, Y.; Liang, H. Thermostable enzyme-immobilized magnetic responsive Ni-based metal-organic framework nanorods as recyclable biocatalysts for efficient biosynthesis of S-adenosylmethionine. *Dalton Trans.* **2019**, *48*, 2077–2085. [CrossRef] [PubMed]

35. Tong, X.; Guo, P.; Liao, S.; Xue, S.; Zhang, H. Nickel-Catalyzed intelligent reductive transformation of the aldehyde group using hydrogen. *Green Chem.* **2019**, *3*, 7. [CrossRef]

36. Cao, Y.; Wu, Z.; Wang, T.; Xiao, Y.; Huo, Q.; Liu, Y. Immobilization of Bacillus subtilis lipase on a Cu-BTC based hierarchically porous metal-organic framework material: A biocatalyst for esterification. *Dalton Trans.* **2016**, *45*, 6998–7003. [CrossRef]

37. Monnard, F.W.; Nogueira, E.S.; Heinisch, T.; Schirmer, T.; Ward, T.R. Human carbonic anhydrase II as host protein for the creation of artificial metalloenzymes: The asymmetric transfer hydrogenation of imines. *Chem. Sci.* **2013**, *4*, 3269–3274. [CrossRef]

38. Bradford, M.M. A rapid method for the quantitation of microgram quantities of protein utilizing the principle of protein-dye binding. *Anal. Biochem.* **1976**, *72*, 248–254. [CrossRef]

 *catalysts*

Communication

Enhancing the Water Resistance of Mn-MOF-74 by Modification in Low Temperature NH$_3$-SCR

Sheng Wang [1,2,†], Qiang Gao [1,2,†], Xiuqin Dong [1,2], Qianyun Wang [1,2], Ying Niu [1,2], Yifei Chen [1,2] and Haoxi Jiang [1,2,*]

[1] Key Laboratory for Green Chemical Technology of Ministry of Education, R&D Center for Petrochemical Technology, Tianjin University, Tianjin 300072, China; shw@tju.edu.cn (S.W.); qgao@tju.edu.cn (Q.G.); xqdong@tju.edu.cn (X.D.); qywang@tju.edu.cn (Q.W.); yniu@tju.edu.cn (Y.N.); yfchen@tju.edu.cn (Y.C.)

[2] Collaborative Innovation Center of Chemical Science and Engineering, Tianjin 300072, China

* Correspondence: hxjiang@tju.edu.cn; Tel.: +86-22-2740-6119

† S.W. and Q.G. contributed equally.

Received: 4 November 2019; Accepted: 27 November 2019; Published: 29 November 2019

Abstract: In this study, Mn-MOF-74 was successfully synthesized and further modified via two paths for enhanced water resistance. The structure and morphology of the modified samples were investigated by a series of characterization methods. The results of selective catalytic reduction (SCR) performance tests showed that polyethylene oxide-polypropylene-polyethylene oxide (P123)-modified Mn-MOF-74 exhibited outstanding NO conversion of up to 92.1% in the presence of 5 vol.% water at 250 °C, compared to 52% for Mn-MOF-74 under the same conditions. It was concluded that the water resistance of Mn-MOF-74 was significantly promoted after the introduction of P123 and that the unmodified P123-Mn-MOF-74 was proven to be a potential low-temperature SCR catalyst.

Keywords: Mn-MOF-74; modification; water resistance; NH$_3$-SCR performance

1. Introduction

As one of the major air pollutants, nitrogen oxides (NO$_x$) are considered to cause a series of environmental problems, such as acid rain, smog, and greenhouse effects [1]. The selective catalytic reduction (SCR) of NO$_x$ with NH$_3$ (NH$_3$-SCR) is one of the most effective techniques for reducing NO$_x$ from stationary resources caused by fossil fuel combustion (e.g., coal-burning power plants) [2]. In the past few years, considerable research work has been performed on the development of high efficiency catalysts for NH$_3$-SCR [3]. V$_2$O$_5$-WO$_3$-TiO$_2$ catalysts have been widely employed as commercial catalysts; nevertheless, the disadvantages of these catalysts are their high operating temperatures (300-400 °C) and the toxicity of vanadium [4]. Hence, it is of great importance to find a new type of catalyst which is able to effectively remove NO$_x$ at low temperature since the catalyst is located behind the desulfurizer and electrostatic precipitator system to reduce the cost of NH$_3$-SCR. A series of transition metal oxides such as MnO$_x$, FeO$_x$, CoO$_x$, and CeO$_x$ supported on different carriers have been studied to raise low temperature activity [2,5,6]. Notably, manganese-oxide-based catalysts have shown promising catalytic activity among the studied catalysts. However, the relatively lower specific surface area of traditional carriers might hinder the further application of manganese-oxide-based catalysts. Thus, it is especially important to find a carrier substitute with a large specific surface area.

Among the potential candidates, metal–organic frameworks (MOFs) have attracted significant attention due to their large specific surface area, low density, high chemical tenability, and controlled structure. It has been reported that a Ni-MOF activated at 220 °C achieved a NO conversion efficiency of over 92% for a large operating-temperature range of 275 to 440 °C [7]. MOF-74 is a potential support for the low-temperature NH$_3$-SCR process owing to its coordinatively unsaturated metal sites (CUSs), highly dispersed and absolute exposed metal sites, large specific surface areas, and high porosity.

According to our previous work [8], the Mn-MOF-74 catalyst has good catalytic performance for the low-temperature reaction of NH_3-SCR, and we found the NO conversion to be nearly 99% at 220 °C. However, the NO conversion of Mn-MOF-74 was observed to drop by 44% after the adding of 5% water, which hinders the further application of these materials. It has been reported that the presence of water could reduce the capacity of gas adsorption significantly and destroy the crystal structure of MOF-74 [9]. It is therefore necessary to improve the water stability of the prepared MOF materials under the premise of maintaining good denitrification performance.

At present, two major strategies are being employed to raise the water-resistance of MOFs to expand their applications. The most effective strategy for preparing MOFs with water stability is to introduce strong coordination bonds [10]; another method is to install a hydrophobic moiety around the coordination sites or on the working surface of the crystal to prevent corrosion from water molecules [11–13]. It has been proven that the latter is an efficient way to enhance the water resistance of MOF materials. Wu and his co-authors enhanced the water resistance of IRMOF1 by adding water-repellent functional groups [12].

In this work, as-synthesized Mn-MOF-74 catalysts were modified via two methods to promote water resistance. One strategy used was to cover water-in-oil surfactants polyethylene oxide-polypropylene-polyethylene oxide (P123) or polyvinylpyrrolidone (PVP) on the surface of Mn-MOF-74 to increase the external surface hydrophobicity of MOFs, producing P123-Mn-MOF-74 and PVP-Mn-MOF-74. The other was to introduce hydrophobic groups (-CH_3) to the ligand of Mn-MOF-74 to synthesize Mn-MOF-74-CH_3 through a post-synthesis modification (PSM) method. In addition, the effects of these groups on crystal structure, morphology, and thermal stability were investigated by powder X-ray diffraction (PXRD), FT-IR, SEM, TEM, and thermogravimetry mass spectrometry (TG-MS), et al. In addition, NH_3-SCR performances with low temperature and water resistance for the prepared catalysts were studied.

2. Results and Discussion

2.1. Characterization of MOFs

Figure 1 shows the PXRD patterns of Mn-MOF-74, P123-Mn-MOF-74, PVP-Mn-MOF-74, and Mn-MOF-74-CH_3. As can be seen from this figure, the two strongest peaks at 2θ = 6.5–7° and 11.5–12° of all the samples correspond to the crystal faces (2-10) and (300), respectively, which are characteristic diffraction peaks of MOF-74. The relative strength of the two diffraction peaks of P123-Mn-MOF-74 and PVP-Mn-MOF-74 accord with the standard pattern while the peak intensity of Mn-MOF-74-CH_3 at 2θ = 11.5–12° is higher than that at 2θ = 6.5–7°. The patterns of -CH_3-functionalized Mn-MOF-74 are inconsistent with the standard pattern, which was probably caused by the synthesis method. In other words, the harsher preparation conditions of Mn-MOF-74-CH_3 did have an influence on the structure of MOF-74, but the main crystal structure was maintained. In summary, the introduction of surfactants P123 and PVP and methyl had little effect on the crystal structure of MOF-74.

To further investigate the influence of modification on catalyst structure, FT-IR tests were carried out on Mn-MOF-74, P123/PVP-Mn-MOF-74, and Mn-MOF-74-CH_3 (the spectra are displayed in Figure S1). The peak at 1625 cm^{-1} can be assigned to ν(-COO) of DHTP (2,5-dihydroxyterephthalic acid) [13]. The bands at 1558 cm^{-1} and 1400 cm^{-1} can be assigned to ν(C=C), which belongs to the skeleton vibration of the benzene ring [14,15] of DHTP. The peak at 1245 cm^{-1} can be assigned to ν(C-N) of DMF (N, N-Dimethyl formamide) [16,17], demonstrating that the solvent DMF exists on the catalysts' surface or in the channel of the catalysts. The bands at 888 cm^{-1} and 811 cm^{-1} can be attributed to a ν(C-H) oscillatory vibration in and out of the plane of the benzene ring [18]. The peaks at 579 cm^{-1} and 473 cm^{-1} belong to the vibration absorption ν(Mn-O) [19,20], which confirms the existence of a metal–ligand bond. Notably, as shown in Figure S1b, Mn-MOF-74-CH_3 has a weak absorption peak at 2960 cm^{-1}, which belongs to the stretching vibration absorption of ν(C-H) in alkanes [19], proving that

-CH$_3$ had been introduced into the structure of Mn-MOF-74. According to the results of FT-IR and XRD, it can be concluded that the three catalysts were synthesized successfully.

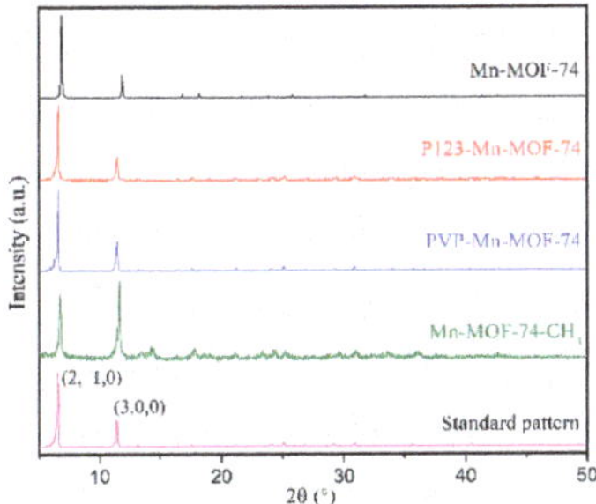

Figure 1. The powder X-ray diffraction (PXRD) patterns of Mn-MOF-74, polyethylene oxide-polypropylene-polyethylene oxide (P123)/polyvinylpyrrolidone (PVP)-Mn-MOF-74, and Mn-MOF-74-CH$_3$.

As can be seen from Figure S2, the four samples showed spherical particles. The morphology of Mn-MOF-74-CH$_3$ displayed almost no change except for there being a few tiny block crystals on the surface; the surfaces of P123-Mn-MOF-74 and PVP-Mn-MOF-74 were almost wrapped with a mass of tiny particles compared to Mn-MOF-74. The results suggest that the original spherical morphology of Mn-MOF-74 remained unchanged whether the water-in-oil surfactants P123 and PVP were introduced or ligand methyl functionalization was performed.

The edge of each sample was observed using TEM characterization techniques and the results are shown in Figure 2. It may be noted that there is no coating on the edges of Mn-MOF-74 and Mn-MOF-74-CH$_3$, while the edges of P123-Mn-MOF-74 and PVP-Mn-MOF-74 are obvious, which should be due to the coating of water-in-oil surfactants P123 and PVP, respectively. Compared with PVP-Mn-MOF-74, the P123-modified Mn-MOF-74 showed better surfactant dispersion and the particle size of the latter was more uniform. In summary, it can be speculated that the surfactants P123 and PVP were coated on the surface of the Mn-MOF-74 successfully.

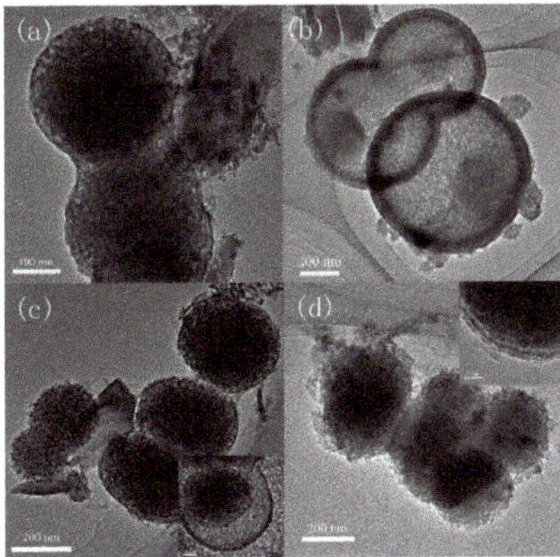

Figure 2. TEM images of (**a**) Mn-MOF-74, (**b**) P123-Mn-MOF-74, (**c**) PVP-Mn-MOF-74, and (**d**) Mn-MOF-74-CH$_3$. Illustration: edge of the related samples.

The prepared catalysts were analyzed by thermogravimetric analysis in air and under a nitrogen atmosphere and the results are shown in Figure S3a,b. Taking Mn-MOF-74-CH$_3$ as an example, the ion fragments of decomposition products in the air atmosphere and N_2 atmosphere were detected by mass spectrometry, as shown in Figure S3c,d.

From Figure S3a, it can be observed that the weight loss of the three catalysts showed two significant stages under air conditions. In the first stage, the mass loss ratio of P123-Mn-MOF-74 can be seen to be about 20% near 235 °C, which can be attributed to the removal of $CHCl_3$, while that of PVP-Mn-MOF-74 and Mn-MOF-74-CH_3 are about 20% at 275 °C, which could be ascribed to methanol and $CHCl_3$ (as can be seen in Figure S3c), respectively. The second weight loss of all the samples is about 35% at about 320 °C, which is due to the total collapse of the skeleton structure. To sum up, it can be seen from the results that the three catalysts can maintain the integrity of the crystal structure when stored in air.

As shown in Figure S3b, there are three clear mass losses for the three catalysts in the N_2 atmosphere. The first stage is about 10% near 105 °C for the Mn-MOF-74-CH_3 sample, which can be attributed to the removal of $CHCl_3$ only (no water, methanol, or DMF ion fragmentations can be detected from the MS results shown in Figure S3d), suggesting that solvent molecules in the tunnel were almost discharged in the pre-preparation evacuation process. For P123-Mn-MOF-74 and PVP-Mn-MOF-74, it is known from the preparation method that the mass loss can be ascribed to the removal of $CHCl_3$ and methanol. The second stage is in the vicinity of 317 °C and is caused by the partial collapse of the skeleton structure from the MS results of Figure S3d. The third stage appears above 535 °C, suggesting the complete collapse of the MOF-74 skeleton structure. Combined with TG and MS results, the activation conditions of the samples should be set as N_2 atmosphere, 200 °C for 3 h, and a heating rate of 2 °C/min.

2.2. The Low-Temperature SCR Performance

Low-temperature SCR catalytic performance was tested to investigate the effects of the surfactants P123 and PVP and ligand methyl functionalization on the catalytic performance of Mn-MOF-74 at low temperature. As shown in Figure 3, Mn-MOF-74 performed well with regard to low-temperature NH_3-SCR activity, but the water resistance still needs further study. As the temperature increased, the NO conversion of all the samples exhibited a rising trend; when the temperature rose to 280 °C, the NO conversion of Mn-MOF-74-CH_3 reached a maximum of 93.2%. The NO conversion of P123-Mn-MOF-74 and PVP-Mn-MOF-74 reached maxima of 92.1% and 71.8% at 265 °C and 250 °C, respectively, and then decreased with the continuous increase in temperature, which was caused by the collapse of the skeletal structure. Compared with the low-temperature SCR catalytic performance curve of Mn-MOF-74, the performance order was Mn-MOF-74 > Mn-MOF-74-CH_3 > P123-Mn-MOF-74 > PVP-Mn-MOF-74. The surfactants P123 and PVP and the methyl ligands existing in the pores may even have been partially covered in the metal active center, hindering the approaches of reactant molecules and thus depressing the SCR catalytic performance.

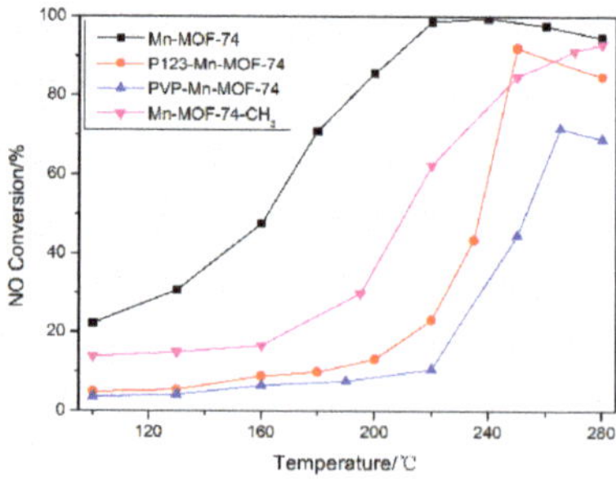

Figure 3. Low-temperature selective catalytic reduction (SCR) activities of P123-Mn-MOF-74, PVP-Mn-MOF-74, and Mn-MOF-74-CH_3 (gas flow rate: 100 mL/min; gas composition: NO 500 ppm, NH_3 500 ppm, O_2 5%, and Ar as balance gas).

2.3. Stability and Water Resistance Study

In this work, stability tests were carried out at temperatures corresponding to the maximum NO conversion of catalysts (MOF-74 240 °C, P123-Mn-MOF-74 250 °C, PVP-Mn-MOF-74 265 °C, and Mn-MOF-74-CH$_3$ 280 °C) for 12 h; the results are displayed in Figure 4. It was concluded that all the samples showed good stability at their own optimal temperature during the 12 h tests.

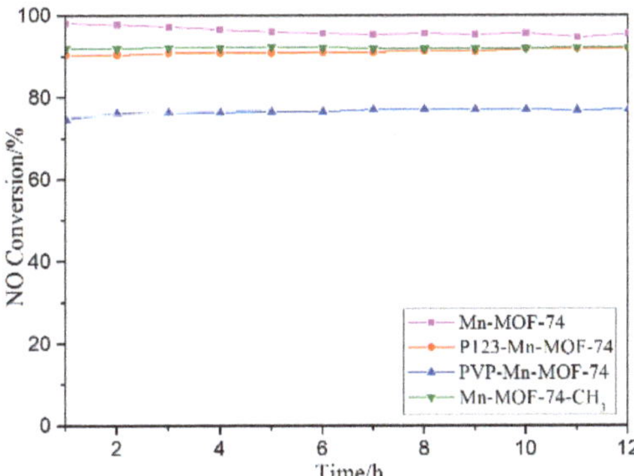

Figure 4. Stability test results of Mn-MOF-74, P123-Mn-MOF-74, PVP-Mn-MOF-74, and Mn-MOF-74-CH$_3$ under an SCR atmosphere at their own optimal conditions (gas flow rate: 100 mL/min; gas composition: NO 500 ppm, NH$_3$ 500 ppm, O$_2$ 5%, H$_2$O (g) 5%, and Ar as balance gas).

To study the water resistance of the prepared samples, 5 vol.% H$_2$O was added into the feed gas and low-temperature catalytic activities were investigated (as shown in Figure 5). For a clearer explanation, the maximum NO conversion and the corresponding temperature, stability, and water resistance of Mn-MOF-74, P123-Mn-MOF-74, PVP-Mn-MOF-74, and Mn-MOF-74-CH$_3$ have been summarized in Table 1.

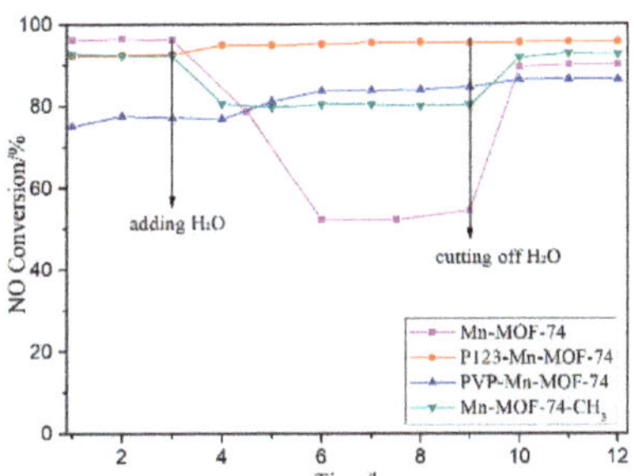

Figure 5. Stability test results of Mn-MOF-74, P123-Mn-MOF-74, PVP-Mn-MOF-74, and Mn-MOF-74-CH$_3$ under an SCR-H$_2$O atmosphere at their own optimal conditions (gas flow rate: 100 mL/min; gas composition: NO 500 ppm, NH$_3$ 500 ppm, O$_2$ 5%, H$_2$O (g) 5%, and Ar as balance gas).

According to the results given in Figure 5 and Table 1, the NO conversion of Mn-MOF-74 decreased by 44% after adding H$_2$O. Combined with the SCR activity results, Mn-MOF-74 did show better catalytic activity, while the water resistance of it was very poor. The reason for this might have been that OH species of the dissolved H$_2$O were bonded to the exposed CUSs, thereby reducing the number of active sites and reducing the activity [21]. After modification, the NO conversion of 71.8% for PVP-Mn-MOF-74 and 81.2% for Mn-MOF-74-CH$_3$ can be observed to be higher than that of 55% for Mn-MOF-74 in the SCR-H$_2$O atmosphere, and then be recovered to the original level when removing H_2O, suggesting that the water resistance of PVP-Mn-MOF-74 and Mn-MOF-74-CH$_3$ increased greatly. Notably, the NO conversion of P123-modified Mn-MOF-74 remained at 92.1% after the addition of 5 vol.% H_2O, which indicates that the introduction of P123 did enhance the water resistance of Mn-MOF-74.

Table 1. The effects of adding H_2O on the low-temperature SCR catalytic activities of Mn-MOF-74, P123-Mn-MOF-74, PVP-Mn-MOF-74, and Mn-MOF-74-CH$_3$.

Sample	Temperature (°C)	NO Conversion (%) (SCR)	Change in NO Conversion (%) (SCR and H_2O)
Mn-MOF-74	240	99.6	−44
P123-Mn-MOF-74	250	92.1	0
PVP- Mn-MOF-74	265	71.8	+4
Mn-MOF-74-CH$_3$	280	93.2	−12

3. Materials and Methods

3.1. Materials

Manganese chloride (MnCl$_2$·4H$_2$O, purity ≥ 99%, Guangfu Fine Chemical Research Institute, Tianjin, China), 2,5-dihydroxyterephthalic acid (C$_8$O$_6$H$_6$, 98%, DHTP, Heowns Biochem LLC, Tianjin, China), (N, N-Dimethyl formamide (HCON(CH$_3$)$_2$, 99.5%, DMF, Jiangtian Chemical Technology Co., Ltd., Tianjin, China), ethanol (CH$_3$CH$_2$OH, ≥ 99.7%, Jiangtian Chemical Technology Co, Ltd., Tianjin, China), methanol (CH$_3$OH, ≥ 99.7%, Jiangtian Chemical Technology Co., Ltd., Tianjin, China), chloroform (CHCl$_3$, ≥ 99%, Jiangtian Chemical Technology Co., Ltd., Tianjin, China), polyvinylpyrrolidone ((C$_6$H$_9$NO)$_n$, Sigma-Alorich, Shanghai, China), and polyethylene oxide-polypropylene-polyethylene oxide (PEO-PPO-PEP, Sigma-Aldrich, Shanghai, China) were used. The gas mixtures used in this work, including NO/Ar, NH$_3$/Ar, SO$_2$/Ar, and O$_2$/Ar, were supported by Beijing AP BAIF Gases Industry Co., Ltd. All the materials and gas mixtures were used without further processing unless noted.

3.2. Catalyst Preparation

According to the procedure described by Zhou et al. [22], Mn-MOF-74 was synthesized using a hydrothermal method. Briefly, 53 mL of DMF, 3.5 mL of ethanol, and 3.5 mL of H_2O were added to a 100 mL beaker and then the ligands DHTP was added into the above mixture. MnCl$_2$·4H$_2$O (0.4396 g, 2.22 mmol) was added into the solution, dispersed by ultrasound for 0.5 h, and then transferred to a Teflon-lined hydrothermal reactor. The hydrothermal reactor was placed in an oven at 135 °C for 24 h. The red–brown precipitates produced were cooled to room temperature, centrifuged, and washed three times with 60 mL of methanol. Finally, the product was dried and preserved for use.

The Mn-MOF-74 samples were pretreated for 3 h at a heating rate of 2 °C/min in a nitrogen atmosphere at 200 °C to remove the residual solvent. P123-Mn-MOF-74 and PVP-Mn-MOF-74 were synthesized by an impregnation method reported in the literature [23]. One gram of P123 or PVP was dissolved in solvent and then 0.2 g of pre-treated Mn-MOF-74 was added into the solution. The mixture was allowed to rest under static conditions at room temperature for 24 h, and then was washed with CHCl$_3$. The initial product was filtered and dried naturally.

Mn-MOF-74-CH$_3$ was synthesized by a post-synthesis modification method referred to in the synthesis method of CH$_3$-MOF-5 [24] but with some modification. Generally, a mixture of 15 mL CHCl$_3$ and 0.2 g pre-treated Mn-MOF-74 was placed in a Teflon-lined hydrothermal reactor and then placed in an oven at 65 °C for 12 h. The initial product was washed with CHCl$_3$ three times to remove unreacted species. The solids were dried at room temperature and stored in a desiccator for use.

3.3. Characterization

PXRD analysis was carried out to study the crystal phases. The instrument used in the PXRD tests was a D/MAX 2500 X-ray diffraction analyzer manufactured by RIGAKU. Analysis conditions were Cu target, scintillation counter tube (SC) detector, tube voltage 40 kV, tube current 200 mA, continuous scanning measurement, time constant 2, scan range 5~90°, and scan speed 5°/min. TG-MS was carried out under air conditions and a nitrogen atmosphere with a flow rate of 50 mL/min, and the temperature ranged from 30 °C to 700 °C with a heating rate of 10 °C/min.

In this study, the morphology of the catalysts was observed and analyzed using a Nanosem 430 scanning electron microscope manufactured by FEI. The instrument parameters were resolution 1 nm, acceleration voltage 0.1~30 kV, magnification 20~800,000, low vacuum, and low voltage deceleration mode. TEM was carried out on a Tecnai G2 F20 instrument made in America. The instrument parameters were accelerating voltage 200 kV, point resolution 0.248 nm, line resolution 0.102 nm, and magnification 1.05×10^6 times. FTIR was utilized to study the surface group of the catalysts on a Nicolet 6700 FTIR infrared spectrometer. The scans were taken 32 times and the scanning wave number was 400~4000 cm^{-1} with a resolution of 4 cm^{-1}.

3.4. Catalytic Activity

Low-temperature SCR catalytic performance of the as-prepared samples was tested in a fixed-bed reactor at atmosphere pressure. In a typical reaction test, 0.2 g of the samples and 0.4 g of quartz sand were loaded into the reactor tube. Before the activity test, the samples had to be pretreated under an N_2 atmosphere to remove the solvent and get activated. The pretreated conditions were as follows: 200 °C, 3 h, and heating rate 2 °C/min. After the pretreatment, the samples were tested for a temperature range of 80–280 °C under ambient conditions. The total flow rate of the feed gas was 100 mL/min, while the gas hourly space velocity (GHSV) was about 50,000 h^{-1}. The composition of the gas flow was as follows: 500 ppm NO, 500 ppm NH_3, 5% O_2, 5 vol.% H_2O (when used), and Ar as a balance gas.

The concentration of outlet NO and NO_2 were determined using KM9106 Quintox Kane International Limited, a flue gas analyzer. The concentrations of the products N_2 and N_2O were monitored by gas chromatography (Agilent GC-6820) at 30 °C when effluent gases passed through it. A 5 Å molecular sieve column and Porapak Q column were used, respectively. The calculation methods used were

$$X = \frac{[NO_x]_{in} - [NO_x]_{out}}{[NO_x]_{in}} \times 100\%$$

$$S = \frac{[N_2]_{out}}{[N_2]_{out} + [N_2O]_{out}} \times 100\%$$

In the formulas, X represents the conversion of NO_x, S represents the N_2 selectivity, $[x]$ represents the concentration of each gas, in represents the inlet gas, and out represents the outlet gas. In our work, the N_2 selectivity of our catalysts was 100%.

4. Conclusions

In this study, Mn-MOF-74 catalysts were modified by introducing surfactants P123 and PVP and by ligand methyl functionalization to enhance their water resistance. The structure and morphology of the samples were investigated by PXRD, FT-IR, SEM, TEM and TG-MA while low-temperature SCR performance was tested in a fix-bed reactor. It was concluded that the modified samples were successfully synthesized and that the water resistance of the samples was significantly enhanced after modification. Moreover, all the catalysts were able to remain stable for at least 12 h at their own optimal temperature in an SCR atmosphere. After the addition of 5% H_2O in the feed gas, the NO conversion of P123-Mn-MOF-74 remained at 92.1% at 250 °C while that of unmodified Mn-MOF-74 dropped by 44%. In summary, P123-Mn-MOF-74 has superior low temperature SCR activity, SCR stability, and water resistance, and its NO conversion rate can be maintained above 90%, which proves it to be a potential low-temperature NH_3-SCR catalyst.

Supplementary Materials: The following are available online at http://www.mdpi.com/2073-4344/9/12/1004/s1, Figure S1: FT-IR spectra of P123/PVP-Mn-MOF-74 and Mn-MOF-74-CH$_3$ at (**a**) 4000~400 cm^{-1} and (**b**) 4000–2500 cm^{-1}; Figure S2: SEM images of (**a**) Mn-MOF-74, (**b**) P123-Mn-MOF-74, (**c**) PVP-Mn-MOF-74, and (**d**) Mn-MOF-74-CH$_3$; Figure S3: TG curves of Mn-MOF-74, P123-Mn-MOF-74, PVP-Mn-MOF-74, and Mn-MOF-74-CH$_3$ under atmospheres of (**a**) air and (**b**) N$_2$. MS results of Mn-MOF-74-CH$_3$ under atmospheres of (**c**) air and (**d**) N$_2$.

Author Contributions: Conceptualization and methodology, S.W. and Q.G.; writing—original draft preparation, Q.W. and Y.N.; writing—review and editing, X.D. and Y.C.; supervision, H.J.; S.W., and Q.G. (these authors made equal contributions).

Funding: This research was funded by the National Natural Science Foundation of China, grant number 21506150.

Acknowledgments: The authors are grateful for financial support from the National Natural Science Foundation of China (no. 21506150). This project was sponsored by the Scientific Research Foundation for Returned Overseas Chinese Scholars, State Education Ministry.

Conflicts of Interest: The authors declare no conflict of interest.

References

1. Chen, H.Y.; Wei, Z.; Kollar, M.; Gao, F.; Wang, Y.; Szanyi, J.; Peden, C.H.F. A comparative study of N_2O formation during the selective catalytic reduction of NOx with NH_3 on zeolite supported Cu catalysts. *J. Catal.* **2015**, *329*, 490–498. [CrossRef]

2. Hu, H.; Cai, S.; Li, H.; Huang, L.; Shi, L.; Zhang, D. In Situ DRIFTs Investigation of the Low-Temperature Reaction Mechanism over Mn-Doped Co_3O_4 for the Selective Catalytic Reduction of NOx with NH_3. *J. Phys. Chem. C* **2015**, *119*, 22924–22933. [CrossRef]

3. Shan, W.; Song, H. Catalysts for the selective catalytic reduction of NOx with NH_3 at low temperature. *Catal. Sci. Technol.* **2015**, *5*, 4280–4288. [CrossRef]

4. Li, J.; Chang, H.; Ma, L.; Hao, J.; Yang, R.T. Low-temperature selective catalytic reduction of NOx with NH_3 over metal oxide and zeolite catalysts-A review. *Catal. Today* **2011**, *175*, 147–156. [CrossRef]

5. Casapu, M.; Cher, O.K.; Elsener, M. Screening of doped MnO_x-CeO_2 catalysts for low-temperature NO-SCR. *Appl. Catal. B* **2009**, *88*, 413–419. [CrossRef]

6. Chen, Z.H.; Li, X.-H.; Yang, Q.; Li, H.; Gao, X.; Jiang, Y.-B.; Wang, F.-R.; Wang, L.-F. Removal of NOx Using Novel Fe-Mn Mixed-Oxide Catalysts at Low Temperature. *Acta Phys.-Chim. Sin.* **2009**, *25*, 601–605.

7. Sun, X.; Shi, Y.; Zhang, W.; Li, C.; Zhao, Q.; Gao, J.; Li, X. A new type Ni-MOF catalyst with high stability for selective catalytic reduction of NOx with NH_3. *Catal. Commun.* **2018**, *114*, 104–108. [CrossRef]

8. Jiang, H.; Wang, C.; Wang, H.; Zhang, M. Synthesis of highly efficient MnO_x catalyst for low-temperature NH_3-SCR prepared from Mn-MOF-74 template. *Mater. Lett.* **2016**, *168*, 17–19. [CrossRef]

9. Zuluaga, S.; Fuentes-Fernandez, E.M.A.; Tan, K.; Xu, F.; Li, J.; Chabal, Y.J.; Thonhauser, T. Understanding and controlling water stability of MOF-74. *J. Mater. Chem. A* **2016**, *4*, 5176–5183. [CrossRef]

10. Bai, Y.; Dou, Y.; Xie, L.; Rutledge, W.; Li, J.; Zhou, H. Zr-based metal-organic frameworks: Design, synthesis, structure, and applications. *Chem. Soc. Rev.* **2016**, *45*, 2327–2367. [CrossRef]

11. Yang, C.; Kaipa, U.; Mather, Q.Z.; Wang, X.; Nesterov, V.; Venero, A.F.; Omary, M.A. Fluorous Metal-Organic Frameworks with Superior Adsorption and Hydrophobic Properties toward Oil Spill Cleanup and Hydrocarbon Storage. *J. Am. Chem. Soc.* **2011**, *133*, 18094–18097. [CrossRef] [PubMed]

12. Wu, T.; Shen, L.; Luebbers, M.; Hu, C.; Chen, Q.; Ni, Z.; Masel, R.I. Enhancing the stability of metal-organic frameworks in humid air by incorporating water repellent functional groups. *Chem. Commun.* **2010**, *46*, 6120–6122. [CrossRef]

13. Makal, T.A.; Wang, X.; Zhou, H.C. Tuning the Moisture and Thermal Stability of Metal-Organic Frameworks through Incorporation of Pendant Hydrophobic Groups. *Cryst. Growth Des.* **2013**, *13*, 4760–4768. [CrossRef]

14. Ranjbar, M.; Taher, M.A. Preparation of Ni-metal organic framework-74 nanospheres by hydrothermal method for SO_2 gas adsorption. *J. Porous Mater.* **2016**, *23*, 1249–1254. [CrossRef]

15. Chavan, S.; Bonino, F.; Valenzano, L.; Civalleri, B.; Lamberti, C.; Acerbi, N.; Cavka, J.H.; Leistner, M.; Bordiga, S. Fundamental Aspects of H_2S Adsorption on CPO-27-Ni. *J. Phys. Chem. C* **2013**, *117*, 15615–15622. [CrossRef]

16. Jiang, H.; Wang, Q.; Wang, H.; Chen, Y.; Zhang, M. MOF-74 as an Efficient Catalyst for the Low-Temperature Selective Catalytic Reduction of NOx with NH$_3$. *ACS Appl. Mater. Interfaces* **2016**, *8*, 26817–26826. [CrossRef]
17. Jiang, H.; Wang, Q.; Wang, H.; Chen, Y.; Zhang, M. Temperature effect on the morphology and catalytic performance of Co-MOF-74 in low-temperature NH$_3$-SCR process. *Catal. Commun.* **2016**, *80*, 24–27. [CrossRef]
18. Bernini, M.C.; García Blanco, A.A.; Villarroelrocha, J.; Fairenjimenez, D.; Sapag, K.; Ramirezpastor, A.J.; Narda, G.E. Tuning the target composition of amine-grafted CPO-27-Mg for capture of CO$_2$ under post-combustion and air filtering conditions: A combined experimental and computational study. *Dalton Trans.* **2015**, *44*, 18970–18982. [CrossRef]
19. Tikhomirov, K.; Kröcher, O.; Elsener, M.; Wokaun, A. MnOx-CeO$_2$ mixed oxides for the low-temperature oxidation of diesel soot. *Appl. Catal. B* **2006**, *64*, 72–78. [CrossRef]
20. Wang, X.; Li, X.; Zhao, Q.; Sun, W.; Tade, M.; Liu, S. Improved activity of W-modified MnOx-TiO$_2$ catalysts for the selective catalytic reduction of NO with NH$_3$. *Chem. Eng. J.* **2016**, *288*, 216–222. [CrossRef]
21. Tan, K.; Zuluaga, S.; Gong, Q.; Canepa, P.; Wang, H.; Li, J.; Chabal, Y.J.; Thonhauser, T. Water Reaction Mechanism in Metal Organic Frameworks with Coordinatively Unsaturated Metal Ions: MOF-74. *Chem. Mater.* **2014**, *26*, 6886–6895. [CrossRef]
22. Wei, Z.; Hui, W.; Taner, Y. Enhanced H$_2$ adsorption in isostructural metal-organic frameworks with open metal sites: Strong dependence of the binding strength on metal ions. *J. Am. Chem. Soc.* **2008**, *130*, 15268.
23. Fernandez, C.A.; Nune, S.K.; Annapureddy, H.V.; Dang, L.X.; Mcgrail, B.P.; Zheng, F.; Polikarpov, E.; King, D.L.; Freeman, C.; Brooks, K.P. Hydrophobic and moisture-stable metal-organic frameworks. *Dalton Trans.* **2015**, *44*, 13490–13497. [CrossRef]
24. Jie, Y.; Anna, G.; Mulder, F.M.; Dingemans, T.J. Methyl modified MOF-5: A water stable hydrogen storage material. *Chem. Commun.* **2011**, *47*, 5244–5246.

Article

Surface Modification of a MOF-based Catalyst with Lewis Metal Salts for Improved Catalytic Activity in the Fixation of CO_2 into Polymers

Sudakar Padmanaban and Sungho Yoon *

Department of Chemistry, Chung-Ang University, Seoul 156-756, Korea; sudakar.p@outlook.com
* Correspondence: sunghoyoon@cau.ac.kr

Received: 10 September 2019; Accepted: 23 October 2019; Published: 26 October 2019

Abstract: The catalyst zinc glutarate (ZnGA) is widely used in the industry for the alternating copolymerization of CO_2 with epoxides. However, the activity of this heterogeneous catalyst is restricted to the outer surface of its particles. Consequently, in the current study, to increase the number of active surface metal centers, ZnGA was treated with diverse metal salts to form heterogeneous, surface-modified ZnGA-Metal chloride (ZnGA-M) composite catalysts. These catalysts were found to be highly active for the copolymerization of CO_2 and propylene oxide. Among the different metal salts, the catalysts treated with $ZnCl_2$ (ZnGA-Zn) and $FeCl_3$ (ZnGA-Fe) exhibited ~38% and ~25% increased productivities, respectively, compared to untreated ZnGA catalysts. In addition, these surface-modified catalysts are capable of producing high-molecular-weight polymers; thus, this simple and industrially viable surface modification method is beneficial from an environmental and industrial perspective.

Keywords: heterogeneous catalysis; metal organic framework; surface modification; Zinc glutarate; CO_2 fixation; polycarbonate

---

1. Introduction

In recent decades, anthropogenic activities have dramatically increased the concentration of atmospheric CO_2; this concentration was found to be higher than 400 ppm in 2017 [1]. The rapid increase in the CO_2 content in the atmosphere makes CO_2 a major greenhouse gas. Consequently, the development of efficient and safe methods for capturing and sequestering CO_2 is garnering increased attention. Furthermore, the development of methods and processes for converting CO_2 into value-added chemicals is of paramount interest because CO_2 is a cheap, non-toxic, and abundant carbon feedstock [2,3]. Recently, a great deal of research has been devoted to capturing and utilizing CO_2 to synthesize a variety of value-added products [4–17].

One of the most sustainable strategies for utilizing CO_2 is the copolymerization of CO_2 with epoxides to produce poly(alkylene carbonates). These materials are commercially viable owing to their vast number of applications, such as in adhesives, packing and coating materials, and ceramic binders [18–22]. Furthermore, these polycarbonates are biodegradable and are useful in biomedical applications [19]. The alternating copolymerization of CO_2 with epoxides was first reported by Inoue et al., who used a diethylzinc–water system as a catalyst [23,24]. Subsequently, numerous homogeneous and heterogeneous catalytic systems have been developed. For example, homogeneous catalysts, such as metalloporphyrins, β-diiminate Zn complexes, and metal salen complexes have been found to be highly active for the copolymerization of CO_2 and epoxides [25–27]. Nevertheless, the industrial utilization of these homogeneous catalysts is limited because of their complicated syntheses, the use of toxic metals like chromium, and the difficulties in separating the catalyst/product mixtures. However, heterogeneous catalysts are preferred for industrial scale applications owing to their low cost, easy synthesis, facile separation from products, and reusability [28–31]. Among the heterogeneous catalysts, zinc dicarboxylates, Zn-Co double metal cyanide complexes, and ternary rare-earth complexes are found to be particularly active for the copolymerization of CO_2 and epoxides. Among them, the zinc glutarate (ZnGA) is widely applied in industry as a catalyst for the copolymerization of propylene oxide (PO) and CO_2 because it is economic, non-toxic, easy to synthesize, and, most beneficially, it yields copolymers with high molecular weights [18,32]. Based on a recent life cycle assessment study, for each one kg of CO_2 incorporated into polycarbonate polyols, up to three kg of CO_2-equivalent greenhouse gas emissions could be reduced [33]. Therefore, improving the productivity of ZnGA becomes necessary in order to meet the increasing economic and environmental requirements.

Despite a growing need and the continued efforts by researchers, the catalytic activity of ZnGA has improved only marginally over the past few decades. Rieger et al. reported on the considerable improvement of the catalytic activity of ZnGA via the introduction of zinc-ethylsulfinate initiator groups to its surface [34]. However, this post-modification has limited industrial viability because of its procedural complexity and use of expensive precursors. Ree et al., demonstrated the effects of different zinc and glutarate sources in the synthesis of ZnGA and consequently the effects of different morphologies of ZnGA on its catalytic activity for CO_2/PO copolymerization [35]. Through continuous research effort, the catalytic activity of ZnGA has been found to be dependent on its surface area and crystallinity [18,36–38]. Recently, the precise structure of ZnGA was obtained by single-crystal X-ray diffraction studies, thus revealing that its catalytic activity mainly originates on the outer surfaces of the Zn-dicarboxylate particles [39]. It should nevertheless be noted that the crystal structure of ZnGA features a 3D-network structure of glutarate ligands and Zn atom of the molecule in which, each Zn atoms tetrahedrally bind to the oxygen atoms of four different glutarate units.

According to the recent definition of metal organic framework (MOF) by Seth and Matzger, ZnGA can be considered a MOF [40]. It is noteworthy that most other Zn-based MOF systems have been reported to yield cyclic carbonate as the predominant product in the reaction of CO_2 and an epoxide, whereas the ZnGA system produces the poly(alkylene carbonate) as the major product [41–47]. In recent years, most studies on ZnGA have focused on increasing its surface area and crystallinity using different supporters, amphiphilic templates, and ultrasonic treatment [48–51]. Furthermore, a recent mechanistic study on CO_2/PO copolymerization using ZnGA indicated a bimetallic mechanism involving sequential insertions of CO_2 and epoxide into Zn-alkoxide and Zn-carboxylate initiator groups on the surface of the catalyst [52]. The optimal separation between two adjacent Zn atoms was suggested to be in the range 4.3–5.0 Å, which results in the optimal activation energy required for copolymerization. Recently, we explored the catalytic ability of ZnGA for the copolymerization of CO_2 with a relatively reluctant epoxide, epichlorohydrin, and the application of ZnGA as a catalyst for the terpolymerization of CO_2, PO, and β-butyrolactone [53,54].

As a part of our continued research effort to develop heterogeneous catalysts for CO_2 conversions, this study reports a facile method for preparing a surface modified ZnGA and its enhanced catalytic activity in the CO_2/PO polymerization. In order to improve the catalytic activity of ZnGA, the number of active sites on the surfaces of the catalyst particles must be increased. It has been reported that ZnGA has protruding glutarate and hydroxyl groups on its surface that act as initiators for CO_2/PO copolymerization [55,56]. These glutarate and hydroxyl groups can ligate additional incoming metal ions and create oxygen-bound metal centers on the surface of the ZnGA. Therefore, we hypothesized that surface modification of ZnGA by treatment with Lewis metal ions to form metal-treated ZnGA catalysts (ZnGA-M) would increase the number of active metal centers on the surface of the ZnGA, as depicted in Figure 1. This would provide catalysts with improved cooperative bimetallic properties for CO_2/PO copolymerization (Scheme 1). Indeed, the pore structure of some MOFs were modified by the installation of additional ligand motif to coordinate additional metal centers in a step called "post synthetic modification" (PSM) [57–61]. However, in the case of ZnGA frameworks, the protruding glutarate groups may be considered for this purpose. To test this hypothesis, different metal chlorides, i.e., $FeCl_3$, $AlCl_3$, $ZnCl_2$, and $CoCl_2$, were selected for the preparation of ZnGA-M catalysts based on their activities in homogeneous complexes species. The resulting catalysts were subsequently assessed for their activities in CO_2/PO copolymerization.

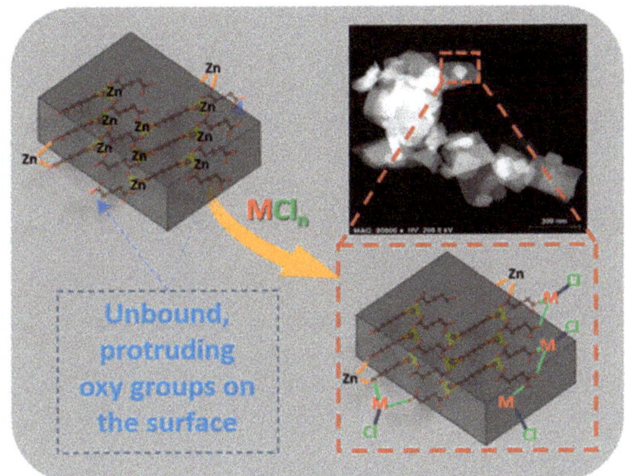

Figure 1. Schematic representation of the formation of ZnGA-M.

Scheme 1. Copolymerization of CO_2 and PO using ZnGA-M.

2. Results and Discussion

2.1. Synthesis and Characterization of Catalysts

Initially, standard ZnGA (std-ZnGA) was prepared according to a published procedure with slight modifications [62]. Figure S1 compares the powder X-ray diffraction (PXRD) pattern of the resultant white precipitate with the pattern calculated from the crystal structure and confirms the formation of std-ZnGA in its pure form with relatively high crystallinity (Figure S1). Fourier-transform infrared (FT-IR) spectroscopic analysis shows typical peaks for ZnGA. The CH_2 scissoring and CH stretching bands were observed at 1445 cm^{-1} and 2955 cm^{-1}. The bands at ~1585 cm^{-1} and ~1405 cm^{-1} correspond to COO$^-$ antisymmetric stretching frequencies. The COO$^-$ symmetric stretching band was observed at 1538 cm^{-1} (Figure S2a). Figure S3a shows a scanning electron microscopy (SEM) image in which the std-ZnGA has taken the form of typical platelet-shaped particles. These analyses collectively confirm the formation of std-ZnGA according to the previous reports in the literature.

The ZnGA-M catalysts were then prepared in anhydrous THF through treatment with different metal chloride solutions at different ratios, as shown in Table 1. The copolymerization of CO_2 and PO requires sequential insertion of CO_2 and PO into the Zn-alkoxide and Zn-carboxylate initiator groups on the surface of the catalyst. Therefore, the metal treatment should be kept as mild as possible to effect synchronized cooperative catalysis because a thick coating of metal chloride would block the monomers from approaching the catalytic sites. Thus, the ratio of metal to Zn was maintained in the range of 10^{-3}–10^{-4} equivalents in the metal treatment step.

The metal-treated ZnGAs were analyzed via PXRD, FT-IR spectroscopy, and SEM and TEM analyses. The FT-IR spectra of the ZnGA-M catalysts are shown in Figure S2a. The FT-IR analysis shows that the metal treatment does not affect the original glutarate binding, since the amount of each metal salt used is too small. As shown in Figure S2b, the amounts of metal ions used are too low to cause any phase changes in the crystal lattice or structural deformations. Thus, the PXRD patterns of all the catalysts are similar. Figure S3 shows the SEM images of the ZnGA-M samples and reveals that the metal treatments do not significantly change the morphology. The metal ions cannot be detected via SEM energy-dispersive X-ray spectroscopy (SEM-EDS) analysis when fewer than 10^{-2} equivalents of the metal ions are used. However, the scanning transmission electron microscopy (STEM) and energy-dispersive X-ray spectroscopy (EDS)-assisted elemental mapping of Al and Zn in ZnGA-Al-10^{-3} reveal the homogeneous distribution of Zn and Al metals in the ZnGA-Al-10^{-3} (Figure 2). In addition, the actual coated amount of Al in ZnGA-Al-10^{-3} was obtained via inductively coupled plasma optical emission spectroscopy (ICPOES), in which the ratio of Zn to Al was found to be 1:1.1 × 10^{-3}. Similarly, the ratio between Zn and Fe in ZnGA-Fe-10^{-3} was found to be 1:1.0 × 10^{-3}. These analyses confirmed the attachment of the different metal ions after metal treatment.

Table 1. Copolymerization of CO_2 with PO using ZnGA-M catalysts [a].

Entry	Catalyst	Zn:MCl$_n$ [b]	TON [c]	Productivity Increment [d] (%)	Fco$_2$ [e]	Selectivity (%) [f]		M$_n$ [g] (kg/mol)	PDI [g]	T$_g$ [h] (°C)
						PPC	PC			
1	Std-ZnGA	1:0	72.4	-	94.9	96.0	4.0	156.4	3.5	42
2	ZnGA-Fe-10^{-3}	1:10^{-3}	82.5	13.9	94.8	98.0	2.0	137.6	2.8	38
3	ZnGA-Fe-10^{-4}	1:10^{-4}	90.9	25.6	94.1	94.0	6.0	262.4	2.0	40
4	ZnGA-Al-10^{-3}	1:10^{-3}	88.7	22.5	93.5	96.0	4.0	196.3	1.8	36
5	ZnGA-Al-10^{-4}	1:10^{-4}	77.5	1.5	95.1	98.0	2.0	231.8	2.2	35
6	ZnGA-Zn-10^{-2}	1:10^{-2}	72.0	−0.5	93.7	98.0	2.0	99.8	3.9	42
7	ZnGA-Zn-10^{-3}	1:10^{-3}	100.1	38.3	94.9	97.0	3.0	208.0	1.8	38
8	ZnGA-Zn-10^{-4}	1:10^{-4}	80.7	11.5	91.7	96.0	4.0	147.8	2.2	37
9	ZnGA-Co-10^{-3}	1:10^{-3}	72.0	−0.5	92.2	95.0	5.0	144.4	1.9	34
10	ZnGA-Co-10^{-4}	1:10^{-4}	73.6	1.7	93.5	95.0	5.0	70.7	2.4	35
11	ZnGA-Al-1	1:1	8.1	-	36.2	91.0	9.0	-	-	-

[a] Conditions: 0.20 g catalyst; 20.0 mL PO; 2.0 MPa CO_2; 60 °C and 40 h. [b] Treated mol ratio of ZnGA and metal salt. [c] TON = weight of polymer formed per gram of catalyst. [d] Values represent the increment in productivity of the metal-treated catalysts in comparison to std-ZnGA. [e] Fco$_2$ of the polymers was determined from ^{1}H NMR spectra of the polymers. [f] Obtained from ^{1}H NMR spectra of the polymers; PC = propylene carbonate. [g] M$_n$, M$_w$, and PDI values of the polymers were determined using GPC with polystyrene standards in THF. [h] Determined from DSC.

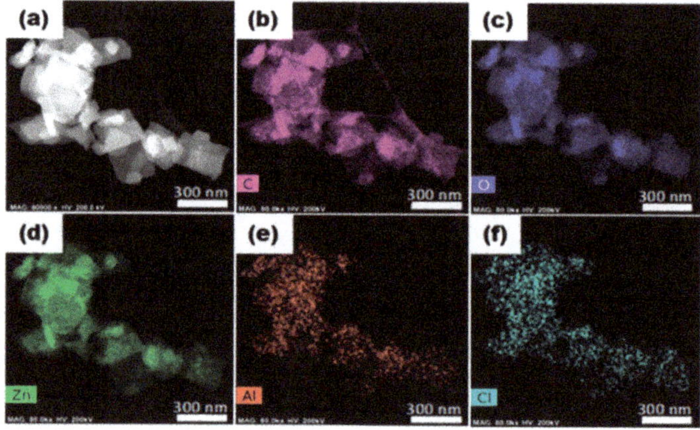

Figure 2. STEM image and EDS mapping. (**a**) STEM image of ZnGA-Al-10^{-3}; STEM-EDS mapping of (**b**) C (purple), (**c**) O (blue), (**d**) Zn (green), (**e**) Al (red), and (**f**) Cl (cyan) elements in ZnGA-Al-10^{-3}; (Scale bar: = 300 nm).

2.2. Catalytic Activity Studies

To see the effect of metal treatment in the copolymerization of CO_2 and PO, the catalytic activities of the metal-treated ZnGAs were assessed and compared with those of std-ZnGA. All the copolymerization reactions were performed using 0.20 g of catalyst and 20.0 mL PO under 2.0 MPa CO_2 at 60 °C for 40 h. The results are summarized in Table 1. The productivity or the turnover number (TON) is given as grams of PPC formed per gram of catalyst (g PPC/g catalyst). The carbonate content (F_{CO_2}) values of the poly(propylene carbonate)-co-polyethers produced were determined using ^{1}H NMR spectroscopy according to the following equation:

$$F_{CO_2} = [(A_{5.0} + A_{4.2})/(A_{5.0} + A_{4.2} + A_{3.8-3.5})] \times 100 \tag{1}$$

where A represents the integral area of the corresponding protons in the ^{1}H NMR spectrum.

In an initial trial, the std-ZnGA produced poly(propylene carbonate) (PPC) with 93.9% F_{CO_2} and a turnover number (TON) of 72.4 g of PPC/g of catalyst. The resultant polymer has a molecular weight of 156.4 kg/mol and a polydispersity index (PDI) of 3.5 (entry 1, Table 1). Then the ZnGA-M catalysts were used as catalysts for the copolymerization of CO_2 and PPO. The results are summarized in Table 1. From Table 1, it is evident that the metal-treated catalysts show a significant increment in TON for CO_2/PO copolymerization, and the activity increments are in the range of 1.6–38.3%. Among the metal-treated ZnGA catalysts, ZnGA-Al, ZnGA-Fe, and ZnGA-Zn show dramatic improvements in productivity compared to that of std-ZnGA, with TONs ranging from 80.7 to 100.1 (entry 2–5, 7, and 8, Table 1). It is noteworthy that treatment with 10^{-3} equivalents of $ZnCl_2$ results in a TON of 100.1, which is ~38% higher than that of std-ZnGA (entry 7, Table 1).

For $AlCl_3$ and $ZnCl_2$, the catalytic activities increase following a rise in the amount of metal from 10^{-4} equivalents to 10^{-3} equivalents (entry 4, 5, 7, and 8, Table 1). Conversely, $FeCl_3$ shows a slight decline in the TON after an increase in the metal treatment amount from 10^{-4} equivalents to 10^{-3} equivalents (entry 2 and 3, Table 1). In addition, when increasing the $ZnCl_2$ amount to 10^{-2} equivalents, the observed productivity is lower than that of std-ZnGA (entry 6, Table 1). These results suggest that there is a threshold limit for the amount of metal ions employed, beyond which the metal ions cover the whole surface of the ZnGA and block the monomers from approaching the active Zn-metal bimetallic sites. This masking effect is repeated when using $CoCl_2$ as the metal salt, because Co^{2+} ions are inactive when used in homogeneous salen complexes and they need to be oxidized to Co^{3+} for higher catalytic performance [63]. Importantly, increasing the amount of $CoCl_2$ has a negative effect on copolymerization, and the productivity of ZnGA-Co-10^{-3} is decreased by ~0.5% (TON = 72.0).

As an extreme example, std-ZnGA was treated with 1.0 equivalent of $AlCl_3$ for 24 h and the resultant white particles were probed using SEM analysis. The SEM image shows that the surface of the ZnGA is completely covered with a thick layer of $AlCl_3$, with the $AlCl_3$ also having round-shaped edges (Figure 3). TEM-EDS mapping had shown the homogeneous distribution of Zn and Al atoms throughout the sample (Figure S4). The ratio of Zn to Al as shown by TEM-EDS is 1 to 7.4, suggesting a thick layer of Al-motif covered the surface. Additionally, the FT-IR spectrum of ZnGA-Al-1 shows a broad hydroxyl stretching in the range of 3000–3500 cm^{-1}. This is different from other ZnGA-M samples (Figure S5). Therefore, as shown in Figure S6, when increasing the amount of metal salt beyond a specific limit, the excess metal salt form a thick layer resembling the bulk metal chloride, thus rendering the catalyst less active (entry 11, Table 1). This fact accounts for the reduced activity of ZnGA-Zn-10^{-2} (entry 6, Table 1). These results clearly show that the presence of other metal ions influences the reactivity of the Zn atoms on the surface of the ZnGA catalyst. Additionally, the added metal on the surface may create a Zn-O-M bimetallic site with optimal distance between the surface metal centers. As reported for a number of homogeneous di-nuclear or bimetallic catalysts, the reaction pathway may follow a bimetallic-cooperative mechanism, wherein one metal may selectively bind with epoxide and ease the ring opening, while the other metal activates the CO_2 and attacks the activated epoxide to form the metal carbonate bond [52]. Alternate additions of epoxide and CO_2 results in a polymer chain growth.

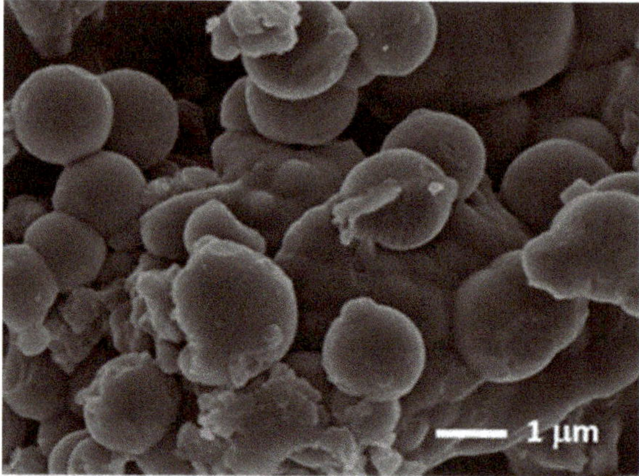

Figure 3. SEM image of ZnGA-Al-1 showing thick layers of $AlCl_3$ on the ZnGA.

2.3. Properties of Polymers

The polymers formed using the different ZnGA-M samples were then characterized with ^{1}H NMR spectroscopy, differential scanning calorimetry (DSC), thermogravimetric analysis (TGA), and Gel Permeation Chromatography (GPC). The ^{1}H NMR spectrum of the polymer exhibits the methyl, methylene, and methine peaks of the polycarbonate at 1.3 ppm, 4.2 ppm, and 5.1–4.9 ppm, respectively (Figure S7).

It is worth mentioning that all these catalysts produce copolymers with significantly high TONs at a relatively low CO_2 pressure of 2.0 MPa, while the F_{CO_2} values are in the range 90–95% with a very small amount of polyether linkages. The microstructure of the polymers was analyzed by ^{13}C NMR analysis. This revealed that the polymers are formed with a predominantly head-to-tail (HT) connectivity (Figure S8). The molecular weight distributions of the prepared polycarbonates were then analyzed using GPC with polystyrene standards in THF and the GPC elugrams. Some of the selected polymers are shown in Figure S9. The results demonstrate that the cooperative bimetallic catalysts actually help to produce polymers with very high molecular weights. As seen in the Table 1, all the catalysts except ZnGA-Co-10^{-4} and ZnGA-Zn-10^{-2} produce polymers with high M_n values and PDI values close to 2.0. Interestingly, ZnGA-Fe-10^{-4} affords a polymer with the very high M_n of 262.4 kg/mol. The PDIs of the polymers varied from 1.8 to 3.5.

The TGA results of the polycarbonates show that the 5% weight loss temperatures ($T_{5\%}$) of the polymers are around 230 °C and complete decomposition occurs in the range 330–350 °C. However, the glass transition temperatures (T_g) of the resulting polymers are slightly lower than that of the polycarbonate produced using std-ZnGA and vary from 33–38 °C (Figure S10). These values are in the accepted range typically reported for PPC prepared from ZnGA under different conditions.

3. Experimental Section

3.1. Materials and Methods

Glutaric acid (≥99.0%) was obtained from Tokyo Chemical Industry Co., Ltd. and used without further purification. Anhydrous $ZnCl_2$, ZnO, $FeCl_3$, $CoCl_2$, $AlCl_3$, and anhydrous tetrahydrofuran (≥99.9%) were purchased from Sigma-Aldrich (Seoul, South Korea) and used as received. Propylene oxide (≥99.9%, PO) was received from Sigma-Aldrich and was distilled over CaH_2 before use. CO_2 gas (99.999) was received from Shinyang gas Industries, Korea. Powder X-ray diffraction

(PXRD) measurements were performed using a Bruker D8 Focus X-ray powder diffractometer (Billerica, MA, USA) using CuKα radiation at room temperature. A Hitachi (Tokyo, Japan.) FE-SEM S-4800 and TEM-Talos; F 200X system were used to study the morphologies of the catalysts. The Fourier-transform-infrared (FT-IR) spectra were measured on Nicolet iS 50 (Thermo Fisher Scientific, Waltham, MA, USA) spectrometer. Metal contents in the catalysts were analyzed using inductively coupled plasma optical emission spectroscopy (ICP-OES) (iCAP-Q, Thermo Fisher Scientific, Waltham, MA, USA) and a microwave-assisted acid digestion system (MARS6, CEM/U.S.A). Bruker Ascend 400 MHz spectrometer was used for measuring the ^{1}H and ^{13}C NMR spectra of the products. Gel Permeation Chromatography (GPC) analysis was performed using a Waters 717 plus instrument equipped with a Waters 515 HPLC Pump (Milford, MA, USA). The columns were eluted with THF at a flow rate of 1.00 mL/min at 35 °C. GPC curves were calibrated using polystyrene standard with molecular weight ranges from 580 to 660,500. A 2960 Simultaneous DSC-TGA instrument (TA instruments, New Castle, DE, USA) was used for the Thermogravimetric Analysis (TGA) with a heating rate of 10 °C/min from 25 °C to 500 °C under Nitrogen atmosphere. A PerkinElmer DSC 4000 instrument (Waltham, MA, USA) was used to perform the Differential Scanning Calorimetric (DSC) tests with a heating rate of 10 °C/min from 20 °C to 120 °C under Nitrogen atmosphere.

3.2. Synthesis of std-ZnGA

std-ZnGA was synthesized by following a published report with slight modification [62]. ZnO (100.0 mmol) was suspended in toluene (150 mL) in a 250-mL round-bottom flask equipped with a Dean-Stark trap and a reflux condenser. Glutaric acid (98.0 mmol) was added to this mixture and refluxed at 60 °C with vigorous stirring for 4 h. After 4 h, heating was stopped and the reaction mixture cooled to room temperature. The white precipitate was filtered and washed with excess of acetone. The resulting product was dried under vacuum at 130 °C, delivering 19.00 g of ZnGA (99.2%). The elemental analysis result, calculated (observed) for $C_5H_6O_4Zn$ (%) was C, 30.72(29.89); H, 3.09(3.05); O, 32.74(33.10).

3.3. General Procedure for Preparing Metal Treated Catalysts

3.3.1. Preparation of Metal Chloride Stock Solutions

The metal chloride stock solutions were prepared by dissolving about 40 to 50 µmol of the corresponding metal chloride in anhydrous THF under argon atmosphere.

3.3.2. Metal Treatment of std-ZnGA

To a dispersion of std-ZnGA (0.5 g, 2.55 mmol) in 20.0 mL of anhydrous THF was added a desired amount of the selected metal chloride as a solution in THF under Ar atm. The white suspension was stirred at ambient temperature for 30 min. After that, the white solid was separated by filtration followed by washing with THF (30.0 mL × 2) and acetone (30.0 mL × 2). The resulting solid was dried under vacuum at 60 °C for 10 h.

3.4. General Procedure for the Copolymerization of CO_2 and PO

All copolymerization reactions were carried out in a pre-dried 100 mL stainless steel autoclave reactor equipped with a magnetic stirrer and a programmable temperature controller. In a typical reaction, 20.0 mL of PO was added to 200.0 mg of the desired catalyst under Ar atmosphere and then pressurized with CO_2 to 2.0 MPa at room temperature. The mixture was stirred at 60 °C for 40 h. After cooling the reactor to room temperature, CO_2 was slowly released. A small fraction was taken for ^{1}H NMR analysis and the remaining mass was dissolved in dichloromethane (DCM) and treated with 1.25 M methanolic HCl solution (1.0 mL × 3). The addition of excess methanol to the solution afforded the polymer as a white precipitate and was dried under vacuum at 60 °C.

4. Conclusions

In this study, surface-modified ZnGA samples were prepared by treatment with different metal chloride salts and used as new, highly active catalysts in the copolymerization of CO_2 and PO under a relatively low CO_2 pressure of 2.0 MPa. Among the various metal-treated ZnGA catalysts, ZnGA-Zn-10^{-3} was found to be highly active with a TON of 100.1 g PPC/g catalyst, which is 38.3% higher than that of std-ZnGA (TON = 72.4). The $FeCl_3$_ treated catalyst ZnGA-Fe-10^{-4} produced the polymer with the highest molecular weight (262 kg/mol) with a TON of 90.9. These results indicate that this simple and industrially viable procedure effectively increases the catalytic activity of ZnGA. This economically beneficial method promotes the use of the ZnGA-M catalytic system for the industrial applications in the production of biodegradable thermoplastics from PO and CO_2.

Supplementary Materials: The following are available online at http://www.mdpi.com/2073-4344/9/11/892/s1, Figure S1: PXRD pattern of std-ZnGA and PXRD pattern calculated from the crystal structure of ZnGA via Mercury 3.7, Figure S2: (a) FT-IR spectra of std-ZnGA and ZnGA-M catalysts, (b) PXRD patterns of std-ZnGA and ZnGA-M catalysts, Figure S3: SEM images of std-ZnGA and ZnGA-MCl_n, Figure S4: STEM image and EDS mapping. (a) STEM image of ZnGA-Al-1; STEM-EDS mapping of (b) Zn (red)and (c) Al (green) elements in ZnGA-Al-1; (Scale bar: = 200 nm), Figure S5: FT-IR spectra of std-ZnGA and ZnGA-Al-1, Figure S6: Expected coordination modes of MCl_n on the std-ZnGA surface, Figure S7: ^{1}H NMR spectrum of PPC from entry 3 in Table 1, Figure S8: ^{13}C NMR spectrum of PPC from entry 3 in Table 1, Figure S9: GPC elugrams for some selected polymers from Table 1, Figure S10: DSC curves of selected polymers from Table 1.

Author Contributions: S.Y. and S.P. have designed the experiments. S.P. conducted the experiments. S.P. and S.Y. wrote the manuscript and S.Y. supervised the project. All authors reviewed the manuscript.

Funding: This research was supported by a Korea CCS R&D Center (KCRC) grant funded by the Korea government (Ministry of Science, ICT & Future Planning) (No. 2014M1A8A1049300).

Acknowledgments: We acknowledge the financial support provided by the Korea CCS R&D Center (KCRC) grant funded by the Korea government (Ministry of Science, ICT & Future Planning) (no. 2014M1A8A1049300).

Conflicts of Interest: The authors declare that the research was conducted in the absence of any commercial or financial relationships that could be construed as a potential conflict of interest.

References

1. The Keeling Curve. Scripps Institution of Oceanography at University of California at San Diego. Available online: https://scripps.ucsd.edu/programs/keelingcurve/ (accessed on 13 September 2017).
2. Scott, D.A.; Christopher, M.B.; Geoffrey, W.C. Carbon Dioxide as a Renewable C1 Feedstock: Synthesis and Characterization of Polycarbonates from the Alternating Copolymerization of Epoxides and CO_2. In *Feedstocks for the Future*; American Chemical Society: Washington, DC, USA, 2006; Volume 921, pp. 116–129.
3. Aresta, M.; Dibenedetto, A. Utilisation of CO_2 as a chemical feedstock: Opportunities and challenges. *Dalton Trans.* **2007**, *28*, 2975–2992. [CrossRef] [PubMed]
4. Song, C. Global challenges and strategies for control, conversion and utilization of CO_2 for sustainable development involving energy, catalysis, adsorption and chemical processing. *Catal. Today* **2006**, *115*, 2–32. [CrossRef]
5. Spinner, N.S.; Vega, J.A.; Mustain, W.E. Recent progress in the electrochemical conversion and utilization of CO_2. *Catal. Sci. Technol.* **2012**, *2*, 19–28. [CrossRef]
6. Kumar, S.; Verma, S.; Shawat, E.; Nessim, G.D.; Jain, S.L. Amino-functionalized carbon nanofibres as an efficient metal free catalyst for the synthesis of quinazoline-2,4(1H,3H)-diones from CO_2 and 2-aminobenzonitriles. *RSC Adv.* **2015**, *5*, 24670–24674. [CrossRef]
7. Park, K.; Gunasekar, G.H.; Prakash, N.; Jung, K.-D.; Yoon, S. A Highly Efficient Heterogenized Iridium Complex for the Catalytic Hydrogenation of Carbon Dioxide to Formate. *ChemSusChem* **2015**, *8*, 3410–3413. [CrossRef]
8. Gunasekar, G.H.; Park, K.; Jung, K.-D.; Yoon, S. Recent developments in the catalytic hydrogenation of CO_2 to formic acid/formate using heterogeneous catalysts. *Inorg. Chem. Front.* **2016**, *3*, 882–895. [CrossRef]
9. Kim, S.-H.; Chung, G.-Y.; Kim, S.-H.; Vinothkumar, G.; Yoon, S.-H.; Jung, K.-D. Electrochemical NADH regeneration and electroenzymatic CO_2 reduction on Cu nanorods/glassy carbon electrode prepared by cyclic deposition. *Electrochim. Acta* **2016**, *210*, 837–845. [CrossRef]

10. Gunasekar, G.H.; Park, K.; Ganesan, V.; Lee, K.; Kim, N.-K.; Jung, K.-D.; Yoon, S. A Covalent Triazine Framework, Functionalized with Ir/N-Heterocyclic Carbene Sites, for the Efficient Hydrogenation of CO_2 to Formate. *Chem. Mater.* **2017**, *29*, 6740–6748. [CrossRef]

11. Park, K.; Lee, K.; Kim, H.; Ganesan, V.; Cho, K.; Jeong, S.K.; Yoon, S. Preparation of covalent triazine frameworks with imidazolium cations embedded in basic sites and their application for CO_2 capture. *J. Mater. Chem. A* **2017**, *5*, 8576–8582. [CrossRef]

12. Gunasekar, G.H.; Park, K.; Jeong, H.; Jung, K.D.; Park, K.; Yoon, S. Molecular Rh(III) and Ir(III) Catalysts Immobilized on Bipyridine-Based Covalent Triazine Frameworks for the Hydrogenation of CO_2 to Formate. *Catalysts* **2018**, *8*, 295. [CrossRef]

13. Gunasekar, G.H.; Shin, J.; Jung, K.-D.; Park, K.; Yoon, S. Design Strategy toward Recyclable and Highly Efficient Heterogeneous Catalysts for the Hydrogenation of CO_2 to Formate. *ACS Catal.* **2018**, *8*, 4346–4353. [CrossRef]

14. Abbas, I.; Kim, H.; Shin, C.-H.; Yoon, S.; Jung, K.-D. Differences in bifunctionality of ZnO and ZrO_2 in $Cu/ZnO/ZrO_2/Al_2O_3$ catalysts in hydrogenation of carbon oxides for methanol synthesis. *Appl. Catal. B Environ.* **2019**, *258*, 117971. [CrossRef]

15. Gunasekar, G.H.; Jung, K.-D.; Yoon, S. Hydrogenation of CO_2 to Formate using a Simple, Recyclable, and Efficient Heterogeneous Catalyst. *Inorg. Chem.* **2019**, *58*, 3717–3723. [CrossRef] [PubMed]

16. Gunasekar, G.H.; Yoon, S. A phenanthroline-based porous organic polymer for the iridium-catalyzed hydrogenation of carbon dioxide to formate. *J. Mater. Chem. A* **2019**, *7*, 14019–14026. [CrossRef]

17. Kim, C.; Choe, Y.-K.; Won, D.H.; Lee, U.; Oh, H.-S.; Lee, D.K.; Choi, C.H.; Yoon, S.; Kim, W.; Hwang, Y.J.; et al. Turning Harmful Deposition of Metal Impurities into Activation of Nitrogen-Doped Carbon Catalyst toward Durable Electrochemical CO_2 Reduction. *ACS Energy Lett.* **2019**, *4*, 2343–2350. [CrossRef]

18. Luinstra, G.A. Poly(propylene carbonate), old copolymers of propylene oxide and carbon dioxide with new interests: Catalysis and material properties. *Polym. Rev.* **2008**, *48*, 192–219. [CrossRef]

19. Scharfenberg, M.; Hilf, J.; Frey, H. Functional Polycarbonates from Carbon Dioxide and Tailored Epoxide Monomers: Degradable Materials and Their Application Potential. *Adv. Funct. Mater.* **2018**, *28*, 1704302. [CrossRef]

20. Luinstra, G.A.; Borchardt, E. Material Properties of Poly(Propylene Carbonates). *Adv. Polym. Sci.* **2012**, *245*, 29–48. [CrossRef]

21. Alessandra, Q.E.; Gabriele, C.; Jean-Luc, D.; Siglinda, P. Carbon Dioxide Recycling: Emerging Large-Scale Technologies with Industrial Potential. *ChemSusChem* **2011**, *4*, 1194–1215.

22. Peters, M.; Koehler, B.; Kuckshinrichs, W.; Leitner, W.; Markewitz, P.; Müller, T.E. Chemical Technologies for Exploiting and Recycling Carbon Dioxide into the Value Chain. *ChemSusChem* **2011**, *4*, 1216–1240. [CrossRef]

23. Inoue, S.; Koinuma, H.; Tsuruta, T. Copolymerization of Carbon Dioxide and Epoxide. *J. Polym. Sci. Pol. Lett.* **1969**, *7*, 287–292. [CrossRef]

24. Inoue, S.; Koinuma, H.; Tsuruta, T. Copolymerization of Carbon Dioxide and Epoxide with Organometallic Compounds. *Makromol. Chem.* **1969**, *130*, 210–220. [CrossRef]

25. Darensbourg, D.J. Making plastics from carbon dioxide: Salen metal complexes as catalysts for the production of polycarbonates from epoxides and CO_2. *Chem. Rev.* **2007**, *107*, 2388–2410. [CrossRef] [PubMed]

26. Ang, R.R.; Sin, L.T.; Bee, S.T.; Tee, T.T.; Kadhum, A.A.H.; Rahmat, A.R.; Wasmi, B.A. A review of copolymerization of green house gas carbon dioxide and oxiranes to produce polycarbonate. *J. Clean. Prod.* **2015**, *102*, 1–17. [CrossRef]

27. Trott, G.; Saini, P.K.; Williams, C.K. Catalysts for CO_2/epoxide ring-opening copolymerization. *Philos. Trans. R. Soc. A* **2016**, *374*, 20150085. [CrossRef]

28. Sudakar, P.; Gunasekar, G.H.; Baek, I.H.; Yoon, S. Recyclable and efficient heterogenized Rh and Ir catalysts for the transfer hydrogenation of carbonyl compounds in aqueous medium. *Green Chem.* **2016**, *18*, 6456–6461. [CrossRef]

29. Ganesan, V.; Yoon, S. Hyper-Cross-Linked Porous Porphyrin Aluminum(III) Tetracarbonylcobaltate as a Highly Active Heterogeneous Bimetallic Catalyst for the Ring-Expansion Carbonylation of Epoxides. *ACS Appl. Mater. Interfaces* **2019**, *11*, 18609–18616. [CrossRef]

30. Padmanaban, S.; Gunasekar, G.H.; Lee, M.; Yoon, S. Recyclable Covalent Triazine Framework-based Ru Catalyst for Transfer Hydrogenation of Carbonyl Compounds in Water. *ACS Sustain. Chem. Eng.* **2019**, *7*, 8893–8899. [CrossRef]

31. Friend, C.M.; Xu, B. Heterogeneous Catalysis: A Central Science for a Sustainable Future. *Acc. Chem. Res.* **2017**, *50*, 517–521. [CrossRef]

32. Padmanaban, S.; Kim, M.; Yoon, S. Acid-mediated surface etching of a nano-sized metal-organic framework for improved reactivity in the fixation of CO_2 into polymers. *J. Ind. Eng. Chem.* **2019**, *71*, 336–344. [CrossRef]

33. von der Assen, N.; Bardow, A. Life cycle assessment of polyols for polyurethane production using CO_2 as feedstock: Insights from an industrial case study. *Green Chem.* **2014**, *16*, 3272–3280. [CrossRef]

34. Eberhardt, R.; Allmendinger, M.; Zintl, M.; Troll, C.; Luinstra, G.A.; Rieger, B. New zinc dicarboxylate catalysts for the CO_2/propylene oxide copolymerization reaction: Activity enhancement through Zn(II)-ethylsulfinate initiating groups. *Macromol. Chem. Phys.* **2004**, *205*, 42–47. [CrossRef]

35. Ree, M.; Hwang, Y.; Kim, J.S.; Kim, H.; Kim, G.; Kim, H. New findings in the catalytic activity of zinc glutarate and its application in the chemical fixation of CO_2 into polycarbonates and their derivatives. *Catal. Today* **2006**, *115*, 134–145. [CrossRef]

36. Wang, S.J.; Du, L.C.; Zhao, X.S.; Meng, Y.Z.; Tjong, S.C. Synthesis and characterization of alternating copolymer from carbon dioxide and propylene oxide. *J. Appl. Polym. Sci.* **2002**, *85*, 2327–2334. [CrossRef]

37. Zhong, X.; Dehghani, F. Solvent free synthesis of organometallic catalysts for the copolymerisation of carbon dioxide and propylene oxide. *Appl. Catal. B Environ.* **2010**, *98*, 101–111. [CrossRef]

38. Ang, R.R.; Sin, L.T.; Bee, S.T.; Tee, T.T.; Kadhum, A.A.H.; Rahmat, A.R.; Wasmi, B.A. Determination of zinc glutarate complexes synthesis factors affecting production of propylene carbonate from carbon dioxide and propylene oxide. *Chem. Eng. J.* **2017**, *327*, 120–127. [CrossRef]

39. Kim, J.S.; Kim, H.; Ree, M. Hydrothermal synthesis of single-crystalline zinc glutarate and its structural determination. *Chem. Mater.* **2004**, *16*, 2981–2983. [CrossRef]

40. Seth, S.; Matzger, A.J. Metal–Organic Frameworks: Examples, Counterexamples, and an Actionable Definition. *Cryst. Growth Des.* **2017**, *17*, 4043–4048. [CrossRef]

41. Miralda, C.M.; Macias, E.E.; Zhu, M.; Ratnasamy, P.; Carreon, M.A. Zeolitic Imidazole Framework-8 Catalysts in the Conversion of CO_2 to Chloropropene Carbonate. *ACS Catal.* **2012**, *2*, 180–183. [CrossRef]

42. Kim, J.; Kim, S.-N.; Jang, H.-G.; Seo, G.; Ahn, W.-S. CO_2 cycloaddition of styrene oxide over MOF catalysts. *Appl. Catal. A Gen.* **2013**, *453*, 175–180. [CrossRef]

43. Kumar, S.; Verma, G.; Gao, W.-Y.; Niu, Z.; Wojtas, L.; Ma, S. Anionic Metal–Organic Framework for Selective Dye Removal and CO_2 Fixation. *Eur. J. Inorg. Chem.* **2016**, *2016*, 4373–4377. [CrossRef]

44. Kuruppathparambil, R.R.; Babu, R.; Jeong, H.M.; Hwang, G.-Y.; Jeong, G.S.; Kim, M.-I.; Kim, D.-W.; Park, D.-W. A solid solution zeolitic imidazolate framework as a room temperature efficient catalyst for the chemical fixation of CO_2. *Green Chem.* **2016**, *18*, 6349–6356. [CrossRef]

45. Babu, R.; Kim, S.-H.; Kathalikkattil, A.C.; Kuruppathparambil, R.R.; Kim, D.W.; Cho, S.J.; Park, D.-W. Aqueous microwave-assisted synthesis of non-interpenetrated metal-organic framework for room temperature cycloaddition of CO_2 and epoxides. *Appl. Catal. A Gen.* **2017**, *544*, 126–136. [CrossRef]

46. Lin, Y.-F.; Huang, K.-W.; Ko, B.-T.; Lin, K.-Y.A. Bifunctional ZIF-78 heterogeneous catalyst with dual Lewis acidic and basic sites for carbon dioxide fixation via cyclic carbonate synthesis. *J. CO_2 Util.* **2017**, *22*, 178–183. [CrossRef]

47. Liang, J.; Huang, Y.-B.; Cao, R. Metal–organic frameworks and porous organic polymers for sustainable fixation of carbon dioxide into cyclic carbonates. *Coord. Chem. Rev.* **2019**, *378*, 32–65. [CrossRef]

48. Zhu, Q.; Meng, Y.Z.; Tjong, S.C.; Zhao, X.S.; Chen, Y.L. Thermally stable and high molecular weight poly(propylene carbonate)s from carbon dioxide and propylene oxide. *Polym. Int.* **2002**, *51*, 1079–1085. [CrossRef]

49. Kim, J.S.; Kim, H.; Yoon, J.; Heo, K.; Ree, M. Synthesis of zinc glutarates with various morphologies using an amphiphilic template and their catalytic activities in the copolymerization of carbon dioxide and propylene oxide. *J. Polym. Sci. Pol. Chem.* **2005**, *43*, 4079–4088. [CrossRef]

50. Wang, J.T.; Zhu, Q.; Lu, X.L.; Meng, Y.Z. ZnGA-MMT catalyzed the copolymerization of carbon dioxide with propylene oxide. *Eur. Polym. J.* **2005**, *41*, 1108–1114. [CrossRef]

51. Gao, L.J.; Luo, Y.C.; Lin, Y.J.; Su, T.; Su, R.P.; Feng, J.Y. Silica-supported zinc glutarate catalyst synthesized by rheological phase reaction used in the copolymerization of carbon dioxide and propylene oxide. *J. Polym. Res.* **2015**, *22*, 220. [CrossRef]

52. Klaus, S.; Lehenmeier, M.W.; Herdtweck, E.; Deglmann, P.; Ott, A.K.; Rieger, B. Mechanistic Insights into Heterogeneous Zinc Dicarboxylates and Theoretical Considerations for CO_2-Epoxide Copolymerization. *J. Am. Chem. Soc.* **2011**, *133*, 13151–13161. [CrossRef]

53. Sudakar, P.; Sivanesan, D.; Yoon, S. Copolymerization of Epichlorohydrin and CO_2 Using Zinc Glutarate: An Additional Application of ZnGA in Polycarbonate Synthesis. *Macromol. Rapid Commun.* **2016**, *37*, 788–793. [CrossRef] [PubMed]

54. Padmanaban, S.; Dharmalingam, S.; Yoon, S. A Zn-MOF-Catalyzed Terpolymerization of Propylene Oxide, CO_2, and β-butyrolactone. *Catalysts* **2018**, *8*, 393. [CrossRef]

55. Chisholm, M.H.; Navarro-Llobet, D.; Zhou, Z.P. Poly(propylene carbonate). 1. More about poly(propylene carbonate) formed from the copolymerization of propylene oxide and carbon dioxide employing a zinc glutarate catalyst. *Macromolecules* **2002**, *35*, 6494–6504. [CrossRef]

56. Kim, J.S.; Ree, M.; Shin, T.J.; Han, O.H.; Cho, S.J.; Hwang, Y.T.; Bae, J.Y.; Lee, J.M.; Ryoo, R.; Kim, H. X-ray absorption and NMR spectroscopic investigations of zinc glutarates prepared from various zinc sources and their catalytic activities in the copolymerization of carbon dioxide and propylene oxide. *J. Catal.* **2003**, *218*, 209–219. [CrossRef]

57. Rieter, W.J.; Taylor, K.M.L.; Lin, W. Surface Modification and Functionalization of Nanoscale Metal-Organic Frameworks for Controlled Release and Luminescence Sensing. *J. Am. Chem. Soc.* **2007**, *129*, 9852–9853. [CrossRef] [PubMed]

58. Cohen, S.M. Postsynthetic Methods for the Functionalization of Metal–Organic Frameworks. *Chem. Rev.* **2012**, *112*, 970–1000. [CrossRef] [PubMed]

59. Bhattacharya, B.; Haldar, R.; Dey, R.; Maji, T.K.; Ghoshal, D. Porous coordination polymers based on functionalized Schiff base linkers: Enhanced CO_2 uptake by pore surface modification. *Dalton Trans.* **2014**, *43*, 2272–2282. [CrossRef]

60. McGuire, C.V.; Forgan, R.S. The surface chemistry of metal–organic frameworks. *Chem. Commun.* **2015**, *51*, 5199–5217. [CrossRef]

61. Yin, Z.; Wan, S.; Yang, J.; Kurmoo, M.; Zeng, M.-H. Recent advances in post-synthetic modification of metal–organic frameworks: New types and tandem reactions. *Coord. Chem. Rev.* **2019**, *378*, 500–512. [CrossRef]

62. Ree, M.; Bae, J.Y.; Jung, J.H.; Shin, T.J. A new copolymerization process leading to poly(propylene carbonate) with a highly enhanced yield from carbon dioxide and propylene oxide. *J. Polym. Sci. Pol. Chem.* **1999**, *37*, 1863–1876. [CrossRef]

63. Ren, W.-M.; Liu, Z.-W.; Wen, Y.-Q.; Zhang, R.; Lu, X.-B. Mechanistic Aspects of the Copolymerization of CO_2 with Epoxides Using a Thermally Stable Single-Site Cobalt(III) Catalyst. *J. Am. Chem. Soc.* **2009**, *131*, 11509–11518. [CrossRef] [PubMed]

Article

Replacement of Chromium by Non-Toxic Metals in Lewis-Acid MOFs: Assessment of Stability as Glucose Conversion Catalysts

Ralentri Pertiwi [1,2,3,†], Ryan Oozeerally [1,†], David L. Burnett [2], Thomas W. Chamberlain [1,2], Nikolay Cherkasov [1], Marc Walker [4], Reza J. Kashtiban [4], Yuni K. Krisnandi [3], Volkan Degirmenci [1,*] and Richard I. Walton [2,*]

[1] School of Engineering, University of Warwick, Coventry CV4 7AL, UK; ralentri.pertiwi@gmail.com (R.P.); R.Oozeerally@warwick.ac.uk (R.O.); T.W.Chamberlain@warwick.ac.uk (T.W.C.); N.Cherkasov@warwick.ac.uk (N.C.)

[2] Department of Chemistry, University of Warwick, Coventry CV4 7AL, UK; D.Burnett.1@warwick.ac.uk

[3] Department of Chemistry, Universitas Indonesia, Depok 16424, Indonesia; yuni.krisnandi@sci.ui.ac.id

[4] Department of Physics, University of Warwick, Coventry CV4 7AL, UK; M.Walker@warwick.ac.uk (M.W.); R.Jalilikashtiban@warwick.ac.uk (R.J.K.)

* Correspondence: V.Degirmenci@warwick.ac.uk (V.D.); R.I.Walton@warwick.ac.uk (R.I.W.)

† These authors contributed equally to the work.

Received: 16 April 2019; Accepted: 6 May 2019; Published: 9 May 2019

Abstract: The metal–organic framework MIL-101(Cr) is known as a solid–acid catalyst for the solution conversion of biomass-derived glucose to 5-hydroxymethyl furfural (5-HMF). We study the substitution of Cr^{3+} by Fe^{3+} and Sc^{3+} in the MIL-101 structure in order to prepare more environmentally benign catalysts. MIL-101(Fe) can be prepared, and the inclusion of Sc is possible at low levels (10% of Fe replaced). On extended synthesis times the polymorphic MIL-88B structure instead forms.Increasing the amount of Sc also only yields MIL-88B, even at short crystallisation times. The MIL-88B structure is unstable under hydrothermal conditions, but in dimethylsulfoxide solvent, it provides 5-HMF from glucose as the major product. The optimum material is a bimetallic (Fe,Sc) form of MIL-88B, which provides ~70% conversion of glucose with 35% selectivity towards 5-HMF after 3 hours at 140 °C: this offers high conversion compared to other heterogeneous catalysts reported in the same solvent.

Keywords: Metal–organic framework; Lewis acid; fructose; 5-hydroxymethyl furfural; biomass

---

1. Introduction

Biomass-derived glucose has received considerable attention as a potential feedstock for the production of useful organic molecules. One attractive target of glucose conversion is the molecule 5-hydroxymethyl furfural (5-HMF): this is regarded as a platform molecule that can be used as a precursor for an extensive range of plastics, polymers, and fuels [1–4]. Acid catalysts are needed to bring about the transformation of glucose, via fructose, into 5-HMF, and both Lewis and Brønsted acids have been implicated in the transformation mechanism [5–8]. Figure 1 shows the most common reaction pathways that have been proposed in the literature [9,10]. It is well established that the isomerisation of glucose to fructose requires a Lewis acid catalyst, and mannose can be produced as an alternative product, while the dehydration of fructose to 5-HMF is catalysed by Brønsted acids that may also result in the further transformation (hydrolysis) to levulinic acid and formic acid [10]. Humins are insoluble carbon-rich materials that are commonly formed as a side-product, and other small molecules may be possible as intermediates and byproducts, depending on the choice of catalyst and solvent [2].

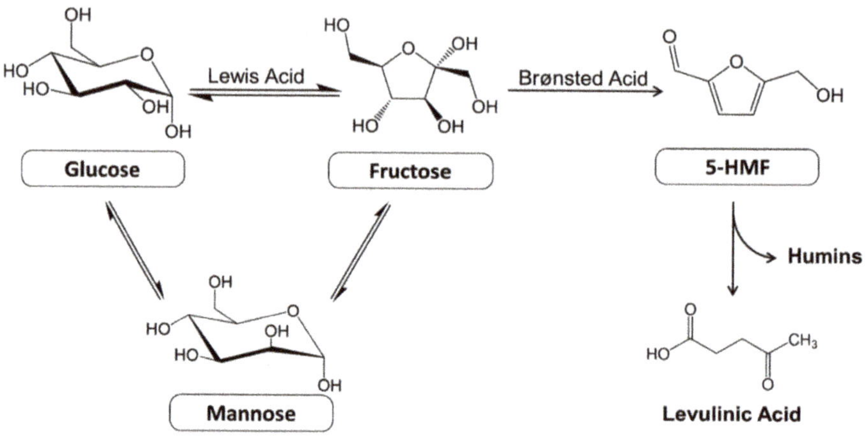

Figure 1. Pathway for the conversion of glucose to 5-HMF, showing possible byproducts, as described in the literature [9,10]. Glucose and mannose are shown in their α-pyranose forms and fructose in its α-furanose form.

Heterogeneous catalysts are desirable for the conversion of glucose, and Davis and co-workers pioneered the use of zeolite-based catalysts that contain both Lewis and Brønsted acid sites; in particular, tin-substituted zeolite beta (Sn-beta) has been shown as an effective catalyst for the isomerisation of glucose into fructose and mannose in water [9], or the direct formation of 5-HMF from glucose when a biphasic water/tetrahydrofuran reactor system was used [11]. The stability of the heterogeneous catalyst in the reaction medium is a major challenge in the conversion of glucose [12]. Although Sn-beta is hydrothermally stable at low pH and the solid can be recycled, a significant disadvantage in its use, particularly on an industrial scale, is that the synthesis of the catalyst is a lengthy process requiring a 40 day hydrothermal reaction, and furthermore requires hydrofluoric acid, an extremely toxic and corrosive substance [9]. A challenge is to identify new solution-stable solid–acid catalysts that provide glucose conversion with selectivity towards fructose and 5-HMF; the catalysts must also be easy to prepare, using environmentally benign reagents and processes. In recent years, porous metal–organic frameworks (MOFs) have been studied for wide-ranging applications [13], such as adsorbents [14], gas storage and separation [15], and heterogeneous catalysts [16–20]. MOF materials have many attractive features for catalysis, such as large surface areas and pore sizes, as well as controllable properties, since the wide choice of ligands and metals provides a large variety of structures, porosities, and chemical functionalities. The scope for adding extra functionality, for example, by post-synthesis modification of the organic ligands, offers the possibility of multifunctional heterogeneous catalysts that contain more than one active site for tandem or multi-step conversions [21].

Brønsted acid functionality in MOFs has been introduced in a number of ways [22], but particularly useful is the addition of sulfonic acid (-SO₃H) functionalities to the organic ligands, such as by their addition to aromatic polycarboxylate linkers that make up many common MOF structures [23,24]. Lewis acidity may be present in MOF structures in the form of coordinatively unsaturated metal sites, such as metals associated with loosely bound solvent, in addition to the coordinating atoms of the ligands making up the framework [25]. The dehydration of glucose to 5-HMF using MIL-101(Cr) functionalised with sulfonic acid groups was reported by Herbst and Janiak in 2016 [10]. The highest 5-HMF yield (29%) was obtained at 403 K with a 5 hour reaction time in a THF:water (39:1 v:v) solvent. Su et al. studied the same catalyst and found the highest conversions were possible in mixed organic:aqueous solvents, and demonstrated the effectiveness of the solid in a fixed-bed continuous reactor [26]. MIL-101(Cr), UiO-66, and MIL-53(Al) functionalised with sulfonic acid groups have also been studied as catalysts for glucose conversion, and MIL-101(Cr)-SO₃H showed high efficiency, with more than 90% yield of 5-HMF from fructose when dimethyl sulfoxide (DMSO) was used as solvent [27]. Previously, MIL-101(Cr) had been used as a host for phosphotungstic acid, thereby adding Brønsted acidity for carbohydrate dehydration [28], while most recently, Guo et al. found that a composite of MIL-101(Cr) and chromium hydroxide gave superior isomerisation of glucose to fructose in ethanol solvent via ethyl fructoside that required hydrolysis in a second step [29]. Other MOFs that have been used for glucose conversion include the zirconium-based NU-1000, with phosphate modification to induce Lewis acidity [30], and the zirconium material UiO-66 with sulfo-modified ligands and inherent Lewis acidity from defects notably allowing conversion of glucose to fructose, along with a significant amount of 5-HMF in water alone [31], and selective conversion of glucose to fructose with iso-propanol solvent [32]. Sulfonated UiO-66 has been further tuned to optimise activity towards glucose conversion [33], compared with other zirconium MOFs [34], and has also been used to effect the esterification of levulinic acid with ethanol [35].

Based on the efficiency of MIL-101(Cr) as a catalyst for glucose conversion, we have now investigated the modification of this material by substitution of the metals making up the structure. MIL-101 is constructed from the cheap and benign ligand benzene-1,4-dicarboxylate (BDC), but can be prepared with various other trivalent cations [36]. The metal Cr is known to be toxic to humans and harmful to the environment, and although these detrimental properties are largely associated with the +6 oxidation state, which is established to be carcinogenic, Cr^{3+} is considered an irritant, with evidence also for it being an allergen [37]; therefore its release into the environment is clearly undesirable, where it might encounter oxidising conditions. Herein, we report a study of variants of MIL-101 using Fe and Sc as replacements for Cr and their potential as catalysts for glucose conversion. Homogeneous scandium triflate is a powerful Lewis acid catalyst [38], and the synthesis of a range of porous scandium carboxylate MOFs with structures analogous to MIL-88B, MIL-100, and MIL-53 was described previously [39], with MIL-100 (Sc) proven to be an effective Lewis-acid catalyst [40]. Mixed-metal MIL-100 (Sc,Fe) materials have been used as dual acid-redox catalysts [41]. Figure 2 shows the structures of two of the MOFs that are relevant to the current work: MIL-101 and MIL-88B. Both are constructed from the same trimeric building unit, consisting of three octahedrally coordinated metal centres with a common shared oxide, and linked by benzene-1,4-dicarboxylate (BDC), but the connectivity gives different structures [42]. Each octahedral metal centre contains a bound solvent molecule, as indicated by the terminal atom on Figure 2a, which may provide a Lewis acid site.

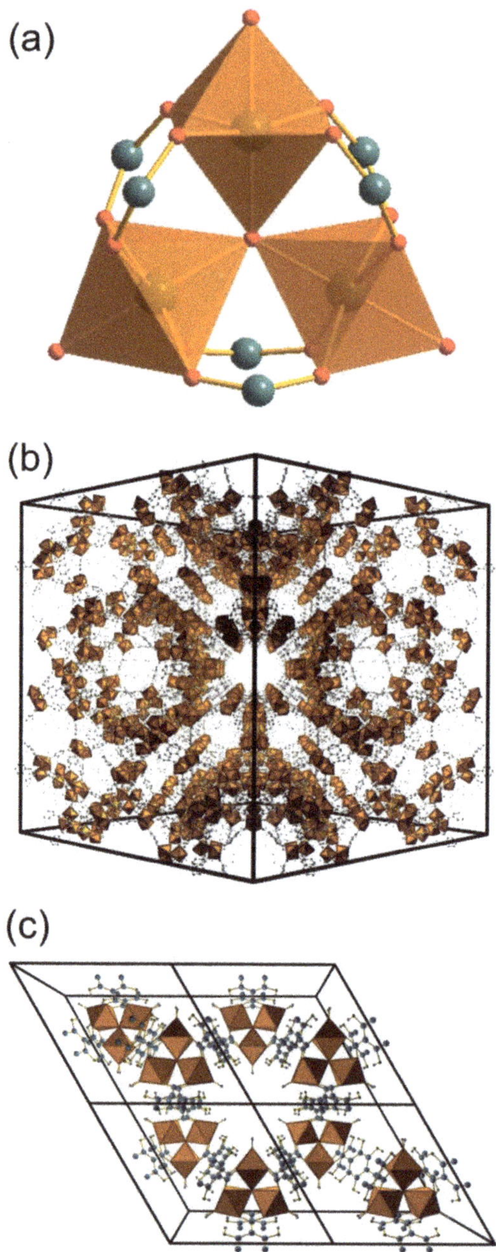

Figure 2. The structures of the MOFs investigated in this work: (**a**) the trimeric building unit showing three {FeO$_6$} octahedral units linked by a common oxygen atom, (**b**) the cubic unit cell of MIL-101, and (**c**) four unit cells of hexagonal MIL-88B. Carbon atoms of the benzene-1,4-dicarboxylate (BDC) linker are shown as grey spheres.

2. Results

MIL-101(Fe) was obtained by a solvothermal method from benzene-1,4-dicarboxylic acid and iron (III) chloride, according to the method of Zhu et al. [43] The powder X-ray diffraction (PXRD) pattern of the material prepared at 3 hours reaction time confirms the identity of the material by comparison with the simulated pattern from the literature (Figure 3). The differences in relative intensity between the low angle peaks (at less than 6 $°2\theta$) and those at higher angle (greater than 8 $°2\theta$) in the measured pattern compared to the simulated pattern is accounted for by the presence of solvent in the as-prepared sample; in contrast, the simulated pattern is for an idealised guest-free framework. These X-ray powder patterns are typical of MIL-101 samples reported in the literature [44–46]. There is also evidence in the PXRD for the presence of an impurity phase in the sample, which we identify below as being MIL-88B (Figure 1) [47]. It is already known that MIL-88B and MIL-101 are "framework isomers" and may form in competition to each other [4,48,49].

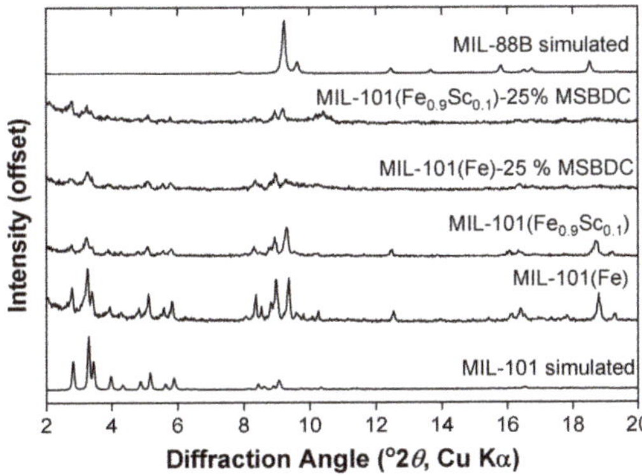

Figure 3. Powder X-ray diffraction patterns of samples of MIL-101 prepared by solvothermal synthesis at 3 hour reaction times. The simulated pattern of MIL-101 was produced from the published structure of Férey et al. [36] and that of MIL-88B from the refined diffraction profile in this work for MIL-88B(Fe).

Figure 3 also shows PXRD patterns of materials prepared where small amounts of Sc^{3+} are used to replace some of the Fe^{3+}, or some of the BDC linker is replaced by mono-2-sulfobenzene-1,4-dicarboxylate (MSBDC). In both cases, the crystallinity of the resultant product is considerably less than for the Fe material, and the MIL-88B impurity remains in the samples. SEM was used to detect the presence of the sulfo-modified ligand in the material, but no signal from sulfur was observed. Since we used this method successfully to image sulfur in sulfo-modified UiO-66 [31], we conclude that for MIL-101(Fe) prepared by this method, the incorporation of the sulfo-modified ligand is negligible. Fourier transform infrared (FT-IR) spectra also provided no evidence for any characteristic sulfo bands. Owing to the lack of any sulfo-functionalisation, we excluded these samples from any further study.

The highest scandium amount that can be introduced to MIL-101(Fe) is only 10%. If 50% of the Fe is replaced by Sc (note that this is the nominal metals ratio used in the synthesis) then the material formed is solely MIL-88B under the synthesis conditions that we have used, labelled MIL-88B(Fe$_{0.5}$Sc$_{0.5}$), and indeed if longer synthesis times are used then, even for the iron material, only MIL-88B is found (see Figure 4). The powder X-ray pattern of MIL-88B matches that reported by Ma et al. who prepared the iron form by microwave synthesis in the same solvent [50]. Figure 4 also demonstrates the stability of MIL-88B in DMSO upon prolonged heating: although the powder pattern shows some broadening of the peaks, the material clearly has the same structure. The materials are, however, unstable in water: upon heating to 140 °C for 3 hours, the MOF structure collapses and the only crystalline product recovered is the acid of the BDC linker, along with an orange powder, presumably amorphous iron oxide.

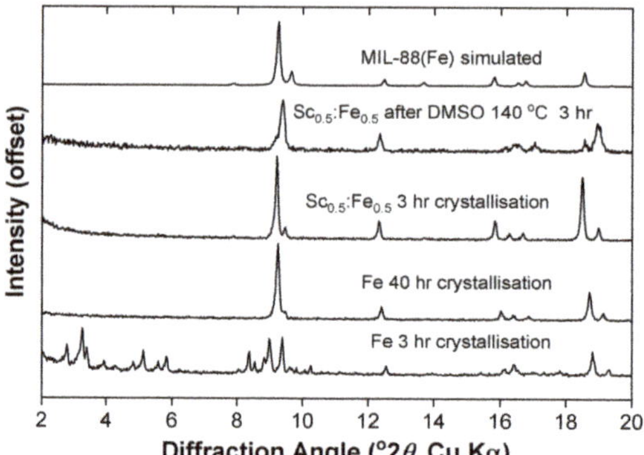

Figure 4. Powder X-ray diffraction patterns of MOFs transformation from MIL-101(Fe), 3 hour crystallisation) to MIL-88B(Fe), 40 hour crystallisation, along with mixed (Fe,Sc) (nominal metals ratio 1:1), 3 hour crystallisation. The second from top pattern illustrates the stability of the mixed-metal phase on heating in DMSO.

Confirmation of the identity of the more stable MOF structure in the synthetic conditions used was provided by a more detailed analysis of higher resolution powder X-ray diffraction patterns, and Figure 5 shows the profile fits to the MIL-88B(Fe), MIL-88B(Sc$_{0.5}$Fe$_{0.5}$) and MIL-88B(Sc) materials. The refined lattice parameters (Figure 5) fitted using the hexagonal space-group $P\bar{6}2c$ are consistent with those reported in the literature for MIL-88B materials: for example, Ma et al. found $a \sim 12.7$ Å, $c \sim 18.4$ Å and $V \sim 2570.2$ Å^3 for MIL-88B(Fe) isolated from N,N-dimethylformamide (DMF) [50], and Mowat et al. found $a = 11.2190$ Å, $c = 19.373$ Å, $V = 2112.27$ Å^3 for MIL-88B(Sc) containing a mixture of water and DMF [39]. MIL-88B, however, is an example of a breathing MOF and the lattice parameters can vary dramatically with the amount, or type, of solvent present [51], so it is difficult to draw conclusions in the trends in the values of the fitted lattice parameters for the three materials that we have studied. Larger unit cell volumes are observed for both the Sc-containing samples we have prepared compared to the iron form, which is consistent with the larger ionic radius of octahedral Sc^{3+} compared to Fe^{3+} [52]. However, the presence of different amounts of solvents (or mixtures of water and DMF) cannot be ruled out, which may also adjust the unit cell volume.

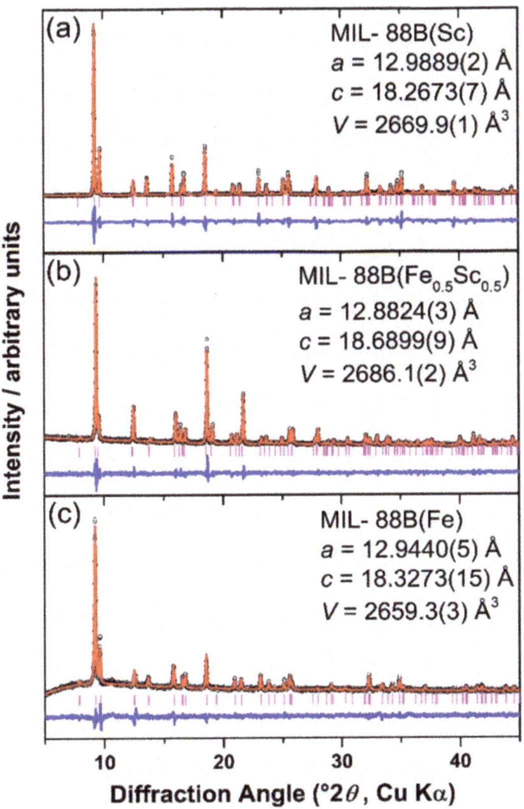

Figure 5. Profile fits to powder X-ray diffraction patterns of (**a**) MIL-88B(Sc), (**b**) MIL-88B(Fe$_{0.5}$Sc$_{0.5}$), and (**c**) MIL-88B(Fe) (Cu Kα_1 radiation) fitted using space-group $P\bar{6}2c$. The points are the measured data, the red line is the final fit, the blue line is the difference curve and the pink ticks are the positions of the allowed reflections. Refined lattice parameters and unit cell volumes are provided on each panel.

In addition to the different unit cell of the mixed-metal material, more substantive evidence for the formation of a mixed-metal sample of the MOF, rather than a phase-separated mixture of materials, comes from EDX mapping on the SEM, Figure 6. This shows how the two metals are both present in micro-sized crystallites of MIL-88B(Sc$_{0.5}$Fe$_{0.5}$). Proving the degree of mixing of the two metals on smaller length-scales is challenging; in the previous work on MIL-100(Fe,Sc), which contains the same trimeric building units as found in MIL-88B, it was impossible to determine, using extended X-ray absorption fine structure (EXAFS) spectroscopy, whether the individual clusters contained mixtures of metals or whether single metal clusters were present [41].

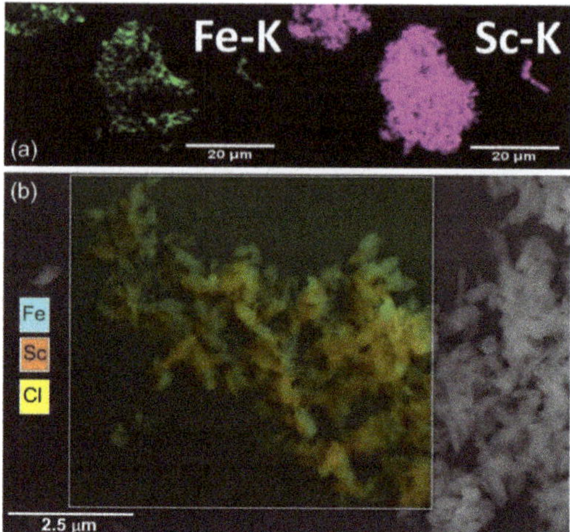

Figure 6. Elemental maps of MIL-88B(Fe$_{0.5}$Sc$_{0.5}$) at two magnifications showing the distribution of the two metals, and the presence of chlorine.

The elemental mapping suggests that the concentration of iron in the MIL-88B(Fe,Sc) sample is lower than that of scandium, despite the initial 1:1 molar ratio used in synthesis. This was verified by bulk analysis for metals using ICP-OES, which gave a Sc:Fe molar ratio of 3.1:1. X-ray photoelectron spectroscopy (XPS, Figure 7), was used to quantify the metal ratio at the surface of the material, and integration of the signal gave a Sc:Fe ratio of 6.0:1. The typical depth of the XPS analysis for these elements is around 5–10 nm, which shows how the surface of the particles is scandium-rich. First, this suggests that most of the Fe is present within the crystallites of the MIL-88B, and, second, shows how the surface is terminated substantially by Sc-containing species, which are strongly bound, since they are not removed by the washing used in the sample preparation. One possible explanation for the excess scandium at the surface is the presence of a scandium chloride species, necessary for charge-balance of the terminated crystal surfaces. Some of the XPS signal can indeed be accounted for as ScCl$_3$ (Figure 7a), and this result is also consistent with the presence of chlorine detected in the EDXA mapping on the SEM. The majority of signal (>80%) in the Sc $2p_{3/2}$ region (Figure 7a), however, can be accounted for by the presence of octahedrally coordinated scandium, assigned using ScOOH as a reference, as expected for the MIL-88B structure. The Fe $2p_{3/2}$ region of the XPS (Figure 7b) is more complex due to multiplet splitting [53–56]. Careful analysis of the overall Fe $2p_{3/2}$ envelope necessitated a component at around 709 eV, indicating the presence of a significant amount of Fe^{2+} as well as Fe^{3+}. This is unlike the spectrum of the single metal MIL-88B(Fe) (Supplementary Materials) that shows >90% Fe^{3+}, as expected, and would suggest that the mixed-metal material contains defects (see below for a further discussion of this).

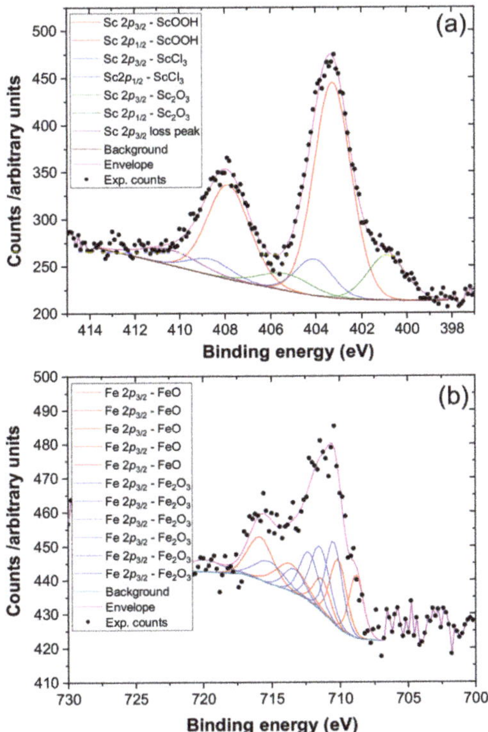

Figure 7. X-ray photoelectron spectra of MIL-88B(Fe,Sc) in (**a**) the Sc *2p* region and (**b**) the Fe *2p*$_{3/2}$
region. The fitted contributions are shown with reference spectra in the literature used for guidance
(see text).

The FT-IR spectra of the MIL-88B materials (Figure 8) shows all the vibrational bands characteristic
of framework-coordinated carboxylate groups expected in the MIL-88B structure, and further evidence
for the isostructural nature of the samples (see Supplementary Materials for assignment of the spectra).
Some weak additional bands in the spectrum of the MIL-88(Fe,Sc) sample suggest the presence of
some uncoordinated (protonated) carboxylic acid groups. This is not due to the presence of crystalline
H_2BDC (not seen in the powder XRD, Figure 5, and which also has a different FT-IR spectrum), and
therefore suggests a defective structure for the mixed-metal material, with some linkers that do not
bridge pairs of clusters. The thermogravimetric analysis (TGA) data (Supplementary Materials) show
that all the solvents and water molecules in MIL-88B materials are removed at above ~150 °C. On
continued heating, a decomposition of the MIL-88B network occurs above 450 °C, forming Fe_2O_3 and/or
Sc_2O_3 as the final product, which is consistent with the literature [47]. The TGA of the mixed-metal
material (Supplementary Materials) shows an extra inflection in mass loss in the region corresponding
to the combustion of the linker, which is consistent with the presence of some partially coordinated
benzene-1,4-dicarboxylate groups. This is consistent with the more defective structure of this material,
also indicated by IR, and charge balance in the material may be explained by the presence of some Fe^{2+}
that was seen by XPS, as discussed above.

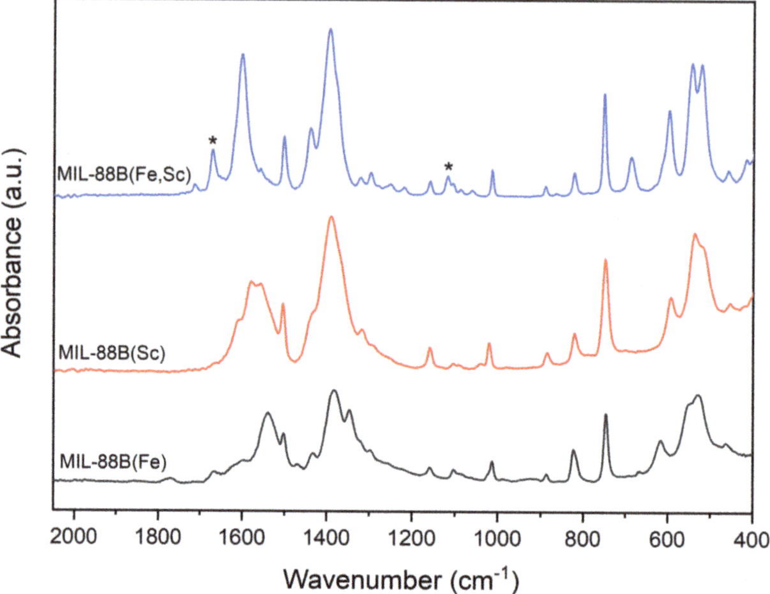

Figure 8. Infrared spectra of MIL-88B materials. The features labelled * are assigned as due to uncoordinated carboxylic acid groups.

Three samples were selected to examine their effectiveness as catalysts for glucose conversion: MIL-88B(Fe), MIL-88B(Fe,Sc), and MIL-88B(Sc). Given the instability of the materials in water, DMSO was used as the reaction solvent, in which the materials proved to be stable (Figure 4). The catalytic results are summarised in Table 1, which shows the total glucose conversion as well as the three desired products analysed for: fructose, mannose, and 5-HMF. We report results from two different catalyst loadings (based on mass, 30:1 and 7.5:1, glucose to catalyst ratio) and at two different temperatures (120 °C and 140 °C). Firstly, it is important to note that DMSO alone, in the absence of any catalyst, results in a conversion of just 1.6% at 120 °C and 60.3% at 140 °C. However, the product mixture contains only a minor amount of 5-HMF at the upper reaction temperature and neither of the intermediate sugars, fructose or mannose, are present, which suggests a decomposition pathway that does not yield these desired products. Indeed, Jia et al. found that anhydrous DMSO alone gave cellobiose as the main dehydration product of glucose at 130 °C for 2 hours [57]. In the presence of the MIL-88B solids, a much more significant formation of 5-HMF is seen, along with the intermediate fructose and mannose sugars. DMSO is known to act as a catalyst for the dehydration of fructose to 5-HMF in the absence of any Brønsted acid at 150 °C [58]. This implies that the MIL-88B acts as a Lewis acid catalyst that produce the fructose by glucose isomerisation, which can then be converted to 5-HMF by the DMSO solvent.

The mixed-metal MIL-88B(Fe,Sc) gives the highest conversion (70.7%) and 5-HMF yield (24.9%) at 140 °C with a 7.5 substrate–catalyst ratio. This catalyst thus produces significant amounts of 5-HMF from glucose. The mixed-metal material is more effective than either the MIL-88B(Fe) or MIL-88B(Sc), which may be due to greater structural disorder leading to a greater concentration of defects, such as coordinatively unsaturated metal sites, implying more Lewis acid sites. As noted above, the degree of mixing of the metals is difficult to prove unambiguously, but since the mixed-metal material always gives a higher glucose conversion than either of the single metal materials, there must be a cooperative effect, which in turn suggests that the two metals are in close proximity. In previous work on MIL-100(Sc), it was noted that upon addition of excess Fe, the presence of small amounts of amorphous Fe_2O_3 enhanced the Lewis acidity [41], so it is possible that we are seeing the same beneficial effect here, although the presence of small amounts of an amorphous second phase is difficult to determine.

The effect of temperature is also significant on the catalysis. At the temperature of 120 °C and the catalyst-to-substrate ratio of 7.5, the selectivity towards 5-HMF is notably higher over the mixed-metal material compared to Fe-only material (16.8% vs 1.7%). At the lower catalyst-to-substrate ratio of 30:1, the mixed-metal material shows a slightly lower activity than the Fe-only material, but a notably higher 5-HMF selectivity. The varying response of the two materials to temperature at both catalyst loadings is consistent with a different, and more favourable, mode of action in the mixed-metal material. Interestingly, the scandium material has a higher selectivity to 5-HMF than the iron at the catalyst low loadings, yet lower total conversion, providing evidence that each metal has a distinct role in the catalysis. Vlachos and co-workers studied the kinetics and mechanism of tandem Lewis–Brønsted acid catalysis of the same reaction, and reported that relative Lewis–Brønsted acid concentration has an optimum and pronounced effect on 5-HMF yield, which leads to 5-HMF yield being maximised due to a change in the rate-limiting step from fructose dehydration to glucose isomerisation [59]. In this sense, the MIL-88B(Fe) catalyst shows the formation of fructose both at 120 °C and 140 °C; however, the MIL-88B(Sc) catalyst yields higher 5-HMF yields. On the other hand, the mixed-metal MIL-88B(Fe,Sc) yields the highest 5-HMF yields, which is more clearly pronounced at 140 °C. Hence, the origin of the behaviour of our catalysts may be due to the distribution and availability of acid sites.

We considered the recyclability of the MIL-88B(Fe,Sc) catalyst by isolating the solid after a run, washing with DMSO, and adding fresh reagents (Figure 9). This was done at a larger scale than the screening reactions reported in Table 1 so mass transport may be different; however, the material maintains activity after three cycles and only after the fourth use shows a notable drop in glucose conversion. The recycling of MOF catalysts has not always been reported in the literature, but for the results that are available, we note that, although the recyclability of MIL-88B(Fe,Sc) is not as complete as for some sulfonated MIL-101(Cr) materials (albeit under different reaction conditions) [27,29], others have noted that the build-up of humins can make recycling difficult [10]. Nevertheless, the observation of recyclability points towards the use of the materials in continuous reactors, as has been achieved for sulfonated MIL-101(Cr) in a fixed-bed system [26]; here, conversion need not be high to generate useful quantities of the desired product.

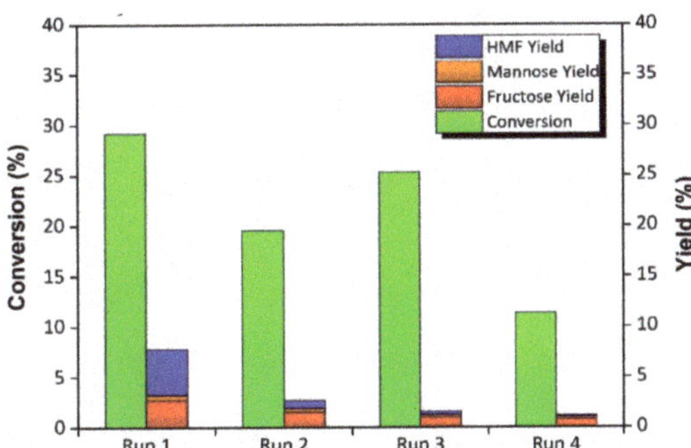

Figure 9. Recyclability of MIL-88B(Fe,Sc) as a catalyst for glucose conversion (120 °C in DMSO for 3 hours).

In comparison to other MOF-based solid acids that have been recently investigated as catalysts for glucose conversion, the MIL-88B(Fe,Sc) shows favourable properties. Tsapatsis and co-workers have recently showed that a composite of MIL-101(Cr) and nanoparticles of $Cr(OH)_3$ gave the highest conversion of glucose in a two-step process with ethanol as the solvent for isomerisation to ethyl fructoside, followed by the addition of water to allow hydrolysis to fructose, each performed at 100 °C for 24 hours: the catalyst was more effective than a physical mixture of the two component solids and various other MIL-101(Cr) materials, but also outperformed the zeolite Sn-beta giving >75% conversion, with almost 60% of the product fructose [29]. Yabushita et al. used the zirconium MOF NU-1000 modified with phosphate to optimise the 5-HMF yield from glucose and also screened various solvent mixtures: for their catalysis, however, water-DMSO gave the lowest overall glucose conversion in comparable conditions to those we have used (at 140 °C for 5 hours), and THF or propanols as solvents gave almost complete conversion [30]. Looking at other solid catalysts applied in DMSO solvent, the recent work of Li et al. on MnO_2 particles surface modified with Sn^{4+}-containing coordination polymers provides a useful comparison [60]. Their catalyst achieved ~56% conversion at 150 °C in 5 hours, with a 60% selectivity towards 5-HMF, which was superior to other Sn-based catalysts reported in the literature in DMSO solvent. Therefore, the MIL-88B(Fe,Sc) catalyst is comparable to the Sn-based catalyst (with higher conversion, although lower selectivity), but has the advantage of being prepared in a single-step, rather than being a surface-modified composite material. Other reported systems that show a higher conversion of glucose to 5-HMF in DMSO involve a multi-step synthesis of the catalyst and strong mineral acids. This includes Sn-containing vanadium phosphate, prepared in three steps starting from a phosphoric acid treatment of vanadium oxide [61]; sulfated mesoporous niobium oxide prepared in several steps, involving ultrasound and hydrothermal treatments, with a final sulfuric acid treatment [62]; and sulfated zirconia materials that are activated using chlorosulfonic acid in ethylene dichloride [63].

The nature and location of the Lewis acid sites in MOF catalysts have not been extensively studied in glucose isomerisation reactions, and the accessibility of active sites will depend on the substrates used. Ammonia temperature-programmed desorption (TPD) was used to characterise the MIL-88B(Fe,Sc) material (Figure 10), and this confirms the presence of surface acid sites at a concentration of 0.36 mmol g^{-1}. The results are complicated by the onset of collapse of the MOF at higher temperatures (see TGA in Supplementary Materials), but a desorption event between 100 and 300 °C is characteristic of the interaction of ammonia with surface acid sites. This is similar behaviour to that reported for the Cu-containing MOF-74 by Jiang et al., who observed desorption of ammonia around 100–250 °C under similar conditions [64].

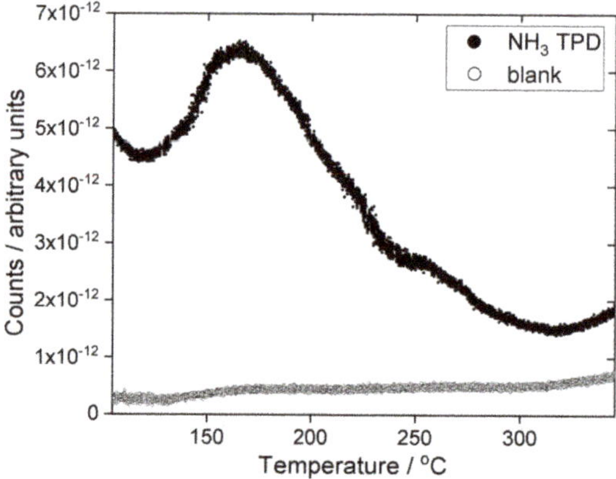

Figure 10. Temperature-programmed desorption of ammonia from MIL-88B(Fe,Sc), with data from a fresh sample of the material (without ammonia adsorbed) heated under the same conditions as a blank.

Table 1. Results of catalytic conversion of glucose using MIL-88B catalysts. All reactions were performed in DMSO and the results reported are after 3 hours of reaction time. Product selectivity is defined as 100 × (product yield/conversion).

Temperature (°C)	Substrate to Catalyst Ratio (wt./wt.)	Catalyst	Conversion (%)	Product Yield (% Mole)			Product Selectivity (% Mole)		
				Fructose	Mannose	5-HMF	Fructose	Mannose	5-HMF
120	∞	none	1.6	0.0	0.0	0.01	0.0	0.0	0.69
	30:1	MIL-88B(Fe)	18.9	6.3	0.3	0.0	33.3	1.6	0.0
		MIL-88B(Sc)	9.1	1.0	0.4	0.1	11.0	4.4	1.1
		MIL-88B(Fe,Sc)	11.5	1.1	0.2	0.7	9.6	1.7	6.1
140	∞	none	60.3	0.0	0.0	0.6	0.0	0.0	1.0
	30:1	MIL-88B(Fe)	54.2	7.6	1.2	0.9	14.0	2.2	1.7
		MIL-88B(Sc)	28.8	2.1	1.1	3.2	7.3	3.8	11.1
		MIL-88B(Fe,Sc)	53.3	3.6	2.0	17.8	6.8	3.8	33.4
120	7.5:1	MIL-88B(Fe)	24.2	3.2	0.0	0.4	13.2	0.0	1.7
		MIL-88B(Fe,Sc)	19.7	0.8	2.3	3.3	4.1	11.7	16.8
140	7.5:1	MIL-88B(Fe)	55.6	3.6	1.0	3.8	6.5	1.8	6.8
		MIL-88B(Fe,Sc)	70.7	3.3	1.4	24.9	4.7	2.0	35.2

While MIL-88B is a porous, breathing framework in the presence of certain solvents [51], it is likely that the bulky glucose does not penetrate into the MOF structure, but the activity is instead associated with the surface of the crystallites. Additional PXRD measurements made on samples immersed in DMSO show no evidence for pore-opening in this solvent at room temperature (Supplementary Materials), so it is probable that the material remains in a contracted form during the catalysis. Indeed, after catalysis and ammonia TPD, there is no evidence for pore swelling in the PXRD (Supplementary Materials). Our measured BET surface area of 16 m^2 g^{-1} for MIL-88B(Fe,Sc) activated in vacuum at 100 °C corroborates this, showing the lack of porosity and a surface area as expected for micron-sized powders. This result is similar to that reported for other MIL-88B(Fe) materials: for example, Rahmani and Rahmani measured a BET surface area of 23 m^2 g^{-1} [65], while Vuong et al. found a surface area of 30 m^2 g^{-1} for a hydrated, mixed (Fe, Ni) form of the material with amino-functionalised linkers [66]. Previous work on dense (non-porous) scandium-organic frameworks has shown how surface Lewis acidity is sufficient for the acetalization of aldehydes or the cyanosilylation of carbonyl compounds [67,68]. Although Mitchell et al. found that MIL-88B(Sc) was an inferior Lewis acid catalyst compared to porous MIL-100(Sc) in a variety of organic transformations [41], which they attributed to its low surface area, in DMSO it is clear that the Lewis acidity is sufficient for the glucose isomerisation.

3. Discussion

The MIL-88B material provides the Lewis acidity that catalyses isomerisation of glucose to fructose, and a cooperative effect of the Fe and Sc in the mixed-metal MIL-88B MOF provides optimum catalytic activity to maximise the 5-HMF yields, while the DMSO solvent catalyses the fructose dehydration to 5-HMF. It is notable that the mixed-metal material shows enhanced reactivity compared to the iron or scandium catalysts, and the presence of a locally disordered structure, such as uncoordinated ligands, may be partly responsible for this. Further work is needed to elucidate the mechanism of reaction, such as NMR studies with isotopic substitution, especially since it is likely that the surface of the MOF crystallites provides the active sites for catalysis and that solvent effects may be at play in this chemistry, but our work shows how one-step synthesis of MOF catalysts under moderate conditions provides a convenient route to acid catalysts for biomass transformations. The long-term stability of the catalysts and their implementation in flow reactors is the next step towards realistic use.

4. Materials and Methods

4.1. Materials and Chemicals

All reagents were purchased and used without further purification. $FeCl_3 \cdot 6H_2O$ (99.9%) was obtained from Sigma Aldrich (Gillingham, United Kingdom). $ScCl_3 \cdot 6H_2O$ (99.9%) and benzene-1,4-dicarboxylic acid (H_2BDC, 98%) were obtained from Alfa Aesar (Heysham, United Kingdom). Monosodium 2-sulfobenzene-1,4-dicarboxylate (NaH_2MSBDC) was obtained from Tokyo Chemical Industry (TCI, >98%, Tokyo, Japan). DMF was purchased from Fischer Scientific (Loughborough, United Kingdom) and absolute ethanol was purchased from VWR Chemicals.

4.2. Synthesis of MIL-101(Fe,Sc)

MIL-101(Fe) was prepared following the method of Zhu et al. [43], $FeCl_3 \cdot 6H_2O$ (0.894 g, 3.3 mmol), H_2BDC (0.280 g, 1.65 mmol), and DMF (20 ml) were mixed with stirring at room temperature for 2 h. The mixture was transferred to a Teflon-lined autoclave (45 mL) and heated at 383 K for either 3 or 40 hours (see below). After cooling to room temperature, the mixture was filtered and the solid product was washed with ethanol (20 mL) at 60 °C for 3 h. The product was dried at 60 °C overnight. Variation of metal ratio (Fe:Sc) and ligand using MSBDC in place of BDC was investigated by following the same synthesis method, but replacing the $FeCl_3 \cdot 6H_2O$ with $ScCl_3 \cdot 6H_2O$ and/or the H_2BDC with NaH_2MSBDC.

4.3. Synthesis of MIL-88B(Sc)

For comparison, MIL-88B(Sc) was prepared using a method based on that of Mowat et al. [39], where 0.22 g (1.45 mmol), scandium (III) chloride, 0.14 g (8.43 mmol) H_2BDC, and 9 mL DMF (Fisher Scientific) were stirred for 5 minutes prior to transfer into a 45 mL Teflon -lined autoclave and heated to 140 °C for 48 h before cooling to room temperature. The solid product was isolated by suction filtration and washed with 25 mL ethanol by stirring at room temperature for 24 h. The recovered solid was finally dried at 70 °C in the air for 24 h.

4.4. Materials Characterisation

For sample identification, PXRD patterns were recorded on a D5000 Siemens diffractometer with Cu K$\alpha_{1/2}$ radiation (λ = 1.54184 Å) and 2θ= 2–30° (step size 0.02° in 2θ), operated at 40 kV and 40 mA. High resolution PXRD patterns were recorded using a Panalytical X'Pert Pro MPD (Malvern Panalytical, Malvern, United Kingdom), equipped with monochromatic Cu K$_{\alpha1}$ radiation (λ = 1.54056 Å) and a PIXcel solid state detector. Profile fitting of the powder patterns was performed using the GSAS software (revision 1188, Los Alamos, USA) [69] implemented using the EXPGUI interface [70]. FT-IR spectra were measured at room temperature in the range 400–4000 cm^{-1} using a Platinum-ATR Bruker Alfa instrument (Bruker Optics GmbH, Ettlingen, Germany). The stability of the catalysts was investigated by TGA using a Mettler Toledo TGA/DSC1 instrument (Leicester, United Kingdom) under ambient air pressure and a heating rate of 10 °C min^{-1}. Samples were heated in air from 25 °C to 1000 °C. Selected elemental compositions of materials were determined by energy dispersive X-ray (EDX) analysis using a FEI scanning electron microscope (SEM, Fei UK Limited, Altrincham, United Kingdom). Elemental analysis for metals was performed by MEDAC Ltd, Surrey, United Kingdom, using the ICP-OES method after acid digestion. XPS was measured using a Kratos AXIS Ultra DLD (Manchester, United Kingdom). XPS measurements were carried out in a UHV system with a base pressure of 5×10^{-11} mbar. The sample was excited with X-rays from a monochromated Al Kα source (1486.7 eV), with the photo-electrons being detected at a 90° take-off angle with respect to the sample surface. Curve fitting was performed using the CasaXPS package, incorporating Voigt (mixed Gaussian–Lorentzian) line shapes and Shirley backgrounds for all regions except the Sc 2p region, where a U 2 Tougaard background was found to be more appropriate.

The acidity of materials was characterised using ammonia TPD. An excess of 0.02 vol% ammonia in helium was dosed onto 50–70 mg of a catalyst contained in a quartz tube at 100 °C (to minimise physisorption). The ammonia was then desorbed from the catalysts by heating the material to 350 °C at a ramp rate of 2 °C min^{-1}. To ensure the complete desorption of ammonia from the material, the temperature was then maintained at 350 °C for 6 hours. The amount of ammonia desorbed from the catalyst was measured using a mass spectrometer at m/z = 17 with the interference with water vapour taken into account. A blank experiment (used as a baseline) was performed with a fresh catalyst, but without ammonia pre-adsorption. Surface area measurements were performed using nitrogen adsorption via the Brunauer–Emmett–Teller (BET) method using a QUADRASORB (gas sorption surface area analyzer) (Quantachrome UK, Hook, United Kingdom) after degassing samples under vacuum at 100 °C.

4.5. Catalytic Tests

Catalytic screening was carried out in 4 mL batch reactors at 393 K or 413 K. 3 mL of 10 wt. % glucose solution (in DMSO) was heated to the desired temperature together with the MOF catalyst and a magnetic stirring bar (either 10 or 40 mg, corresponding to a 30 or 7.5 substrate:catalyst ratio, respectively) for 3 hours. Blank experiments were also carried out without catalyst. The products were analysed by high performance liquid chromatography (HPLC) equipped with a Bio-Rad HPX 87P column; a photo diode array detector and evaporative light scattering detector (ELSD) were used to monitor 5-HMF and sugars, respectively. The mobile phase was water with 0.6 mL min^{-1} flow rate. The products and the reactant (glucose) were quantified by calibration with external standard solutions. Recycle reactions were conducted in a 25 mL reactor with PTFE lining (Berghof, BR-25). In a typical reaction, 200 mg of catalyst and a magnetic stirring bar was placed into the reactor. 15 mL of a solution of 10 wt. % glucose in DMSO was then added. The reactor was sealed and pressurized to 10 bar with helium. The reactor was brought to reaction temperature (120 °C) by placing it into a preheated aluminium block heated via a heating/stirring plate. At the end of the reaction (3 hours), the reactor was removed from the heating block and quenched in an ice bath at 0 °C to stop the reaction. The reactor was then depressurised and opened. The solid catalyst was recovered from the reaction solution using a centrifuge and washed with DMSO. The reaction solution was filtered and analysed using a Shimadzu HPLC (Shimadzu UK Ltd, Milton Keynes, United Kingdom), as described above. In the subsequent reaction tests, the recovered catalyst was added back into the 25 mL reactor along with fresh stock solution. The reaction procedure was then repeated under the same conditions in order to test the recyclability of the catalyst, and products were analysed as described above.

5. Conclusions

The replacement of Cr^{3+} in MIL-101 by more benign metal cations, Fe^{3+} and Sc^{3+}, leads to an instability of the structure and the formation of the polymorphic MOF MIL-88B structure. Although MIL-88B(Fe,Sc) materials are unstable under aqueous hydrothermal conditions, they are stable in DMSO at 140 °C and efficiently transform glucose to 5-HMF with high conversion and selectivity. While DMSO is less desirable as a solvent than water, the simplicity of the synthesis of the MIL-88B catalyst materials in a single-step is a significant advantage over the more conventional solid–acid catalysts that have so far been proposed for this reaction.

Supplementary Materials: Additional characterisation data. The following are available online at http://www.mdpi.com/2073-4344/9/5/43/s1. Figure S1: XPS of MIL-88B(Fe), Table S1: Assignment of key bands in the IR spectra, Figure S2: Thermogravimetric analysis of MIL-88B catalysts, Figure S3: Powder on samples immersed in DMSO, Figure S4: Powder XRD of MIL-88(Fe,Sc) as-made, after catalysis, and after ammonia TPD.

Author Contributions: Conceptualization, V.D., Y.K.K., and R.I.W.; Formal Analysis, R.P., R.O., D.L.B, T.W.C, N.C., M.W., R.J.K.; Investigation, R.P., R.O., D.L.B, T.W.C, N.C., M.W., R.J.K.; Writing—Original Draft Preparation, R.P, and R.I.W.; Writing—Review and Editing, R.P., R.O., D.L.B., Y.K.K, V.D.; Supervision, V.D., Y.K.K., and R.I.W.; Project Administration, V.D., Y.K.K., and R.I.W.; Funding Acquisition, V.D., Y.K.K., and R.I.W.

Acknowledgments: This work was supported by the Royal Society [Challenge Grant CH160099], the UK Engineering and Physical Sciences Research Council (EPSRC) [grant EP/P511432/1 Global Challenge Research Fund (GCRF) Institutional Award for the University of Warwick], and a Research Grants for Higher Education 2016 from DGHE, Indonesia [No.1136/UN2.R12/HKP/05.00/2016]. We thank Baiwen Zhao for his assistance with measuring surface area. The data supporting this article are available at https://wrap.warwick.ac.uk/116888.

Conflicts of Interest: The authors declare no conflict of interest. The sponsors had no role in the design, execution, interpretation, or writing of the study.

References

1. Rosatella, A.A.; Simeonov, S.P.; Frade, R.F.M.; Afonso, C.A.M. 5-Hydroxymethylfurfural (HMF) as a building block platform: Biological properties, synthesis and synthetic applications. *Green Chem.* **2011**, *13*, 754–793. [CrossRef]

2. Van Putten, R.J.; van der Waal, J.C.; de Jong, E.; Rasrendra, C.B.; Heeres, H.J.; de Vries, J.G. Hydroxymethylfurfural, a Versatile Platform Chemical Made from Renewable Resources. *Chem. Rev.* **2013**, *113*, 1499–1597. [CrossRef]

3. Teong, S.P.; Yi, G.S.; Zhang, Y.G. Hydroxymethylfurfural production from bioresources: Past, present and future. *Green Chem.* **2014**, *16*, 2015–2026. [CrossRef]

4. Carson, F.; Su, J.; Platero-Prats, A.E.; Wan, W.; Yun, Y.F.; Samain, L.; Zou, X.D. Framework Isomerism in Vanadium Metal-Organic Frameworks: MIL-88B(V) and MIL-101(V). *Cryst. Growth Des.* **2013**, *13*, 5036–5044. [CrossRef]

5. Roman-Leshkov, Y.; Moliner, M.; Labinger, J.A.; Davis, M.E. Mechanism of Glucose Isomerization Using a Solid Lewis Acid Catalyst in Water. *Angew. Chem. Int. Edit.* **2010**, *49*, 8954–8957. [CrossRef] [PubMed]

6. Pidko, E.A.; Degirmenci, V.; van Santen, R.A.; Hensen, E.J.M. Glucose Activation by Transient Cr^{2+} Dimers. *Angew. Chem. Int. Edit.* **2010**, *49*, 2530–2534. [CrossRef] [PubMed]

7. Pagan-Torres, Y.J.; Wang, T.F.; Gallo, J.M.R.; Shanks, B.H.; Dumesic, J.A. Production of 5-Hydroxymethylfurfural from Glucose Using a Combination of Lewis and Bronsted Acid Catalysts in Water in a Biphasic Reactor with an Alkylphenol Solvent. *ACS Catal.* **2012**, *2*, 930–934. [CrossRef]

8. Choudhary, V.; Mushrif, S.H.; Ho, C.; Anderko, A.; Nikolakis, V.; Marinkovic, N.S.; Frenkel, A.I.; Sandler, S.I.; Vlachos, D.G. Insights into the Interplay of Lewis and Brønsted Acid Catalysts in Glucose and Fructose Conversion to 5-(Hydroxymethyl)furfural and Levulinic Acid in Aqueous Media. *J. Am. Chem. Soc.* **2013**, *135*, 3997–4006. [CrossRef]

9. Moliner, M.; Roman-Leshkov, Y.; Davis, M.E. Tin-containing zeolites are highly active catalysts for the isomerization of glucose in water. *Proc. Nat. Acad. Sci. USA* **2010**, *107*, 6164–6168. [CrossRef] [PubMed]

10. Herbst, A.; Janiak, C. Selective glucose conversion to 5-hydroxymethylfurfural (5-HMF) instead of levulinic acid with MIL-101Cr MOF-derivatives. *New J. Chem.* **2016**, *40*, 7958–7967. [CrossRef]

11. Nikolla, E.; Roman-Leshkov, Y.; Moliner, M.; Davis, M.E. "One-Pot" Synthesis of 5-(Hydroxymethyl)furfural from Carbohydrates using Tin-Beta Zeolite. *ACS Catal.* **2011**, *1*, 408–410. [CrossRef]

12. Degirmenci, V.; Hensen, E.J.M. Development of a Heterogeneous Catalyst for Lignocellulosic Biomass Conversion: Glucose Dehydration by Metal Chlorides in a Silica-Supported Ionic Liquid Layer. *Environ. Prog. Sustainable Energy* **2014**, *33*, 657–662. [CrossRef]

13. Furukawa, H.; Cordova, K.E.; O'Keeffe, M.; Yaghi, O.M. The Chemistry and Applications of Metal-Organic Frameworks. *Science* **2013**, *341*, 1230444. [CrossRef] [PubMed]

14. Seo, P.W.; Bhadra, B.N.; Ahmed, I.; Khan, N.A.; Jhung, S.H. Adsorptive Removal of Pharmaceuticals and Personal Care Products from Water with Functionalized Metal-organic Frameworks: Remarkable Adsorbents with Hydrogen-bonding Abilities. *Sci. Rep.* **2016**, *6*, 34462. [CrossRef] [PubMed]

15. Li, B.; Wen, H.M.; Zhou, W.; Chen, B.L. Porous Metal-Organic Frameworks for Gas Storage and Separation: What, How, and Why? *J. Phys. Chem. Lett.* **2014**, *5*, 3468–3479. [CrossRef] [PubMed]

16. Farrusseng, D.; Aguado, S.; Pinel, C. Metal-Organic Frameworks: Opportunities for Catalysis. *Angew. Chem. Int. Edit.* **2009**, *48*, 7502–7513. [CrossRef] [PubMed]

17. Corma, A.; Garcia, H.; Xamena, F. Engineering Metal Organic Frameworks for Heterogeneous Catalysis. *Chem. Rev.* **2010**, *110*, 4606–4655. [CrossRef]

18. Isaeva, V.I.; Kustov, L.M. The application of metal-organic frameworks in catalysis (Review). *Petrol. Chem.* **2010**, *50*, 167–180. [CrossRef]

19. Chughtai, A.H.; Ahmad, N.; Younus, H.A.; Laypkov, A.; Verpoort, F. Metal-organic frameworks: versatile heterogeneous catalysts for efficient catalytic organic transformations. *Chem. Soc. Rev.* **2015**, *44*, 6804–6849. [CrossRef]

20. Herbst, A.; Janiak, C. MOF catalysts in biomass upgrading towards value-added fine chemicals. *CrystEngCommun* **2017**, *19*, 4092–4117. [CrossRef]

21. Huang, Y.B.; Liang, J.; Wang, X.S.; Cao, R. Multifunctional metal-organic framework catalysts: synergistic catalysis and tandem reactions. *Chem. Soc. Rev.* **2017**, *46*, 126–157. [CrossRef] [PubMed]

22. Jiang, J.C.; Yaghi, O.M. Brønsted Acidity in Metal-Organic Frameworks. *Chem. Rev.* **2015**, *115*, 6966–6997. [CrossRef]

23. Goesten, M.G.; Juan-Alcaniz, J.; Ramos-Fernandez, E.V.; Gupta, K.; Stavitski, E.; van Bekkum, H.; Gascon, J.; Kapteijn, F. Sulfation of metal-organic frameworks: Opportunities for acid catalysis and proton conductivity. *J. Catal.* **2011**, *281*, 177–187. [CrossRef]

24. Juan-Alcaniz, J.; Gielisse, R.; Lago, A.B.; Ramos-Fernandez, E.V.; Serra-Crespo, P.; Devic, T.; Guillou, N.; Serre, C.; Kapteijn, F.; Gascon, J. Towards acid MOFs—catalytic performance of sulfonic acid functionalized architectures. *Catal. Sci. Technol.* **2013**, *3*, 2311–2318. [CrossRef]

25. Hu, Z.G.; Zhao, D. Metal-organic frameworks with Lewis acidity: Synthesis, characterization, and catalytic applications. *CrystEngCommun* **2017**, *19*, 4066–4081. [CrossRef]

26. Su, Y.; Chang, G.G.; Zhang, Z.G.; Xing, H.B.; Su, B.G.; Yang, Q.W.; Ren, Q.L.; Yang, Y.W.; Bao, Z.B. Catalytic dehydration of glucose to 5-hydroxymethylfurfural with a bifunctional metal-organic framework. *AICHE J.* **2016**, *62*, 4403–4417. [CrossRef]

27. Chen, J.Z.; Li, K.G.; Chen, L.M.; Liu, R.L.; Huang, X.; Ye, D.Q. Conversion of fructose into 5-hydroxymethylfurfural catalyzed by recyclable sulfonic acid-functionalized metal-organic frameworks. *Green Chem.* **2014**, *16*, 2490–2499. [CrossRef]

28. Zhang, Y.M.; Degirmenci, V.; Li, C.; Hensen, E.J.M. Phosphotungstic Acid Encapsulated in Metal-Organic Framework as Catalysts for Carbohydrate Dehydration to 5-Hydroxymethylfurfural. *ChemSusChem* **2011**, *4*, 59–64. [CrossRef] [PubMed]

29. Guo, Q.; Ren, L.M.; Kumar, P.; Cybulskis, V.J.; Mkhoyan, K.A.; Davis, M.E.; Tsapatsis, M. A Chromium Hydroxide/MIL-101(Cr) MOF Composite Catalyst and Its Use for the Selective Isomerization of Glucose to Fructose. *Angew. Chem. Int. Edit.* **2018**, *57*, 4926–4930. [CrossRef] [PubMed]

30. Yabushita, M.; Li, P.; Islamoglu, T.; Kobayashi, H.; Fukuoka, A.; Farha, O.K.; Katz, A. Selective Metal-Organic Framework Catalysis of Glucose to 5-Hydroxymethylfurfural Using Phosphate-Modified NU-1000. *Ind. Eng. Chem. Res.* **2017**, *56*, 7141–7148. [CrossRef]

31. Oozeerally, R.; Burnett, D.L.; Chamberlain, T.W.; Walton, R.I.; Degirmenci, V. Exceptionally Efficient and Recylable Heterogeneous Metal-Organic Framework Catalyst for Glucose Isomerisation in Water. *ChemCatChem* **2018**, *10*, 706–709. [CrossRef]

32. De Mello, M.D.; Tsapatsis, M. Selective Glucose-to-Fructose Isomerization over Modified Zirconium UiO-66 in Alcohol Media. *ChemCatChem* **2018**, *10*, 2417–2423. [CrossRef]

33. Zhang, Y.L.; Jin, P.; Meng, M.J.; Gao, L.; Liu, M.; Yan, Y.S. Acid-Base Bifunctional Metal-Organic Frameworks: Green Synthesis and Application in One-Pot Glucose to 5-HMF Conversion. *Nano* **2018**, 13. [CrossRef]

34. Gong, J.; Katz, M.J.; Kerton, F.M. Catalytic conversion of glucose to 5-hydroxymethylfurfural using zirconium-containing metal-organic frameworks using microwave heating. *RSC Adv.* **2018**, *8*, 31618–31627. [CrossRef]

35. Desidery, L.; Yusubov, M.S.; Zhuiykov, S.; Verpoort, F. Fully-sulfonated hydrated UiO-66 as efficient catalyst for ethyl levulinate production by esterification. *Catal. Commun.* **2018**, *117*, 33–37. [CrossRef]

36. Férey, G.; Mellot-Draznieks, C.; Serre, C.; Millange, F.; Dutour, J.; Surblé, S.; Margiolaki, I. A chromium terephthalate-based solid with unusually large pore volumes and surface area. *Science* **2005**, *309*, 2040–2042. [CrossRef]

37. Dayan, A.D.; Paine, A.J. Mechanisms of chromium toxicity, carcinogenicity and allergenicity: Review of the literature from 1985 to 2000. *Hum. Exp. Toxicol.* **2001**, *20*, 439–451. [CrossRef] [PubMed]

38. Ishihara, K.; Kubota, M.; Kurihara, H.; Yamamoto, H. Scandium trifluoromethanesulfonate as an extremely active Lewis acid catalyst in acylation of alcohols with acid anhydrides and mixed anhydrides. *J. Org. Chem.* **1996**, *61*, 4560–4567. [CrossRef]

39. Mowat, J.P.S.; Miller, S.R.; Slawin, A.M.Z.; Seymour, V.R.; Ashbrook, S.E.; Wright, P.A. Synthesis, characterisation and adsorption properties of microporous scandium carboxylates with rigid and flexible frameworks. *Micropor. Mesopor. Mater.* **2011**, *142*, 322–333. [CrossRef]

40. Mitchell, L.; Gonzalez-Santiago, B.; Mowat, J.P.S.; Gunn, M.E.; Williamson, P.; Acerbi, N.; Clarke, M.L.; Wright, P.A. Remarkable Lewis acid catalytic performance of the scandium trimesate metal organic framework MIL-100(Sc) for C-C and C=N bond-forming reactions. *Catal. Sci. Technol.* **2013**, *3*, 606–617. [CrossRef]

41. Mitchell, L.; Williamson, P.; Ehrlichova, B.; Anderson, A.E.; Seymour, V.R.; Ashbrook, S.E.; Acerbi, N.; Daniels, L.M.; Walton, R.I.; Clarke, M.L.; Wright, P.A. Mixed-Metal MIL-100(Sc,M) (M=Al, Cr, Fe) for Lewis Acid Catalysis and Tandem C-C Bond Formation and Alcohol Oxidation. *Chem. Eur. J.* **2014**, *20*, 17185–17197. [CrossRef] [PubMed]

42. Yang, L.; Zhao, T.; Boldog, I.; Janiak, C.; Yang, X.Y.; Li, Q.; Zhou, Y.J.; Xia, Y.; Lai, D.W.; Liu, Y.J. Benzoic acid as a selector-modulator in the synthesis of MIL-88B(Cr) and nano-MIL-101(Cr). *Dalton Trans.* **2019**, *48*, 989–996. [CrossRef] [PubMed]

43. Zhu, T.T.; Zhang, Z.M.; Chen, W.L.; Liu, Z.J.; Wang, E.B. Encapsulation of tungstophosphoric acid into harmless MIL-101(Fe) for effectively removing cationic dye from aqueous solution. *RSC Adv.* **2016**, *6*, 81622–81630. [CrossRef]

44. Bromberg, L.; Diao, Y.; Wu, H.M.; Speakman, S.A.; Hatton, T.A. Chromium(III) Terephthalate Metal Organic Framework (MIL-101): HF-Free Synthesis, Structure, Polyoxometalate Composites, and Catalytic Properties. *Chem. Mater.* **2012**, *24*, 1664–1675. [CrossRef]

45. Wang, T.; Wang, J.; Zhang, C.L.; Yang, Z.; Dai, X.P.; Cheng, M.S.; Hou, X.H. Metal-organic framework MIL-101(Cr) as a sorbent of porous membrane-protected micro-solid-phase extraction for the analysis of six phthalate esters from drinking water: a combination of experimental and computational study. *Analyst* **2015**, *140*, 5308–5316. [CrossRef] [PubMed]

46. Montazerolghaem, M.; Aghamiri, S.F.; Tangestaninejad, S.; Talaie, M.R. A metal-organic framework MIL-101 doped with metal nanoparticles (Ni & Cu) and its effect on CO_2 adsorption properties. *RSC Adv.* **2016**, *6*, 632–640.

47. Surblé, S.; Serre, C.; Mellot-Draznieks, C.; Millange, F.; Férey, G. A new isoreticular class of metal-organic-frameworks with the MIL-88 topology. *Chem. Commun.* **2006**, 284–286. [CrossRef]

48. Bauer, S.; Serre, C.; Devic, T.; Horcajada, P.; Marrot, J.; Férey, G.; Stock, N. High-throughput assisted rationalization of the formation of metal organic frameworks in the iron(III) aminoterephthalate solvothermal system. *Inorg. Chem.* **2008**, *47*, 7568–7576. [CrossRef]

49. Tanasaro, T.; Adpakpang, K.; Ittisanronnachai, S.; Faungnawakij, K.; Butburee, T.; Wannapaiboon, S.; Ogawa, M.; Bureekaew, S. Control of Polymorphism of Metal-Organic Frameworks Using Mixed-Metal Approach. *Cryst. Growth Des.* **2018**, *18*, 16–21. [CrossRef]

50. Ma, M.Y.; Betard, A.; Weber, I.; Al-Hokbany, N.S.; Fischer, R.A.; Metzler-Nolte, N. Iron-Based Metal-Organic Frameworks MIL-88B and NH_2-MIL-88B: High Quality Microwave Synthesis and Solvent-Induced Lattice "Breathing". *Cryst. Growth Des.* **2013**, *13*, 2286–2291. [CrossRef]

51. Serre, C.; Mellot-Draznieks, C.; Surblé, S.; Audebrand, N.; Filinchuk, Y.; Férey, G. Role of solvent-host interactions that lead to very large swelling of hybrid frameworks. *Science* **2007**, *315*, 1828–1831. [CrossRef]

52. Shannon, R.D. Revised effective ionic radii and systematic studies of interatomic distances in halides and chalcogenides. *Acta Cryst.* **1976**, *32*, 751–767. [CrossRef]

53. Gupta, R.P.; Sen, S.K. Calculation of multiplet structure of core p -vacancy levels. II. *Phys. Rev. B* **1975**, *12*, 15–19. [CrossRef]

54. Gota, S.; Guiot, E.; Henriot, M.; Gautier-Soyer, M. Atomic-oxygen-assisted MBE growth of a-Fe_2O_3 on a-Al_2O_3 (0001): Metastable FeO(111)-like phase at subnanometer thicknesses. *Phys. Rev. B* **1999**, *60*, 14387–14395. [CrossRef]

55. Biesinger, M.C.; Payne, B.P.; Grosvenor, A.P.; Lau, L.W.M.; Gerson, A.R.; Smart, R.S. Resolving surface chemical states in XPS analysis of first row transition metals, oxides and hydroxides: Cr, Mn, Fe, Co and Ni. *Appl. Surf. Sci.* **2011**, *257*, 2717–2730. [CrossRef]

56. Jimenez-Cavero, P.; Lucas, I.; Anadon, A.; Ramos, R.; Niizeki, T.; Aguirre, M.H.; Algarabel, P.A.; Uchida, K.; Ibarra, M.R.; Saitoh, E.; Morellon, L. Spin Seebeck effect in insulating epitaxial gamma-Fe2O3 thin films. *APL Mater.* **2017**, *5*.

57. Jia, S.Y.; Xu, Z.W.; Zhang, Z.C. Catalytic conversion of glucose in dimethylsulfoxide/water binary mix with chromium trichloride: Role of water on the product distribution. *Chem. Eng. J.* **2014**, *254*, 333–339. [CrossRef]

58. Amarasekara, A.S.; Williams, L.D.; Ebede, C.C. Mechanism of the dehydration of D-fructose to 5-hydroxymethylfurfural in dimethyl sulfoxide at 150 °C: An NMR study. *Carbohydr. Res.* **2008**, *343*, 3021–3024. [CrossRef] [PubMed]

59. Swift, T.D.; Nguyen, H.; Anderko, A.; Nikolakis, V.; Vlachos, D.G. Tandem Lewis/Brønsted homogeneous acid catalysis: Conversion of glucose to 5-hydoxymethylfurfural in an aqueous chromium(III) chloride and hydrochloric acid solution. *Green Chem.* **2015**, *17*, 4725–4735. [CrossRef]

60. Li, K.; Du, M.M.; Ji, P.J. Multifunctional Tin-Based Heterogeneous Catalyst for Catalytic Conversion of Glucose to 5-Hydroxymethylfurfural. *ACS Sust. Chem. Eng.* **2018**, *6*, 5636–5644. [CrossRef]

61. Behera, G.C.; Parida, K.M. One-pot synthesis of 5-hydroxymethylfurfural: a significant biomass conversion over tin-promoted vanadium phosphate (Sn-VPO) catalyst. *Catal. Sci. Technol.* **2013**, *3*, 3278–3285. [CrossRef]

62. Ngee, E.L.S.; Gao, Y.J.; Chen, X.; Lee, T.M.; Hu, Z.G.; Zhao, D.; Yan, N. Sulfated Mesoporous Niobium Oxide Catalyzed 5-Hydroxymethylfurfural Formation from Sugars. *Ind. Eng. Chem. Res.* **2014**, *53*, 14225–14233. [CrossRef]

63. Yan, H.P.; Yang, Y.; Tong, D.M.; Xiang, X.; Hu, C.W. Catalytic conversion of glucose to 5-hydroxymethylfurfural over SO_4^{2-}/ZrO_2 and SO_4^{2-}/ZrO_2-Al_2O_3 solid acid catalysts. *Catal. Commun.* **2009**, *10*, 1558–1563. [CrossRef]

64. Jiang, H.X.; Zhou, J.L.; Wang, C.X.; Li, Y.H.; Chen, Y.F.; Zhang, M.H. Effect of Cosolvent and Temperature on the Structures and Properties of Cu-MOF-74 in Low-temperature NH_3-SCR. *Ind. Eng. Chem. Res.* **2017**, *56*, 3542–3550. [CrossRef]

65. Rahmani, E.; Rahmani, M. Alkylation of benzene over Fe-based metal organic frameworks (MOFs) at low temperature condition. *Micropor. Mesopor. Mater.* **2017**, *249*, 118–127. [CrossRef]
66. Vuong, G.T.; Pham, M.H.; Do, T.O. Synthesis and engineering porosity of a mixed metal Fe_2Ni MIL-88B metal-organic framework. *Dalton Trans.* **2013**, *42*, 550–557. [CrossRef] [PubMed]
67. Perles, J.; Iglesias, M.; Ruiz-Valero, C.; Snejko, N. Rare-earths as catalytic centres in organo-inorganic polymeric frameworks. *J. Mater. Chem.* **2004**, *14*, 2683–2689. [CrossRef]
68. Gandara, F.; Gomez-Lor, B.; Iglesias, M.; Snejko, N.; Gutierrez-Puebla, E.; Monge, A. A new scandium metal organic framework built up from octadecasil zeolitic cages as heterogeneous catalyst. *Chem. Commun.* **2009**, 2393–2395. [CrossRef]
69. Larson, A.C.; Dreele, R.B.V. *General Structure Analysis System (GSAS)*; Los Alamos National Laboratory Report 1994, LAUR 86–748; Los Alamos National Laboratory: Los Alamos, NM, USA, 1994.
70. Toby, B.H. EXPGUI, a graphical user interface for GSAS. *J. Appl. Cryst.* **2001**, *34*, 210–213. [CrossRef]

Review

Metal-Organic Frameworks as a Platform for CO_2 Capture and Chemical Processes: Adsorption, Membrane Separation, Catalytic-Conversion, and Electrochemical Reduction of CO_2

Salma Ehab Mohamed Elhenawy [1], Majeda Khraisheh [1,*], Fares AlMomani [1] and Gavin Walker [2]

[1]   Department of Chemical Engineering, College of Engineering, Qatar University, Doha P.O. Box 2713, Qatar; se1105821@student.qu.edu.qa (S.E.M.E.); falmomani@qu.edu.qa (F.A.)

[2]   Bernal Institute, Department of Chemical Sciences, University of Limerick, V94 T9PX Limerick, Ireland; Gavin.Walker@ul.ie

*   Correspondence: m.khraisheh@qu.edu.qa; Tel.: +974-4403-4990 (ext. 4138); Fax: +974-4403-4001

Received: 17 September 2020; Accepted: 28 October 2020; Published: 9 November 2020

Abstract: The continuous rise in the atmospheric concentration of carbon dioxide gas (CO_2) is of significant global concern. Several methodologies and technologies are proposed and applied by the industries to mitigate the emissions of CO_2 into the atmosphere. This review article offers a large number of studies that aim to capture, convert, or reduce CO_2 by using a superb porous class of materials (metal-organic frameworks, MOFs), aiming to tackle this worldwide issue. MOFs possess several remarkable features ranging from high surface area and porosity to functionality and morphology. As a result of these unique features, MOFs were selected as the main class of porous material in this review article. MOFs act as an ideal candidate for the CO_2 capture process. The main approaches for capturing CO_2 are pre-combustion capture, post-combustion capture, and oxy-fuel combustion capture. The applications of MOFs in the carbon capture processes were extensively overviewed. In addition, the applications of MOFs in the adsorption, membrane separation, catalytic conversion, and electrochemical reduction processes of CO_2 were also studied in order to provide new practical and efficient techniques for CO_2 mitigation.

Keywords: metal organic frame works; CO_2 adsorption; pre combustion; gas membrane separation

1. Introduction

The Earth's global climate system is continually facing devastating changes due to various human-made and natural factors. Smithson [1] mentioned that the increase in greenhouse gas concentrations in the atmosphere directly impacts the global climate system, which is known as global warming [1]. These greenhouse gases trap the sun's radiation in the Earth's atmosphere; this phenomenon is known as the greenhouse effect, causing global warming. Carbon dioxide gas is blamed for being the main factor that causes the greenhouse effect because it is the most important anthropogenic greenhouse gas (Intergovernmental Panel on Climate Change (IPCC), 2007) [2]. Rochelle [3] stated that more than 85% of the world's energy demand is based on burning fossil fuels; this will result in massive emissions of CO_2 into the atmosphere [3]. There are natural and anthropogenic sources for carbon dioxide gas emissions into the atmosphere. Chemical engineering industries are considered one of the primary anthropogenic sources of CO_2 into the atmosphere in which natural gas and fossil fuels are burned for various purposes. As a result of the industrial revolution and the rapid increase in the population growth rate, more fossil fuels are burnt to satisfy the population's needs and demands. Hence, more carbon dioxide is emitted into the atmosphere. Consequently, the separation and capture of CO_2 became a necessity.

CO_2 can be separated and captured using five leading technologies, including absorption, adsorption, cryogenics, membrane, and microbial or algae [4]. In the meantime, the research trend has been focusing on three main types of technology for carbon capture, namely oxy-fuel combustion, pre-combustion, and post-combustion. Omoregbe (2020) investigated those main types of technology by using publications retrieved from the Web of Science database from the year 1998 to 2018. The results of the authors' investigations presented that from the year 1998 to 2007 there was almost no research output on carbon capture, until the year 2008, in which climate change abatement was first introduced, and the industrial and public awareness of clean, greener fossil energy options grew. The authors also stated that among the commonly studied carbon capture technologies, the post-combustion capture technology was the most referenced technology for carbon capture, with approximately 80.9% of total publications retrieved. On the other hand, the technology with the lowest number of publications is oxy-fuel combustion, with approximately 3.4% of total publications retrieved [5]. Several porous materials can be incorporated into these carbon capture technologies to allow and enhance the separation or the capture of CO_2. One of the best-used porous materials in carbon capture technologies are metal-organic frameworks (MOFs).

During the last two decades, a new crystalline porous materials class has emerged [6]. This class of materials is known as MOFs. As a result of the MOFs' unique properties, this class has gained remarkable attention across the globe. The main limitation of the application of MOFs in the carbon capture processes is the high cost. The synthesis process of MOFs is very costly, which makes them economically unviable. This review paper investigates MOFs' applications in the CO_2 capture, adsorption, separation, conversion, and reduction processes. It aims to draw and provide general guidelines and conclusions for the MOFs' importance as a porous material for carbon dioxide gas-related process.

2. Fundamentals of Metal-Organic Frameworks (MOFs)

MOFs are made-up of metal-containing nodes linked by organic ligand bridges and assembled primarily by strong coordination bonds; this can be shown in Figure 1 below. MOFs have well defined crystallographic and geometric three-dimensional (3D) microporous structures [7]. These structures are sturdy and durable, allowing the removal of the included guest species, which results in permanent porosity.

MOFs can be easily designed, synthesized, and tuned. By comparing MOFs to other porous materials like zeolites and activated carbons, MOFs allows a facile optimization of the structures of their pores, surface functions, and other properties, making them applicable for several specific and precise applications as porous materials. MOFs can be categorized into two classes flexible/dynamic [8] and rigid. Flexible MOFs hold a dynamic and soft framework with a fast response to external stimuli, for example, guest molecules, temperature, and pressure. This extraordinary and superb sensitivity to external stimuli allows the MOFs to possess special properties such as temperature/pressure-dependent molecular sieving, which puts them ahead of the traditional adsorbents, including activated carbons and zeolites.

On the other hand, rigid MOFs possess a stable and strong porous framework with an enduring porosity that is similar to zeolites. In the meantime, rigid MOFs have been used extensively for the selective gas adsorption processes. The selective adsorption mechanism in rigid MOFs is quite similar to zeolites; hence selective adsorption can be achieved based on the molecular sieving effect. Also, it is probably achieved according to the strength of the different interactions between adsorbate–adsorbate and adsorbate–adsorbent. Li, et al. [9] mentioned that selective adsorption in rigid MOFs depends on three main factors: adsorbate–surface interactions, size/shape exclusion, and simultaneous corporation of both factors.

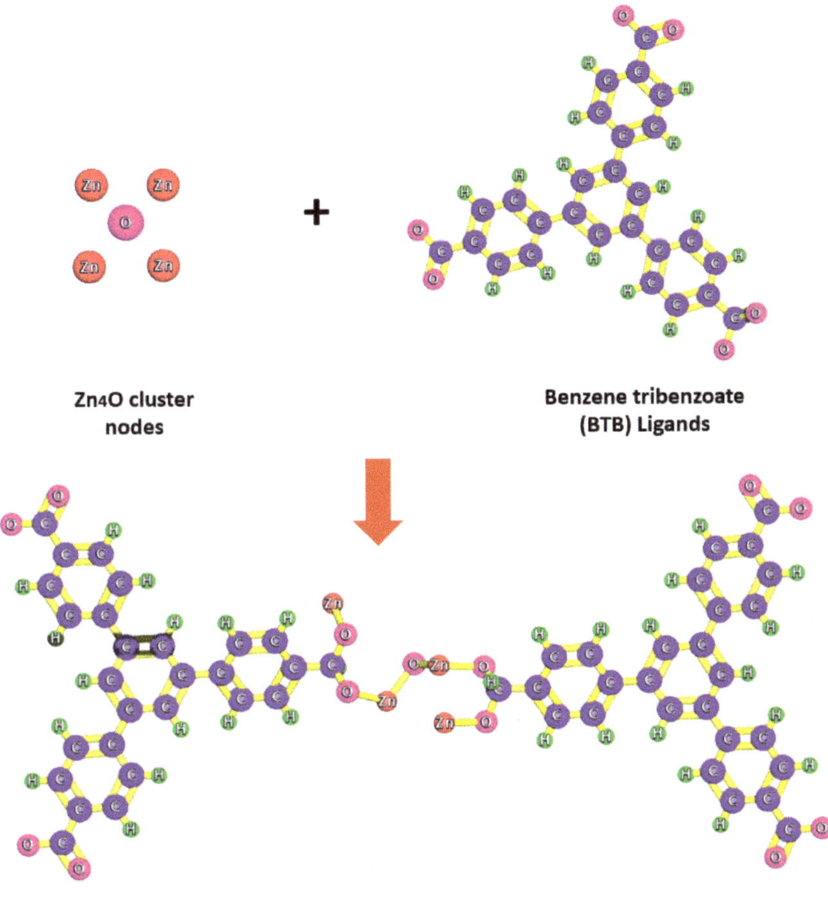

Figure 1. Typical synthesis approach for metal-organic framework MOF-177.

Based on the literature, many MOFs are used to selectively adsorb different gases relying on the molecular sieving effect [10–13]. This implies that only the molecules with appropriate pore kinetic diameters can pass through the MOF pores (Figure 2). Table 1 below shows the kinetic diameters of several gases.

CU-BTC MOF

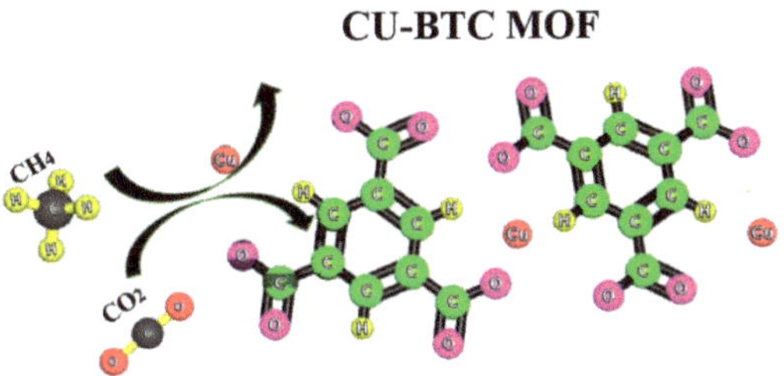

Figure 2. Schematic diagram of selective gas adsorption in Cu-BTC MOF (copper benzene-1,3,5 -tricarboxylate) based on molecular sieving effect.

Table 1. Kinetic diameters of several gases [14].

Molecule	Kinetic Diameter (Å)
CO_2	3.3
O_2	3.46
N_2	3.64
H_2O	2.65
CH_4	3.8
H_2	2.89

MOFs can be synthesized with an exceptionally high porosity under mild conditions via self-assembly reaction between organic linkers and several metal ions. Table 2 below provides chemical formulas and chemical structures of some organic ligands in some frequently used MOFs.

Table 2. Chemical formulas and chemical structures of some organic ligands in some frequently used MOFs [15].

MOF Name	Organic Ligand	Organic Ligand Structure
MOF-200	BBC: 4,4′,4″-benzene-1,3,5-triyl-tris(benzene-4,1-diyl)tribenzoate	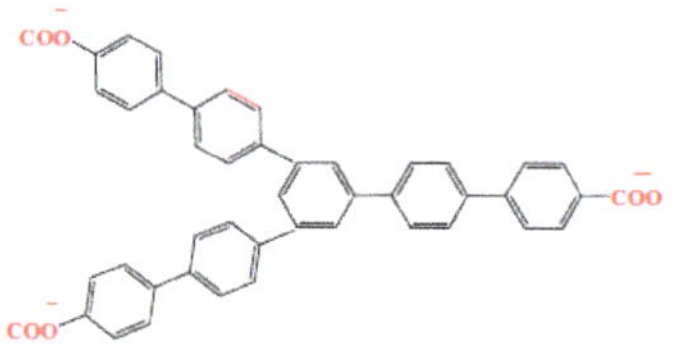
MOF-177	BTB: 4,4′,4″-benzene-1,3,5-triyl-tribenzoate	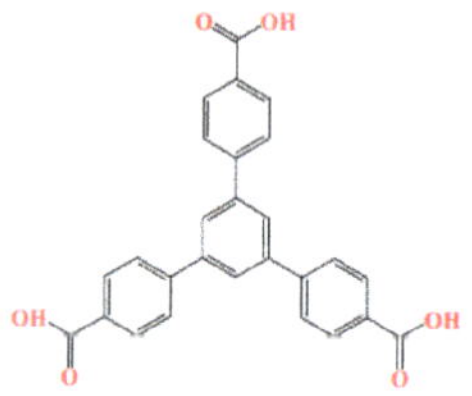
MOF-180	BTE: 4,4′,4″-benzene-1,3,5-triyl-tris(ethyne-2,1-diyl)tribenzoate	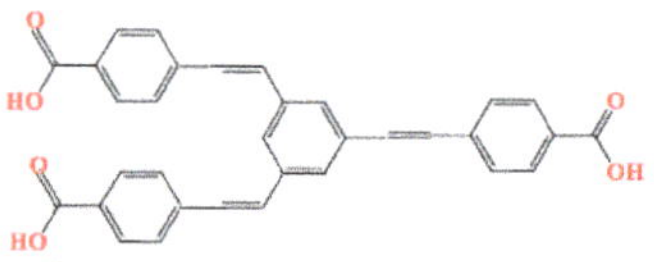
MOF-205	BTB + 2,6-naphtalenedicarboxylate (NDC)	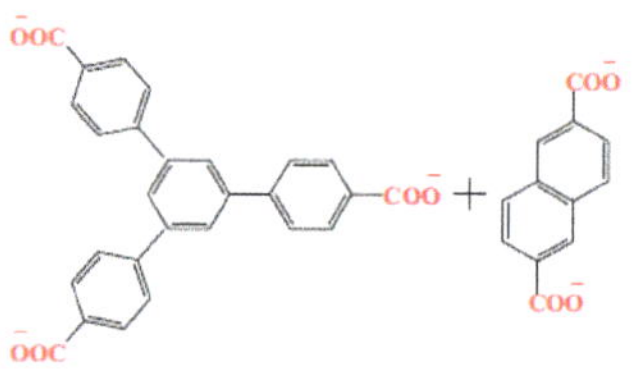
MOF-210	BTE + biphenyl-4,40-dicarboxilate (BPDC)	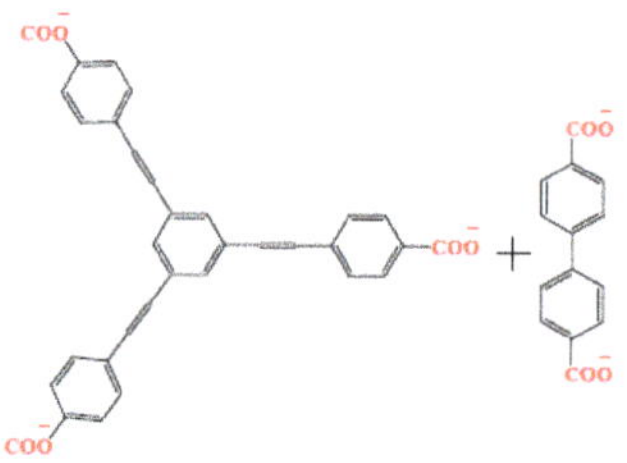

In the field of porous materials, MOFs have excelled and surpassed the traditional porous materials in the following properties: they possess very high CO_2 [16] and methane storage [17], uptake of hydrogen-based on physical adsorption [18], and surface area [19]. Hence, MOFs are extensively used in the carbon capture processes. Table 3 below shows a comparison between the strengths and weaknesses of carbon capture materials. It can be seen from Table 3 that MOFs have the highest working capacity among all other carbon capture materials.

Table 3. Strengths and weaknesses comparison of selected carbon-capture materials [20].

	MOFs	Liquid Amines	Amine Grafted MOFs	Zeolites	Ionic Liquids	Hybrid Ultraporous Materials (HUMs)	Soda Lime	Amine Grafted Inorganics
Selectivity	Low	High	High	Low	High	Very high	High	High
Stability	Low	Low	Medium	High	High	Medium	High	High
Humidity effect	High	Low	Medium	High	Low	Medium	Low	Low
Material cost	Medium/high	Low	High	Low	Low	Low	Low	Medium
Process cost	Medium	Low	High	Low	Medium	Low	Low	Medium
Recycling cost	High	High	Medium	High	Medium/high	Low	Very high	Medium
Working capacity	High	Medium	Medium	Medium	Low	Medium	Medium	Medium
Kinetics	Medium	Fast	Medium	Medium	Fast	Fast	Fast	Medium
Upside potential	High	Low	Medium	Low	Medium	High	Low	Medium

An extensive number of review articles have been published recently to highlight the extreme developments in the synthesis, design, and application of MOFs in the carbon dioxide capture and storage (CCS) field [21–23]. Simmons, et al. [24] stated that MOFs had been displayed as excellent materials for carbon dioxide storage, and also they are useful in the removal of carbon dioxide from flue gas stacks [24]. MOFs can be consequently considered ideal membrane-filling materials and adsorbents for CO_2 gas storage, separation, and capture due to their pore surface controllable properties, adjustable pore sizes, and large surface area [25]. In synthetic MOFs, the pore size and channels can be trivial, reaching nanometers and angstrom. At high pressure, these small pores and channels of the MOFs can store CO_2 gas up to 10 to 12 times greater than an empty container [26,27]. It is crucial to choose MOFs with high CO_2 uptake at low pressures to facilitate an effective CO_2 capture process for enclosed localities. Thus, MOFs provide a solid platform for many carbon capture processes. Table 4 below shows the list of abbreviations and acronyms used in this article.

Table 4. List of the abbreviations and acronyms used in this review article.

Abbreviation	Name
MOF	Metal-organic frameworks
HUMs	Hybrid ultra-microporous materials
CCS	Carbon capture and storage
HKUST	Hong Kong university of science and technology
ZIF	Zeolitic imidazolate framework
MIL	Materials of institut lavoisier
TEPA	Tetraethylenepentamine
PIM	Polymer of intrinsic microporosity
TFC	Thin film composite
PDMS	Polydimethylsiloxane
PEBA	Polyether-block-amide
KAUST	King Abdullah university of science and technology
MMM	Mixed-matrix membranes
NPs	Nanoparticles
CNFs	Chitosan nanofibers
GO	Graphene oxide
PSF	Polysulfone
MUF	Massey university framework
PDA	Polydopamine

Table 4. *Cont.*

Abbreviation	Name
DABCO	Diazabicyclo octane
HOMO	Highest occupied molecular orbital
LUMO	Lowest unoccupied molecular orbital
PL	Photoluminescence
SHE	Standard hydrogen electrode
PIC	Porous interconnected carbon
FE	Faradaic efficiency
ECR	Electrochemical reduction
TMOS	Tetramethyl orthosilicate
CO_2	Carbon dioxide
N_2	Nitrogen
CH_4	Methane
H_2	Hydrogen
CO	Carbon monoxide
CH_3OH	methanol
HCOOH	Formic acid
C_2H_4	Ethylene
HCHO	Formaldehyde
$HCOO^-$	Formate

3. Carbon Dioxide (CO_2) Capture Using Metal-Organic Frameworks

Carbon dioxide capture and storage are now being vigorously investigated to simultaneously combat global climate change while producing more sustainable synthetic fuels to use for several purposes [28–33]. The key approaches for capturing CO_2 using MOFs for the mitigation of the CO_2 emissions resulting from fuel combustion power plants are pre-combustion capture, post-combustion capture, and oxy-fuel combustion capture. The CO_2 capture selection method is mostly based on its advantages, disadvantages, and CO_2 feed input conditions such as partial pressure and concentration of CO_2 in the flue gas. Table 5 below shows a comparison between oxy-fuel combustion capture, pre-combustion capture, and post-combustion capture based on the advantages and disadvantages of each method.

Wang, et al. [34] mentioned that for newly-built power plants, oxy-fuel combustion could be adopted, whereas, for gasification plants, pre-combustion capture is the most appropriate carbon capture approach [34]. Moreover, post-combustion carbon capture is frequently favored for retrofitting power plants, since direct fuel combustion occurs in the boiler of all coal-fired power generation. Zou and Zhu [35] mentioned that the recognition ability of a CO_2 porous adsorbent is usually evaluated by two main key factors: the adsorption capacity of CO_2 and the selectivity of the material. The authors also mentioned that the ideal MOF materials with the high capturing ability of CO_2 are expected to exhibit both high adsorption and high uptake for CO_2 gas over other gases, like CH_4 and N_2 [35]. The effective CO_2 capture ability of MOF materials is owing to their distinguishable chemical and structural features. These unique features include pore size, unsaturated or open metal sites, function control, polar functional groups into the pore channels, and alkylamine incorporation. Kang, et al. [36] stated that the gas separation process is one of the most challenging and critical steps for the industrial processes, and MOFs are potential candidates for this separation application [36]. Table 6 below shows a summary of MOF-based materials for CO_2 capture.

Table 5. Advantages and disadvantages of oxy-fuel combustion capture, pre-combustion capture, and post-combustion capture.

Oxy-Combustion Carbon Capture	Pre-Combustion Carbon Capture	Post-Combustion Carbon Capture
Advantages		
Produce high efficiency steam cycles	Frequently used in the industrial processes	Applicable for existing and new coal-fired power plants
Low level of Pollutants emissions at low cost	Lower energy requirements compared to other CO_2 capture methods	Extensive studies are made to improve the sorbents and the capture equipment
Cost effective compared to other CO_2 capture methods. A low cost is required to capture more than 98% of CO_2	Syngas can be used as a fuel for turbine cycle	Future developments of pulverized coal systems will increase the plant efficiency and reduce CO_2 emissions
Easy to retrofit into an existing power plant, and does not require an on-site chemical operation	Requires less amount of water compared to post-combustion capture	Most commonly used technology in CO_2 capture methods
Disadvantages		
High Energy penalty	Significant loss of energy compared to post-combustion capture.	Low CO_2 partial pressure at ambient pressure
High overall cost	High equipment cost	The amine technologies used results in an almost 30% loss of the net power output and an efficiency reduction of 11%
Technology needs to be proved for large scale operations.	Requires extensive supporting systems	The steam extraction decreases the flow to low-pressure turbine; affecting the efficiency and reducing capability
High risk of CO_2 leakage	Mainly applicable to new plants	High performance, circulation volume, and water requirements are needed for high capture levels

Table 6. Summary of MOF-based materials CO_2 capture.

MOF	CO_2 Uptake	T (°C)	P	Ref.
Zn(adc) (4,4$_0$-bpe)$_{0.5}$	130 mmol g^{-1}	−78.15	1p/p	[37]
(MIL-53)	7.5 mmol g^{-1}	30.85	20 bar	[38]
Cu(fam) (4,40-bpe)$_{0.5}$	100 mL g^{-1}	−78.15	760 torr	[39]
Ni$_2$(cyclam)$_2$(mtb)	57 mL g^{-1}	−78.15	1 atm	[40]
MIL-53 M = Al, Cr	10 mmol g^{-1}	30.85	30 bar	[41]
(PCN-5)	210 mg g^{-1}	−78.15	760 torr	[42]
Cu(dhbc)$_2$ (4,40-bpy)	70 mL g^{-1}	24.85	0.4~8 atm	[43]
Cu(bdc) (4,40-bpy)$_{0.5}$	70 mL g^{-1}	24.85	0.1~0.2 MPa	[43]
(ZIF-20)	70 mL g^{-1}	0	760 torr	[44]
[Ni(bpe)$_2$ (N(CN)$_2$)] (N(CN)$_2$)	35 mL g^{-1}	−78.15	1p/p	[45]
Zn$_2$(tcom) (4,40-bpy)	5 wt%	24.85	1 bar	[46]
Cu(pyrdc)(bpp)	Differed adsorption capacity	−78.15	Different pressure	[47]
Ni$_3$(BTC)$_2$	3.0 mmol g^{-1}	40	1 bar	[48]
SNU-110	6.0 mmol g^{-1}	78	1 bar	[49]
1D-MOF	4.0 mmol g^{-1}	78	1 bar	[50]
2D-MOF	2.9 mmol g^{-1}	0	1 bar	[51]
A core–shell MOF	41 mmol g^{-1}	0	1 bar	[52]
NJU-Bai12	23.8 mmol g^{-1}	0	20 bar	[53]
PCN-124	9.1 mmol g^{-1}	0	1 bar	[54]
MOF-5/graphite oxide	1.1 mmol g^{-1}	25	4 bar	[55]
HCM-Cu$_3$(BTC)$_2$-3	2.8 mmol g^{-1}	25	1 bar	[56]
Zn doped Ni-ZIF-8	4.3 mmol g^{-1}	0	1 bar	[57]
Zn(II)-based MOFs	9.2 mmol g^{-1}	25	1 bar	[58]
MOF with PEI	4.2 mmol g^{-1}	78	0.15 bar	[59]
MIL-53 with BNH$_x$	4.5 mmol g^{-1}	0	1 bar	[50]
Mg-MOF-74	8.0 mmol g^{-1}	23	1 bar	[60]
UMCM-1-NH$_2$-MA	19.8 mmol g^{-1}	25	18 bar	[61]

Table 6 represents a summary of MOF based materials for CO_2 capture and it is clearly seen that as the temperature increases, the adsorption capacity of CO_2 decreases. Chen, Jin and Chen [62] mentioned in their study that the adsorption capacity and saturated adsorption capacity decreases with the increase in adsorption temperature. They also added that an exponential function is the best function that describes the relationship between saturated adsorption capacity and temperature. Hence, to obtain a high CO_2 adsorption capacity, the adsorption temperature should be low enough.

3.1. Oxy-Fuel Combustion CO_2 Capture

Oxy-fuel combustion is the process at which hydrocarbon fuel is combusted in a nearly pure oxygen environment, as opposed to air. For controlling the temperature, oxygen is diluted in a portion of the flue gas rather than dilution in nitrogen. In a coal-fired power plant, the oxy-fuel combustion aims to produce flue gas that is enriched with CO_2 and water vapor. This allows the separation or the capture of CO_2 from the flue gas by using low-temperature desulfurization and dehydration processes [63]. Figure 3 below shows a block flow diagram of an oxy-combustion carbon capture system.

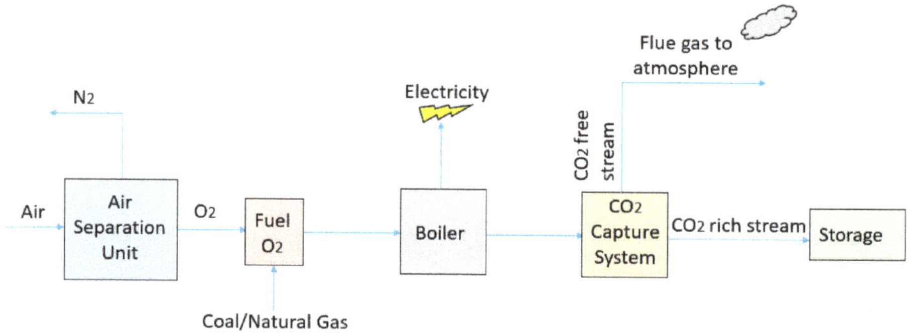

Figure 3. Block flow diagram of an oxy-fuel combustion carbon capture system.

Sumida, et al. [64] stated that oxy-fuel combustion refers to the combustion process in a nearly pure O_2 environment of pulverized coal or other carbonaceous fuel. The significant merit of the oxy-fuel combustion process is based on the fact that the flue gas is almost entirely CO_2. This eases the capture step, and also most of the existing power plants can be easily retrofitted with an oxy-fuel combustion system. Even though there are no full-scale plants that currently adopt oxy-fuel combustion, theoretical studies in combination with pilot-scale and laboratory studies have mentioned some operational issues and important design parameters that aid in studying the oxy-fuel combustion process [64]. The current carbon capture technologies include membranes and MOF-based adsorbents. Hu, et al. [65] mentioned that MOF, activated carbon, and Zeolite adsorbents are all physisorption-based and can be applied to oxyfuel, pre-combustion, post-combustion, and CO_2 capture. The authors also mentioned that as a result of the weak interactions with CO_2, activated carbon, and zeolite adsorbents are unlikely to be used for direct air capture.

On the other hand, MOFs can be tuned to undertake a strong interaction with CO_2 making it suitable for direct air capture processes [65]. The nitrogen-free combustion atmosphere of the oxy-fuel combustion entails a flue gas with a high concentration of CO_2 and water vapors for easier separation. The capture of CO_2 is not required in oxy-fuel combustion since purification can be easily achieved by water vapors condensation. The MOFs' selectivity for other molecules is based on the polarizabilities, diversity of the competing molecules, and the differences between the quadruple moments. Consequently, MOFs have a poor selectivity for O_2/N_2 as a result of the similarity in their molecular nature. Thus, the application of MOFs in the oxy-fuel combustion process is restricted in the carbon capture processes [66].

3.2. Pre-Combustion CO_2 Capture

Pre-combustion carbon capture technology involves capturing the carbon from the fuel before completing the combustion process [67]. A pre-treatment stage is conducted for the fuel, such as natural gas steam reforming, biomass, and coal gasification, before the actual combustion stage [68,69]. Syngas, which is a mixture of CO and H_2, is produced in this pretreatment stage. By using the water gas shift (WGS) reaction, CO in the syngas is then reacted with steam to produce additional CO_2 and H_2 [63]. The separation of H_2 and CO_2 can then be achieved by various technologies. The pre-combustion technology has the merit of lower energy requirements; however, the efficiency and the temperature associated with the H_2-rich gas turbine fuel is considered a big problem. Figure 4 below illustrates a block flow diagram of a pre-combustion carbon capture system.

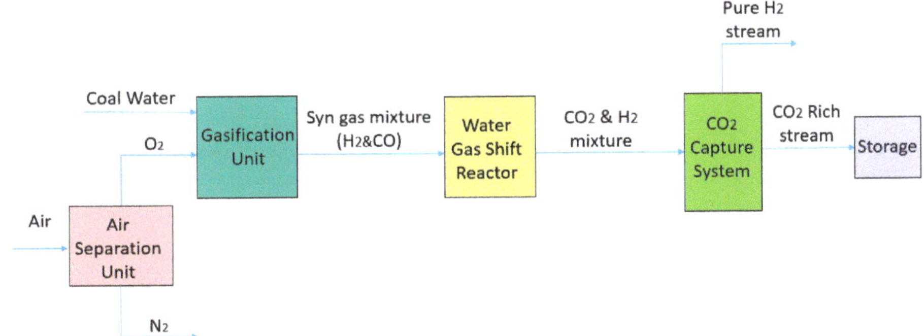

Figure 4. Block flow diagram of pre-combustion carbon capture system.

Lea-Langton and Andrews [70] mentioned that the pre-combustion technology offers a lower-cost CCS since the combined-cycle gas turbine is the base power generator, which has an approximately 62% thermal efficiency in the latest technology plants compared to approximately 50% efficiency in the latest steam-cycle technology [70]. The addition of the MOFs to the pre-combustion processes aids in enhancing the separation of CO_2 from the CO_2/H2 mixture. Zhang, et al. [71] mentioned that MOFs are better than zeolites as adsorbents for the pre-combustion CO_2/H_2 separation because MOFs possess higher porosities and result in vast uptakes of CO_2 at moderately high pressures [71]. Chung, et al. [72] discovered new MOFs adsorbent material that has a high CO_2 working capacity, which might help in the reduction of CO_2 emissions from the newly commissioned power plants that use the pre-combustion carbon capture technique. The authors have reported the in-silico discovery of high-performance adsorbents for the CO_2 pre-combustion capture. A genetic algorithm was used by the authors to efficiently search for a large database of top candidates of MOFs. The MOFs with the highest performance that were identified from the in-silico search were then synthesized and activated. These MOFs in the study have shown a high CO_2/H_2 selectivity and a CO_2 working capacity. The authors' also mentioned that one of the MOFs that they synthesized had shown the highest CO_2 working capacity compared to all the MOFs reported in the literature under the same operating conditions [72]. Nandi, et al. [73] synthesized MOFs with high CO_2/H_2 selectivity that are suitable for the pre-combustion capture. The authors reported an ultra-microporous (3.5 and 4.8 Å pores) Ni-(4-pyridylcarboxylate) with a working capacity of (3.95 mmol/g), which makes it applicable for hydrogen purification under normal CO_2 pre-combustion capture conditions. The MOFs reported in this study exhibited facile adsorption-desorption CO_2 cycling and have a CO_2 self-diffusivity of approximately 3×10^{-9} m²/s, which is approximately double zeolite 13X [73]. Ding, et al. [74] stated that the usage of MOFs in the CO_2 conversion and capture applications went through three main development stages. First, the CO_2 selectivity and adsorption capacity of MOFs were tuned. Based on that, a large number of strategies were used and investigated for the enhancement of the MOF-CO_2

interactions. The authors also mentioned that with time, many MOFs with strong interactions with CO_2 were developed. Many researchers focused their attention on the applications of pre-combustion carbon capture for CO_2 separation. The authors of the study mentioned that the MOF (Ni-4PyC) is considered as an ideal candidate for pre-combustion of CO_2 capture since, at a very high pressure reaching 35 bar it adsorbed almost no H_2 [74]. Herm, et al. [75] studied some selected MOFs with high structural flexibility, high surface area, or with open metal cation sites, for the utility in CO_2 separation from H_2 using pressure swing adsorption. The authors measured the single-component H_2 and CO_2 adsorption isotherms at a temperature of 313 K and pressures reaching 40 bar. For the pre-combustion CO_2 capture and H_2 purification, ideal adsorbed solution theory was employed by the authors in order to have a realistic estimation of the isotherms for the 80:20 and 60:40 H_2/CO_2 gas mixtures. The results of the authors' study have shown that the MOFs with high concentrations of the exposed metal cation sites, Cu-BTTri and Mg_2 (dobdc), had significant improvements over the traditional commonly used adsorbents. Thus, those MOFs have promising applications in the CO_2/H_2 separations [75]. Asgari and Queen [76] stated that by considering the limited ability to tune the pore shape, pore size, and surface functionality of the zeolites and activated carbons, only fractional improvements in their separation efficiency can be achieved. On the other hand, the MOFs offer a record-breaking CO_2 adsorption capacity in the pre-combustion CO_2 capture pressure regime. In the meantime, NU-11 has the highest pressure CO_2 adsorption with an absolute uptake of 856 cm^3 per gram of MOF at 25 °C and 30 bar. In addition, as mentioned by the authors, the MOFs facile structural tunability can allow significant improvements in the CO_2 binding strength and hence increases the CO_2 selectivity over H_2. Also, the MOFs' surpassing internal surface area is considered a very important factor for high-pressure separations improvements [76]. Sumida, Rogow, Mason, McDonald, Bloch, Herm, Bae and Long [64] mentioned that MOF-based membranes are a promising strategy for pre-combustion CO_2 capture. This is mainly because the pre-combustion gas mixture high pressure is an outstanding driving force for the membrane separation of H_2 and CO_2 [64].

3.3. Post-Combustion CO_2 Capture

The post-combustion carbon capture (PCC) technology is from the most commonly used technologies in the carbon capture field, but there are certain drawbacks and limitations that govern its usage [77–81]. Zamarripa, Eslick, Matuszewski and Miller [79] mentioned that PCC is an expensive energy-intensive process that is subjected to a considerable number of researches using various types of technologies, including sorbents, solvents, and membranes [79]. In the post-combustion carbon capture technology, the carbon-based fuel first undergoes combustion before the separation of the CO_2 from the flue gas is carried out [80]. A pre-treatment of the flue gas has to be conducted for the removal of all the corrosive substances and impurities. Since the flue gas temperature from the combustion units is more likely to be high, ranging from 120 °C to 180 °C, energy-intensive cooling systems are required before pretreatment [82]. Also, because of the high flue gas volume and low CO_2 partial pressure, large size equipment is needed. Both aspects increase the capture cost considerably. Figure 5 below shows a block flow diagram of a post-combustion carbon capture system.

Hu, et al. [83] stated that the post-combustion capture of CO_2 from the flue gas of a coal-fired power plants is a very important and critical approach because the CO_2 emitted from fossil fuel combustion contributed to approximately 60% of total CO_2 worldwide emission in 2004 [2]. Figueroa, et al. [84] mentioned that the flue gas emitted from post-combustion consists of approximately 75% N_2 and 15% CO_2, balanced by other impurities and moisture at ambient pressures (1 bar) and temperatures (30 °C) and pressures (1 bar) [84]. Consequently, an economically viable post-combustion CO_2 capture approach should efficiently separate the 15% CO_2 from the 75% N_2 with the lowest cost at ambient conditions without being affected by the moisture [83].

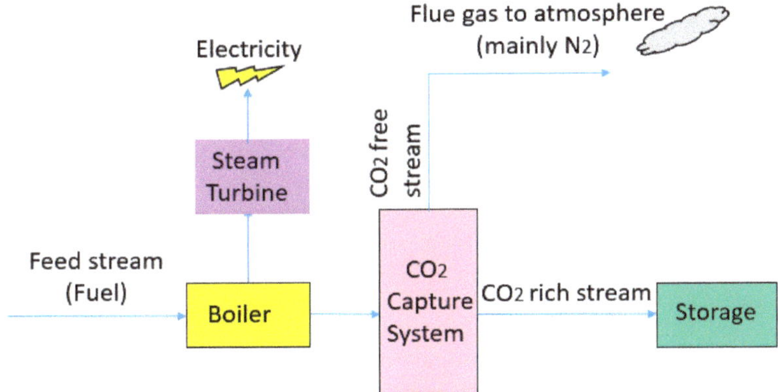

Figure 5. Block flow diagram of post-combustion carbon capture system.

Several studies have been conducted to investigate the efficiency of MOFs in the CO_2 capture under post-combustion conditions. Martínez, et al. [85] investigated three distinct commercial amino-containing MOFs (MIL-53(Al), HKUST-1, and ZIF-8) in the adsorption of CO_2 under post-combustion conditions. The authors' have modified these MOFs by wetness impregnation of tetraethylenepentamine (TEPA) molecules. The results of their study have shown that the amino-impregnated ZIF-8 samples have exhibited higher adsorption capacities by combining the chemical and physical adsorption of CO_2 compared to the TEPA-impregnated microporous framework. Under CO_2 post-combustion capture conditions, the TEPA-impregnated ZIF-8 samples went through a significant increase in the CO_2 uptake reaching 104 mg CO_2/gads, as a result of the moisture present, which governs the CO_2 capture efficiency increase of amino groups [85]. Pai, et al. [86] studied five distinct diamine-appended MOFs that exhibit an S-shaped CO_2 isotherm using a vacuum swing adsorption process in the post-combustion CO_2 capture from dry flue gas. The authors' algorithm to maximize the CO_2 recovery and purity has shown a linkage between the evacuation pressure, feed temperature, and the performance of the process. The MOFs that have achieved a target CO_2 recovery ≥90% and purity ≥95%, namely, mmen-Mg_2 (dobpdc), and mmen-Mn_2 (dobpdc), were optimized by the authors to increase productivity and reduce parasitic energy. The authors also mentioned that the low affinity of N_2 and the distinct shape of the CO_2 isotherm were the main reasons for the lower energy consumption [86].

Hedin, et al. [87] stated that MOFs had been intensively studied as an efficient class of adsorbents for CO_2 capture. Several studies mentioned by the authors have shown that certain members of this class of solid adsorbents can adsorb large quantities of CO_2 while having a higher selectivity of CO_2 over N_2 for the post-combustion capture of CO_2 from flue gas [87]. Furthermore, Samanta, et al. [88] stated that MOFs as a sorbent for the post-combustion capture of CO_2 are expected to have a remarkable and significant adsorption capacity; however, they require substantial, intensive research efforts to be applicable under flue gas conditions. Also, they mentioned that the CO_2/N_2 selectivity further limits the usage of these sorbents for CO_2 adsorption [88]. Maurya and Singh [89] comparatively studied some water-stable microporous adsorbents for post-combustion CO_2 capture. They investigated three metal-organic frameworks (MOFs), a single-wall carbon nanotube (SWCNT) and two covalent organic frameworks (COFs), and a single-wall carbon nanotube (SWCNT) under equal flue gas conditions. The results of the simulation made by the authors depicted that the pure component CO_2 adsorption capacity followed this descending order SWCNT > InOF-1 > COF-300 > UiO-66 > COF-108 > ZIF-8 under post-combustion conditions [89]. Babarao and Jiang [90] reported a computational study for the characterization of the cation and the capture of CO_2 in the Li^+-exchanged metal−organic frameworks (Li^+-MOFs). The authors' have adopted a density functional theory for the cation locations optimization and the evaluation of the atomic charges, and the molecular simulation used to investigate the separation of CO_2/H_2 and CO_2/N_2 gas mixtures for the pre-combustion and post-combustion CO_2 capture. The results of the authors' study show that at ambient conditions, the selectivity is around 60 for CO_2/N_2 mixture and 550 for CO_2/H_2 mixture, higher than the selectivities in other nano-porous adsorbents and the non-ionic MOFs. They also mentioned that the charge of the cations and framework have a remarkable impact on the selectivity, that was found to decrease by a magnitude of 1 order by switching off the charges. Also, the Li^+-MOF cations' hydration leads to a reduction in the free volume, leading to a lower adsorption extent [90]. Further studies have focused their research on the rule of photoresponsive MOFs in the CO_2 capture under post-combustion conditions. Park, et al. [91] proposed a new photoresponsive MOF, namely Mg-IRMOF-74-III structure with azopyridine molecules bonded to its unsaturated metal sites for CO_2 capture. The authors' computational simulations showed that the photochemical MOF had induced the trans-to-cis transition of the material leading to a remarkable alteration in the capacity of CO_2. Their work aimed to provide a blueprint for the computational design of the new photoresponsive MOF before the actual experimental synthesis [91]. Wang, et al. [92] also mentioned that MOFs, with their distinguishable characteristics and their fine-tunable structures, are exceptional porous solid materials that can provide many powerful and efficient platforms for the exploration of high-performance adsorbents for the CO_2 post-combustion capture [92].

Marti [93] mentioned that the post-combustion CO_2 capture technology is currently the most utilized method for power production. The CCS targets for the post-combustion CO_2 capture are to achieve 90% CO_2 capture with less than 20% increase in the electricity cost. This financially translates to a CO_2 separation and compression cost of \$30–50 per ton of CO_2 [93]. Moreover, a very remarkable study has synthesized MOFs with no N_2 adsorption. In this study, Hu, et al. [94] post-synthetically tethered distinct alkylamine molecules to the unsaturated Cr (III) centers in the MOF MIL-101 for post-combustion CO_2 capture. The resulting MOFs of their study showed almost no N_2 adsorption with a remarkably increased CO_2 capture as a result of the interaction between CO_2 molecules and amine groups under ambient conditions. The authors stated that MIL-101-diethylenetriamine extraordinary CO_2 uptake, exceptional stability, very high CO_2/N_2 selectivity, and the mild regeneration energy, makes it very promising for CO_2/N_2 separation and post-combustion CO_2 capture [94].

3.4. MOFs as Filler in Mixed-Matrix Membranes for CO_2 Separation

Membrane separation is one of the most efficient and commonly used techniques in the CCS field. However, it is governed by certain limitations, including selectivity, permeability, pore size, fouling, and high cost of the membranes [95,96]. Most of the time, selectivity and permeability of the membranes are the main drawbacks to the efficiency of the membranes. Certain porous materials like MOFs, zeolites [97–99], and activated carbons are incorporated in the membranes to increase its CO_2 separation efficiency. Figure 6 below describes how a MOF-based mixed matrix membrane (MMM) is used for CO_2 capture. Figure 6 shows that the feed gas stream containing CO_2 and CH_4 moves across a MOF-based mixed matrix membrane surface that has a selective permeability to the CO_2 gas over CH_4 gas. The CO_2 gas molecules diffuse through the membrane's pores via the solution-diffusion mechanism forming the permeate. Moreover, the CH_4 gas molecules that did not diffuse through the membrane pores form the retentate. The MOF filler enhances the selectivity and permeability of the membrane, allowing more CO_2 molecules to diffuse through the pores of the membrane.

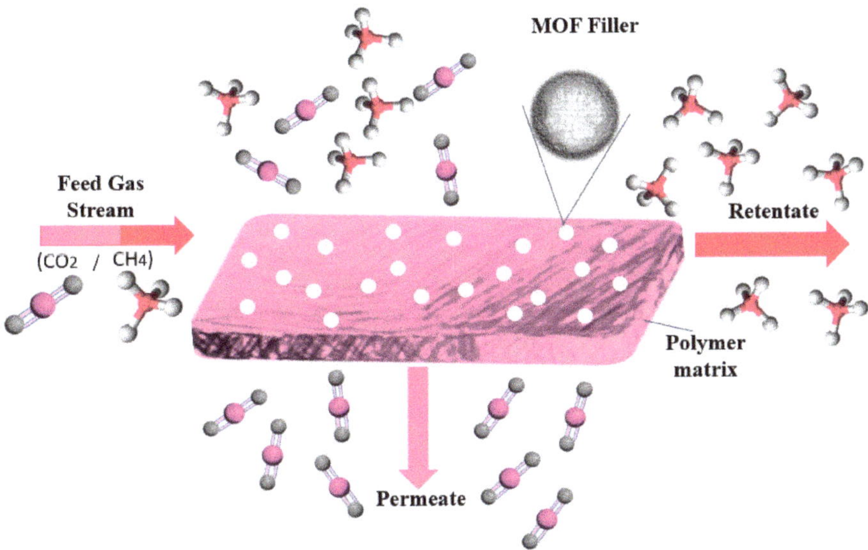

Figure 6. MOF-based mixed matrix membrane (MMM) for CO_2 capture.

With regard to the incorporation of MOFs into the membranes for carbon capture, reference [100] described a thin film composite (TFC) membrane that incorporates MOF nanoparticles and a polymer of intrinsic microporosity (PIM-1) for post-combustion CO_2 capture. The novel TFC membrane design used by the author consists of three layers; the first layer is a CO_2 selective layer synthesized of the mixed matrix PIM-1@MOF. The second layer is an ultra-permeable PDMxS gutter layer that is doped with MOF nanosheets. Moreover, the third layer has a porous polymeric substrate. The results of the study show that the PDMS@MOF gutter layer incorporated with amorphous nanosheets provides a 10,000–11,000 gas permeance units (GPU) permeance of CO_2 permeance. The authors' owes the high CO_2 permeance to the less gas transport resistance compared to the pristine PDMS gutter layers. Furthermore, TFC membrane assembly resulting from a nanosized MOF particles blend (NH_2-UiO-66 and MOF-74-Ni) into PIM-1 improved the permeation of CO_2 4660–7460 GPU and the selectivity of CO_2/N_2 of 26–33, in comparison to the pristine PIM-1 counterpart with aCO_2 permeance of 4320 GPU and a CO_2/N_2 selectivity of 19. This enhancement in the CO_2 separation after the incorporation of MOFs strongly suggests that MOFs' incorporation into the mixed matrix membranes improves CO_2 separation [100]. Sun, et al. [101] introduced a novel metal-organic framework MOF-801 nanocrystal into a polyether-block-amide (PEBA) polymer in order to synthesize a new mixed-matrix material for the separation of CO_2. The author found that the uniform incorporation of the MOF-801 microporous with preferential adsorption of CO_2 provided selective and fast transport channels for CO_2 over N_2, leading to an increase in both the CO_2/N_2 mixed-gas selectivity and the CO_2 permeance compared with the pure PEBA membrane. Based on the study, the MOF-801/PEBA optimized mixed-matrix composite membrane has shown a highly stabilized separation performance with CO_2/N_2 selectivity of 66 and CO_2 permeance of 22.4 GPU under a mixed-gas permeation test. This shows great potential for CO_2 separation and capture [101]. Chen, et al. [102] studied the CO_2 separation performance of a new type of MMM with a microporous filler of MOF-801 and a polymer matrix of PIM-1. The author's experimental results show that the CO_2-philic MOF-801 filler uniform dispersion provided a channel for a fast, selective CO_2 transport; hence, the MOF-801/PIM-1 MMMs show a greater CO_2 permeability and CO_2/N_2 ideal selectivity over the pure polymer membrane [102]. Majumdar, et al. [103] synthesized Mg-MOF-74 crystals for the preparation of a polymer/Mg-MOF-74 MMMs for the separation of CO_2/CH_4 gas mixture. Activation temperature and time of the Mg-MOF-74 crystals were determined to enhance the performance of the polymer/Mg-MOF-74 mixed matrix. The authors used a solvent-casting method to incorporate the Mg-MOF-74 crystals into the polyvinyl acetate (PVAc) matrix to form a dense MMM. The results showed that the MMM's mixed gas permeability measurements improved in both the permeability and the selectivity of CO_2 with the increase in the amount of MOF. This suggests a strong CO_2 adsorption selectivity of Mg-MOF-74. Also, the authors' have found that the incorporation of Mg-MOF-74 has reduced the effect of plasticization [103]. Ahmad, et al. [104] investigated the gas separation properties of the MMMs 6FDA-DAM with three types of zirconium-based MOFs nanoparticles (MOF NPs, ca. 40 nm) up to 20 bar. The authors investigated the separation of CO_2/CH_4 at high feed pressure with different CO_2 concentrations in the feed in a temperature range of 35–55 °C. The results of the study show that incorporating Zr-MOFs in the 6FDA-DAM MMMs increased both the CO_2 permeability and the CO_2/CH_4 selectivity of this polymer. This study suggests that 6FDA-DAM Zr-MOF MMMs have great potential in the carbon capture process [104].

The MMMs that are derived from MOF nanocrystals represent a promising alternative for overcoming the trade-off between selectivity and permeability of the pristine polymeric membrane. Chen, et al. [105] incorporated CO_2-philic KAUST-nanocrystals into 6FDA-durene polyimide membrane. This incorporation made by the author has increased the permeability and the selectivity of the MMMs. The developed MMMs in this study has a promising application in the CO_2 capture from natural gas and biogas [105]. Jiamjirangkul, et al. [106] mentioned that for the development of metal-organic frameworks (MOFs) nanofibrous membranes, chitosan nanofibers are a very promising template because of their high surface area with the presence of functional groups for the cationic/anionic binding. In their study, Cu-BTC-integrated chitosan/PVA nanofibrous membrane (Cu-BTC/CNFs) hybrids were synthesized. The CNFs/Cu/BTC-3 synthesized by the authors show an adsorption capacity of CO_2/N_2 over 14 times. Hence, the membrane has a great potential for selective capture and filtration of CO_2 [106]. Lee, et al. [107] prepared a Ni-MOF-74 continuous and defect-free membrane on α-alumina support by using the technique of a layer-by-layer seeding followed by a secondary growth crystallization. The gas permeation properties of the membranes were investigated by the authors for small gases, including CO_2, CH_4, H_2, and N_2. The results of their study showed that the Ni-MOF-74 membrane exhibited a stronger adsorption affinity to CO_2 compared to the other gases, and hence the Ni-MOF-74 membrane CO_2 permeation was dominated by the surface diffusion [107].

There is increasing attention nowadays to MMMs comprised of inorganic fillers scattered in an organic matrix for the separation of gas mixtures due to the membrane enhancement in the material robustness, separation selectivity, and throughput. Anastasiou, et al. [108] developed ZIF-8/graphene oxide (GO) hybrid nanofillers and ZIF-8 MOFs and incorporated them into a polysulfone (PSF) matrix. The authors then tested the membranes for their selectivity and permeation properties for CO_2, CH_4, and N_2. The results highlighted that the PSF+ (ZIF-8/GO) MMMs showed an enhancement in the CO_2 permeability (up to 87% increase) and the selectivity of the CO_2/CH_4 pair (up to 61% increase), compared to the pristine PSF membrane. Also, the selectivity of the PSF^+ (ZIF-8/GO) MMM was increased up to 7-fold compared to the PSF + ZIF-8 MMM selectivity. Based on the results, the composite fillers that combine MOFs and the GO functionality have a great potential in boosting and tuning the performance of the polymeric membranes for CO_2 separation from flue gas and natural gas [108]. Inorganic fillers can mainly define the optimal performance of the MMMs. Hence, the development and identification of new inorganic fillers are critical for optimizing the MMMs' performance. MMMs incorporated with MOF fillers are extensively investigated. However, MOF fillers with high performance remain scarce and in high demand. Yin, et al. [109] combined for the first time, an emerging MOF that has an exceptional physicochemical property, namely MUF-15 (Massey University Framework-15), into a MMM with PIM-1. As mentioned by the authors, based on the MUF-15 intrinsic ability to discriminate distinct guest molecules, MUF-15 is considered an impressive filler that delivers the MMMs with excellent CO_2 separation property. Hence, MUF-15 can be proposed as a strategy to further increase the MMM performance [109].

The enhancement of MOF membranes' CO_2 separation performance is attracting the attention of several researchers. Wu, et al. [110] reported a versatile post-modification strategy based on polydopamine (PDA) grafting for the improvement of the MOF membranes' CO_2 separation performance. The PDA was deposited by the authors' on the UiO-66 membrane via a simple and mild process. The results of their study show that the modified PDA/UiO-66 membrane exhibited an enhancement in the CO_2/CH_4 and CO_2/N_2 and selectivities of 28.9 and 51.6, respectively. These selectivity results were 2 to 3 times greater than the MOF membranes that are reported with similar permeance. Moreover, under moist conditions and in the 36 h measurement period, the PDA/UiO-66 membrane prepared in their study exhibited superb long-term stability for the capture of CO_2 [110].

The search for effective carbon-capture materials has allowed the disclosure and institution of nanoporous fluorinated MOFs with a contracted pore system as a CO_2-selective benchmark adsorbent. Chernikova, et al. [111] transplanted/integrated SIFSIX-3-M (M = Cu, Zn, and Ni) MOF adsorbent that encompasses a fluorine moieties with the periodic arrangement in a one-dimensional confined channel, showing a remarkable CO_2 adsorption-based selectivity over H_2 and CH_4 and several other industrially related gas mixtures for carbon capture. The single and mixed-gas permeation tests made by the authors showed that the nanoporous MOF membrane is a highly CO_2-selective membrane that shows a greater CO_2-selectivity over H_2, and CH_4 is limited by the selective adsorption of CO_2 in the SIFSIX-3-M functional and contracted channels [111]. The fabrication and design of novel MMMs with a simultaneously enhanced gas selectivity and permeability are greatly demanded by the industries as membrane technology for large-scale CO_2 capture and storage. Traditional fillers consisting of isotropic bulky particles often limit the interfacial compatibility leading to a great loss in the MMMs' selectivity. Cheng, et al. [112] incorporated chemically stable MOF nanosheets into a highly permeable polymer matrix to synthesize defect-free MMMs. The authors homogeneously dispersed the MOF nanosheets within the polymer matrix, owing to their high aspect ratios that enhances the integration of the polymer-filler. The MMMs prepared by the authors showed a high selective separation performance for CO_2, good antiaging, and anti-pressure abilities, thereby offering a new strategy in the development of advanced membranes for the industrial gas separation applications [112].

The efficient separation of CO_2 from CO_2/CH_4 mixtures with membranes has economic, environmental, and industrial importance. Membrane technologies are currently dominated by polymers due to their processing abilities and low manufacturing costs. However, polymeric membranes suffer from either low gas permeabilities or low selectivities. MOFs are suggested as potential membrane candidates that offer both high selectivity and permeability for CO_2/CH_4 separation. Experimental testing of every single synthesized MOF material as a membrane is not practical due to the availability of thousands of different MOF materials. Altintas and Keskin [113] used a multilevel, high-throughput computational screening methodology to examine the MOF database for membrane-based CO_2/CH_4 separation. MOF membranes offering the best combination of CO_2 permeability ($>10^6$ Barrer) and CO_2/CH_4 selectivity (>80) were identified by combining grand canonical Monte Carlo and molecular dynamics simulations. The results revealed that the best MOF membranes are located above the Robeson's upper bound, indicating that they outperform polymeric membranes for CO_2/CH_4 separation. The impact of framework flexibility on the membrane properties of the selected top MOFs was studied by comparing the results of rigid and flexible molecular simulations.

The relationship between the structure of the MOFs and the performance was also investigated to provide atomic-level insights into the design of novel MOFs, which will be useful for CO_2/CH_4 separation processes. Prediction of permeabilities and selectivities of the MMM found the best MOF candidates to incorporate as filler particles into polymers, and it was found that MOF-based MMMs have significantly higher CO_2 permeabilities and moderately higher selectivities than pure polymers [113]. Light-responsive metal-organic frameworks are attracting special attention for their use as a filler in MMMs for CO_2 capture. Prasetya and Ladewig [114] synthesized a new generation-2 light-responsive MOF by using Zn as the metal source and both 1,4-diazabicyclo [2.2.2] octane (DABCO) and 2-phenyldiazenyl terephthalic acid as the ligands. The results showed that Zn-azo-dabco MOF

(Azo-DMOF-1) have exhibited photoresponsive adsorption of CO_2 in both a static and dynamic conditions; this is owed to the abundance of azobenzene functionalities from the ligand. The authors' have also incorporated the MOF as a filler in a mixed matrix membrane with PIM-1 as the polymer matrix and evaluated MMM separation performance for CO_2/N_2 gas mixture. The results of their study showed that azo-DMOF-1 might increase the pristine polymer permeability and selectivity of CO_2. Also, the azo-DMOF-1–PIM-1 composite membranes have a good performance, which has surpassed the 2008 Robeson Upper Bound [114]. Benzaqui, et al. [115] used a microporous Al trimesate-based MOF, namely MIL-96-(Al), as a porous hybrid filler in MMMs for the post-combustion separation of CO_2/N_2. The homogeneous and defect-free MMMs with a high MOF loading (up to 25 wt%) synthesized by the authors have super passed the pure polymer membranes for CO_2/N_2 separation [115]. Maina, et al. [116] reported a new route for the synthesis of hybrid membranes containing inorganic nanoparticles and MOFs with potential applications in several areas, including catalysis, separation, electrochemical, and sensing applications was reported [116]. Zhao, et al. [117] studied at different temperatures, feed compositions and feed pressures, the separation and permeation properties of CO_2/N_2 and CO_2/H_2 mixtures for thin high-quality MOF-5 membranes synthesized by the secondary growth method. The MOF-5 membranes synthesized by the authors under the experimental conditions were studied to offer a selective permeation for CO_2 over N_2 and H_2 in CO_2/N_2 and CO_2/H_2 feed mixture. The results showed that the MOF-5 membranes exhibit high permeance and separation for CO2. It was mentioned by the authors that the sharp increase in the MOF-5 membranes separation factor with the increase in the feed pressure is an unobserved phenomenon for other inorganic microporous membranes [117]. Hu, et al. [118] mentioned that as a result of the MOFs' functionality, easily tunable porosity, and morphology, they are regarded as an ideal filler for MMMs. Fan, et al. [119] incorporated two isomorphous MOFs Ni_2(l-asp)$_2$pz) and (Ni_2(l-asp)$_2$bipy with different pore sizes into a poly(ether-block-amide) (Pebax-1657) to synthesize MMMs with gas permeation properties for CO_2, N_2, H_2, and CH_4. The results of their study show that the two series of MMMs showed an enhanced CO_2/H_2 selectivity and CO_2 permeability compared to the pure polymer membrane. Ni_2 (l-asp)$_2$bipy@Pebax-20 have shown the highest CO_2 permeation property in the study of 120.2 barrers with an enhanced CO_2/H_2 selectivity of 32.88 compared to the pure polymer membrane of, respectively, 55.85 barrers and 1.729. This study shows that the synthesized MMMs with MOF fillers are remarkable candidates for the future applications in CO_2 capturing [119].

3.5. MOFs in Photo-Catalytic Conversion of CO_2

In the road toward a sustainable low-carbon environment, aside from the physical capture and the underground injection and geologic sequestration of the anthropogenic CO_2 emitted from the power plants or industrial processes, catalytic chemical conversion of CO_2 into less harmful valuable chemicals is a very efficient way to undertake the carbon capture process [120–122]. Moreover, CO_2 molecules are very stable as a result of the C=O interactions; multistep reduction via photochemical or electrochemical methods are more imperative than water splitting reactions and have various technical implications. The CO_2 reduction reaction can occur in several distinct pathways that yield a varied range of reduction products, including carbon monoxide (CO) [123–127], methanol (CH_3OH) [128–135], methane (CH_4) [136], ethylene (C_2H_4) [137,138], formic acid (HCOOH) [139], and others [140,141]. Hence, the target product needed is the factor that governs the overall design process for the CO_2 reduction reaction. MOFs have great application as catalysts in the catalytic conversion reactions of CO_2, including conversion of CO_2 to fuels, hydrogenation, cycloaddition, and photo-reduction of CO_2.

The photo-reduction of CO_2 occurs in the presence of ultraviolet (UV) and visible light irradiation. Depending on the reduction potentials, the CO_2 reduction reaction products can include HCHO, CO, CH_3OH, and CH_4. Table 7 below provides the photo-reduction potentials for the CO_2 reduction reaction.

Table 7. [142]: Standard photo-reduction potentials for the CO_2 reduction reaction.

Reduction Potentials of CO_2	Reduction Potential vs. Normal Hydrogen Electrode (NHE) (V)
$CO_2 + e^- \rightarrow CO_2^-$	−1.9
$CO_2 + 2H^+ + 2e^- \rightarrow CO + H_2O$	−0.53
$CO_2 + 2H^+ + 2e^- \rightarrow HCOOH$	−0.61
$CO_2 + 4H^+ + 4e^- \rightarrow HCHO + H_2O$	−0.48
$CO_2 + 2H^+ + 2e^- \rightarrow HCOO^-$	−0.49
$CO_2 + 6H^+ + 6e^- \rightarrow CH_3OH + H_2O$	−0.38
$2H^+ + 2e^- \rightarrow H_2$	−0.41
$H_2O \rightarrow \frac{1}{2} O_2 + 2H^+ + 2e^-$	+0.41
$CO_2 + 8H^+ + 8e^- \rightarrow HCHO + H_2O$	−0.24

The photocatalytic CO_2 reduction process comprises a series of reactions including adsorption of CO_2, charge carrier separation and transportation, electron-hole pair photo-generation, and chemical reactions between the charge carriers and surface species [143–145]. However, certain photo-induced electrons on the surface of the catalyst are specifically utilized for CO_2 reduction. Consequently, catalysts with high redox potential and relatively low bandgap value are favorable. The photo-induced activation of CO_2 on the surface of MOFs includes some main steps. The catalytic material first adsorbs a photon leading to electron-hole pair separation. This separation excites a negative electron (e^-) from the highest occupied molecular orbital (HOMO) to the lowest unoccupied molecular orbital (LUMO), forming a positive hole (H^+) on the HOMO [146]. The CO_2 molecules are then absorbed on the MOFs' catalytic center and accept electrons forming different products such as CH_4, CO, and HCOOH. The mechanism of the CO_2 photo-reduction process on MOFs is shown in Figure 7. Nevertheless, not all MOFs exhibit photocatalytic activity, and their electronic properties are identical. This may be determined based on the HOMO and LUMO of MOF materials.

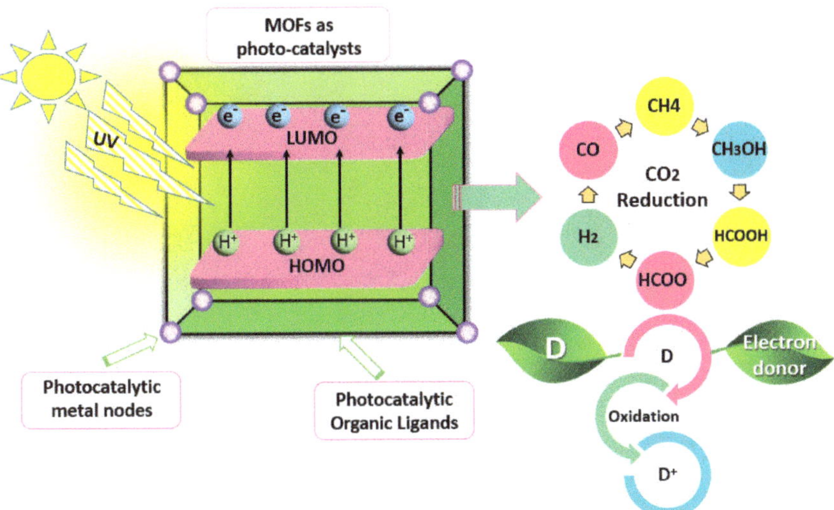

Figure 7. CO_2 photo-reduction process mechanism on MOFs.

The construction of MOF-based photo- and electro-catalysts is very promising as a result of their considerable flexible structures and active sites. Reddy, et al. [147] mentioned that MOFs are an emerging new class of functional materials with a highly porous structure, exceptional specific surface areas, and tunable surface chemistry; hence, they hold great potential as photocatalysts [147].

Wang, et al. [148] stated that the photocatalytic reduction of CO_2 for valuable chemicals is an attractive way to create a better overall environment. The authors have proposed two distinct conversion processes. In the first process, CO_2 is split into CO, and in the second process, CO_2 is converted into organic chemicals (like CH_3OH, CH_4, and $HCOOH$) [148]. It is clear that MOFs have a bright future, prospect, and applications in the field of CO_2 photocatalytic reduction. Table 8 below shows a summary of MOF-based catalysts for photocatalytic CO_2 reduction.

Table 8. Summary of MOF-based catalysts for photocatalytic CO_2 reduction.

Sample ID	Proton Donor	Products and Yield (μ mol/g h)				Light Source	Reference
MOF4	TEA	CO	10.9	–	–	UV	[149]
Zn_2GeO_4/ZIF-8	H_2O	CH_3OH	0.22	–	–	UV	[150]
NH_2-MIL-125(Ti)	TEOA	$HCOO^-$	16.3	–	–	Visible	[151]
$Cu_3(BTC)_2$@TiO_2	H_2O	CH_4 [a]	2.64	–	–	UV	[152]
Copper porphyrin MOF [b]	TEOA	CH_3OH [c]	262.6	–	–	Visible	[153]
Pt-NH_2-MIL-125(Ti) Au-NH_2-MIL-125(Ti)	TEOA	HCOO	32.4 16.3	–	–		[154]
NH_2-UiO-66(Zr) NH_2-UiO-66(Zr/Ti)	TEOA	HCOO [d]	3.4 5.8	–	–	Visible	[155]
Ui-66-CrCAT Ui-66-GaCAT	TEOA	HCOOH	1724 959	–	–		[156]
Co-ZIF-9 Co-MOF-74 Mn-MOF-74 Zn-ZIF-8	TEOA	CO	12.6 9.9 0.3 0.2	H_2	2.8 1.9 0.5 0.2	Visible	[157]
CPO-27-Mg/TiO_2 TiO_2 CPO-27-Mg	H_2O	CO	4.09 2.25 0	CH_4	2.35 1.37 0	UV	[158]
Co-ZIF-9/TiO_2	H_2O	CO	8.8	H_2	2.6	UV-Vis	[159]
Zn/PMOF	H_2O	CH_4	8.7	–	–	UV-Vis	[160]
PCN-22	TEOA	HCOO	52.8	–	–	Visible	[161]
2Cu/ZIF-8N_2	Na_2SO_3	CH_3OH [e]	35.82	–	–	Visible	[162]
Ag@Co-ZIF-9	TEOA	CO [f]	28.4	H_2	22.9	Visible	[163]
Ni MOLs	TEOA	CO	12.5	H_2	0.28	Visible	[164]
TiO_2/Cu_2O/$Cu_3(BTC)_2$	H_2O	CO	210	CH_4	160	Visible	[165]
CdS/UiO-bpy/Co	TEOA	CO	235	–	–	Visible	[165]
NH_2-rGO (5 wt%)/Al-PMOF	TEOA	HCOO	685.6	–	–	Visible	[166]
Zn-MOF nanosheets/ [CO_2 (OH)L](ClO_4)$_3$	TEOA	CO	14.45	H_2	2.6	Visible	[167]

[a] Average after 4-h operation. [b] (5,10,15,20-tetrakis(4-carboxyphenyl) porphyrin. [c] Production in (ppm/gcat) after 1-h operation. [d] Production in (mmol/molcat) after 10 h operation. [e] Production in (mmol/L g) after 6-h operation. [f] Production in (mmol after) 0.5-h operation.

By looking at Table 8 above, which summarizes MOF-based catalysts for photocatalytic CO_2 reduction, the highest and lowest yields for CO are produced by CdS/UiO-bpy/Co and TiO_2, yielding 235 and 2.25 μ mol/g h, respectively. In addition, the highest and lowest yields for CH_3OH are produced by TEOA and Zn_2GeO_4/ZIF-8 yielding 262.6, and 0.22 μ mol/g h, respectively. Furthermore, the highest and lowest yield for HCOO is produced by TEOA and NH_2-UiO-66(Zr), yielding 685.6 and 3.4 μ mol/g h, respectively. These results prove that MOFs are successfully capable of photo catalytically, reducing CO_2 into other useful products.

Photocatalytic reduction of CO_2 into highly valuable added chemicals via clean, renewable solar energy sources is a remarkable pathway to address the environmental and energy issues. In recent days, MOFs have been exploited intensively as catalysts for the photocatalytic reduction of CO_2 owing to their extinguishable CO_2 capture abilities, photochemical and structural properties. Li, et al. [168] studied the recent progress made in the MOF-based photocatalysts for the reduction of CO_2 on the basis of the products reduced, including the CO_2 photocatalytic conversion into CO and other organic chemicals (methanol, formic acid, and methane). The authors' have also mentioned several modification techniques for the relevant improvements in the photocatalytic performance and the structural activity-corresponding relationships. The authors have mainly focused on the CO_2 capture capacity role for the CO_2 photocatalytic reduction performance over the MOF-based materials [168]. Li, et al. [169] constructed a MOF that incorporates unsaturated metal sites and accessible nitrogen-rich groups by using a solvothermal assembly of an acylamide-containing Cu(II) ions and tetracarboxylate ligand. The results showed that the MOFs synthesized had a high CO_2-adsorbing capability, and high porosity exposed Lewis acid metal sites. The inherent structural features of the MOFs synthesized in the study makes them very promising candidates as heterogeneous catalysts for the chemical conversion of CO_2; this was confirmed by their highly efficient CO_2 cycloaddition with the small-sized epoxides in the study.

The high efficiency and remarkable size selectivity on the CO_2 catalytic conversion allows the synthesized MOFs in the study to act as an advanced heterogeneous catalyst for the carbon fixation process [169]. Ding, et al. [170] presented an in-situ-growth strategy and a facile double solvent to integrate CdS NPs with MIL-101(Cr) to synthesize CdS/MIL-101(Cr) composite photocatalyst. The results show that under visible light irradiation, the CdS/MIL-101(Cr) exhibited a remarkable activity enhancement for the conversion of CO_2 to CO [170]. Mu, Zhu, Li, Zhang, Su, Lian, Qi, Deng, Zhang, Wang, Zhu and Peng [144] assembled a composite thin film comprised of 2D/2D MOF/rGO heterostructures by using a Coulomb interaction for the use as a co-catalyst for the photocatalytic reduction of CO_2 for the first time. The study results showed that the best thin-film catalyst with an optimal MOF/rGO ratio exhibited a high evolution rate of CO 3.8×10^4 μmol h^{-1} g$_{film}^{-1}$ (0.46 min^{-1} in TOF) and an acceptable selectivity of 91.74% [144]. Li and Zhu [171] focused their work on the MOF-based materials active sites that achieve efficient charge separation for conductivity in the electrocatalytic CO_2 reduction and visible-light absorption for photocatalytic CO_2 reduction. The authors showed the distinguishable characteristics of the MOF-based material for catalytic CO_2 reduction, the recent progress and development in the MOF-based material for the CO_2 catalytic reduction, and the challenges and future perspectives for the development of MOF-based materials for CO_2 reduction [171].

One of the most effective techniques in the preparation of the catalysts for the CO_2 photocatalytic reduction into high value-added chemicals is by the use of metalloporphyrin as a light-harvesting mixed ligand to optimize the MOF. This method is valuable because it can improve the dispersibility of the prophyrin, thus inhibiting its potential agglomeration. Wang, et al. [172] incorporated through coordination mode a one-pot synthetic strategy to immobilize chemically Cu (II) tetra (4-carboxylphenyl) porphyrin (CuTCPP) into a UiO-66 MOF structure. Also, in-situ growth of the TiO_2 nanoparticles onto the MOF was actualized with the composite's generation of the CuTCPP⊂UiO-66/TiO_2 (CTU/TiO_2). The results showed that the catalytic results represented an optimal value of 31.32 μmol g^{-1} h^{-1} CO evolution amount, which was approximately 7 times greater than that of the pure TiO_2 obtained using photo-catalysis under Xe lamp irradiation ($\lambda > 300$ nm) Wang, Jin, Duan, She, Huang and Wang [172]. Dong, et al. [173] improved the CO_2 conversion activity of a synthesized MOF by the regulation of the metal species in the MOFs metal-cluster nodes, and the MOF's best photocatalytic activity for CO_2 conversion was obtained. The authors' synthesized a stable MOF, PCN-250-Fe$_3$ with open metal sites and Fe$_2^{III}$Fe^{II} metal-cluster nodes, and they further improved the CO_2 reduction activity by tuning the M^{II} metal ions species in the cluster. The results of their study showed that all bi-metallic PCN-250-Fe$_2$M (M = Mn, Ni, Zn, Co) exhibited higher catalytic activity and selectivity for stable reduction of CO_2 into CO, compared to the mono-metallic PCN-250-Fe$_3$. Further investigations revealed that introducing a second M^{II} metal ions can enhance the migration of the photogenerated electrons to the active sites and enhance the CO_2 activation and adsorption by favoring the rout of CO_2 reduction and limiting the production of hydrogen as an intermediate [173].

CO_2 conversion into clean energy by using photocatalysts with a porous hollow structure that has a superb activity is of worldwide interest. Chen, et al. [174] prepared a porous hollow spheres ZnO/NiO with sheet-like subunits by calcination of Ni–Zn bimetallic organic frameworks. The prepared ZnO/NiO composites showed an improved photocatalytic activity for the CO_2 recondition and an outstanding photocatalytic CO_2 reduction performance [174]. Ye, Gao, Cao, Chen, Yao, Hou and Sun [167] synthesized and used an ultrathin two-dimensional Zn porphyrin-based metal-organic framework (Zn-MOF nanosheets) in the photoreduction of CO_2 to CO. The two novelty noble-metal-free hybrid photocatalytic systems displayed outstanding selectivity and photocatalytic activity for CO emissions under mild photocatalytic reaction conditions. These studies highlight that the development of noble-metal-free photocatalytic systems and MOF-based materials for photocatalytic applications are promising [167]. Crake, et al. [175] effectively coupled under ultraviolet–visible (UV–vis) light irradiation TiO_2 nanosheets and metal-organic framework (NH_2-UiO-66) using an *in-situ* growth strategy for the formation of bifunctional materials for the combined photocatalytic reduction and capture of CO_2. The results of their study showed that the nanocomposites were durable and dramatically more efficient in the reduction of CO_2 to CO than their single components. Furthermore, the photocatalytic activity was significantly altered by the composition of the nanocomposites with the optimal TiO_2 content doubling the evolution rate of CO compared to the pure TiO_2 [175].

3.6. MOF-Based Materials for Electrochemical and Electrocatalytic Conversion of CO_2

Electrochemical and electrocatalytic reduction of CO_2 into hydrocarbons and value-added chemicals is a remarkable and clean way to mitigate greenhouse gas emissions as a result of our over-dependence on fossil fuels. The electrocatalytic CO_2 reduction reaction consists of two half-reactions that can occur by two to fourteen-electron exchange process. These reactions are shown in Table 9, along with various standard electrode potentials vs. the standard hydrogen electrode (SHE).

Table 9. Standard electrochemical potentials for CO_2 reduction [142].

Reduction Potentials of CO_2	Standard Electrode Potentials vs. SHE (V)
$CO_2 + 2H^+ + 2e^- \rightarrow CO + H_2O$	−0.106
$2CO_2 + 2H^+ + 2e^- \rightarrow H_2C_2O_4$	−0.500
$CO_2 + 2H^+ + 2e^- \rightarrow HCOOH + H_2O$	−0.250
$CO_2 + 4H^+ + 4e^- \rightarrow CH_2O + 2H_2O$	−0.070
$CO_2 + 4H^+ + 4e^- \rightarrow C + 2H_2O$	0.210
$CO_2 + 8H^+ + 8e^- \rightarrow CH_4 + 2H_2O$	0.169
$CO_2 + 6H^+ + 6e^- \rightarrow CH_3OH + H_2O$	0.016
$CO_2 + 14H^+ + 14e^- \rightarrow C_2H_6 + 4H_2O$	0.084
$CO_2 + 12H^+ + 12e^- \rightarrow C_2H_4 + 4H_2O$	0.064

Ma, et al. [176] employed Zn–Ni bimetal MOFs as precursors for the synthesis of Ni-N-doped porous interconnected carbon (NiNPIC) catalysts to enhance CO_2RR electrocatalytic activity and selectivity. The interconnected porous structures and the high surface area of the catalyst have provided a convenient channel for mass diffusion and highly accessible Ni-N sites that lead to a higher electron transfer, less interface resistance, and greater electrolyte/gas transport in the CO_2RR. The results demonstrated that the synthesized catalyst has a high conversion efficiency of CO_2 into CO and excellent electrochemical stability at a moderate over-potential [176]. Table 10 below shows a summary of MOF-based catalysts for electrocatalytic CO_2 reduction.

Table 10. Summary of MOF-based catalysts for electrocatalytic CO_2 reduction.

Sample ID	Product	FE (%)	Potential	Reference
Zn-BTC	CH_4	80.1 ± 6.6	−2.2 V vs. Ag/AgCl	[177]
M-PMOF	CO	98.7	−0.8 V vs. RHE [1]	[178]
Re-SURMOF	CO	93 ± 5	−1.6 V vs. NHE	[179]
ZIF-8	CO	65.5	−1.8 V vs. SCE	[180]
ZIF-CNT-FA-p	CO	100	−0.86 V vs. RHE	[181]
$Al_2(OH)_2$TCPP-Co	CO	76	−0.7 V vs. RHE	[182]
CR-MOF	HCOOH	98	−1.2 V vs. SHE	[183]
Ru(III)-doped HKUST1	CH_3OH, C_2H_5OH	47.2	20 mA cm^{-2}	[184]
Ag_2O/layer ZIF	CO	80.5	−1.2 V vs. RHE	[184]
C-AFC@ZIF-8	CO	93	−0.6 V vs. RHE	[185]
ZIF-8 derived Fe-N-C	CO	91	−0.6 V vs. RHE	[186]

[1] Reversible Hydrogen Electrode.

By looking at Table 10 above, CO_2 is mostly electro-catalytically reduced into CO. Furthermore, ZIF-CNT-FA-p shows the greatest FE (%) with a value of 100 for CO. On the other hand, the lowest FE (%) for CO is given by ZIF-8, with a value of 65.

The Cu-based catalysts exhibit distinguishable superiorities; however, achieving high selectivity for hydrocarbon is still a great challenge. Tan, et al. [187] reported a multifunction-coupled Cu–MOF tailor-made electrocatalyst by using time-resolved controllable restructuration from Cu_2O to Cu_2O@Cu-MOF. The restructured electrocatalyst from their study had a high CO_2 adsorption capacity and electrocatalytic activity. The results showed that their synthesized MOF exhibited high performance towards hydrocarbons, with a hydrocarbon Faradaic efficiency (FE) of 79.4% [187]. Li, et al. [188] reported a remarkable 2D bismuth metal–organic framework (Bi-MOF) that exhibits an accessible permanent porosity for a highly efficient CO_2 electrochemical reduction (ECR) to HCOOH. The results of the author's study show that the 2D Bi-MOF open-framework structure synthesized shows excellent Faradaic efficiency for the formation of HCOOH over a large potential window, reaching 92.2% at approximately −0.9 V [188].

In sustainable energy research, it is well known that metallic copper acts as an electrocatalyst for the CO_2 reduction to multicarbon products like hydrocarbons and alcohols. However, a great challenge remains in the development of a selective, cost-effective, and stable catalyst/electrode material for this reduction reaction. Rayer, et al. [189] studied the potentials of copper carbonized MOF-derived electrocatalysts as catalytic materials for CO_2 electrochemical reduction. The author used two copper-decorated commercial MOFs, PCN-62, and HKUST-1, pyrolyzed at a variable temperature range (400 to 800 °C), were coated on both copper and metallic nickel supports as inks. The authors' study shows that the MOF-derived coatings produce electrodes with higher selectivity and current density towards isopropanol compared with the uncoated copper electrodes. Also, the best-performing electrocatalyst in their study exhibits an isopropanol Faradaic efficiency (FE) of over 72% [189]. Zhang, et al. [190] synthesized a novel mixed-metallic MOF $[Ag_4Co_2 (pyz)PDC_4][Ag_2Co(pyz)_2PDC_2]$ and transformed it into an Ag-doped Co_3O_4 catalyst that exhibits excellent electrocatalytic performance for CO_2 reduction in water to syngas (H_2 + CO). The as-prepared Ag/Co3O4 material showed a high CO selectivity in a 0.1 M $KHCO_3$ aqueous solution (CO_2 saturated) with an approximately 55.6% corresponding Faradaic efficiency. The results showed that the presence of Ag can increase the efficiency of CO greatly, hence inhibiting the H_2 production [190]. Cao, et al. [191] used a nitrogen-rich Cu–BTT MOF as a catalyst for the electrochemical reduction of CO2. The results showed that the high-temperature pyrolysis product of Cu–N–C_{1100} has the best catalytic activity for production of CO and HCOOH [191]. Sun, et al. [192] synthesized nitrogen-doped mesoporous carbon nanoparticles that have atomically dispersed iron sites (namely mesoNC-Fe) using high-temperature pyrolysis of a Fe that contains ZIF-8 MOF. The hydrolysis of tetramethyl orthosilicate (TMOS) in the MOF framework made by the author prior to pyrolysis has a fundamental role in the maintenance of a high surface area in the formation phase of the carbon structure, impeding the iron (oxide) nanoparticles' formation.

The results showed that the combination of such a distinguishable coordination environment that has a high surface area in the mesoNC-Fe carbon structure makes more accessible active sites during catalysis and promotes CO_2 electro-reduction [192].

Transforming CO_2 into a broad range of chemicals, including methanol, is a high priority field of study owing to the direct link between CO_2 emissions and global warming. There is an environmental and industrial need for substituting non-renewable energy fuels with renewable and sustainable energy sources. Electrochemical reduction acts as a superb approach in the conversion of CO_2 to methanol by the employment of alternative energy sources at which an electrocatalyst plays a fundamental role. Many efforts are being made by several researchers to understand and increase the catalytic efficiency of electrocatalysts. MOFs, composite materials, and metal oxide are employed for CO_2 electrochemical reduction to methanol. However, MOFs catch most of the researchers' attention in CO_2 conversion as a result of their high surface area, simplicity, and exquisite structural features. In recent decades, there have been significant applications of MOFs and their derivatives in CO_2 reduction. Al-Rowaili, et al. [193] focused their work on the electro-reduction of CO_2 to methanol by coalescing MOFs' vantages, and their composite materials. The authors highlighted the challenges in achieving CO_2 electro-reduction with high efficiency and selectivity [193]. Dong, et al. [194] introduced a highly stable 3D porphyrin-based MOF of PCN-222(Fe) into a heterogeneous catalysis by using a simple dip-coating method. Their study shows that their composite catalyst PCN-222(Fe)/C exhibited a high catalytic performance for the electrochemical CO_2 conversion to CO with an overpotential of 494 mV and a maximum of 91% FE_{CO} in a CO_2-saturated aqueous solution of 0.5 M $KHCO_3$ [194].

Hod, et al. [195] demonstrated that MOF material thin-film electrophoretic deposition is an effective method for immobilizing the required quantity of a catalyst. The authors used in their study for electrocatalytic CO_2 reduction a material that consists of functionalized Fe-porphyrins as catalytically competent, redox-conductive linkers. Their method yielded an electrochemically addressable catalytic site with a highly effective surface coverage. The chemical products of their reduction contain mixtures of CO and H_2. These results show that the MOFs are very promising as catalysts for electrochemical reactions [195]. Wang, et al. [196] synthesized a nitrogen-doped carbon via the pyrolysis of a well-known MOF, namely ZIF-8, for the use as a catalyst in the electrochemical reduction of CO_2, and a subsequent acid treatment was then applied. Their study's resulting electrode exhibited a Faradaic efficiency to CO of approximately 78%, with hydrogen the only byproduct [196].

4. Conclusions and Future Perspective

MOFs' applications in the CO_2 capture, adsorption, membrane separation, catalytic conversion, and electrochemical reduction processes were analyzed thoroughly in this paper. As an emerging new class of crystalline porous materials, MOFs have attracted great attention over recent decades. The high surface area, high porosity, well-defined structures, and spectacular CO_2 adsorption of MOFs are in great demand for CO_2 capture, separation, conversion, and reduction processes.

There are three primary approaches in carbon dioxide capture using MOFs: pre-combustion capture, post-combustion capture, and oxy-fuel combustion capture. The post-combustion CO_2 capture is the most adopted technology in the carbon capture field. However, a large number of aspects should be considered before implementing MOFs in these technologies. These aspects include MOFs' adsorption capacity, thermal stability, selectivity, and life cycle in several operational conditions. Also, for large-scale processes, further details should be considered; an economic analysis must be conducted, the supply chain of the raw materials, including the MOFs, should be analyzed, and environmental impact analysis should be produced. In addition, the thermal stability for the MOFs should be increased for large-scale processes by regenerating the MOFs. This increase in thermal stability will allow the MOFs to be used under high operational conditions. Optimistically, MOFs' properties are continually improving and developing over time, which allows them to serve as the next generation material class for CO_2 capture.

In the membrane separation process of CO_2, MOFs act as a filler in mixed matrix membranes, enhancing the membranes' separation efficiency. However, the membrane separation process is usually hindered by several factors, including the cost, permeability, and selectivity of the membrane. The greatest challenge in the membrane separation with MOFs is the cost. Unfortunately, membranes are usually expensive to develop and maintain; with MOFs' incorporation, they will be even more expensive. Hence, economic MOFs should be chosen as a filler for the mixed matrix membranes for an economically viable process.

In the reduction and conversion processes of CO_2, MOFs acts as a catalyst for the processes. The applications of MOFs as a catalyst for these chemical processes are extensively overviewed in this article. Various parameters should be considered before choosing the MOFs, such as photocatalytic activity, electrical conductivity, and stability. The cost of the MOFs is also the major issue that faces their practical application.

Based on the above studies, MOFs are very promising materials for CO_2 capture, separation, adsorption, and chemical processes. However, their application is mainly suppressed by their high cost. MOFs have a high cost as a result of the costly synthesis of their raw materials. The choice of the MOFs' synthesis approach plays a major role in the economic aspect of the process. The high cost of the MOFs' raw materials is mainly owing to the lack of industrial-scale manufacturing facilities [197]. The MOFs' synthesis process involves sophisticated and time-consuming batch operations with complicated separation techniques and costly organic solvents. Various studies have been conducted to find an efficient approach for reducing the cost of the MOFs. DeSantis, et al. [198] performed a techno-economic analysis to identify the primary factors of MOF adsorbents' high production cost (Mg-MOF-74, Ni-MOF-74, HKAUST-1 and MOF-5) and find approaches for cost reduction. The authors found the cost of the solvent used in the MOFs' synthesis to be the main factor for the high cost. The authors also mentioned that by changing from solvothermal synthesis to liquid assisted grinding and aqueous synthesis it is estimated to decrease the cost by up to 83% [198]. Hence, a detailed economic analysis should be performed before choosing the MOF for any process. With their remarkable properties, MOFs are predicted to be further developed for more economical and efficient applications.

Funding: This research was funded by Qatar Foundation NPRP10-0107-170119.

Acknowledgments: The work was made possible by a grant from the Qatar National Research Fund under the National Priorities Research Program award number NPRP10-0107-170119. Its content is solely the responsibility of the authors and does not necessarily represent the official views of QNRF.

References

1. Smithson, P.A. IPCC, 2001: Climate Change 2001: The Scientific Basis. In *Contribution of Working Group 1 to the Third Assessment Report of the Intergovernmental Panel on Climate Change*; Houghton, J.T., Ding, Y., Griggs, D.J., Noguer, M., van der Linden, P.J., Dai, X., Maskell, K., Johnson, C.A., Eds.; Cambridge University Press: Cambridge, UK; New York, NY, USA, 2001; Volume 22, p. 1144, ISBN 0-521-01495-6. [CrossRef]
2. Lemke, P.; Ren, J.F.; Alley, R.; Allison, I.; Carrasco, J.; Flato, G.; Fujii, Y.; Kaser, G.; Mote, P.; Thomas, R.; et al. *IPCC, 2007. Climate Change 2007. Synthesis Report. Contribution of Working Groups I, II & III to the Fourth Assessment Report of the Intergovernmental Panel on Climate Change. Geneva*; Cambridge University Press: Cambridge, UK; New York, NY, USA, 2007. [CrossRef]
3. Rochelle, G.T. Amine scrubbing for CO_2 capture. *Science* **2009**, *325*, 1652–1654. [CrossRef]
4. Thiruvenkatachari, R.; Su, S.; An, H.; Yu, X.X. Post combustion CO_2 capture by carbon fibre monolithic adsorbents. *Prog. Energy Combust. Sci.* **2009**, *35*, 438–455. [CrossRef]
5. Omoregbe, O.; Mustapha, A.N.; Steinberger-Wilckens, R.; El-Kharouf, A.; Onyeaka, H. Carbon capture technologies for climate change mitigation: A bibliometric analysis of the scientific discourse during 1998–2018. *Energy Rep.* **2020**, *6*, 1200–1212. [CrossRef]

6. Ghoufi, A.; Maurin, G. Hybrid Monte Carlo Simulations Combined with a Phase Mixture Model to Predict the Structural Transitions of a Porous Metal–Organic Framework Material upon Adsorption of Guest Molecules. *J. Phys. Chem. C* **2010**, *114*, 6496–6502. [CrossRef]

7. Younas, M.; Rezakazemi, M.; Daud, M.; Wazir, M.B.; Ahmad, S.; Ullah, N.; Inamuddin; Ramakrishna, S. Recent progress and remaining challenges in post-combustion CO_2 capture using metal-organic frameworks (MOFs). *Prog. Energy Combust. Sci.* **2020**, *80*, 100849. [CrossRef]

8. Qin, J.-S.; Yuan, S.; Alsalme, A.; Zhou, H.-C. Flexible Zirconium MOF as the Crystalline Sponge for Coordinative Alignment of Dicarboxylates. *ACS Appl. Mater. Interfaces* **2017**, *9*, 33408–33412. [CrossRef] [PubMed]

9. Li, J.-R.; Kuppler, R.J.; Zhou, H.-C. Selective gas adsorption and separation in metal–organic frameworks. *Chem. Soc. Rev.* **2009**, *38*, 1477–1504. [CrossRef] [PubMed]

10. Dybtsev, D.N.; Chun, H.; Yoon, S.H.; Kim, D.; Kim, K. Microporous Manganese Formate: A Simple Metal–Organic Porous Material with High Framework Stability and Highly Selective Gas Sorption Properties. *J. Am. Chem. Soc.* **2004**, *126*, 32–33. [CrossRef]

11. Loiseau, T.; Lecroq, L.; Volkringer, C.; Marrot, J.; Férey, G.; Haouas, M.; Taulelle, F.; Bourrelly, S.; Llewellyn, P.L.; Latroche, M. MIL-96, a Porous Aluminum Trimesate 3D Structure Constructed from a Hexagonal Network of 18-Membered Rings and μ3-Oxo-Centered Trinuclear Units. *J. Am. Chem. Soc.* **2006**, *128*, 10223–10230. [CrossRef]

12. Xue, M.; Ma, S.; Jin, Z.; Schaffino, R.M.; Zhu, G.-S.; Lobkovsky, E.B.; Qiu, S.-L.; Chen, B. Robust Metal–Organic Framework Enforced by Triple-Framework Interpenetration Exhibiting High H2 Storage Density. *Inorg. Chem.* **2008**, *47*, 6825–6828. [CrossRef]

13. Zhuang, W.; Yuan, D.; Liu, D.; Zhong, C.; Li, J.-R.; Zhou, H.-C. Robust Metal–Organic Framework with An Octatopic Ligand for Gas Adsorption and Separation: Combined Characterization by Experiments and Molecular Simulation. *Chem. Mater.* **2012**, *24*, 18–25. [CrossRef]

14. Scholes, C.; Kentish, S.; Stevens, G. Carbon Dioxide Separation through Polymeric Membrane Systems for Flue Gas Applications. *Recent Pat. Chem. Eng.* **2010**, *1*. [CrossRef]

15. Furukawa, H.; Ko, N.; Go, Y.B.; Aratani, N.; Choi, S.B.; Choi, E.; Yazaydin, A.Ö.; Snurr, R.Q.; O'Keeffe, M.; Kim, J.; et al. Ultrahigh Porosity in Metal-Organic Frameworks. *Science* **2010**, *329*, 424. [CrossRef]

16. Qasem, N.A.A.; Ben-Mansour, R.; Habib, M.A. An efficient CO_2 adsorptive storage using MOF-5 and MOF-177. *Appl. Energy* **2018**, *210*, 317–326. [CrossRef]

17. Kayal, S.; Sun, B.; Chakraborty, A. Study of metal-organic framework MIL-101(Cr) for natural gas (methane) storage and compare with other MOFs (metal-organic frameworks). *Energy* **2015**, *91*, 772–781. [CrossRef]

18. Wei, W.; Xia, Z.; Wei, Q.; Xie, G.; Chen, S.; Qiao, C.; Zhang, G.; Zhou, C. A heterometallic microporous MOF exhibiting high hydrogen uptake. *Microporous Mesoporous Mater.* **2013**, *165*, 20–26. [CrossRef]

19. Joharian, M.; Morsali, A. Ultrasound-assisted synthesis of two new fluorinated metal-organic frameworks (F-MOFs) with the high surface area to improve the catalytic activity. *J. Solid State Chem.* **2019**, *270*, 135–146. [CrossRef]

20. Mukherjee, S.; Kumar, A.; Zaworotko, M.J. Metal-organic framework based carbon capture and purification technologies for clean environment. In *Metal-Organic Frameworks (MOFs) for Environmental Applications*; Ghosh, S.K., Ed.; Elsevier: Amsterdam, The Netherlands, 2019; pp. 5–61. [CrossRef]

21. Ghanbari, T.; Abnisa, F.; Wan Daud, W.M.A. A review on production of metal organic frameworks (MOF) for CO_2 adsorption. *Sci. Total Environ.* **2020**, *707*, 135090. [CrossRef]

22. Kazemi, S.; Safarifard, V. Carbon dioxide capture in MOFs: The effect of ligand functionalization. *Polyhedron* **2018**, *154*, 236–251. [CrossRef]

23. Duan, C.; Yu, Y.; Xiao, J.; Li, Y.; Yang, P.; Hu, F.; Xi, H. Recent advancements in metal–organic frameworks for green applications. *Green Energy Environ.* **2020**. [CrossRef]

24. Simmons, J.M.; Wu, H.; Zhou, W.; Yildirim, T. Carbon capture in metal–organic frameworks—A comparative study. *Energy Environ. Sci.* **2011**, *4*, 2177–2185. [CrossRef]

25. Li, J.-R.; Ma, Y.; McCarthy, M.C.; Sculley, J.; Yu, J.; Jeong, H.-K.; Balbuena, P.B.; Zhou, H.-C. Carbon dioxide capture-related gas adsorption and separation in metal-organic frameworks. *Coord. Chem. Rev.* **2011**, *255*, 1791–1823. [CrossRef]

26. Li, H.; Eddaoudi, M.; O'Keeffe, M.; Yaghi, O.M. Design and synthesis of an exceptionally stable and highly porous metal-organic framework. *Nature* **1999**, *402*, 276–279. [CrossRef]

27. Koppens, F.H.L.; Folk, J.A.; Elzerman, J.M.; Hanson, R.; van Beveren, L.H.W.; Vink, I.T.; Tranitz, H.P.; Wegscheider, W.; Kouwenhoven, L.P.; Vandersypen, L.M.K. Control and Detection of Singlet-Triplet Mixing in a Random Nuclear Field. *Science* **2005**, *309*, 1346. [CrossRef]

28. Ludig, S.; Haller, M.; Bauer, N. Tackling long-term climate change together: The case of flexible CCS and fluctuating renewable energy. *Energy Procedia* **2011**, *4*, 2580–2587. [CrossRef]

29. Miyagawa, T.; Matsuhashi, R.; Murai, S.; Muraoka, M. Comparative assessment of CCS with other technologies mitigating climate change. *Energy Procedia* **2011**, *4*, 5710–5714. [CrossRef]

30. Koljonen, T.; Flyktman, M.; Lehtilä, A.; Pahkala, K.; Peltola, E.; Savolainen, I. The role of CCS and renewables in tackling climate change. *Energy Procedia* **2009**, *1*, 4323–4330. [CrossRef]

31. Freund, P. 1 - Anthropogenic climate change and the role of CO_2 capture and storage (CCS). In *Geological Storage of Carbon Dioxide (CO_2)*; Gluyas, J., Mathias, S., Eds.; Woodhead Publishing: Cambridge, UK, 2013; pp. 3–25. [CrossRef]

32. Hanaoka, T.; Masui, T. Exploring the 2 °C Target Scenarios by Considering Climate Benefits and Health Benefits—Role of Biomass and CCS. *Energy Procedia* **2017**, *114*, 2618–2630. [CrossRef]

33. Gibbins, J.; Chalmers, H. Is all CCS equal? Classifying CCS applications by their potential climate benefit. *Energy Procedia* **2011**, *4*, 5715–5720. [CrossRef]

34. Wang, Y.; Zhao, L.; Otto, A.; Robinius, M.; Stolten, D. A Review of Post-combustion CO_2 Capture Technologies from Coal-fired Power Plants. *Energy Procedia* **2017**, *114*, 650–665. [CrossRef]

35. Zou, X.; Zhu, G. CO_2 Capture with MOF Membranes. In *Microporous Materials for Separation Membranes*; John Wiley & Sons: Hoboken, NJ, USA, 2019; pp. 323–359. [CrossRef]

36. Kang, Z.; Fan, L.; Sun, D. Recent advances and challenges of metal–organic framework membranes for gas separation. *J. Mater. Chem. A* **2017**, *5*, 10073–10091. [CrossRef]

37. Chen, B.; Ma, S.; Hurtado, E.J.; Lobkovsky, E.B.; Zhou, H.-C. A Triply Interpenetrated Microporous Metal–Organic Framework for Selective Sorption of Gas Molecules. *Inorg. Chem.* **2007**, *46*, 8490–8492. [CrossRef]

38. Llewellyn, P.L.; Bourrelly, S.; Serre, C.; Filinchuk, Y.; Férey, G. How hydration drastically improves adsorption selectivity for CO(2) over CH(4) in the flexible chromium terephthalate MIL-53. *Angew. Chem. Int. Ed. Engl.* **2006**, *45*, 7751–7754. [CrossRef]

39. Chen, B.; Ma, S.; Zapata, F.; Fronczek, F.R.; Lobkovsky, E.B.; Zhou, H.-C. Rationally Designed Micropores within a Metal–Organic Framework for Selective Sorption of Gas Molecules. *Inorg. Chem.* **2007**, *46*, 1233–1236. [CrossRef]

40. Cheon, Y.E.; Suh, M.P. Multifunctional Fourfold Interpenetrating Diamondoid Network: Gas Separation and Fabrication of Palladium Nanoparticles. *Chem. A Eur. J.* **2008**, *14*, 3961–3967. [CrossRef]

41. Bourrelly, S.; Llewellyn, P.L.; Serre, C.; Millange, F.; Loiseau, T.; Férey, G. Different Adsorption Behaviors of Methane and Carbon Dioxide in the Isotypic Nanoporous Metal Terephthalates MIL-53 and MIL-47. *J. Am. Chem. Soc.* **2005**, *127*, 13519–13521. [CrossRef]

42. Ma, S.; Wang, X.-S.; Manis, E.S.; Collier, C.D.; Zhou, H.-C. Metal–Organic Framework Based on a Trinickel Secondary Building Unit Exhibiting Gas-Sorption Hysteresis. *Inorg. Chem.* **2007**, *46*, 3432–3434. [CrossRef]

43. Kitaura, R.; Seki, K.; Akiyama, G.; Kitagawa, S. Porous Coordination-Polymer Crystals with Gated Channels Specific for Supercritical Gases. *Angew. Chem. Int. Ed.* **2003**, *42*, 428–431. [CrossRef]

44. Hayashi, H.; Côté, A.P.; Furukawa, H.; O'Keeffe, M.; Yaghi, O.M. Zeolite A imidazolate frameworks. *Nat. Mater.* **2007**, *6*, 501–506. [CrossRef]

45. Maji, T.K.; Matsuda, R.; Kitagawa, S. A flexible interpenetrating coordination framework with a bimodal porous functionality. *Nat. Mater.* **2007**, *6*, 142–148. [CrossRef]

46. Thallapally, P.K.; Tian, J.; Radha Kishan, M.; Fernandez, C.A.; Dalgarno, S.J.; McGrail, P.B.; Warren, J.E.; Atwood, J.L. Flexible (breathing) interpenetrated metal-organic frameworks for CO_2 separation applications. *J. Am. Chem. Soc.* **2008**, *130*, 16842–16843. [CrossRef]

47. Maji, T.K.; Mostafa, G.; Matsuda, R.; Kitagawa, S. Guest-induced asymmetry in a metal-organic porous solid with reversible single-crystal-to-single-crystal structural transformation. *J. Am. Chem. Soc.* **2005**, *127*, 17152–17153. [CrossRef]

48. Wade, C.R.; Dincă, M. Investigation of the synthesis, activation, and isosteric heats of CO_2 adsorption of the isostructural series of metal–organic frameworks M3(BTC)2 (M = Cr, Fe, Ni, Cu, Mo, Ru). *Dalton Trans.* **2012**, *41*, 7931–7938. [CrossRef]

49. Hong, D.H.; Suh, M.P. Selective CO_2 adsorption in a metal–organic framework constructed from an organic ligand with flexible joints. *Chem. Commun.* **2012**, *48*, 9168–9170. [CrossRef]

50. Bataille, T.; Bracco, S.; Comotti, A.; Costantino, F.; Guerri, A.; Ienco, A.; Marmottini, F. Solvent dependent synthesis of micro- and nano- crystalline phosphinate based 1D tubular MOF: Structure and CO_2 adsorption selectivity. *CrystEngComm* **2012**, *14*, 7170–7173. [CrossRef]

51. Yan, Q.; Lin, Y.; Wu, P.; Zhao, L.; Cao, L.; Peng, L.; Kong, C.; Chen, L. Designed Synthesis of Functionalized Two-Dimensional Metal–Organic Frameworks with Preferential CO_2 Capture. *ChemPlusChem* **2013**, *78*, 86–91. [CrossRef]

52. Li, T.; Sullivan, J.E.; Rosi, N.L. Design and Preparation of a Core–Shell Metal–Organic Framework for Selective CO_2 Capture. *J. Am. Chem. Soc.* **2013**, *135*, 9984–9987. [CrossRef] [PubMed]

53. Zheng, B.; Yun, R.; Bai, J.; Lu, Z.; Du, L.; Li, Y. Expanded porous MOF-505 analogue exhibiting large hydrogen storage capacity and selective carbon dioxide adsorption. *Inorg. Chem.* **2013**, *52*, 2823–2829. [CrossRef] [PubMed]

54. Park, J.; Li, J.-R.; Chen, Y.-P.; Yu, J.; Yakovenko, A.A.; Wang, Z.U.; Sun, L.-B.; Balbuena, P.B.; Zhou, H.-C. A versatile metal–organic framework for carbon dioxide capture and cooperative catalysis. *Chem. Commun.* **2012**, *48*, 9995–9997. [CrossRef]

55. Zhao, Y.; Ding, H.; Zhong, Q. Synthesis and characterization of MOF-aminated graphite oxide composites for CO_2 capture. *Appl. Surf. Sci.* **2013**, *284*, 138–144. [CrossRef]

56. Qian, D.; Lei, C.; Hao, G.-P.; Li, W.-C.; Lu, A.-H. Synthesis of Hierarchical Porous Carbon Monoliths with Incorporated Metal–Organic Frameworks for Enhancing Volumetric Based CO_2 Capture Capability. *ACS Appl. Mater. Interfaces* **2012**, *4*, 6125–6132. [CrossRef]

57. Yu, J.; Balbuena, P.B. Water Effects on Postcombustion CO_2 Capture in Mg-MOF-74. *J. Phys. Chem. C* **2013**, *117*, 3383–3388. [CrossRef]

58. Masoomi, M.Y.; Stylianou, K.C.; Morsali, A.; Retailleau, P.; Maspoch, D. Selective CO_2 Capture in Metal–Organic Frameworks with Azine-Functionalized Pores Generated by Mechanosynthesis. *Cryst. Growth Des.* **2014**, *14*, 2092–2096. [CrossRef]

59. Lin, Y.; Yan, Q.; Kong, C.; Chen, L. Polyethyleneimine incorporated metal-organic frameworks adsorbent for highly selective CO_2 capture. *Sci. Rep.* **2013**, *3*, 1859. [CrossRef]

60. Caskey, S.R.; Wong-Foy, A.G.; Matzger, A.J. Dramatic Tuning of Carbon Dioxide Uptake via Metal Substitution in a Coordination Polymer with Cylindrical Pores. *J. Am. Chem. Soc.* **2008**, *130*, 10870–10871. [CrossRef] [PubMed]

61. Xiang, Z.; Peng, X.; Cheng, X.; Li, X.; Cao, D. CNT@Cu3(BTC)2 and Metal–Organic Frameworks for Separation of CO_2/CH4 Mixture. *J. Phys. Chem. C* **2011**, *115*, 19864–19871. [CrossRef]

62. Chen, S.; Jin, L.; Chen, X. The effect and prediction of temperature on adsorption capability of coal/CH4. *Procedia Eng.* **2011**, *26*, 126–131. [CrossRef]

63. Carpenter, S.M.; Long, H.A. 13 - Integration of carbon capture in IGCC systems. In *Integrated Gasification Combined Cycle (IGCC) Technologies*; Wang, T., Stiegel, G., Eds.; Woodhead Publishing: Southorn, UK; Cambridge, UK, 2017; pp. 445–463. [CrossRef]

64. Sumida, K.; Rogow, D.L.; Mason, J.A.; McDonald, T.M.; Bloch, E.D.; Herm, Z.R.; Bae, T.-H.; Long, J.R. Carbon Dioxide Capture in Metal–Organic Frameworks. *Chem. Rev.* **2012**, *112*, 724–781. [CrossRef]

65. Hu, Z.; Wang, Y.; Shah, B.B.; Zhao, D. CO_2 Capture in Metal–Organic Framework Adsorbents: An Engineering Perspective. *Adv. Sustain. Syst.* **2019**, *3*, 1800080. [CrossRef]

66. Günther, C.; Weng, M.; Kather, A. Restrictions and Limitations for the Design of a Steam Generator for a Coal-fired Oxyfuel Power Plant with Circulating Fluidised Bed Combustion. *Energy Procedia* **2013**, *37*, 1312–1321. [CrossRef]

67. Jansen, D.; Gazzani, M.; Manzolini, G.; Dijk, E.v.; Carbo, M. Pre-combustion CO_2 capture. *Int. J. Greenh. Gas Control* **2015**, *40*, 167–187. [CrossRef]

68. Looyd, P.J.D. Precombustion technologies to aid carbon capture. In *Greenhouse Gas Control Technologies 7*; Rubin, E.S., Keith, D.W., Gilboy, C.F., Wilson, M., Morris, T., Gale, J., Thambimuthu, K., Eds.; Elsevier Science Ltd.: Oxford, UK, 2005; pp. 1957–1961. [CrossRef]

69. Zhong, D.-L.; Wang, J.-L.; Lu, Y.-Y.; Li, Z.; Yan, J. Precombustion CO_2 capture using a hybrid process of adsorption and gas hydrate formation. *Energy* **2016**, *102*, 621–629. [CrossRef]

70. Lea-Langton, A.; Andrews, G. Pre-combustion Technologies. In *Biomass Energy with Carbon Capture and Storage (BECCS): Unlocking Negative Emissions*; Wiley: Hoboken, NJ, USA, 2018; pp. 67–91.

71. Zhang, Z.; Yao, Z.-Z.; Xiang, S.; Chen, B. Perspective of microporous metal–organic frameworks for CO_2 capture and separation. *Energy Environ. Sci.* **2014**, *7*, 2868–2899. [CrossRef]

72. Chung, Y.G.; Gómez-Gualdrón, D.A.; Li, P.; Leperi, K.T.; Deria, P.; Zhang, H.; Vermeulen, N.A.; Stoddart, J.F.; You, F.; Hupp, J.T.; et al. In silico discovery of metal-organic frameworks for precombustion CO_2 capture using a genetic algorithm. *Sci. Adv.* **2016**, *2*, e1600909. [CrossRef]

73. Nandi, S.; De Luna, P.; Daff, T.D.; Rother, J.; Liu, M.; Buchanan, W.; Hawari, A.I.; Woo, T.K.; Vaidhyanathan, R. A single-ligand ultra-microporous MOF for precombustion CO_2 capture and hydrogen purification. *Sci. Adv.* **2015**, *1*, e1500421. [CrossRef]

74. Ding, M.; Flaig, R.W.; Jiang, H.-L.; Yaghi, O.M. Carbon capture and conversion using metal–organic frameworks and MOF-based materials. *Chem. Soc. Rev.* **2019**, *48*, 2783–2828. [CrossRef] [PubMed]

75. Herm, Z.R.; Swisher, J.A.; Smit, B.; Krishna, R.; Long, J.R. Metal–Organic Frameworks as Adsorbents for Hydrogen Purification and Precombustion Carbon Dioxide Capture. *J. Am. Chem. Soc.* **2011**, *133*, 5664–5667. [CrossRef]

76. Asgari, M.; Queen, W. *Carbon Capture in Metal–Organic Frameworks*; Wiley: Hoboken, NJ, USA, 2018; pp. 1–78.

77. Favre, E. Membrane processes and postcombustion carbon dioxide capture: Challenges and prospects. *Chem. Eng. J.* **2011**, *171*, 782–793. [CrossRef]

78. Tillman, D.A. Chapter Nine—The Development of Postcombustion Control Technology. In *Coal-Fired Electricity and Emissions Control*; Tillman, D.A., Ed.; Butterworth-Heinemann: Oxford, UK, 2018; pp. 237–276. [CrossRef]

79. Zamarripa, M.A.; Eslick, J.C.; Matuszewski, M.S.; Miller, D.C. Multi-objective Optimization of Membrane-based CO_2 Capture. In *Computer Aided Chemical Engineering*; Eden, M.R., Ierapetritou, M.G., Towler, G.P., Eds.; Elsevier: Amsterdam, The Netherlands, 2018; Volume 44, pp. 1117–1122.

80. Breeze, P. Chapter 7—Carbon Capture and Storage. In *Coal-Fired Generation*; Breeze, P., Ed.; Academic Press: Boston, MA, USA, 2015; pp. 73–86. [CrossRef]

81. Ghoshal, S.; Zeman, F. Carbon dioxide (CO_2) capture and storage technology in the cement and concrete industry. In *Developments and Innovation in Carbon Dioxide (CO_2) Capture and Storage Technology*; Maroto-Valer, M.M., Ed.; Woodhead Publishing: Cambridge, UK, 2010; Volume 1, pp. 469–491.

82. Spigarelli, B.P.; Kawatra, S.K. Opportunities and challenges in carbon dioxide capture. *J. CO_2 Util.* **2013**, *1*, 69–87. [CrossRef]

83. Hu, Z.; Khurana, M.; Seah, Y.H.; Zhang, M.; Guo, Z.; Zhao, D. Ionized Zr-MOFs for highly efficient post-combustion CO_2 capture. *Chem. Eng. Sci.* **2015**, *124*, 61–69. [CrossRef]

84. Figueroa, J.D.; Fout, T.; Plasynski, S.; McIlvried, H.; Srivastava, R.D. Advances in CO_2 capture technology—The U.S. Department of Energy's Carbon Sequestration Program. *Int. J. Greenh. Gas Control* **2008**, *2*, 9–20. [CrossRef]

85. Martínez, F.; Sanz, R.; Orcajo, G.; Briones, D.; Yángüez, V. Amino-impregnated MOF materials for CO_2 capture at post-combustion conditions. *Chem. Eng. Sci.* **2016**, *142*, 55–61. [CrossRef]

86. Pai, K.N.; Baboolal, J.D.; Sharp, D.A.; Rajendran, A. Evaluation of diamine-appended metal-organic frameworks for post-combustion CO_2 capture by vacuum swing adsorption. *Sep. Purif. Technol.* **2019**, *211*, 540–550. [CrossRef]

87. Hedin, N.; Andersson, L.; Bergström, L.; Yan, J. Adsorbents for the post-combustion capture of CO_2 using rapid temperature swing or vacuum swing adsorption. *Appl. Energy* **2013**, *104*, 418–433. [CrossRef]

88. Samanta, A.; Zhao, A.; Shimizu, G.K.H.; Sarkar, P.; Gupta, R. Post-Combustion CO_2 Capture Using Solid Sorbents: A Review. *Ind. Eng. Chem. Res.* **2012**, *51*, 1438–1463. [CrossRef]

89. Maurya, M.; Singh, J.K. Effect of Ionic Liquid Impregnation in Highly Water-Stable Metal–Organic Frameworks, Covalent Organic Frameworks, and Carbon-Based Adsorbents for Post-combustion Flue Gas Treatment. *Energy Fuels* **2019**, *33*, 3421–3428. [CrossRef]

90. Babarao, R.; Jiang, J.W. Cation Characterization and CO_2 Capture in Li+-Exchanged Metal–Organic Frameworks: From First-Principles Modeling to Molecular Simulation. *Ind. Eng. Chem. Res.* **2011**, *50*, 62–68. [CrossRef]

91. Park, J.; Suh, B.L.; Kim, J. Computational Design of a Photoresponsive Metal–Organic Framework for Post Combustion Carbon Capture. *J. Phys. Chem. C* **2020**, *124*, 13162–13167. [CrossRef]

92. Wang, Q.; Bai, J.; Lu, Z.; Pan, Y.; You, X. Finely tuning MOFs towards high-performance post-combustion CO_2 capture materials. *Chem. Commun.* **2016**, *52*, 443–452. [CrossRef]

93. Marti, A. *Metal-Organic Frameworks Materials for Post-Combustion CO 2 Capture*; Wiley: Hoboken, NJ, USA, 2018; pp. 79–111.

94. Hu, Y.; Verdegaal, W.M.; Yu, S.-H.; Jiang, H.-L. Alkylamine-Tethered Stable Metal–Organic Framework for CO_2 Capture from Flue Gas. *ChemSusChem* **2014**, *7*, 734–737. [CrossRef]

95. Siagian, U.W.R.; Raksajati, A.; Himma, N.F.; Khoiruddin, K.; Wenten, I.G. Membrane-based carbon capture technologies: Membrane gas separation vs. membrane contactor. *J. Nat. Gas Sci. Eng.* **2019**, *67*, 172–195. [CrossRef]

96. Khalilpour, R.; Mumford, K.; Zhai, H.; Abbas, A.; Stevens, G.; Rubin, E.S. Membrane-based carbon capture from flue gas: A review. *J. Clean. Prod.* **2015**, *103*, 286–300. [CrossRef]

97. Krishna, R.; van Baten, J.M. In silico screening of zeolite membranes for CO_2 capture. *J. Membr. Sci.* **2010**, *360*, 323–333. [CrossRef]

98. Chen, Y.; Wang, B.; Zhao, L.; Dutta, P.; Winston Ho, W.S. New Pebax®/zeolite Y composite membranes for CO_2 capture from flue gas. *J. Membr. Sci.* **2015**, *495*, 415–423. [CrossRef]

99. Zhao, L.; Chen, Y.; Wang, B.; Sun, C.; Chakraborty, S.; Ramasubramanian, K.; Dutta, P.K.; Ho, W.S.W. Multilayer polymer/zeolite Y composite membrane structure for CO_2 capture from flue gas. *J. Membr. Sci.* **2016**, *498*, 1–13. [CrossRef]

100. Liu, M.; Nothling, M.D.; Webley, P.A.; Jin, J.; Fu, Q.; Qiao, G.G. High-throughput CO_2 capture using PIM-1@MOF based thin film composite membranes. *Chem. Eng. J.* **2020**, *396*, 125328. [CrossRef]

101. Sun, J.; Li, Q.; Chen, G.; Duan, J.; Liu, G.; Jin, W. MOF-801 incorporated PEBA mixed-matrix composite membranes for CO_2 capture. *Sep. Purif. Technol.* **2019**, *217*, 229–239. [CrossRef]

102. Chen, W.; Zhang, Z.; Hou, L.; Yang, C.; Shen, H.; Yang, K.; Wang, Z. Metal-organic framework MOF-801/PIM-1 mixed-matrix membranes for enhanced CO_2/N2 separation performance. *Sep. Purif. Technol.* **2020**, *250*, 117198. [CrossRef]

103. Majumdar, S.; Tokay, B.; Martin-Gil, V.; Campbell, J.; Castro-Muñoz, R.; Ahmad, M.Z.; Fila, V. Mg-MOF-74/Polyvinyl acetate (PVAc) mixed matrix membranes for CO_2 separation. *Sep. Purif. Technol.* **2020**, *238*, 116411. [CrossRef]

104. Ahmad, M.Z.; Peters, T.A.; Konnertz, N.M.; Visser, T.; Téllez, C.; Coronas, J.; Fila, V.; de Vos, W.M.; Benes, N.E. High-pressure CO_2/CH_4 separation of Zr-MOFs based mixed matrix membranes. *Sep. Purif. Technol.* **2020**, *230*, 115858. [CrossRef]

105. Chen, K.; Xu, K.; Xiang, L.; Dong, X.; Han, Y.; Wang, C.; Sun, L.-B.; Pan, Y. Enhanced CO_2/CH4 separation performance of mixed-matrix membranes through dispersion of sorption-selective MOF nanocrystals. *J. Membr. Sci.* **2018**, *563*, 360–370. [CrossRef]

106. Jiamjirangkul, P.; Inprasit, T.; Intasanta, V.; Pangon, A. Metal organic framework-integrated chitosan/poly(vinyl alcohol) (PVA) nanofibrous membrane hybrids from green process for selective CO_2 capture and filtration. *Chem. Eng. Sci.* **2020**, *221*, 115650. [CrossRef]

107. Lee, D.-J.; Li, Q.; Kim, H.; Lee, K. Preparation of Ni-MOF-74 membrane for CO_2 separation by layer-by-layer seeding technique. *Microporous Mesoporous Mater.* **2012**, *163*, 169–177. [CrossRef]

108. Anastasiou, S.; Bhoria, N.; Pokhrel, J.; Kumar Reddy, K.S.; Srinivasakannan, C.; Wang, K.; Karanikolos, G.N. Metal-organic framework/graphene oxide composite fillers in mixed-matrix membranes for CO_2 separation. *Mater. Chem. Phys.* **2018**, *212*, 513–522. [CrossRef]

109. Yin, H.; Alkaş, A.; Zhang, Y.; Zhang, Y.; Telfer, S.G. Mixed matrix membranes (MMMs) using an emerging metal-organic framework (MUF-15) for CO_2 separation. *J. Membr. Sci.* **2020**, *609*, 118245. [CrossRef]

110. Wu, W.; Li, Z.; Chen, Y.; Li, W. Polydopamine-Modified Metal–Organic Framework Membrane with Enhanced Selectivity for Carbon Capture. *Environ. Sci. Technol.* **2019**, *53*, 3764–3772. [CrossRef] [PubMed]

111. Chernikova, V.; Shekhah, O.; Belmabkhout, Y.; Eddaoudi, M. Nanoporous Fluorinated Metal–Organic Framework-Based Membranes for CO_2 Capture. *ACS Appl. Nano Mater.* **2020**, *3*, 6432–6439. [CrossRef]

112. Cheng, Y.; Tavares, S.R.; Doherty, C.M.; Ying, Y.; Sarnello, E.; Maurin, G.; Hill, M.R.; Li, T.; Zhao, D. Enhanced Polymer Crystallinity in Mixed-Matrix Membranes Induced by Metal–Organic Framework Nanosheets for Efficient CO_2 Capture. *ACS Appl. Mater. Interfaces* **2018**, *10*, 43095–43103. [CrossRef] [PubMed]

113. Altintas, C.; Keskin, S. Molecular Simulations of MOF Membranes and Performance Predictions of MOF/Polymer Mixed Matrix Membranes for CO_2/CH4 Separations. *ACS Sustain. Chem. Eng.* **2019**, *7*, 2739–2750. [CrossRef]

114. Prasetya, N.; Ladewig, B.P. New Azo-DMOF-1 MOF as a Photoresponsive Low-Energy CO_2 Adsorbent and Its Exceptional CO_2/N2 Separation Performance in Mixed Matrix Membranes. *ACS Appl. Mater. Interfaces* **2018**, *10*, 34291–34301. [CrossRef]

115. Benzaqui, M.; Pillai, R.S.; Sabetghadam, A.; Benoit, V.; Normand, P.; Marrot, J.; Menguy, N.; Montero, D.; Shepard, W.; Tissot, A.; et al. Revisiting the Aluminum Trimesate-Based MOF (MIL-96): From Structure Determination to the Processing of Mixed Matrix Membranes for CO_2 Capture. *Chem. Mater.* **2017**, *29*, 10326–10338. [CrossRef]

116. Maina, J.W.; Schütz, J.A.; Grundy, L.; Des Ligneris, E.; Yi, Z.; Kong, L.; Pozo-Gonzalo, C.; Ionescu, M.; Dumée, L.F. Inorganic Nanoparticles/Metal Organic Framework Hybrid Membrane Reactors for Efficient Photocatalytic Conversion of CO_2. *ACS Appl. Mater. Interfaces* **2017**, *9*, 35010–35017. [CrossRef] [PubMed]

117. Zhao, Z.; Ma, X.; Kasik, A.; Li, Z.; Lin, Y.S. Gas Separation Properties of Metal Organic Framework (MOF-5) Membranes. *Ind. Eng. Chem. Res.* **2013**, *52*, 1102–1108. [CrossRef]

118. Hu, Z.; Kang, Z.; Qian, Y.; Peng, Y.; Wang, X.; Chi, C.; Zhao, D. Mixed Matrix Membranes Containing UiO-66(Hf)-(OH)2 Metal–Organic Framework Nanoparticles for Efficient H2/CO_2 Separation. *Ind. Eng. Chem. Res.* **2016**, *55*, 7933–7940. [CrossRef]

119. Fan, L.; Kang, Z.; Shen, Y.; Wang, S.; Zhao, H.; Sun, H.; Hu, X.; Sun, H.; Wang, R.; Sun, D. Mixed Matrix Membranes Based on Metal–Organic Frameworks with Tunable Pore Size for CO_2 Separation. *Cryst. Growth Des.* **2018**, *18*, 4365–4371. [CrossRef]

120. Takht Ravanchi, M.; Sahebdelfar, S. Catalytic conversions of CO_2 to help mitigate climate change: Recent process developments. *Process Saf. Environ. Prot.* **2021**, *145*, 172–194. [CrossRef]

121. Dewangan, N.; Hui, W.M.; Jayaprakash, S.; Bawah, A.-R.; Poerjoto, A.J.; Jie, T.; Jangam, A.; Hidajat, K.; Kawi, S. Recent progress on layered double hydroxide (LDH) derived metal-based catalysts for CO_2 conversion to valuable chemicals. *Catal. Today* **2020**. [CrossRef]

122. Galadima, A.; Muraza, O. Catalytic thermal conversion of CO_2 into fuels: Perspective and challenges. *Renew. Sustain. Energy Rev.* **2019**, *115*, 109333. [CrossRef]

123. Yadav, D.K.; Singh, D.K.; Ganesan, V. Recent strategy(ies) for the electrocatalytic reduction of CO_2: Ni single-atom catalysts for the selective electrochemical formation of CO in aqueous electrolytes. *Curr. Opin. Electrochem.* **2020**, *22*, 87–93. [CrossRef]

124. Guo, W.; Shim, K.; Odongo Ngome, F.O.; Moon, Y.H.; Choi, S.-Y.; Kim, Y.-T. Highly active coral-like porous silver for electrochemical reduction of CO_2 to CO. *Appl. Surf. Sci.* **2020**, *41*, 101242. [CrossRef]

125. Guo, C.; Zhang, T.; Liang, X.; Deng, X.; Guo, W.; Wang, Z.; Lu, X.; Wu, C.-M.L. Single transition metal atoms on nitrogen-doped carbon for CO_2 electrocatalytic reduction: CO production or further CO reduction? *Appl. Surf. Sci.* **2020**, *533*, 147466. [CrossRef]

126. Guo, W.; Shim, K.; Kim, Y.-T. Ag layer deposited on Zn by physical vapor deposition with enhanced CO selectivity for electrochemical CO_2 reduction. *Appl. Surf. Sci.* **2020**, *526*, 146651. [CrossRef]

127. Tahir, B.; Tahir, M.; Nawawi, M.G.M. Highly stable 3D/2D WO3/g-C3N4 Z-scheme heterojunction for stimulating photocatalytic CO_2 reduction by H2O/H2 to CO and CH4 under visible light. *J. CO_2 Util.* **2020**, *41*, 101270. [CrossRef]

128. Kazemi Movahed, S.; Najinasab, A.; Nikbakht, R.; Dabiri, M. Visible light assisted photocatalytic reduction of CO_2 to methanol using Fe_3O_4@N-C/Cu_2O nanostructure photocatalyst. *J. Photochem. Photobiol. A Chem.* **2020**, *401*, 112763. [CrossRef]

129. Jiang, X.X.; De Hu, X.; Tarek, M.; Saravanan, P.; Alqadhi, R.; Chin, S.Y.; Rahman Khan, M.M. Tailoring the properties of g-C_3N_4 with CuO for enhanced photoelectrocatalytic CO_2 reduction to methanol. *J. CO_2 Util.* **2020**, *40*, 101222. [CrossRef]

130. Aranda-Aguirre, A.; Ojeda, J.; Ferreira de Brito, J.; Garcia-Segura, S.; Boldrin Zanoni, M.V.; Alarcon, H. Photoelectrodes of Cu2O with interfacial structure of topological insulator Bi2Se3 contributes to selective photoelectrocatalytic reduction of CO_2 towards methanol. *J. CO_2 Util.* **2020**, *39*, 101154. [CrossRef]

131. Karim, K.M.R.; Tarek, M.; Sarkar, S.M.; Mouras, R.; Ong, H.R.; Abdullah, H.; Cheng, C.K.; Khan, M.M.R. Photoelectrocatalytic reduction of CO_2 to methanol over $CuFe_2O_4$@PANI photocathode. *Int. J. Hydrog. Energy* **2020**. [CrossRef]

132. Huang, W.; Yuan, G. A composite heterogeneous catalyst C-Py-Sn-Zn for selective electrochemical reduction of CO_2 to methanol. *Electrochem. Commun.* **2020**, *118*, 106789. [CrossRef]

133. Cheng, J.; Yang, X.; Xuan, X.; Liu, N.; Zhou, J. Development of an efficient catalyst with controlled sulfur vacancies and high pyridine nitrogen content for the photoelectrochemical reduction of CO_2 into methanol. *Sci. Total Environ.* **2020**, *702*, 134981. [CrossRef]

134. Liu, Y.; Li, F.; Zhang, X.; Ji, X. Recent progress on electrochemical reduction of CO_2 to methanol. *Curr. Opin. Green Sustain. Chem.* **2020**, *23*, 10–17. [CrossRef]

135. Tarek, M.; Rezaul Karim, K.M.; Sarkar, S.M.; Deb, A.; Ong, H.R.; Abdullah, H.; Cheng, C.K.; Rahman Khan, M.M. Hetero-structure CdS–$CuFe_2O_4$ as an efficient visible light active photocatalyst for photoelectrochemical reduction of CO_2 to methanol. *Int. J. Hydrogen Energy* **2019**, *44*, 26271–26284. [CrossRef]

136. Feng, S.; Zhao, J.; Bai, Y.; Liang, X.; Wang, T.; Wang, C. Facile synthesis of Mo-doped TiO2 for selective photocatalytic CO_2 reduction to methane: Promoted H2O dissociation by Mo doping. *J. CO_2 Util.* **2020**, *38*, 1–9. [CrossRef]

137. Merino-Garcia, I.; Albo, J.; Solla-Gullón, J.; Montiel, V.; Irabien, A. Cu oxide/ZnO-based surfaces for a selective ethylene production from gas-phase CO_2 electroconversion. *J. CO_2 Util.* **2019**, *31*, 135–142. [CrossRef]

138. Qin, T.; Qian, Y.; Zhang, F.; Lin, B.-L. Cloride-derived copper electrode for efficient electrochemical reduction of CO_2 to ethylene. *Chin. Chem. Lett.* **2019**, *30*, 314–318. [CrossRef]

139. Fan, L.; Xia, C.; Zhu, P.; Lu, Y.; Wang, H. Electrochemical CO_2 reduction to high-concentration pure formic acid solutions in an all-solid-state reactor. *Nat. Commun.* **2020**, *11*, 3633. [CrossRef]

140. Zhao, Y.; Wang, T.; Wang, Y.; Hao, R.; Wang, H.; Han, Y. Catalytic reduction of CO_2 to HCO_2^- by nanoscale nickel-based bimetallic alloy under atmospheric pressure. *J. Ind. Eng. Chem.* **2019**, *77*, 291–302. [CrossRef]

141. Li, P.; Liu, L.; An, W.; Wang, H.; Guo, H.; Liang, Y.; Cui, W. Ultrathin porous g-C3N4 nanosheets modified with AuCu alloy nanoparticles and C-C coupling photothermal catalytic reduction of CO_2 to ethanol. *Appl. Catal. B Environ.* **2020**, *266*, 118618. [CrossRef]

142. Tekalgne, M.A.; Do, H.H.; Hasani, A.; Van Le, Q.; Jang, H.W.; Ahn, S.H.; Kim, S.Y. Two-dimensional materials and metal-organic frameworks for the CO_2 reduction reaction. *Mater. Today Adv.* **2020**, *5*, 100038. [CrossRef]

143. Billo, T.; Shown, I.; Anbalagan, A.k.; Effendi, T.A.; Sabbah, A.; Fu, F.-Y.; Chu, C.-M.; Woon, W.-Y.; Chen, R.-S.; Lee, C.-H.; et al. A mechanistic study of molecular CO_2 interaction and adsorption on carbon implanted SnS2 thin film for photocatalytic CO_2 reduction activity. *Nano Energy* **2020**, *72*, 104717. [CrossRef]

144. Mu, Q.; Zhu, W.; Li, X.; Zhang, C.; Su, Y.; Lian, Y.; Qi, P.; Deng, Z.; Zhang, D.; Wang, S.; et al. Electrostatic charge transfer for boosting the photocatalytic CO_2 reduction on metal centers of 2D MOF/rGO heterostructure. *Appl. Catal. B Environ.* **2020**, *262*, 118144. [CrossRef]

145. Ješić, D.; Lašič Jurković, D.; Pohar, A.; Suhadolnik, L.; Likozar, B. Engineering Photocatalytic and Photoelectrocatalytic CO_2 Reduction Reactions: Mechanisms, Intrinsic Kinetics, Mass Transfer Resistances, Reactors and Multi-scale Modelling Simulations. *Chem. Eng. J.* **2020**, 126799. [CrossRef]

146. Omar, S.; Shkir, M.; Ajmal Khan, M.; Ahmad, Z.; AlFaify, S. A comprehensive study on molecular geometry, optical, HOMO-LUMO, and nonlinear properties of 1,3-diphenyl-2-propen-1-ones chalcone and its derivatives for optoelectronic applications: A computational approach. *Optik* **2020**, *204*, 164172. [CrossRef]

147. Reddy, C.V.; Reddy, K.R.; Harish, V.V.N.; Shim, J.; Shankar, M.V.; Shetti, N.P.; Aminabhavi, T.M. Metal-organic frameworks (MOFs)-based efficient heterogeneous photocatalysts: Synthesis, properties and its applications in photocatalytic hydrogen generation, CO_2 reduction and photodegradation of organic dyes. *Int. J. Hydrog. Energy* **2020**, *45*, 7656–7679. [CrossRef]

148. Wang, C.-C.; Zhang, Y.-Q.; Li, J.; Wang, P. Photocatalytic CO_2 reduction in metal–organic frameworks: A mini review. *J. Mol. Struct.* **2015**, *1083*, 127–136. [CrossRef]

149. Wang, C.; Xie, Z.; deKrafft, K.E.; Lin, W. Doping Metal–Organic Frameworks for Water Oxidation, Carbon Dioxide Reduction, and Organic Photocatalysis. *J. Am. Chem. Soc.* **2011**, *133*, 13445–13454. [CrossRef]

150. Liu, Q.; Low, Z.-X.; Li, L.; Razmjou, A.; Wang, K.; Yao, J.; Wang, H. ZIF-8/Zn2GeO4 nanorods with an enhanced CO_2 adsorption property in an aqueous medium for photocatalytic synthesis of liquid fuel. *J. Mater. Chem. A* **2013**, *1*, 11563–11569. [CrossRef]

151. Fu, Y.; Sun, D.; Chen, Y.; Huang, R.; Ding, Z.; Fu, X.; Li, Z. An Amine-Functionalized Titanium Metal–Organic Framework Photocatalyst with Visible-Light-Induced Activity for CO_2 Reduction. *Angew. Chem. Int. Ed.* **2012**, *51*, 3364–3367. [CrossRef] [PubMed]

152. Li, R.; Hu, J.; Deng, M.; Wang, H.; Wang, X.; Hu, Y.; Jiang, H.L.; Jiang, J.; Zhang, Q.; Xie, Y.; et al. Integration of an inorganic semiconductor with a metal-organic framework: A platform for enhanced gaseous photocatalytic reactions. *Adv. Mater.* **2014**, *26*, 4783–4788. [CrossRef]

153. Liu, Y.; Yang, Y.; Sun, Q.; Wang, Z.; Huang, B.; Dai, Y.; Qin, X.; Zhang, X. Chemical Adsorption Enhanced CO_2 Capture and Photoreduction over a Copper Porphyrin Based Metal Organic Framework. *ACS Appl. Mater. Interfaces* **2013**, *5*, 7654–7658. [CrossRef] [PubMed]

154. Sun, D.; Liu, W.; Fu, Y.; Fang, Z.; Sun, F.; Fu, X.; Zhang, Y.; Li, Z. Noble Metals Can Have Different Effects on Photocatalysis Over Metal–Organic Frameworks (MOFs): A Case Study on M/NH2-MIL-125(Ti) (M=Pt and Au). *Chem. A Eur. J.* **2014**, *20*, 4780–4788. [CrossRef]

155. Sun, D.; Liu, W.; Qiu, M.; Zhang, Y.; Li, Z. Introduction of a mediator for enhancing photocatalytic performance via post-synthetic metal exchange in metal–organic frameworks (MOFs). *Chem. Commun.* **2015**, *51*, 2056–2059. [CrossRef] [PubMed]

156. Lee, Y.; Kim, S.; Fei, H.; Kang, J.K.; Cohen, S.M. Photocatalytic CO_2 reduction using visible light by metal-monocatecholato species in a metal–organic framework. *Chem. Commun.* **2015**, *51*, 16549–16552. [CrossRef]

157. Wang, S.; Wang, X. Photocatalytic CO_2 reduction by CdS promoted with a zeolitic imidazolate framework. *Appl. Catal. B Environ.* **2015**, *162*, 494–500. [CrossRef]

158. Wang, M.; Wang, D.; Li, Z. Self-assembly of CPO-27-Mg/TiO2 nanocomposite with enhanced performance for photocatalytic CO_2 reduction. *Appl. Catal. B Environ.* **2016**, *183*, 47–52. [CrossRef]

159. Yan, S.; Ouyang, S.; Xu, H.; Zhao, M.; Zhang, X.; Ye, J. Co-ZIF-9/TiO2 nanostructure for superior CO_2 photoreduction activity. *J. Mater. Chem. A* **2016**, *4*, 15126–15133. [CrossRef]

160. Sadeghi, N.; Sharifnia, S.; Sheikh Arabi, M. A porphyrin-based metal organic framework for high rate photoreduction of CO_2 to CH4 in gas phase. *J. CO$_2$ Util.* **2016**, *16*, 450–457. [CrossRef]

161. Xu, H.-Q.; Hu, J.; Wang, D.; Li, Z.; Zhang, Q.; Luo, Y.; Yu, S.-H.; Jiang, H.-L. Visible-Light Photoreduction of CO_2 in a Metal–Organic Framework: Boosting Electron–Hole Separation via Electron Trap States. *J. Am. Chem. Soc.* **2015**, *137*, 13440–13443. [CrossRef]

162. Goyal, S.; Shaharun, M.S.; Kait, C.F.; Abdullah, B.; Ameen, M. Photoreduction of Carbon Dioxide to Methanol over Copper Based Zeolitic Imidazolate Framework-8: A New Generation Photocatalyst. *Catalysts* **2018**, *8*, 581. [CrossRef]

163. Chen, M.; Han, L.; Zhou, J.; Sun, C.; Hu, C.; Wang, X.; Su, Z. Photoreduction of carbon dioxide under visible light by ultra-small Ag nanoparticles doped into Co-ZIF-9. *Nanotechnology* **2018**, *29*, 284003. [CrossRef]

164. Han, B.; Ou, X.; Deng, Z.; Song, Y.; Tian, C.; Deng, H.; Xu, Y.-J.; Lin, Z. Nickel Metal–Organic Framework Monolayers for Photoreduction of Diluted CO_2: Metal-Node-Dependent Activity and Selectivity. *Angew. Chem. Int. Ed.* **2018**, *57*, 16811–16815. [CrossRef]

165. Chen, C.; Wu, T.; Wu, H.; Liu, H.; Qian, Q.; Liu, Z.; Yang, G.; Han, B. Highly effective photoreduction of CO_2 to CO promoted by integration of CdS with molecular redox catalysts through metal–organic frameworks. *Chem. Sci.* **2018**, *9*, 8890–8894. [CrossRef]

166. Sadeghi, N.; Sharifnia, S.; Do, T.-O. Enhanced CO_2 photoreduction by a graphene–porphyrin metal–organic framework under visible light irradiation. *J. Mater. Chem. A* **2018**, *6*, 18031–18035. [CrossRef]

167. Ye, L.; Gao, Y.; Cao, S.; Chen, H.; Yao, Y.; Hou, J.; Sun, L. Assembly of highly efficient photocatalytic CO_2 conversion systems with ultrathin two-dimensional metal–organic framework nanosheets. *Appl. Catal. B Environ.* **2018**, *227*, 54–60. [CrossRef]

168. Li, D.; Kassymova, M.; Cai, X.; Zang, S.-Q.; Jiang, H.-L. Photocatalytic CO_2 reduction over metal-organic framework-based materials. *Coord. Chem. Rev.* **2020**, *412*, 213262. [CrossRef]

169. Li, P.-Z.; Wang, X.-J.; Liu, J.; Phang, H.S.; Li, Y.; Zhao, Y. Highly Effective Carbon Fixation via Catalytic Conversion of CO_2 by an Acylamide-Containing Metal–Organic Framework. *Chem. Mater.* **2017**, *29*, 9256–9261. [CrossRef]

170. Ding, D.; Jiang, Z.; Jin, J.; Li, J.; Ji, D.; Zhang, Y.; Zan, L. Impregnation of semiconductor CdS NPs in MOFs cavities via double solvent method for effective photocatalytic CO_2 conversion. *J. Catal.* **2019**, *375*, 21–31. [CrossRef]

171. Li, X.; Zhu, Q.-L. MOF-based materials for photo- and electrocatalytic CO_2 reduction. *EnergyChem* **2020**, *2*, 100033. [CrossRef]

172. Wang, L.; Jin, P.; Duan, S.; She, H.; Huang, J.; Wang, Q. In-situ incorporation of Copper(II) porphyrin functionalized zirconium MOF and TiO2 for efficient photocatalytic CO_2 reduction. *Sci. Bull.* **2019**, *64*, 926–933. [CrossRef]

173. Dong, H.; Zhang, X.; Lu, Y.; Yang, Y.; Zhang, Y.-P.; Tang, H.-L.; Zhang, F.-M.; Yang, Z.-D.; Sun, X.; Feng, Y. Regulation of metal ions in smart metal-cluster nodes of metal-organic frameworks with open metal sites for improved photocatalytic CO_2 reduction reaction. *Appl. Catal. B Environ.* **2020**, *276*, 119173. [CrossRef]

174. Chen, S.; Yu, J.; Zhang, J. Enhanced photocatalytic CO_2 reduction activity of MOF-derived ZnO/NiO porous hollow spheres. *J. CO_2 Util.* **2018**, *24*, 548–554. [CrossRef]

175. Crake, A.; Christoforidis, K.C.; Kafizas, A.; Zafeiratos, S.; Petit, C. CO_2 capture and photocatalytic reduction using bifunctional TiO2/MOF nanocomposites under UV–vis irradiation. *Appl. Catal. B Environ.* **2017**, *210*, 131–140. [CrossRef]

176. Ma, Z.; Wu, D.; Han, X.; Wang, H.; Zhang, L.; Gao, Z.; Xu, F.; Jiang, K. Ultrasonic assisted synthesis of Zn-Ni bi-metal MOFs for interconnected Ni-N-C materials with enhanced electrochemical reduction of CO_2. *J. CO_2 Util.* **2019**, *32*, 251–258. [CrossRef]

177. Kang, X.; Gollan, R.J.; Jacobs, P.A.; Veeraragavan, A. Suppression of instabilities in a premixed methane–air flame in a narrow channel via hydrogen/carbon monoxide addition. *Combust. Flame* **2016**, *173*, 266–275. [CrossRef]

178. Wang, Y.-R.; Huang, Q.; He, C.-T.; Chen, Y.; Liu, J.; Shen, F.-C.; Lan, Y.-Q. Oriented electron transmission in polyoxometalate-metalloporphyrin organic framework for highly selective electroreduction of CO_2. *Nat. Commun.* **2018**, *9*, 4466. [CrossRef]

179. Ye, L.; Liu, J.; Gao, Y.; Gong, C.; Addicoat, M.; Heine, T.; Wöll, C.; Sun, L. Highly oriented MOF thin film-based electrocatalytic device for the reduction of CO_2 to CO exhibiting high faradaic efficiency. *J. Mater. Chem. A* **2016**, *4*, 15320–15326. [CrossRef]

180. Wang, Y.; Hou, P.; Wang, Z.; Kang, P. Zinc Imidazolate Metal–Organic Frameworks (ZIF-8) for Electrochemical Reduction of CO_2 to CO. *ChemPhysChem* **2017**, *18*, 3142–3147. [CrossRef] [PubMed]

181. Guo, Y.; Yang, H.; Zhou, X.; Liu, K.; Zhang, C.; Zhou, Z.; Wang, C.; Lin, W. Electrocatalytic reduction of CO_2 to CO with 100% faradaic efficiency by using pyrolyzed zeolitic imidazolate frameworks supported on carbon nanotube networks. *J. Mater. Chem. A* **2017**, *5*, 24867–24873. [CrossRef]

182. Kornienko, N.; Zhao, Y.; Kley, C.S.; Zhu, C.; Kim, D.; Lin, S.; Chang, C.J.; Yaghi, O.M.; Yang, P. Metal–Organic Frameworks for Electrocatalytic Reduction of Carbon Dioxide. *J. Am. Chem. Soc.* **2015**, *137*, 14129–14135. [CrossRef]

183. Hinogami, R.; Yotsuhashi, S.; Deguchi, M.; Zenitani, Y.; Hashiba, H.; Yamada, Y. Electrochemical Reduction of Carbon Dioxide Using a Copper Rubeanate Metal Organic Framework. *ECS Electrochem. Lett.* **2012**, *1*, H17–H19. [CrossRef]

184. Perfecto-Irigaray, M.; Albo, J.; Beobide, G.; Castillo, O.; Irabien, A.; Pérez-Yáñez, S. Synthesis of heterometallic metal–organic frameworks and their performance as electrocatalyst for CO_2 reduction. *RSC Adv.* **2018**, *8*, 21092–21099. [CrossRef]

185. Ye, Y.; Cai, F.; Li, H.; Wu, H.; Wang, G.; Li, Y.; Miao, S.; Xie, S.; Si, R.; Wang, J.; et al. Surface functionalization of ZIF-8 with ammonium ferric citrate toward high exposure of Fe-N active sites for efficient oxygen and carbon dioxide electroreduction. *Nano Energy* **2017**, *38*, 281–289. [CrossRef]

186. Huan, T.N.; Ranjbar, N.; Rousse, G.; Sougrati, M.; Zitolo, A.; Mougel, V.; Jaouen, F.; Fontecave, M. Electrochemical Reduction of CO_2 Catalyzed by Fe-N-C Materials: A Structure–Selectivity Study. *ACS Catal.* **2017**, *7*, 1520–1525. [CrossRef]

187. Tan, X.; Yu, C.; Zhao, C.; Huang, H.; Yao, X.; Han, X.; Guo, W.; Cui, S.; Huang, H.; Qiu, J. Restructuring of Cu2O to Cu2O@Cu-Metal–Organic Frameworks for Selective Electrochemical Reduction of CO_2. *ACS Appl. Mater. Interfaces* **2019**, *11*, 9904–9910. [CrossRef]

188. Li, F.; Gu, G.H.; Choi, C.; Kolla, P.; Hong, S.; Wu, T.-S.; Soo, Y.-L.; Masa, J.; Mukerjee, S.; Jung, Y.; et al. Highly stable two-dimensional bismuth metal-organic frameworks for efficient electrochemical reduction of CO_2. *Appl. Catal. B Environ.* **2020**, *277*, 119241. [CrossRef]

189. Rayer, A.V.; Reid, E.; Kataria, A.; Luz, I.; Thompson, S.J.; Lail, M.; Zhou, J.; Soukri, M. Electrochemical carbon dioxide reduction to isopropanol using novel carbonized copper metal organic framework derived electrodes. *J. CO_2 Util.* **2020**, *39*, 101159. [CrossRef]

190. Zhang, S.-Y.; Yang, Y.-Y.; Zheng, Y.-Q.; Zhu, H.-L. Ag-doped Co3O4 catalyst derived from heterometallic MOF for syngas production by electrocatalytic reduction of CO_2 in water. *J. Solid State Chem.* **2018**, *263*, 44–51. [CrossRef]

191. Cao, S.-M.; Chen, H.-B.; Dong, B.-X.; Zheng, Q.-H.; Ding, Y.-X.; Liu, M.-J.; Qian, S.-L.; Teng, Y.-L.; Li, Z.-W.; Liu, W.-L. Nitrogen-rich metal-organic framework mediated Cu–N–C composite catalysts for the electrochemical reduction of CO_2. *J. Energy Chem.* **2021**, *54*, 555–563. [CrossRef]

192. Sun, X.; Wang, R.; Ould-Chikh, S.; Osadchii, D.; Li, G.; Aguilar, A.; Hazemann, J.-l.; Kapteijn, F.; Gascon, J. Structure-activity relationships in metal organic framework derived mesoporous nitrogen-doped carbon containing atomically dispersed iron sites for CO_2 electrochemical reduction. *J. Catal.* **2019**, *378*, 320–330. [CrossRef]

193. Al-Rowaili, F.N.; Jamal, A.; Ba Shammakh, M.S.; Rana, A. A Review on Recent Advances for Electrochemical Reduction of Carbon Dioxide to Methanol Using Metal–Organic Framework (MOF) and Non-MOF Catalysts: Challenges and Future Prospects. *Acs Sustain. Chem. Eng.* **2018**, *6*, 15895–15914. [CrossRef]

194. Dong, B.-X.; Qian, S.-L.; Bu, F.-Y.; Wu, Y.-C.; Feng, L.-G.; Teng, Y.-L.; Liu, W.-L.; Li, Z.-W. Electrochemical Reduction of CO_2 to CO by a Heterogeneous Catalyst of Fe–Porphyrin-Based Metal–Organic Framework. *ACS Appl. Energy Mater.* **2018**, *1*, 4662–4669. [CrossRef]

195. Hod, I.; Sampson, M.D.; Deria, P.; Kubiak, C.P.; Farha, O.K.; Hupp, J.T. Fe-Porphyrin-Based Metal–Organic Framework Films as High-Surface Concentration, Heterogeneous Catalysts for Electrochemical Reduction of CO_2. *ACS Catal.* **2015**, *5*, 6302–6309. [CrossRef]

196. Wang, R.; Sun, X.; Ould-Chikh, S.; Osadchii, D.; Bai, F.; Kapteijn, F.; Gascon, J. Metal-Organic-Framework-Mediated Nitrogen-Doped Carbon for CO_2 Electrochemical Reduction. *ACS Appl. Mater. Interfaces* **2018**, *10*, 14751–14758. [CrossRef]

197. Witman, M.; Ling, S.; Gladysiak, A.; Stylianou, K.C.; Smit, B.; Slater, B.; Haranczyk, M. Rational Design of a Low-Cost, High-Performance Metal–Organic Framework for Hydrogen Storage and Carbon Capture. *J. Phys. Chem. C* **2017**, *121*, 1171–1181. [CrossRef] [PubMed]

198. DeSantis, D.; Mason, J.A.; James, B.D.; Houchins, C.; Long, J.R.; Veenstra, M. Techno-economic Analysis of Metal–Organic Frameworks for Hydrogen and Natural Gas Storage. *Energy Fuels* **2017**, *31*, 2024–2032. [CrossRef]

Publisher's Note: MDPI stays neutral with regard to jurisdictional claims in published maps and institutional affiliations.

Review

POM@MOF Hybrids: Synthesis and Applications

Jiamin Sun [1], Sara Abednatanzi [1], Pascal Van Der Voort [1,*], Ying-Ya Liu [2] and Karen Leus [1]

[1] COMOC—Center for Ordered Materials, Organometallics and Catalysis, Department of Chemistry, Ghent University, Krijgslaan 281, Building S3, 9000 Ghent, Belgium; jiamin.sun@ugent.be (J.S.); sara.abednatanzi@ugent.be (S.A.); Karen.Leus@ugent.be (K.L.)

[2] State Key Laboratory of Fine Chemicals, School of Chemical Engineering, Dalian University of Technology, Dalian 116024, China; yingya.liu@dlut.edu.cn

* Correspondence: pascal.vandervoort@ugent.be; Tel.: +32-(0)9-264-44-42

Received: 4 May 2020; Accepted: 19 May 2020; Published: 21 May 2020

Abstract: The hybrid materials that are created by supporting or incorporating polyoxometalates (POMs) into/onto metal–organic frameworks (MOFs) have a unique set of properties. They combine the strong acidity, oxygen-rich surface, and redox capability of POMs, while overcoming their drawbacks, such as difficult handling, a low surface area, and a high solubility. MOFs are ideal hosts because of their high surface area, long-range ordered structure, and high tunability in terms of the pore size and channels. In some cases, MOFs add an extra dimension to the functionality of hybrids. This review summarizes the recent developments in the field of POM@MOF hybrids. The most common applied synthesis strategies are discussed, together with major applications, such as their use in catalysis (organocatalysis, electrocatalysis, and photocatalysis). The more than 100 papers on this topic have been systematically summarized in a handy table, which covers almost all of the work conducted in this field up to now.

Keywords: metal–organic frameworks; polyoxometalates; hybrid materials; synthesis; catalysis

---

1. Introduction

Polyoxometalates (POMs), a class of metal oxide clustered anions, have already been investigated for more than 200 years. Their history dates back to 1826, when Berzelius reported the discovery of the first POM cluster $(NH_4)_3[PMo_{12}O_{40}] nH_2O$ [1]. However, due to difficulties achieving insights into the POM structure, no significant progress was made until Keggin determined the structure of $H_3PW_{12}O_{40}$ in 1934 [2]. Since then, the interest of scientists in POMs has increased drastically, not only in the development of new POM structures, but also towards their use in various applications, such as catalysis, optics, magnetism, biological medicine, environmental science, life science, and technology [3–7]. In particular, their use in catalysis is one of the most examined fields because of their strong acidity, oxygen-rich surface, photoactivity, and redox capability. Despite these interesting characteristics, POMs still exhibit some drawbacks for their use in catalysis. First, POMs possess a low surface area (<10 m^2 g^{-1}), which consequently hinders the accessibility of reactants and secondly, their high solubility in aqueous solutions and polar organic solvents results in a low recyclability [4]. The immobilization of POMs into/onto porous solids has been proposed to overcome these shortcomings and to achieve catalysts with a high catalytic performance. In the past few decades, many porous materials have been examined as supports for the immobilization of POMs, e.g., silica, ion-exchange resin, zeolites, and activated carbon [5–8]. Since the discovery of metal–organic frameworks (MOFs), much effort has been dedicated to use these porous materials as potential supports for POMs. MOFs are inorganic–organic hybrid crystalline materials that are constructed from metal ions or clusters and organic linkers through coordination bonds. These materials have attracted considerable interest in recent years due to their large surface areas, tunable pore size, and designable functionalities.

So far, MOFs have shown great potential in gas storage and separation, catalysis, sensing, drug delivery, proton conductivity, solar cells, supercapacitors, and biomedicine [8–13]. Moreover, MOFs are regarded as an outstanding platform for introducing guest molecules because of the high accessibility of their internal surface area and long-range ordered channels. So far, several active sites have been successfully embedded in the pores or cages of MOFs, such as noble metals, metal oxides, enzymes, and POMs [14–17].

The first report of a POM@MOF hybrid was reported in 2005 by Férey and co-workers [18]. In this seminal work, the POM, $K_7PW_{11}O_{39}$ (van der Waals radius, 13.1 Å), was successfully encapsulated into the big cages of the highly stable Cr-based MOF, MIL-101, by using an impregnation method. To date, several other thermal and chemical stable MOFs have been applied as supports to host POMs for their use in catalysis, including MIL, UiO, ZIF series, NU-1000, and Cu-BTC frameworks (see Table 1). The most examined POMs that have been encapsulated into MOFs are the well-known Keggin $[XM_{12}O_{40}]^{n-}$ and Dawson $[X_2M_{18}O_{62}]^{n-}$ (X = Si, P, V, Bi, etc.; M = V, Mo, W, etc.) POMs and their derivatives. These POMs are of significant interest because their structure and properties can be easily varied by removing one or more MO^{4+} units, leading to lacunary POMs such as $[PW_9O_{34}]^{9-}$, or by the substitution of X and M by different metals or a combination of two fragments of the Keggin structure, leading to sandwich-type POMs such as $[Tb(PW_{11}O_{39})_2]^{11-}$.

There are several advantages of using MOFs as a host matrix to encapsulate POMs. First of all, their exceptionally high surface areas and confined cages/channels make it possible to ensure a homogeneous distribution of the POM in the MOF host. This not only prevents the agglomeration of POMs, but also improves their stability and recyclability and ensures a fast diffusion of substrates and products. Secondly, the highly regular cages and windows of MOFs ensure a high substrate selectivity, or, in other words, only specific substrates/products are able to reach the active POM sites. Thirdly, owing to the good interaction and electron transfer between the MOF and POM, an increased synergistic catalytic performance is typically observed. Finally, the chemical environment of POMs can easily be adjusted through modification or functionalization of MOFs. Therefore, POM@MOF hybrids not only combine the interesting properties of POMs and MOFs, but also allow the aforementioned disadvantages of POMs to be tackled to afford synergistic catalysis. This review is focused on the synthetic aspects of POM@MOF hybrids, as well as their use in catalysis (organocatalysis, electrocatalysis, and photocatalysis). Alongside the POM@MOF systems discussed here, where the POMs are encapsulated inside a MOF host, POM-based MOFs have also been investigated. In these MOFs, the POMs form the actual metal nodes that are interconnected by organic linkers [19–21]. However, these fall outside the scope of this review.

Table 1. Physical properties of representative metal–organic frameworks (MOFs) used to encapsulate polyoxometalates (POMs) for catalysis.

MOFs	Chemical Formula	Window	Porosity	BET/cm^3 g^{-1}	Ref.
UiO-66	$Zr_6O_4(OH)_4(BDC)_6$, BDC = 1,4-benzenedicarboxylate	6 Å triangular	12 Å (octahedral cages) and 7.5 Å (tetrahedral cages)	1100–2000	[22]
UiO-67	$Zr_6O_4(OH)_4(BPDC)_6$, BPDC = biphenyl-4,4′-dicarboxylate	8 Å triangular	16 Å (octahedral cages) and 12 Å (tetrahedral cages)	2100–2900	[23]
NU-1000	$Zr_6O_4(OH)_4(H_2O)_4(OH)_4(TBAPy)_2$, TBAPy = 1,3,6,8-tetrakis(p-benzoate)pyrene)	10 Å orthogonal	31 Å (hexagonal channels) and 12 Å (triangular channels)	900–2000	[24]
MOF-545	$Zr_6O_8(H_2O)_8(TCPP-H_2)_2$, TCPP = tetrakis (4-carboxyphenyl)porphyrin	-	~16 Å and 36 Å	1900–2500	[25]
ZIF-8/67	$Zn(MeIM)_2/Co(MeIM)_2$ MeIM = imidazolate-2-methyl	3.4 Å hexagonal	~12 Å	1000–1800	[26]
MIL-101	$X_3(F)O(BDC)_3(H_2O)_2$, (X = Cr, Al, Fe) BDC = 1,4-benzenedicarboxylate)	12 Å pentagonal and 16 Å hexagonal	~29 Å and 34 Å	2500–4500	[18]
MIL-100	$Fe_3FO(H_2O)_2(BTC)_3$ BTC = 1,3,5-benzenetricarboxylat	~5.5 Å and 8.6 Å	25 Å and 29 Å	1500–3000	[27]
Cu-BTC	$Cu_3(BTC)_2$,	~9 Å and 4.6 Å	10–13 Å	1000–1500	[28]

2. Synthesis and Design of POM@MOF

To date, several well-known highly stable MOFs have been used to encapsulate POMs, including MIL, UiO, and ZIF series, as well as NU-1000 and Cu-BTC frameworks. One of the most commonly applied methods to embed POMs in MOFs is impregnation. Wet impregnation is a simple and straightforward method, since most of the POMs are well-soluble in polar solvents. Typically, the activated MOF powder is immersed in the POM solution to obtain the composite material. Several POM@MOF hybrids have been successfully synthesized through this wet impregnation method, such as POM@MIL, POM@ZIF, and POM@UN-1000. An important aspect allowing the use of this method is that the size of the POM must be smaller than the windows of the MOF. Moreover, for some POMs and MOFs, it was observed that the POM loading could not be enhanced by increasing the concentration of POM in aqueous solution when a certain POM loading was achieved. For example, for POMs whose size is bigger than the pentagonal windows (12 Å) of MIL-101(Cr), the POMs were only encapsulated into the large cages, while the other cages, which represent 2/3 of the total number of cages of MIL-101(Cr), were unfilled. Naseri et al. demonstrated that the loading of a sandwich-type POM $[(HOSnOH)_3(PW_9O_{34})_2]^{12-}$ (15.2 Å × 10.4 Å) could not be enhanced by increasing the concentration of POM in the aqueous solution [29].

The impregnation method cannot be used for MOFs whose window size is smaller than the POMs, e.g., Cu-BTC, UiO, and ZIF. Therefore, for these MOFs, the one-pot (also known as bottle-around-the-ship) synthesis method has been applied to obtain POM@MOF hybrids. The one-pot method is also often used to obtain POM-encapsulated MOFs in which the anionic form of the POM acts as a structure directing agent to ensure deprotonation of the organic carboxylate ligand. For the preparation of POM@MOF hybrids, typically, the synthesis parameters employed to obtain the parent MOF are used upon addition of the POM. The one-pot method can not only be used to synthesize POM@MOFs that cannot be obtained by impregnation, but can also confine the POMs in the MOF cages to prevent leaching if the size of the POMs is bigger than the windows of the MOFs.

Therefore, in conclusion, the synthesis approaches commonly used to incorporate POMs into MOFs are impregnation and one-pot synthesis. To choose, however, the "correct" methodology, two questions need to be addressed in advance: does the size of the POM fit into the MOF cages and can the pore window of the MOF confine the POM? If both criteria are met, one can expect that the obtained catalyst will work efficiently at a molecular level.

As was mentioned before, the first report on the embedding of a POM into the cages of an MOF was reported by Férey's group. They showed that a Keggin-type POM, $K_7PW_{11}O_{40}$ (van der Waals radius, 13.1 Å), can be confined in MIL-101(Cr) by simple impregnation. The resulting MIL-101-Keggin solid was characterized by XRD, TGA, and N_2 sorption, as well as IR and ^{31}P solid state NMR, which confirmed the presence of Keggin ions within the pores [18]. As summarized in Table 1, MIL-101 has two types of mesoporous cages: a smaller one with an inner size of ~29 Å and pentagonal windows of ~12 Å, and a larger one with an inner size of ~34 Å and hexagonal windows of ~15 Å. Based on the size of the cage windows and the size of the POM, one can conclude that the POM can only diffuse into the largest cages.

In 2010, Gascon and co-workers prepared PW_{12}@MIL-101 ($PW_{12} = [PW_{12}O_{40}]^{3-}$) composites by using a one-pot and wet impregnation method [30]. The authors observed a homogeneous distribution of PW_{12} when the one-pot synthesis was applied under stirring conditions. By using the wet impregnation method, high loadings of PW_{12} in MIL-101 resulted in a drastic decrease in the surface area and pore volume. However, this decrease in surface area and pore volume was smaller for the one-pot synthesis method in comparison to the impregnation method using the same PW_{12} loading. The authors stated that in the one-pot synthesis, both the large- and medium-sized cavities were occupied, while, when using the impregnation method, only the larger cavities were accessible.

Canioni and co-workers compared different synthesis methods for encapsulating POMs in MIL-100(Fe) [31]. The authors observed a good agreement between the experimentally obtained POM loading and the maximum theoretical loading for the PMo_{12}@MIL-100 ($PMo_{12} = [PMo_{12}O_{40}]^{3-}$) obtained by a one-pot solvothermal synthesis. In addition to this, the solvothermally obtained

PMo$_{12}$@MIL-100 showed a good stability in aqueous solution and no POM leaching was observed after 2 months. On the contrary, the PMo$_{12}$@MIL-100 material prepared through impregnation exhibited significant POM leaching after 2 months.

Based on the above examples, POM leaching was observed for the POM@MIL-101 and POM@MIL-100 obtained by impregnation, since immobilization is based on an adsorption equilibrium. One way to circumvent this leaching is to use amino-functionalized MOF structures, e.g., UiO-66-NH$_2$ and MIL-53-NH$_2$, which can ensure a better interaction with the polyanions [32–34]. The formation of complexes such as $-NH_3{}^+[H_2PW_{12}O_{40}]^-$ between primary amines, ammonia, or pyridine and PW$_{12}$ is well-documented [35]. In 2012, Gascon and co-workers used a microwave-assisted one-pot synthesis to obtain PW$_{12}$@MIL-101-NH$_2$(Al) as their attempts to synthesize MIL-101-NH$_2$(Al) containing PW$_{12}$ by one-pot solvothermal synthesis were not successful [36]. One year later, Bromberg et al. examined the encapsulation of POMs in amino-functionalized MOFs (NH$_2$-MIL-101(Al) and NH$_2$-MIL-53(Al)) by immobilization. They concluded that POMs electrostatically interact with the MOF surface to form a stable composite. The thermal stability of the composites PW$_{12}$@NH$_2$-MIL-53(Al) and PW$_{12}$@NH$_2$-MIL-101(Al) was similar to the stability of the parent MOFs [37].

Besides MIL-101, Cu-BTC (namely HKUST-1 or MOF-199) has also been used as a host material to encapsulate Keggin- and Dawson-type POMs [38,39]. As shown in Table 1, the larger cages of Cu-BTC have an inner diameter of 1.3 nm and a pore window of 0.9 nm, which perfectly ensures the stable entrapment of POMs. For example, PW$_{12}$ with a diameter of approximately 1.06 nm was used as a structure directing agent for the self-assembly of Cu-BTC at room temperature [38]. The authors observed an enhanced chemical and thermal stability after the embedding of POM and no POM leaching was noted during catalysis in several studies [40,41]. In one of these studies, Shuxia Liu's group prepared a series of Keggin-type POMs in Cu-BTC, denoted as NENU-n, n = 1~10, and formulated as [Cu$_2$(BTC)$_{4/3}$(H$_2$O)$_2$]$_6$ [H$_n$XM$_{12}$O$_{40}$]·(C$_4$H$_{12}$N)$_2$ (X = Si, Ge, P, As, V, Ti; M = W, Mo), by using a one-pot hydrothermal synthesis [42,43]. The templating effect of the POMs resulted in highly crystalline composite materials which showed an enhanced thermal stability. Moreover, as large crystals were obtained, the structures were elucidated by means of single-crystal X-ray diffraction, demonstrating that the Keggin polyanions were confined in the larger cuboctahedral cages (inner diameter of 1.3 nm) [43].

Besides MIL-101 and Cu-BTC, isostructural imidazolate frameworks, namely ZIF-8 and ZIF-67, have also been frequently used as the host matrix. The sodalite-type cavities of ZIF-8 have a size of approximately 1.1 nm, but the accessible window of the cavity is rather small (0.34 nm). Keggin-based POMs possess a relatively larger particle size up to 1.3 ~ 1.4 nm in comparison to the cavities of ZIF-8, but can fit perfectly in their anionic form (1 nm diameter of PW$_{12}$) [33]. Therefore, the bottle-around-the-ship method is an ideal approach for confining POMs inside ZIF-8 or ZIF-67 [44]. For instance, Malka et al. reported a POM encapsulated in ZIF-8 for its use as an esterification catalyst. The authors were able to obtain a PW$_{12}$ loading of 18 wt% by using a one-pot synthesis strategy at room temperature in aqueous solution. However, after three catalytic cycles, degradation of the MOF occurred and 9% of the POM leached out [45]. A way to overcome the POM leaching in ZIF-8 was demonstrated in the work of Jeon et al. In this study, an impregnation method was used to functionalize the surface of the ZIF-8 nanoparticles with a Keggin-type PW$_{12}$, in order to obtain a core–shell MOF–POM composite. Interestingly, due to the strong interaction, the POM-decorated MOF became insoluble in hydrophilic solvents [46].

Lin and co-workers constructed a POM@MOF molecular catalytic system with a Ni-containing POM [Ni$_4$(H$_2$O)$_2$(PW$_9$O$_{34}$)$_2$]$^{10-}$ (namely Ni$_4$P$_2$) into an [Ir(ppy)$_2$(bpy)]$^+$-derived MOF by one-pot synthesis, and the MOF was isostructural to UiO-66, with extended ligands. Ni$_4$P$_2$ POMs can be encapsulated in the octahedral cages with an inner dimension of 2.2 nm [47].

In the studies mentioned above, the one-pot synthesis and wet impregnation methodology both give a high chance of success in the synthesis of POM-encapsulated MOFs. Although the impregnation method is straightforward, it is only applicable for MOF pore windows larger than the POMs. However, leaching of the POMs might happen unless precautions are taken in advance to ensure a good interaction between POM and MOF supports. The use of POMs as a template might enhance the crystallinity of the MOF framework, which makes a one-pot synthesis very attractive. However, it is important to note here that the size of the POM needs to be larger than the pore window of the MOFs to prevent leaching. In addition to this, the one-pot synthesis method is not applicable to all MOF structures. In most cases, the POM@MOFs materials obtained through a one-pot synthesis or impregnation method have a positive influence on the thermal stability in comparison to the parent non-functionalized MOF.

Besides the commonly used one-pot and impregnation method, there are some other efficient methods for constructing POM-encapsulated MOFs. In 2018, Zhong et al. synthesized NENU-3 (PW_{12}@HKUST-1) by a liquid-assisted grinding method [48]. By using a two-step synthesis, PW_{12} and the Cu salt were first dissolved and evaporated to obtain the copper salt of PW_{12}. Hereafter, the H_3BTC ligand was added in the presence of a small amount of alcohol (MeOH and EtOH), which was used as the grinding liquid. The mixture was ground for 5 min and the color gradually changed to blue. After washing and drying at 60 °C for 24 h, the obtained nanocrystalline, NENU-3, showed a high crystallinity, and the surface area was slightly higher compared to NENU-3 obtained in one-pot solvothermal synthesis. In 2019, G. Li et al. employed an in-situ hot-pressing approach to encapsulate the Keggin-type PW_{12} into an indium-based MOF (MFM-300(In)) [49]. As shown in Figure 1, all the ingredients, including the POMs, were ground in the absence of a solvent, after which they were packed with an aluminum foil and heated on a plate at 80 °C for only 10 min to obtain PW_{12}@MFM-300(In) composites. The resulting materials exhibited a high crystallinity and stability and no PW_{12} aggregates were observed on the MOF surface.

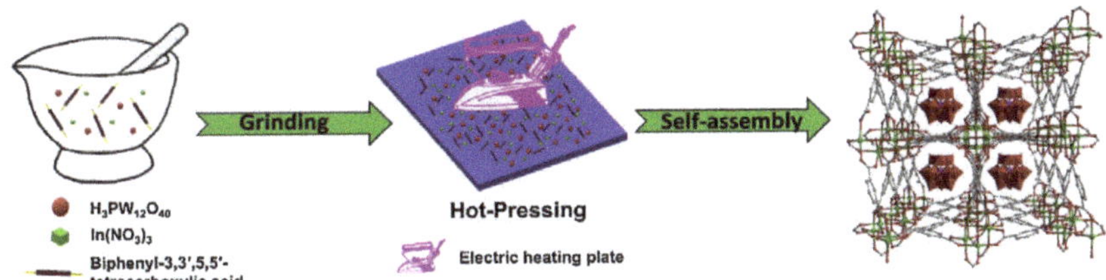

Figure 1. The hot-pressing synthesis process of PW_{12}@MFM-300(In). Reprinted from [49]. Copyright (2019), with permission from Elsevier.

3. Catalytic Applications

3.1. Organocatalysis

3.1.1. Oxidation Reaction

Oxidation reactions are one of the most elementary reactions, and have already been extensively investigated by various catalytic systems. POM@MOF hybrid materials are considered as potential oxidation catalysts due to the presence of acidic sites within MOFs, along with the strong acidity and redox performance of POMs. Accordingly, some well-known MOFs have been reported to encapsulate POMs for their use in oxidation reactions, including MIL(Cr, Fe, or Al), UiO(Zr), and ZIF series, as well as Cu-BTC and NU-1000 frameworks. Among the different oxidation reactions, oxidative desulfurization (ODS) and the selective oxidation of alcohols and alkenes are the most studied reactions using POM@MOF catalysts.

ODS, as one of the promising methods for removing sulfur-containing compounds from fuels, has significant importance in both academic research and industrial chemistry. Since 2012, several Keggin- and sandwich-type POMs, including $[A\text{-}PW_9O_{34}]^{9-}$ [50], $[PW_{12}O_{40}]^{3-}$ [51–54], $[PW_{11}Zn(H_2O)O_{39}]^{5-}$ [55,56], $[Tb(PW_{11}O_{39})_2]^{11-}$ [57], and $[Eu(PW_{11}O_{39})_2]^{11-}$ [58], have been incorporated into the cavities of MIL(Cr, Fe, or Al) for the ODS reaction, using H_2O_2 as the oxidant. The heterogeneous POM@MIL catalysts could not only be easily recycled and reused, but also showed a higher catalytic activity compared to the homogeneous POM counterparts. For example, Balula's group reported that the heterogeneous $Tb(PW_{11})_2$@MIL-101 ($Tb(PW_{11})_2 = [Tb(PW_{11}O_{39})_2]^{11-}$) catalyst exhibits 95% conversion of benzothiophene (BT) at 50 °C after 2 h, whereas the homogeneous $Tb(PW_{11})_2$ catalyst affords a conversion of only 32% under the same reaction conditions [57]. Although the synthesized $Tb(PW_{11})_2$@MIL-101 catalyst showed POM leaching, its structure and morphology remained intact after three consecutive ODS runs. Another study proved that the chemical and thermal stability of POM@MIL-101(Cr) systems could be enhanced compared to individual POMs and MOFs. More specifically, Silva's group demonstrated the high stability of the PW_{11}@MIL-101 ($PW_{11} = [PW_{11}O_{39}]^{7-}$) catalyst in aqueous H_2O_2, while PW_{11} decomposed into peroxo-complexes in the presence of H_2O_2 [59]. In addition, Naseri and co-workers observed that the thermal stability of the synthesized $P_2W_{18}Ce_3$@MIL-101 ($P_2W_{18}Ce_3 = [(OCe^{IV}O)_3(PW_9O_{34})_2]^{12-}$) and $P_2W_{18}Sn_3$@MIL-101 ($P_2W_{18}Sn_3 = [(HOSn^{IV}OH)_3(PW_9O_{34})_2]^{12-}$) materials improved in comparison to the single MIL-101(Cr) framework. The thermally stable POM@MOF materials exhibited >95% conversion of diphenyl sulfide after five cycles [29].

In an interesting study by Cao and co-workers, the effect of the window size within MOFs on the catalytic ODS performance of different POM@MOF materials was investigated [60]. In this work, PW_{12} was encapsulated into three robust MOFs with different window sizes, namely MIL-100(Fe) (8.6 and 5.8 Å), UiO-66 (6 Å), and ZIF-8 (3.4 Å) (see Table 1). Among them, PW_{12}@MIL-100(Fe) exhibited the highest activity (92.8%) for the oxidation of 4,6-dimethyldibenzothiophene ($3.62 \times 6.17 \times 7.86$ Å^3) compared to UiO-66 (39.1%) and ZIF-8 (9.1%). The observed higher activity was attributed to the large window size of MIL-100(Fe), which enabled a fast diffusion of the substrate into the cages. Another important parameter is the influence of the POM loading on the catalytic performance. The conversion of dibenzothiophene (DBT) was at least two times higher when 16%-PW_{12}@MIL-100(Fe) was used as a catalyst in comparison to the 7%-PW_{12}@MIL-100(Fe) catalyst, owing to the higher POM loading. However, when the loading was increased to 35%, the conversion of DBT decreased a lot due to partial pore blocking, which limited the diffusion of reactants to the active sites.

To further enhance the reactivity and recyclability of POM@MOF materials in ODS reactions, amine-functionalized MOFs were employed for encapsulating POMs owing to the strong electrostatic interaction between amine groups and POM anions, including NH_2-MIL-101(Cr) [33], NH_2-MIL-101(Al) [36,56], and NH_2-MIL-53(Al) [58]. For instance, Cao and co-workers reported the incorporation of PW_{12} into NH_2-MIL-101(Cr) as a catalyst for the ODS reaction. The obtained material gave a full conversion of DBT at 50 °C after 1 h [33]. Notably, the reusability tests indicated that the conversion of DBT remained unchanged during six consecutive catalytic cycles using PW_{12}@NH_2-MIL-101(Cr) as a catalyst, due to the strong electrostatic interactions between PW_{12} and the amine groups. Another report by Su and co-workers showed that the PW_{12}@MIL-101(Cr)-diatomite gave 98.6% conversion of DBT at 60 °C for 2 h after three consecutive cycles, which was attributed to the high dispersion of POMs [54].

In addition to ODS reactions, the selective oxidation of alkenes [61–66] and alcohols [34] was evaluated using POM@MIL catalysts. For example, Bo and co-workers synthesized $H_{3+x}PMo_{12-x}V_xO_{40}$@MIL-100(Fe) (x = 0, 1, 2) materials and their catalytic performance were assessed in the oxidation of cyclohexene, using H_2O_2 as the oxidant [66]. Among them, the $H_4PMo_{11}VO_{40}$@MIL-100(Fe) material exhibited 83% conversion of cyclohexene, with an excellent selectivity for 2-cyclohexene-1-one (90%) after five successive catalytic cycles. In 2007, our group developed a new POM@MIL-101 catalyst based on dual amino-functionalized ionic liquid (DAIL) [34]. Firstly, DAIL was introduced onto the coordinatively unsaturated chromium sites of MIL-101(Cr) by a post-synthetic strategy, followed by immobilization of the Keggin-type PW_{12} onto the DAIL-modified MIL-101 through anion exchange (see Figure 2). The PW_{12}/DAIL/MIL-101 catalyst exhibited a very high turnover number (TON: 1900) for the selective oxidation of benzyl alcohol towards benzaldehyde at 100 °C for 6 h. The PW_{12}/DAIL/MIL-101 catalyst demonstrated a higher catalytic activity compared to the PW_{12}/MIL-101 catalyst without DAIL functionalities (TON: 1400). The higher activity was due to the presence of remaining free amino groups anchored on the imidazolium moieties of DAIL, which play a crucial role in enhancing the accessibility of TBHP as the oxidant. Moreover, the PW_{12}/DAIL/MIL-101 catalyst was reused for at least five cycles, with no significant leaching of the tungsten species.

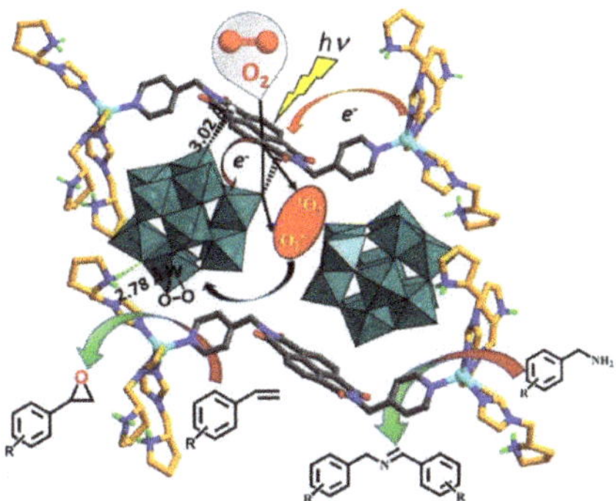

Figure 2. (**a**) Schematic illustration of the preparation of PW/dual amino-functionalized ionic liquid (DAIL)/MIL-101(Cr); (**b**) recyclability of the PW/DAIL/MIL-101(Cr) catalyst. Reprinted with permission from [34]. Copyright (2017), Royal Society of Chemistry.

Another type of MOF, namely Cu-BTC, has also been employed to encapsulate POMs. In 2008, six kind of Keggin-type POMs were encapsulated into Cu-BTC (named NENU-n, NENU = Northeast Normal University) using a one-pot hydrothermal method and their crystal structures were determined [43]. Subsequently, various POMs were encapsulated into the Cu-BTC framework and their catalytic performance was examined in ODS reactions [67,68], the oxidation of alcohols [69,70], olefins [39,71–73], benzene, and H_2S [41,74]. For example, Zheng et al. prepared different sizes of nanocrystal-based catalysts, $[Cu_2(BTC)_{4/3}(H_2O)_2]_6[H_5PV_2Mo_{10}O_{40}]$ (NENU-9N), by using various copper salts and adjusting the pH of the solution for the ODS reaction (see Figure 3) [75]. They proposed that the reaction kinetics can be facilitated by decreasing the size of the nanocrystals. The 550 nm NENU-9 showed a significantly higher conversion of DBT (~90%) in 60 min compared to 300 μm NENU-9 (41%) and the homogeneous POM (2%) in 90 min. To further improve the stability of POM@MOF materials and their catalytic ODS performance, POM@MOF compounds were confined in other porous materials, e.g., MCM-41 [76,77], carbon nanotubes [78], mesoporous SBA-15 [79], and hollow ZSM-5 zeolite [80]. For example, POM@Cu-BTC was confined in the pores of MCM-41 to prevent deactivation of the catalyst [76]. The POM@Cu-BTC@MCM-41 (POM = $Cs_2HPMo_6W_6O_{40}$) exhibited almost full conversion (99.6%) of DBT in 180 min under optimal reaction conditions and could be reused more than 15 times without a significant loss of activity. Lu and co-workers prepared a series of POM@Cu-BTC (POM = PW_{12}, $[PMo_{12-x}V_xO_{40}]^{(3+x)-}$ (x = 0, 1, 2, 3)) catalysts and investigated their performance for the oxidation of benzyl alcohol to benzaldehyde, with H_2O_2 as the oxidant (Figure 4) [70]. The authors observed that the vanadium-containing POMs improved the conversion of benzyl alcohol because of the high redox ability of the POMs. However, when increasing the vanadium content in the POMs, overoxidation to benzoic acid resulted in a lower selectivity towards benzaldehyde. The PMo_{12}@Cu-BTC showed approximately 75% conversion of benzyl alcohol with ~90% selectivity towards benzaldehyde, whereas the PMo_9V_3@Cu-BTC showed ~98% conversion of benzyl alcohol with ~65% selectivity using the same reaction conditions. In other words, the product distribution could be controlled by adjusting the redox capability of the POMs.

Interestingly, in a few studies, a synergistic effect between the POM and Cu-BTC was observed [41,74,81]. For example, Hill prepared $CuPW_{11}$@Cu-BTC ($CuPW_{11}$ = $[CuPW_{11}O_{39}]^{5-}$) for the oxidation of several sulfur compounds and proposed synergistic effects between $PW_{11}Cu$ and Cu-BTC [41]. Not only the hydrolytic stability of the hybrid POM@MOF was improved, but also the

TON (12), as the oxidation of H_2S under ambient conditions increased significantly compared to the individual Cu-BTC (0.02) and POM (no production).

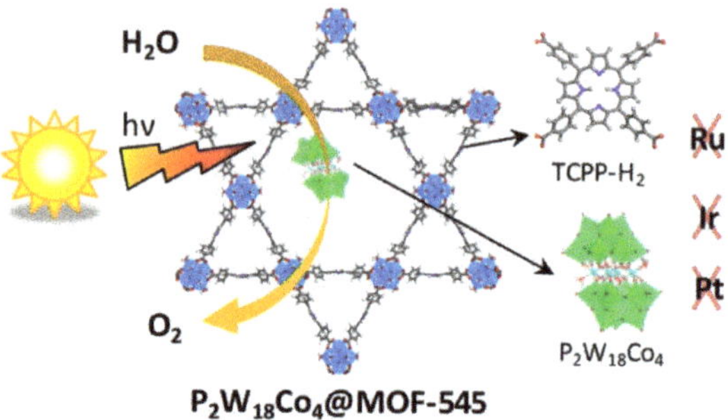

Figure 3. (a) Field emission SEM of Northeast Normal University (NENU)-9N with (**a**) copper nitrate as the metal source at pH 2.5, (**b**) copper acetate as the metal source at pH 2.5, and c) copper acetate as the metal source at pH 4.0. The percentage of DBT-to-DBTO$_2$ conversion versus reaction time by using a) NENU-9N (average diameter = 550 nm), (**b**) NENU-9 (average diameter = 300 mm), and (**c**) POVM (average diameter = 300 mm) as catalysts. Reaction conditions: catalyst (0.01 mmol), DBT (147 mg, 0.8 mmol), and isobutyraldehyde (0.72 mL, 8 mmol) in decalin (50 mL) at 80 °C. Reprinted with permission from [75]. Copyright (2013), John Wiley and Sons.

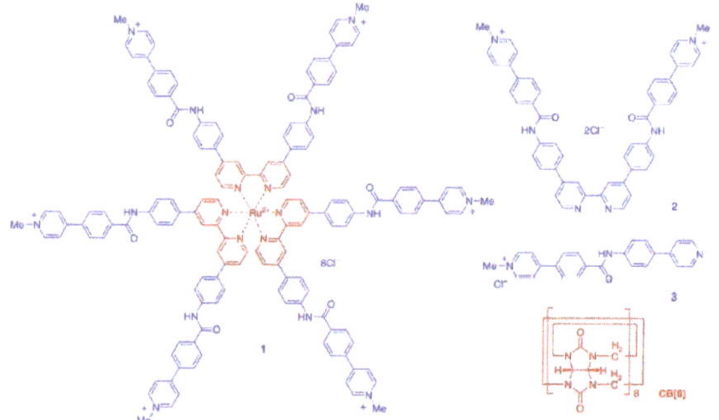

Figure 4. Oxidation of benzyl alcohol by different POM@MOF-199 catalysts. Reprinted with permission from [70]. Copyright (2014), John Wiley and Sons.

The robust Zr-based MOFs have also attracted much attention for hosting POMs for oxidation reactions. The earliest study on the introduction of POMs into a Zr-based MOF was reported by Dolbecq and co-workers in 2015 [82]. Three tungstate POMs ($[PW_{12}O_{40}]^{3-}$ (12 Å), $[PW_{11}O_{39}]^{7-}$, and $[P_2W_{18}O_{62}]^{6-}$ (14 Å)) were encapsulated into the pores of UiO-67. Subsequently, Dai and co-workers examined the catalytic performance of 35%-PW$_{12}$@UiO-66 for the selective oxidation of cyclopentene (CPE) to glutaraldehyde (GA) [83]. The catalyst showed ~95% conversion of CPE, with a ~78% yield for GA at 35 °C after 24 h of reaction. Unfortunately, the catalyst showed PW$_{12}$ leaching (~3 wt%) after three catalytic cycles. To address this POM leaching issue, Yu and co-workers used UiO-bpy (bpy = 2,2′-bipyridine-5,5′-dicarboxylic acid) to encapsulate polyoxomolybdic cobalt (CoPMA) [84]. The bpy sites of the UiO-bpy framework provided an extra interaction with the POM compared to the UiO-67 without bpy moieties. The catalytic activities of CoPMA@UiO-bpy and CoPMA@UiO-67 were assessed in the oxidation of styrene, using O$_2$ as the oxidant. The CoPMA@UiO-bpy exhibited the highest catalytic performance, with 80% conversion of styrene and 59% selectivity towards styrene epoxide.

Another Zr-based MOF, denoted as NU-1000, with small triangular (12 Å) and larger hexagonal (31 Å) channels, has been used to support POMs such as $[PW_{12}O_{40}]^{3-}$ and $[PMo_{10}V_2O_{40}]^{5-}$ [85–87]. For example, Farha's group prepared PW_{12}@NU-1000 through an impregnation method for the oxidation of 2-chloroethyl ethyl sulfide (CEES), using H_2O_2 as the oxidant. The authors demonstrated that the most likely location for PW_{12} clusters is in the small triangular channels, which was further confirmed by means of powder X-ray diffraction, scanning transmission electron microscopy, and difference envelope density analysis. At the same time, PW_{12}@NU-1000 showed a higher conversion of CEES (98% after 20 min) compared to the pristine NU-1000 (77% after 90 min) and homogeneous POM (98% after 90 min). However, the PW_{12}@NU-1000 exhibited only 57% selectivity towards 2-chloroethyl ethyl sulfoxide (CEESO). In a subsequent work, the authors demonstrated that the PW_{12} could migrate from the micropores to the mesopores of NU-1000 under mild thermal activation (see Figure 5). Moreover, the PW_{12}@NU-1000 showed a full conversion of CEES after 5 min, with ~95% selectivity towards CEESO. Recently, this group also prepared the PV_2Mo_{10}@NU-1000 catalyst by using the same method and the synthesized material showed a full conversion of CEES, with O_2 as the oxidant.

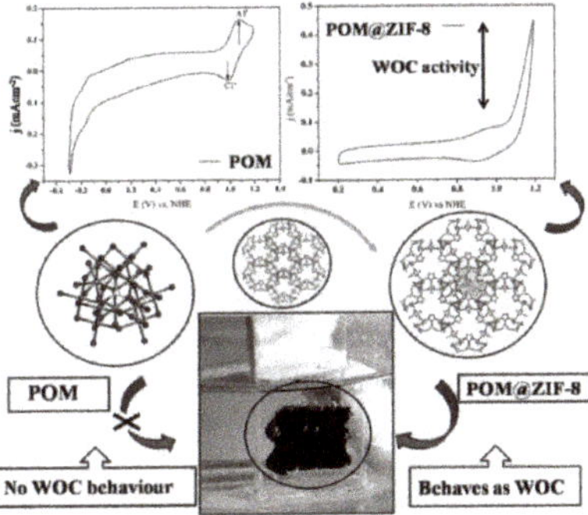

Figure 5. Structural representations of the PW_{12}@NU-1000. Reprinted with permission from [86]. Copyright (2018), Royal Society of Chemistry.

In addition to the well-known MOFs, several other POM@MOF hybrid materials, including $[Co(BBPTZ)_3][HPMo_{12}O_{40}]\cdot24H_2O$ and $[Cu^I_6(trz)_6(PW_{12}O_{40})_2]$, have been synthesized and applied for ODS [88], the oxidation of aryl alkenes [89,90], alkylbenzenes [91], and alcohols [92] (see Table 2).

Besides the use of POMs encapsulated in the cages of MOFs, some POMs have been covered on the surface of MOFs to achieve core–shell structured hybrid materials for oxidation reactions [46,93]. For example, PW_{12} was loaded onto the ZIF-8 surface to obtain a core–shell catalyst for the oxidation of benzyl alcohol. Notably, strong O-N bonding between PW_{12} and the imidazole group of the ZIF-8 was detected through X-ray photoelectron spectroscopy and X-ray absorption near-edge structure measurements. Accordingly, the ZIF-8@PW_{12} material was insoluble in hydrophilic solvents. The ZIF-8@PW_{12} material exhibited a high conversion of benzyl alcohol (>95%), with 90% selectivity towards benzaldehyde, and outperformed the activity of pure PW_{12} (51%) and ZIF-8 (30%).

Table 2. Application of POM@MOF materials in heterogeneous catalysis.

Entry	MOF	POM	Synthesis Approach	Catalytic Reaction	Ref.
		Organocatalysis			
1	MIL-101(Cr)	$TBA_{4.2}H_{0.8}[PW_{11}Zn(H_2O)O_{39}]$	Impregnation	ODS	[55]
2	MIL-101(Cr)	$H_3PW_{12}O_{40}$	One-pot	ODS	[52,54]
3	MIL-101(Cr)	$[Tb(PW_{11}O_{39})_2]^{11-}$	Impregnation	ODS	[57]
4	MIL-101(Cr)	$TBA_3PW_{12}O_{40}$	Impregnation	ODS	[51]
5	MIL-100(Fe) UiO-66 ZIF-8	$H_3PW_{12}O_{40}$	One-pot	ODS	[60]
6	MIL-101(Cr) NH$_2$-MIL-53(Al)	$[Eu(PW_{11}O_{39})_2]^{11-}$	Impregnation	ODS	[58]
7	NH$_2$-MIL-101(Cr)	$H_3PW_{12}O_{40}$	Impregnation	ODS	[33]
8	NH$_2$-MIL-101(Al)	$[PW_{11}Zn(H_2O)O_{39}]^{5-}$	One-pot and Impregnation	ODS	[56]
9	MIL-100(Fe)	$H_{3+x}PMo_{12-x}V_xO_{40}$ (x = 0, 1, 2)	One-pot	Oxidation of cyclohexene	[66]
10	MIL-101(Cr)	$H_5PV_2Mo_{10}O_{40}$	Impregnation	Oxidation of 2-chloroethyl ethyl sulfide	[94]
11	MIL-101(Cr)	$(TBA)_7H_3[Co_4(H_2O)_2(PW_9O_{34})_2]$	Impregnation	Oxidation of alkenes and cyclooctane	[65]
12	MIL-101(Cr)	$[PW_{11}CoO_{39}]^{5-}$ $[PW_{11}TiO_{40}]^{5-}$	Impregnation	Oxidation of alkenes	[62]
13	MIL-101(Cr)	$[(HOSn^{IV}OH)_3(PW_9O_{34})_2]^{12-}$, $[(OCe^{IV}O)_3(PW_9O_{34})_2]^{12-}$	Impregnation	Selective oxidation of various sulfides to sulfones	[29]
14	MIL-101(Cr)	$[PW_4O_{24}]^{3-}$ $[PW_{12}O_{40}]^{3-}$	Impregnation	Epoxidation of various alkenes	[64]
15	MIL-101(Cr)	$H_3PW_{12}O_{40}$	One-pot and Impregnation	Selective oxidation of sulfides to sulfoxides and sulfones	[53]
16	MIL-101(Cr)	$[PZnMo_2W_9O_{39}]^{5-}$	Impregnation	Oxidation of alkenes	[63]
17	NH$_2$-MIL-101(Al)	$H_3PW_{12}O_{40}$	One-pot	CO oxidation	[36]
18	MIL-101(Cr)	$H_3PMo_{12}O_{40}$	One-pot	Epoxidation of propylene	[61]
19	MIL-101(Cr)	$H_3PW_{12}O_{40}$	Impregnation	Oxidation of various alcohols	[34]
20	MIL-101(Cr)	$[A-PW_9O_{34}]^{9-}$	Impregnation	ODS Oxidation of geraniol and R-(+)-limonene	[50]
21	MIL-101(Cr)	$[PW_{11}O_{39}]^{7-}$ $[SiW_{11}O_{39}]^{8-}$	Impregnation	Oxidation of alkenes	[59]
22	Cu-BTC	$H_{3+x}PMo_{12-x}V_xO_{40}$ (x = 1, 2, 3)	One-pot	Synthesis of phenol from benzene	[74]
23	Cu-BTC	$H_3PW_{12}O_{40}$	Liquid-assisted grinding method	Degradation of phenol	[48]
24	Cu-BTC	$H_{6+n}P_2Mo_{18-n}V_nO_{62}\cdot mH_2O$ (n=1-5)	One-pot	ODS	[78]
25	Cu-BTC	Cs^+ ion modified $H_3PMo_6W_6O_{40}$	One-pot	ODS	[77]
26	Cu-BTC	$H_{3+x}PMo_{12-x}V_xO_{40}$ (x = 0, 1, 2, 3) $H_3PW_{12}O_{40}$	One-pot	Selective oxidation of alcohols	[70]
27	Cu-BTC	$H_3PW_{12}O_{40}$ $H_3PMo_{12}O_{40}$ $H_7[P(W_2O_7)(Mo_2O_7)_6]$ $H_4SiW_{12}O_{40}$	One-pot	ODS	[67,68,80]
28	Cu-BTC	$H_3PW_{12}O_{40}$	One-pot	Oxidation of cyclopentene to glutaraldehyde	[72]
29	Cu-BTC	$[CuPW_{11}O_{39}]^{5-}$	One-pot	Oxidation of thiols and H_2S	[41]
30	Cu-BTC	$H_5PMo_{10}V_2O_{40}$	One-pot	ODS	[75]
31	Cu-BTC	$H_6PMo_9V_3O_{40}$	One-pot	Oxidation of benzene to phenol	[79]
32	Cu-BTC	$H_3PMo_6W_6O_{40}$	One-pot	ODS	[76]
33	UiO-66	$H_3PW_{12}O_{40}$	One-pot	Selective oxidation of cyclopentene to glutaraldehyde	[83]

Table 2. *Cont.*

Entry	MOF	POM	Synthesis Approach	Catalytic Reaction	Ref.
34	UiO-67 UiO-bpy	$H_3PMo_{12}O_{40}$	One-pot	Olefins epoxidation	[84]
35	NU-1000	$H_5PV_2Mo_{10}O_{40}H_3PW_{12}O_{40}$	Impregnation	Oxidation of 2-chloroethyl ethyl sulfide	[85–87]
36	$[Co(BBPTZ)_3][HP\ Mo_{12}O_{40}]\cdot24H_2O$	$H_3PMo_{12}O_{40}$	One-pot	ODS	[88]
37	$[Co(BBTZ)_2][H_3BW_{12}O_{40}]\cdot10H_2O$ $[Co_3(H_2O)_6(BBTZ)_4][BW_{12}O_{40}]\cdot NO_3\cdot4H_2O$	$[BW_{12}O_{40}]^{5-}$	One-pot	Oxidation of styrene to benzaldehyde	[90]
38	$[Cu^I_6(trz)_6(PW_{12}O_{40})_2]$ $[Cu^I_3(trz)_3(PMo_{12}O_{40})]$	$H_3PMo_{12}O_{40}$ $H_3PW_{12}O_{40}$	One-pot	Oxidation of alkylbenzenes to aldehydes	[91]
39	$[Cu_3(4,4'\text{-}bpy)_3][HSiW_{12}O_{40}](imi)$ $[Cu_3(4,4'\text{-}bpy)_3][PMo_{12}O_{40}](ampyd),$ $[Cu_2(4,4'\text{-}bpy)_2][HPMo_{12}O_{40}]\ (ampyd)$ $[Cu(Phen)(4,4'\text{-}bpy)(H_2O)]_2[PW_{12}O_{40}](4,4'\text{-}bpy)$	$H_3PW_{12}O_{40}$ $H_3PMo_{12}O_{40}$ $H_4SiW_{12}O_{40}$	One-pot	Oxidation of various alcohols	[69]
40	$Ni(4,4'\text{-}bpy)_2]_2\ [V_7^{IV}V_9^{V}O_{38}Cl]\ (4,4'\text{-}bpy)$ $6H_2O$	$[V_7^{IV}V_9^{V}O_{38}Cl]^{4-}$	One-pot	Oxidation of alkenes	[71]
41	$[H(bpy)Cu_2][PW_{12}O_{40}]$ $[H(bpy)Cu_2][PMo_{12}O_{40}]$	$H_3PMo_{12}O_{40}$ $H_3PW_{12}O_{40}$	One-pot	Oxidation of ethylbenzene, alcohol, and cyclooctene	[89]
42	$H[Cu^I_5Cu^{II}(pzc)_2(pz)_{4.5}(P_2W_{18}O_{62})]\cdot6H_2O$	$[P_2W_{18}O_{62}]^{6-}$	One-pot	Oxidation of alcohols	[92]
43	$[Cu^I(bbi)]_2([Cu^I(bbi)]_2V^{IV}_2V^{V}_8O_{26})\cdot2H_2O$	$[V_{10}O_{26}]^{4-}$	One-pot	Oxidative cleavage of β-O-4 lignin	[81]
44	ZIF-8	Mo_{132}	Impregnation	ODS	[93]
45	ZIF-8	$H_3PW_{12}O_{40}$	Impregnation	Oxidation of benzyl alcohol	[46]
		Condensation reaction			
46	ZIF-8	$Al_{0.66}$-DTP Horic	One-pot	Aldol condensation of 5-hydroxymethylfurfural (HMF) with acetone	[95]
47	MIL-101(Cr)	$H_3PW_{12}O_{40}$	Impregnation	Biginelli condensation reaction	[96]
48	MIL-101(Cr)	$H_3PW_{12}O_{40}$	One-pot	Cyclopentanone self-condensation	[97]
49	MIL-101(Cr)	$H_3PW_{12}O_{40}$	One-pot and Impregnation	Baeyer condensation	[98,99]
50	MIL-101(Cr)	$H_3PW_{12}O_{40}$	One-pot and Impregnation	Knoevenagel condensation of benzaldehyde with ethyl cyanoacetate	[30]
51	NH_2-MIL-101(Al) NH_2-MIL-53(Al)	$H_3PW_{12}O_{40}$	Impregnation and Joint Heating	Aldehyde condensation and polymerization	[37]
52	MIL-100(Fe) MIL-101(Cr)	$H_3PW_{12}O_{40}$	One-pot	Hydroxyalkylation of phenol with formaldehyde	[100]
		Esterification reaction			
53	MIL-100(Fe) Cu-BTC	$H_3PW_{12}O_{40}$	One-pot	Enzymatic esterification of cinnamic acid	[101]
54	Cu-BTC	$H_3PW_{12}O_{40}$	One-pot	Esterification of acetic acid with 1-propanol	[102]
55	Cu-BTC	$H_3PMo_{12}O_{40}$	One-pot	Esterification of levulinic acid (LA) and ethanol	[103]
56	UiO-66	$H_4SiW_{12}O_{40}$	One-pot	Esterification of lauric acid with methanol	[104]
57	MIL-101(Cr)	$K_5[CoW_{12}O_{40}]$	One-pot	Esterification of acetic acid with various alcohols, and cycloaddition of CO_2 with epoxides	[105]
58	Cu-BTC	$H_3PW_{12}O_{40}$ $H_3PMo_{12}O_{40}, H_4PVMo_{11}O_{40},$ $H_5PV_2Mo_{10}O_{40}, H_6PV_3Mo_9O_{40}$	One-pot	Oxidative esterification of glycerol	[106]
59	ZIF-8	$H_3PW_{12}O_{40}$	One-pot	Esterification of benzoic anhydride with cinnamyl alcohol	[45]
60	Fe-BTC	$H_3PMo_{12}O_{40}$	One-pot	Esterification of free fatty acids to biodiesel	[107]

Table 2. *Cont.*

Entry	MOF	POM	Synthesis Approach	Catalytic Reaction	Ref.
61	MIL-100(Fe)	$H_3PW_{12}O_{40}$	One-pot	Esterification of acetic acid with monohydric alcohols, and acetalization of benzaldehyde with ethanediol	[108]
62	Cu-BTC	$H_3PW_{12}O_{40}$	One-pot	Esterification of acetic acid with 1-propanol and salicylic acid with ethanol	[40,109]
63	MIL-101(Cr)	$H_3PW_{12}O_{40}$	One-pot	Esterification of acetic acid with n-hexanol, and hydrolysis of ethyl acetate	[110]
64	MIL-53	$H_3PW_{12}O_{40}$	One-pot by ultrasound irradiation	Esterification of oleic acid by various alcohols	[111]
65	MIL-100(Fe)	$H_3PW_{12}O_{40}$	One-pot	Esterification of oleic acid with ethanol	[112]
66	NENU-3a	$H_3PW_{12}O_{40}$	One-pot	Esterification of long-chain fatty acids	[113]
67	UiO-66-2COOH	$H_3PW_{12}O_{40}$	One-pot	Ransesterification-esterification of acidic vegetable oils	[114]
68	MOF-74	$H_3PW_{12}O_{40}$	One-pot	Hydrogenation–esterification tandem reactions	[115]
			Other organic transformations		
69	MIL-101(Cr)	$H_3PW_{12}O_{40}$	One-pot	Dehydration of fructose to 5-hydroxymethylfurfural	[116]
70	Cu-BTC	$H_3PMo_{12}O_{40}$	One-pot	Transesterification of 5-hydroxymethylfurfural with ethanol	[117]
71	ZIF-8	$H_3PW_{12}O_{40}$	Impregnation	Transesterification of rapeseed oil with methanol	[46]
72	MIL-100(Fe)	$H_3PMo_{12}O_{40}$	One-pot	Transesterification of soybean oil with methanol and esterification of free fatty acids	[118]
73	UiO-66	$H_4SiW_{12}O_{40}$	One-pot	Hydrogenation of methyl levulinate/transesterification of methyl-3-hydroxyvalerate	[119]
74	MIL-100(Fe)	$H_3PW_{12}O_{40}$	One-pot	Hydrogenation of cellobiose/hydrolysis of cellulose	[120]
75	MIL-101(Cr)	$K_5[CoW_{12}O_{40}]$	One-pot	Methanolysis of epoxides	[32]
76	COK-15 MIL-101(Cr)	$H_3PW_{12}O_{40}$	One-pot	Methanolysis of styrene oxide	[121,122]
77	NENU-11	$[PW_{12}O_{40}]^{3-}$	One-pot	Hydrolysis of dimethyl methylphosphonate	[123]
78	NENU-15	$[SiW_{12}O_{40}]^{4-}$	One-pot	Reduction of NO	[124]
79	NENU-1a	$H_4SiW_{12}O_{40}$	One-pot	Dehydration of methanol	[125]
80	Basolite F300	$H_3PW_{12}O_{40}$	Impregnation	Dehydration of ethanol	[126]
81	NENU-3a	$H_3PW_{12}O_{40}$	One-pot	Hydrolysis of esters	[43]
82	NH$_2$-MIL-101(Fe) MIL-101(Cr)	$TBA_4[PW_{11}Fe(H_2O)O_{39}]$	Impregnation	Ring opening of styrene oxide with aniline	[127]
83	MIL-100(Fe)	$[PMo_{11}Mn(H_2O)O_{39}]^{5-}$	One-pot	Reduction of p-nitrophenol	[128]
84	MOF-808	$[H_3PW_{12}O_{40}]$	One-pot	Friedel-Crafts acylation of anisole with benzoyl chloride	[129]
85	MIL-101(Cr)	$H_3PW_{12}O_{40}$	Impregnation	Pechmann, esterification, and Friedel-Crafts acylation	[130]
86	UiO-66	$Cs_{2.5}H_{0.5}PW_{12}O_{40}$	One-pot	Acidolysis of soybean oil	[131]
87	Cu-based MOF	$H_4SiW_{12}O_{40}.xH_2O$	One-pot	Azide-alkyne click reaction	[132]
88	NU-1000	$H_3PW_{12}O_{40}$	Impregnation	Isomerization/disproportionation of o-xylene	[133]
89	MIL-101(Cr)	$H_3PW_{12}O_{40}$	One-pot	Hydroformylation of 1-octene	[134]
90	MIL-101(Cr)	$H_3PW_{12}O_{40}$	One-pot	Cycloaddition of CO_2 to styrene oxide	[135]
			Electrocatalysis		
91	MIL-101(Cr)	$[Co(H_2O)_2(PW_9O_{34})_2]^{10-}$	Ion exchange	H_2O oxidation	[136]
92	ZIF-8	$[CoW_{12}O_{40}]^{6-}$ $H_4SiW_{12}O_{40}$	One-pot	H_2O oxidation	[137,138]
93	ZIF-67	$H_3PW_{12}O_{40}$	Core–shell coating of POM	H_2O oxidation	[139]

Table 2. *Cont.*

Entry	MOF	POM	Synthesis Approach	Catalytic Reaction	Ref.
94	Ag-based metal-organic nanotubes	$H_3PW_{12}O_{40}$ $H_4SiW_{12}O_{40}$	One-pot	H_2 evolution	[140]
95	Cu-based metal-organic nanotubes	$K_6P_2W_{18}O_{62}$ $H_6As_2W_{18}O_{62}$	One-pot	H_2 evolution	[141]
96	MIL-101(Cr)	$[PMo_{10}V_2O_{40}]^{5-}$	Impregnation	Oxidation of ascorbic acid	[142]
		Photocatalysis			
97	NH_2-MIL-101(Al)	$K_6[a\text{-}AgPW_{11}O_{39}]$	Impregnation	Degradation of Rhodamine B	[143]
98	Cu-based MOF	$H_4SiMo_{12}O_{40}$	One-pot	Degradation of Rhodamine B	[144]
99	MFM-300(In)	$H_3PW_{12}O_{40}$	In situ hot-pressing	Degradation of sulfamethazine	[49]
100	MOF-545	$[(PW_9O_{34})_2Co_4(H_2O)_2]^{10-}$	Impregnation	Water oxidation	[145, 146]
101	MIL-100(Fe)	$[Co^{II}Co^{III}W_{11}O_{39}(H_2O)]^{7-}$ $[Co_4(PW_9O_{34})_2(H_2O)_2]^{10-}$	One-pot	Water oxidation	[147]
102	MIL-100(Fe)	$H_3PMo_{12}O_{40}$	One-pot	Oxidation of alcohols and reduction of Cr (VI)	[148]
103	Zn-based MOF	$[BW_{12}O_{40}]^{5-}$	One-pot	Coupling of amines and epoxidation of olefins	[149]
104	UiO-66-NH_2	$H_3PW_{12}O_{40}$	One-pot	H_2 evolution/degradation of Rhodamine B	[150]
105	NH_2-MIL-53	$H_3PW_{12}O_{40}$	Impregnation	H_2 evolution	[151]
106	UiO-67	$[P_2W_{18}O_{62}]^{6-}$ $H_4SiW_{12}O_{40}$	One-pot	H_2 evolution	[152, 153]
107	UiO derived structure	$[Ni_4(H_2O)_2(PW_9O_{34})_2]^{10-}$	One-pot	H_2 evolution	[47]
108	MIL-101(Cr)	$\alpha\text{-}PW_{15}V_3$, $\alpha\text{-}P_2W_{17}Ni$, $\alpha\text{-}P_2W_{17}Co$	One-pot	H_2 evolution	[154]
109	SMOF-1	$[P_2W_{18}O_{62}]^{6-}$	Impregnation	H_2 evolution	[155]
110	Cu-BTC	$[PTi_2W_{10}O_4]^{7-}$	One-pot	CO_2 reduction	[42]

BBPTZ = 4,4'-bis(1,2,4-triazol-1-ylmethyl)biphenyl]; BBTZ = 1,4-bis(1,2,4-triazol-1-ylmethyl)benzene; trz = 1,2,4-triazole; imi = imidazole; ampyd = 2-aminopyridine; bpy = bipyridine; Phen = 1,10-phenanthroline; bipy = 4,4'-bipyridine; Hpzc = pyrazine-2-carboxylic acid, pz = pyrazine; bbi = 1,1'-(1,4-butanediyl)bis(imidazole); DTP = dodecatungstophosp; TBA = tetrabutylammonium.

3.1.2. Condensation Reaction

POM@MOF has revealed potential applications in a range of condensation reactions for producing value-added cyclic organic compounds. Recently, Malkar et al. compared the catalytic performance of three different catalysts, namely 20%-Cs-DTP-K10, 18%-DTP@ZIF-8, and $Al_{0.66}$-DTP@ZIF-8 (DTP = dodecatungstophosp), for the aldol condensation of HMF (5-hydroxymethylfurfural), as shown in Figure 6 [95]. It has been proved that the substitution of protons of heteropolyacids with metal ions increases the mobility of protons, which results in an enhancement of the acidity. Based on NH_3-TPD analysis, Cs-DTP-K10 displays the highest acidity (1.51 mmol g^{-1}), whereas DTP@ZIF-8 and Al-DTP@ZIF-8 possess 0.44 and 0.54 mmol g^{-1} of acidic sites, respectively. Cs-DTP-K10, with the highest acidity, showed the highest activity for the aldol condensation of HMF and acetone to selectively produce the desired C9 product (71.6% after 6 h of reaction), while the selectivity was only 43.1%. Although the total number of acidic sites was much lower in the case of the Al-DTP@ZIF-8 catalyst, a good conversion of 63.1% was still obtained after 6 h of reaction, which is comparable to the former value. Notably, the Al-DTP@ZIF-8 catalyst displayed a much higher selectivity (~92%) towards the C9 product compared to C15. The lowest conversion of HMF was achieved in the case of the 18%-DTP@ZIF-8 material with the lowest acidity. However, a high selectivity towards the C9 product was observed. The higher selectivity towards the C9 adduct, as the desired product in the presence of the DTP@ZIF-8 and Al-DTP@ZIF-8 catalysts, confirms the shape selectivity supplied by the small pore diameter of ZIF-8, which can prevent the production of the C15 adduct.

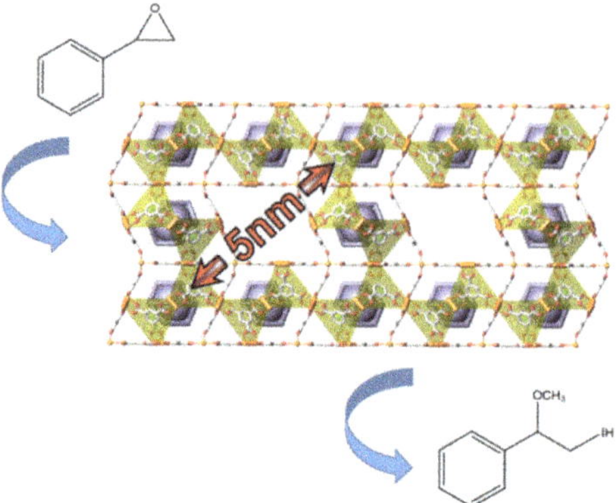

Figure 6. Aldol condensation of 5-hydroxymethylfurfural (HMF) with acetone over Al-DTP@ZIF-8. Reproduced with permission from [95]. Copyright (2019), American Chemical Society.

Another example of the use of MOFs in condensation reactions is the well-known MIL-101. For this purpose, PW_{12}@MIL-101(Cr) composites were synthesized through the direct hydrothermal procedure or post-synthesis modification route [98]. The acidic sites within the MIL-101 and PW_{12}@MIL-101(Cr) materials are desirable for catalyzing the Baeyer condensation of benzaldehyde and 2-naphthol, in the three-component condensation of benzaldehyde, 2-naphthol, and acetamide, as depicted in Figure 7. While no product was produced in the absence of catalysts, a high yield of around 95% was observed for the formation of 1-amidoalkyl-2-naphthol at 130 °C using microwave heating for 5 min. Moreover, no leaching of the active sites was observed, and the catalyst could be reused for four cycles without a notable loss in the product yield.

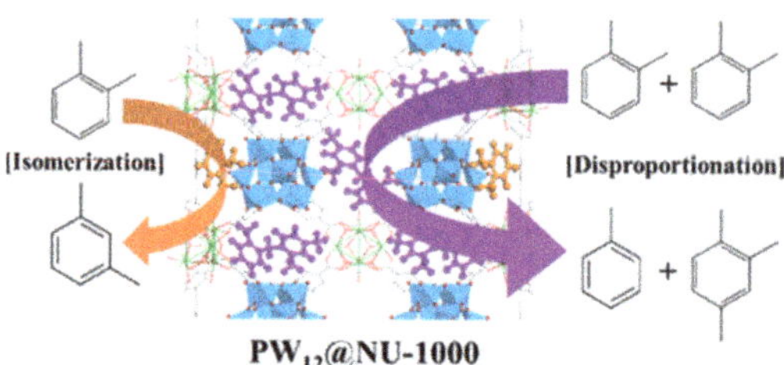

Figure 7. Condensation of benzaldehyde, 2-Naphthol, and acetamide. Reprinted with permission from [98]. Copyright (2012), American Chemical Society.

In addition, PW_{12} clusters were uniformly encapsulated in the cages of MIL-101 as a selective heterogeneous catalyst for the self-condensation of cyclic ketones [97]. As can be observed in Figure 8, the self-condensation of cyclopentanone can result in three different products based on the active sites in the applied catalysts. By using PW_{12} as the catalyst, a conversion of around 78% could be obtained after 24 h reaction to trindane as the main product. However, PW_{12}@MIL-101 exhibits a considerably higher selectivity (>98%) towards the mono-condensed component (2-cyclopentylidenecyclopentanone) as the desired product due to the possibility of shape-selective catalysis. The PW_{12}@MIL-101 catalyst could be recycled up to five cycles, with no obvious reduction in the conversion and selectivity.

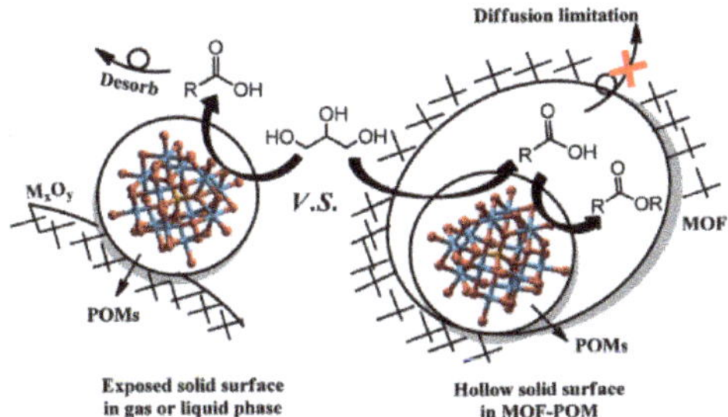

Figure 8. Reaction figure of a cyclopentanone self-condensation reaction. Adapted with permission from [97]. Copyright (2015), Royal Society of Chemistry.

3.1.3. Esterification Reaction

Modified MOFs with POMs can be employed as active catalysts for a wide range of esterification reactions. Biodiesel, as a secure and sustainable energy source, is a promising alternative for fossil fuel-based energy systems [156]. Among the various methods for biodiesel production, transesterification is the most common procedure. Recently, Xie et al. investigated the catalytic one-pot transesterification-esterification of acidic vegetable oil transesterification reaction over a functionalized UiO-66-2COOH with a Keggin-type POM, namely, AILs/POM/UiO-66-2COOH (AIL = sulfonated acidic ionic liquid) (see Figure 9) [114]. The prepared catalyst displayed synergistic benefits arising from the introduction of AIL as Brønsted acid sites. The presence of both Brønsted acid sites of ILs and Lewis acid sites of POM promoted the catalytic reaction for green biodiesel production. The control experiments showed that all of the applied POMs (PW_{12}, SiW_{12}, and PMo_{12}) could convert soybean oil to biodiesel with a high catalytic performance (conversion of ~100%). However, challenges associated with the work-up and recyclability of these homogeneous catalysts limit their potential application. The pristine UiO-66-2COOH material presented a poor activity, with an oil conversion of around 8% because of its insufficient acidic properties. In addition, the PW_{12}@UiO-66-2COOH, SiW_{12}@UiO-66-2COOH, and PMo_{12}@UiO-66-2COOH composites suffered from sluggish reaction kinetics with conversions below 30% towards biodiesel production, which could have been due to the lack of enough acidic sites required to advance the catalytic reaction. Another control experiment was performed by using the sulfonic acid-functionalized IL as the homogeneous catalyst, affording a high catalytic activity of around 99% conversion. It is interesting to note that AILs/POM/UiO-66-2COOH catalysts can combine the advantages of POMs, AILs, and porous MOFs and therefore present the highest catalytic performance in the mentioned reaction (conversion > 90%). Furthermore, the strong interaction between the POMs and AILs was able to hinder the leaching of active components into the reaction media, which further resulted in no notable loss in the catalytic conversion of oil to biodiesel after five consecutive catalytic cycles.

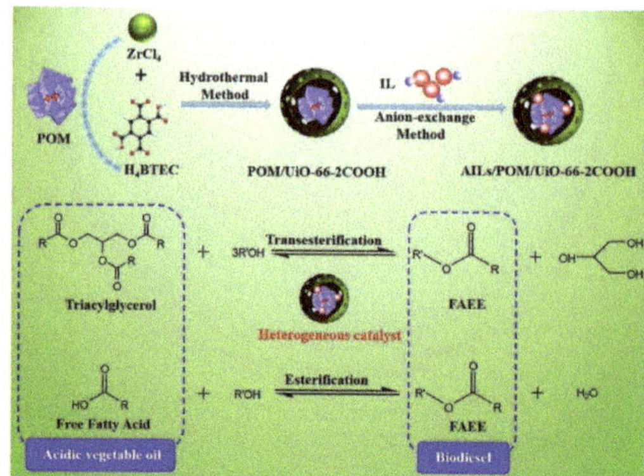

Figure 9. Synthesis procedure of the AILs/HPW/UiO-66-2COOH catalyst, and one-pot transesterification-esterification of acidic vegetable oils. Reprinted from [114]. Copyright (2019), with permission from Elsevier.

In 2015, Liu et al. described an effective procedure for designing NENU-3a with different crystal morphologies (cubic and octahedral) comprised of a Cu-BTC skeleton and encapsulated phosphotungstic acid catalyst [113]. The morphology of this framework was generated by applying the method of coordination modulation, using *P*-toluic acid as the modulator. The NENU-3a with cubic crystals ((100) facets) could effectively promote the conversion of long-chain (C12-C22) fatty acids into corresponding monoalkyl esters (>90% yield) compared to the octahedral counterpart (<22% yield). Moreover, the cubic NENU-3a catalyst was highly robust and could be reused for five reaction runs with a preserved structure and catalytic activity. This report confirms the vital impact of morphological control on MOFs for improving the facet exposure of catalytic sites, which accordingly results in an enhancement of the catalytic performance, especially for bulky substrates with limited access to the catalytic active sites within the pores of MOF catalysts. Another important feature of MOFs is the possibility to control the product selectivity arising from the pore size effect of MOFs. Within this context, Zhu et al. studied the selective esterification of glycerol using a MOF-supported POM catalyst [106]. The catalytic performance of the obtained POM@Cu-BTC catalyst was compared to the metal oxide-supported POMs as the reference materials. Since there was no pore limitation impact using the POM@metal-oxide catalyst, the conversion of glycerol stopped at the acid stage without further reaction and was free to be released from the reaction site (Figure 10). However, when the POM@Cu-BTC catalyst was applied, diffusion of the acid product within the MOF pores was limited and further reaction of the acid product produced the corresponding ester compound.

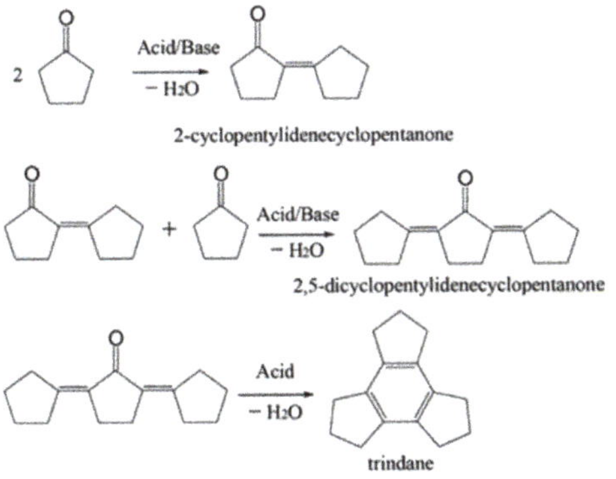

Figure 10. Diffusion limited glycerol transformation on MOF-POMs. Reprinted with permission from [106]. Copyright (2015), Royal Society of Chemistry.

3.1.4. Other Organic Transformations

POMs exhibit great potential as solid acid catalysts because of their strong Brønsted acidity. The first report on a well-defined MOF-supported POM compound, which behaved as a true heterogeneous acid catalyst, was reported by Su et al. [43]. In this work, a series of POM@MOF catalysts were synthesized using a one-pot method. The POM@MOF compound, which contained the strongest Keggin Brønsted acid PW_{12}, was examined in the hydrolysis of ethyl acetate in the presence of an excess amount of water. This catalyst, denoted as NENU-3a, exhibited almost full conversion (>95%) after approximately 7 h of reaction, which is far more superior than most inorganic solid acids and comparable to organic solid acids. More specifically, when the activity was reported per unit of acid, NENU-3a was 3-7 times more active than H_2SO_4, PW_{12}, nafion-H, and Amberlyst-15. In addition to this, no deactivation of the acid sites by water was observed and no leaching of the POM was noted up to at least 15 cycles. Later on, the same group reported the use of POMs as templates for the construction of novel hybrid compounds, for which the properties of the POM could be tailored towards a specific application [123–125]. One of these targeted applications was the adsorption and subsequent hydrolysis of the nerve gas dimethyl methylphosphate to methyl alcohol, for which the conversion increased up to 93% when the temperature was raised to 50 °C [123]. Recycling tests demonstrated that the structural integrity was preserved up until at least 10 cycles. However, it is important to note here that a stabilizing effect of the POM on the MOF will only be obtained when the shape, size, and symmetry of the POM match the MOF host [135]. This stabilizing effect even allowed the application of POM@MOF catalysts in aggressive reactions, as was demonstrated in the very nice work of Hupp, Farha, and Notestein [133]. In this study, the Zr-based MOF, NU-1000, was loaded with $H_3PW_{12}O_{40}$ for its use in the strong acid-catalyzed reaction of o-xylene isomerization/disproportionation at 250 °C (see Figure 11). At low POM loadings (0.3 to 0.7 POM per Zr_6 node), no activity was observed, which was due to the collapse of the POM and/or MOF structure upon activation or at the start of the reaction. However, when the loading was increased to its maximum, with 1 Keggin unit per unit cell of NU-1000, the hybrid catalyst exhibited an initial reactivity in the examined C-C skeletal rearrangement reaction which was even higher than that of the reference WO_x-ZrO_2 catalyst.

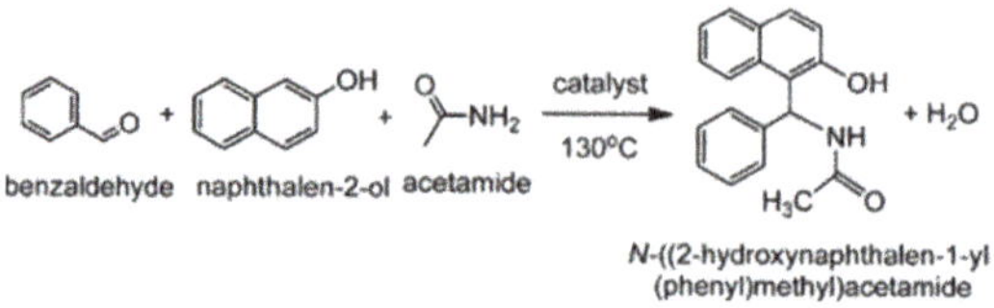

Figure 11. Phosphotungstic acid encapsulated in NU-1000 for its use in the aggressive hydrocarbon isomerization reaction. Reprinted with permission from [133]. Copyright (2018), American Chemistry Society.

While, in the previously discussed studies, the Keggin ion acted as a template to stabilize the microporous/mesoporous structure of the MOF, the group of Martens et al. used this templating mechanism to introduce mesopores separated by uniform microporous walls in a single crystal structure [121] (see Figure 12). More specifically, a hierarchical variant of the Cu-based MOF, Cu-BTC, was synthesized using a dual templating approach in which the Keggin ions served as a molecular template for the structural motif of the MOF, while the surfactant cetyltrimethylammonium bromide was used to introduce mesoporosity. The resulting mesoporous MOF, denoted as COK-15, was investigated in the alcoholysis of styrene oxide, which often suffers from a low selectivity. The COK-15 catalyst not only exhibited a remarkable activity (100% conversion), but also achieved 100% selectivity for 2 methoxy-2 phenylethanol after 3 h of reaction at 40 °C. For comparison, the microporous POM@Cu-BTC and Cu-BTC material only showed 40% and 2% conversion, respectively. The authors addressed the good activity of the COK-15 to the mesoporous feature, which allowed efficient mass transport. Moreover, the catalyst could be recycled for at least four runs, with a negligible loss in activity and selectivity.

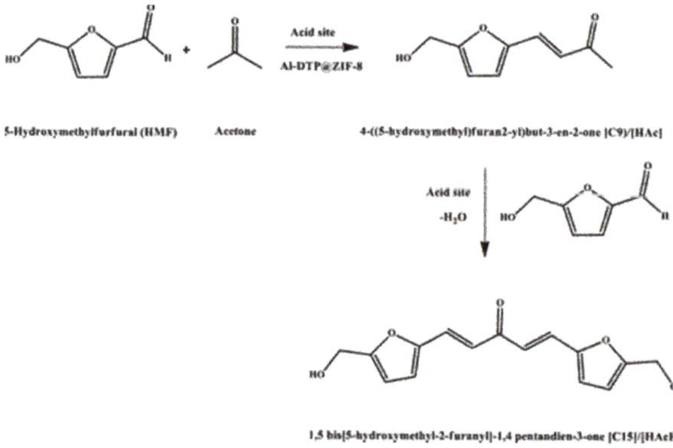

Figure 12. A copper benzene tricarboxylate metal–organic framework, COK-15, with a wide permanent mesoporous feature stabilized by Keggin POM ions for the methanolysis of styrene oxide. Adapted with permission from [121]. Copyright (2012), American Chemistry Society.

Besides this increase in stability after the embedding of the POM in a MOF support, several groups have demonstrated the mutual activation of the POM guest and MOF support [32,119,127]. A very special and extreme example of such a synergism was demonstrated in the work of Kögerler et al. [128]. In this work, a POM@MOF composite was prepared through a hydrothermal reaction in which an Mn-based POM was added to the reaction mixture to synthesize MIL-100. The obtained 30 wt% loaded Mn-POM@MIL-100 was evaluated for its catalytic performance in the reduction of p-nitrophenol to p-aminophenol in the presence of NaBH$_4$. While both the individual compounds exhibited no catalytic activity, the composite showed an excellent performance (the activity and rate constant at 50 °C were 683 L g^{-1} s^{-1} and 0.23 min^{-1}, respectively), which was even comparable to those observed for noble metal-based catalysts. The authors stated that the high catalytic activity originated from the fact that the Mn-POM facilitated the electron transfer from BH_4^- to the Fe^{3+} Lewis acid sites of the MOF, as they assumed that the MIL-100 alone could not accept electrons directly from BH_4^-. Additionally, the group of Shul observed a distinct acid-base synergy upon examination of the core–shell structured heteropoly acid-functionalized ZIF-8 in the transesterification of rapeseed oil with methanol to produce biodiesel [46]. More than 95% of the rapeseed oil was converted to biodiesel due to the simultaneous presence of the acid functionalities of the POM and the basicity of the imidazolate groups of the MOF, whereas the pure POM and ZIF-8 catalysts showed a catalytic performance of 61% and 32%, respectively. Moreover, the strong chemical O-N bonding between the Keggin and the imidazole units ensured a good recyclability, with no noticeable decrease in the catalytic performance after five cycles and no POM leaching.

3.2. Electrocatalysis

Besides the use of POM@MOF hybrids in organocatalysis, POMs also exhibit interesting electrocatalytic properties as they can undergo fast and reversible multi-electron transfers [157]. Within this context, POMs have already shown great potential in the electrochemical oxygen evolution reaction (OER) in a homogenous manner [158]. Despite the remarkable progress in this field, there are only a few reports on the encapsulation of POMs in the cages of MOFs for electrocatalytic water oxidation, as can be seen in Table 2. This is probably due to the fact that the majority of MOFs possess a low electrical conductivity and high hydrophobicity. The first report on the encapsulation of an unsubstituted Keggin POM in a MOF to perform electrocatalytic water oxidation was reported by Das et al. [137]. More specifically, a one-pot synthesis was performed to include the $[CoW_{12}O_{40}]^{6-}$ anion in the size matching cage of ZIF-8 (see Figure 13). During the electrochemical measurements, performed at pH 1.9, the authors observed a clear shift to a less anodic potential for the redox Co^{III}/Co^{II} couple in the cyclic voltammogram of POM@MOF with respect to that of the uncapsulated POM (from 1.14 V for the Keggin POM to 0.97 V for the composite material). In addition to this, the POM@MOF catalyst exhibited an excellent stability as only a very small drop in the catalytic current was observed after 1000 catalytic cycles and no leaching of Co species was observed. It is, however, important to note here that, although the catalyst exhibited a high turnover frequency (TOF = 12.5 s^{-1} based on the quantitative oxygen evolution) and an excellent faradaic efficiency of 95.7%, a rather high overpotential was required (784.19 mV at a current density of 1 mA cm^{-2}).

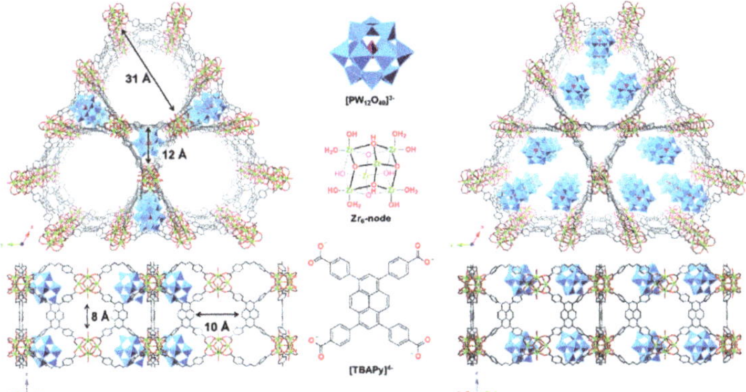

Figure 13. Encapsulation of an inactive Keggin POM in ZIF-8 to become an active oxygen evolution reaction (OER) catalyst. Reprinted with permission from [137]. Copyright (2018), John Wiley and Sons.

In a very recent report of the same group, a redox inactive SiW_{12} POM was used to lower the required overpotential [138]. The co-encapsulation of this POM together with the true catalyst, an Fe(salen) complex, within ZIF-8 resulted in a decrease of the overpotential of more than 150 mV. In the absence of the encapsulated POM, the Fe-salen@ZIF-8 required an overpotential of 672.9 mV to attain a current

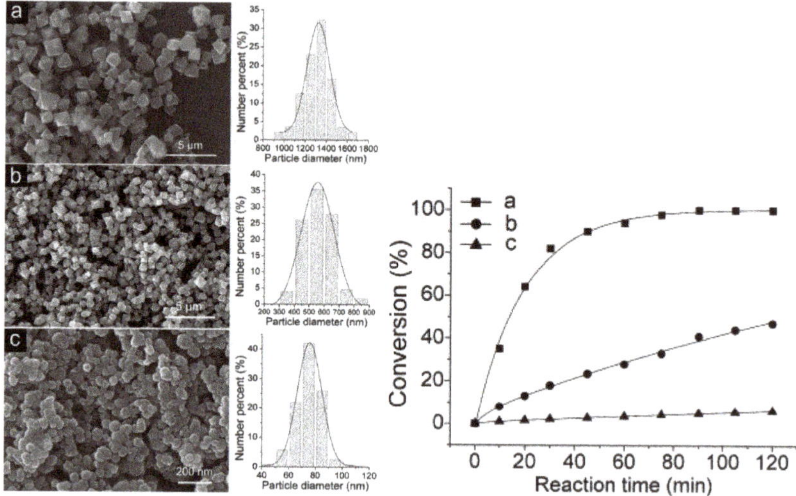

Figure 15. A fully noble metal-free POM@MOF catalyst for the photocatalytic oxidation of water. Reprinted with permission from [145]. Copyright (2018), American Chemistry Society.

However, it is important to note here that in the previously presented studies, a sacrificial donor or acceptor was required for the photocatalytic process. In a very recent work by Niu and co-workers, the assembling of a photosensitizer, electron donor, and acceptor into one single framework was reported [149]. For the synthesis of this Zn-based framework, the photosensitizer *N,N'*-di(4-pyridyl)-1,4,5,8-naphthalenetetracarboxydiimide (DPNDI) was used as the organic ligand, while pyrrolidine-2-yl-imidazole and the $[BW_{12}O_{40}]^{5-}$ anion were introduced, respectively, as an electron donor and electron acceptor (see Figure 16). The resulting Zn-DPNDI-PYI catalyst was examined in the oxidative coupling of benzylamine, exhibiting a conversion of 99% after 16 h of reaction. This high activity was not only the result of the consecutive photo-induced electron transfer (conPET) process, but was also assigned to the long-lived charge separated state.

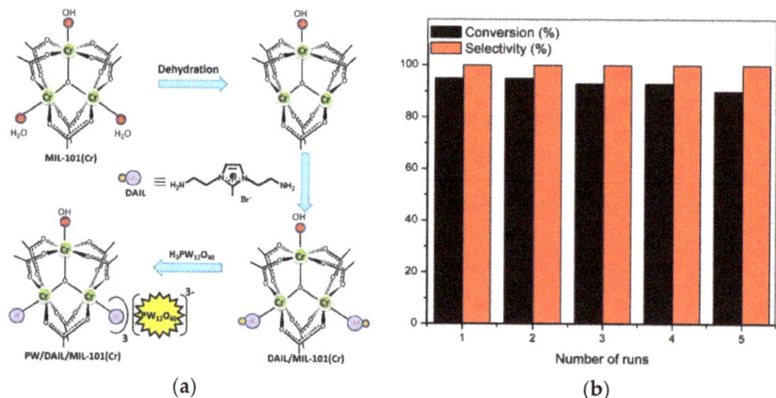

Figure 16. Zn-DPNDI-PYI as photocatalyst for the coupling of primary amines and oxidation of olefins with air under visible light. Reprinted from [149]. Copyright (2019), with permission from Elsevier.

4. Summary and Outlook

Metal–organic frameworks exhibiting well-defined cages, large surface areas, and a high thermal and chemical stability are excellent hosts for encapsulating polyoxometalates. More than 100 studies on such POM@MOF hybrids have appeared in the last decade. In this review, we mainly focused on the common synthetic aspects and catalytic applications of POM@MOF hybrids in organocatalysis, electrocatalysis, and photocatalysis. More specifically, the activity, recyclability, stability, and interesting synergetic functions of POM@MOF were discussed.

The size of the pores and the aperture of the pore windows are very critical parameters in the design of a POM@MOF. The embedding of POMs into MOFs not only allows the shortcomings of POMs to be overcome, but also ensures the use of the unique advantages of MOFs. The rise of POM@MOF systems is mainly attributed to the excellent dispersion and subsequent stability of the POM in the MOF host. The unique cages and windows and the tunable chemical environment of MOFs enable interesting interactions and synergic effects between POM and MOFs, thus creating excellent novel heterogeneous catalysts.

Although POM@MOF hybrid materials have made tremendous progress in recent years, many challenges still need to be addressed. First of all, the interaction between POMs and MOFs is often limited to weak electrostatic interactions, which can result in POM leaching during the catalysis. To this end, stronger interactions, such as covalent bonds between the MOF host and the encapsulated POM, would allow a further increase of the POM@MOF reusability in catalysis. Secondly, at this moment, there is still too much 'trial and error' involved to obtain a good control on the position and distribution of POMs inside MOF cages/channels. Thirdly, very little is known about the synergetic effects and electron transfer mechanism in catalytic reactions. To address this problem, theoretical calculations combined with in-situ and ex-situ characterization techniques would provide a better understanding of the synergetic effects and electron transfer mechanism. Finally, up until now, only some well-known archetypical POMs have been encapsulated in MOFs. New and innovative types of POMs (such as V-centered POMs) with a proven excellent performance in oxidation and photocatalytic reactions should be combined with MOFs to further enhance the application range of these hybrids. We have no doubt that several exciting new (catalytic) applications will be reported in the next months and years in this strongly growing field of research.

Author Contributions: Conceptualization and supervision: P.V.D.V. Writing-review and editing: J.S., S.A., Y.-Y.L. and K.L. All authors have read and agreed to the published version of the manuscript.

Funding: This work was financially supported by Ghent University BOF doctoral grant 01D04318, the Research Foundation Flanders (FWO-Vlaanderen) grant no. G000117N, the International S&T Cooperation Program of China (2016YFE0109800) and the China Scholarship Council (CSC).

Conflicts of Interest: The authors declare no conflicts of interest.

References

1. Berzelius, J.J.A. Beitrag zur näheren kenntniss des molybdäns. *Ann. Phys.* **1826**, *82*, 369–392. [CrossRef]
2. Keggin, J.F. The structure and formula of 12-phosphotungstic acid. *Proc. R. Soc. Lond. Ser. A* **1934**, *144*, 75–100.
3. Omwoma, S.; Gore, C.T.; Ji, Y.C.; Hu, C.W.; Song, Y.F. Environmentally benign polyoxometalate materials. *Coord. Chem. Rev.* **2015**, *286*, 17–29. [CrossRef]
4. Cao, Y.W.; Chen, Q.Y.; Shen, C.R.; He, L. Polyoxometalate-based catalysts for CO_2 conversion. *Molecules* **2019**, *24*, 2069. [CrossRef]
5. Misra, A.; Kozma, K.; Streb, C.; Nyman, M. Beyond charge balance: Counter-cations in polyoxometalate chemistry. *Angew. Chem. Int. Ed.* **2020**, *59*, 596–612. [CrossRef]
6. Huang, B.; Yang, D.H.; Han, B.H. Application of polyoxometalate derivatives in rechargeable batteries. *J. Mater. Chem. A* **2020**, *8*, 4593–4628. [CrossRef]
7. Zhai, L.; Li, H.L. Polyoxometalate-polymer hybrid materials as proton exchange membranes for fuel cell applications. *Molecules* **2019**, *24*, 3425. [CrossRef]

8. Liu, D.X.; Zou, D.T.; Zhu, H.L.; Zhang, J.Y. Mesoporous metal-organic frameworks: Synthetic strategies and emerging applications. *Small* **2018**, *14*, 1801454. [CrossRef]

9. Qin, J.S.; Yuan, S.; Lollar, C.; Pang, J.D.; Alsalme, A.; Zhou, H.C. Stable metal-organic frameworks as a host platform for catalysis and biomimetics. *Chem. Commun.* **2018**, *54*, 4231–4249. [CrossRef]

10. Bedia, J.; Muelas-Ramos, V.; Penas-Garzon, M.; Gomez-Aviles, A.; Rodriguez, J.J.; Belver, C. A review on the synthesis and characterization of metal organic frameworks for photocatalytic water purification. *Catalysts* **2019**, *9*, 52. [CrossRef]

11. Shen, X.; Pan, Y.; Sun, Z.H.; Liu, D.; Xu, H.J.; Yu, Q.; Trivedi, M.; Kumar, A.; Chen, J.X.; Liu, J.Q. Design of metal-organic frameworks for pH-responsive drug delivery application. *Mini Rev. Med. Chem.* **2019**, *19*, 1644–1665. [CrossRef] [PubMed]

12. Yang, J.; Yang, Y.W. Metal-organic frameworks for biomedical applications. *Small* **2020**, *16*, 1906846–1906869. [CrossRef] [PubMed]

13. Amini, A.; Kazemi, S.; Safarifard, V. Metal-organic framework-based nanocomposites for sensing applications—A review. *Polyhedron* **2020**, *177*, 114260. [CrossRef]

14. Li, G.D.; Zhao, S.L.; Zhang, Y.; Tang, Z.Y. Metal-organic frameworks encapsulating active nanoparticles as emerging composites for catalysis: Recent progress and perspectives. *Adv. Mater.* **2018**, *30*, 1800702. [CrossRef] [PubMed]

15. Majewski, M.B.; Howarth, A.J.; Li, P.; Wasielewski, M.R.; Hupp, J.T.; Farha, O.K. Enzyme encapsulation in metal-organic frameworks for applications in catalysis. *CrystEngComm* **2017**, *19*, 4082–4091. [CrossRef]

16. Samaniyan, M.; Mirzaei, M.; Khajavian, R.; Eshtiagh-Hosseini, H.; Streb, C. Heterogeneous catalysis by polyoxometalates in metal-organic frameworks. *ACS Catal.* **2019**, *9*, 10174–10191. [CrossRef]

17. Fang, Y.; Powell, J.A.; Li, E.R.; Wang, Q.; Perry, Z.; Kirchon, A.; Yang, X.Y.; Xiao, Z.F.; Zhu, C.F.; Zhang, L.L.; et al. Catalytic reactions within the cavity of coordination cages. *Chem. Soc. Rev.* **2019**, *48*, 4707–4730. [CrossRef]

18. Ferey, G.; Mellot-Draznieks, C.; Serre, C.; Millange, F.; Dutour, J.; Surble, S.; Margiolaki, I. A chromium terephthalate-based solid with unusually large pore volumes and surface area. *Science* **2005**, *309*, 2040–2042. [CrossRef]

19. Liu, J.X.; Zhang, X.B.; Li, Y.L.; Huang, S.L.; Yang, G.Y. Polyoxometalate functionalized architectures. *Coordin. Chem. Rev.* **2020**, *414*, 213260–213275. [CrossRef]

20. Du, D.Y.; Qin, J.S.; Li, S.L.; Su, Z.M.; Lan, Y.Q. Recent advances in porous polyoxometalate-based metal-organic framework materials. *Chem. Soc. Rev.* **2014**, *43*, 4615–4632. [CrossRef]

21. Li, X.X.; Zhao, D.; Zheng, S.T. Recent advances in POM-organic frameworks and POM-organic polyhedra. *Coordin. Chem. Rev.* **2019**, *397*, 220–240. [CrossRef]

22. Chavan, S.; Vitillo, J.G.; Gianolio, D.; Zavorotynska, O.; Civalleri, B.; Jakobsen, S.; Nilsen, M.H.; Valenzano, L.; Lamberti, C.; Lillerud, K.P.; et al. H_2 storage in isostructural UiO-67 and UiO-66 MOFs. *Phys. Chem. Chem. Phys.* **2012**, *14*, 1614–1626. [CrossRef] [PubMed]

23. Cavka, J.H.; Jakobsen, S.; Olsbye, U.; Guillou, N.; Lamberti, C.; Bordiga, S.; Lillerud, K.P. A new zirconium inorganic building brick forming metal organic frameworks with exceptional stability. *J. Am. Chem. Soc.* **2008**, *130*, 13850–13851. [CrossRef] [PubMed]

24. Islamoglu, T.; Otake, K.; Li, P.; Buru, C.T.; Peters, A.W.; Akpinar, I.; Garibay, S.J.; Farha, O.K. Revisiting the structural homogeneity of NU-1000, a Zr-based metal-organic framework. *CrystEngComm* **2018**, *20*, 5913–5918. [CrossRef]

25. Luo, H.Z.; Zeng, Z.T.; Zeng, G.M.; Zhang, C.; Xiao, R.; Huang, D.L.; Lai, C.; Cheng, M.; Wang, W.J.; Xiong, W.P.; et al. Recent progress on metal-organic frameworks based- and derived-photocatalysts for water splitting. *Chem. Eng. J.* **2020**, *383*, 123196–123214. [CrossRef]

26. Park, K.S.; Ni, Z.; Cote, A.P.; Choi, J.Y.; Huang, R.D.; Uribe-Romo, F.J.; Chae, H.K.; O'Keeffe, M.; Yaghi, O.M. Exceptional chemical and thermal stability of zeolitic imidazolate frameworks. *Proc. Natl. Acad. Sci. USA* **2006**, *103*, 10186–10191. [CrossRef]

27. Horcajada, P.; Surble, S.; Serre, C.; Hong, D.Y.; Seo, Y.K.; Chang, J.S.; Greneche, J.M.; Margiolaki, I.; Ferey, G. Synthesis and catalytic properties of MIL-100(Fe), an iron(III) carboxylate with large pores. *Chem. Commun.* **2007**, 2820–2822. [CrossRef]

28. Krungleviciute, V.; Lask, K.; Migone, A.D.; Lee, J.Y.; Li, J. Kinetics and equilibrium of gas adsorption on RPM1-Co and Cu-BTC metal-organic frameworks: Potential for gas separation applications. *AIChE J.* **2008**, *54*, 918–923. [CrossRef]

29. Naseri, E.; Khoshnavazi, R. Sandwich type polyoxometalates encapsulated into the mesoporous material: Synthesis, characterization and catalytic application in the selective oxidation of sulfides. *RSC Adv.* **2018**, *8*, 28249–28260. [CrossRef]

30. Juan-Alcañiz, J.; Ramos-Fernandez, E.V.; Lafont, U.; Gascon, J.; Kapteijn, F. Building MOF bottles around phosphotungstic acid ships: One-pot synthesis of bi-functional polyoxometalate-MIL-101 catalysts. *J. Catal.* **2010**, *269*, 229–241. [CrossRef]

31. Canioni, R.; Roch-Marchal, C.; Sécheresse, F.; Horcajada, P.; Serre, C.; Hardi-Dan, M.; Férey, G.; Grenèche, J.-M.; Lefebvre, F.; Chang, J.-S.; et al. Stable polyoxometalate insertion within the mesoporous metal organic framework MIL-100(Fe). *J. Mater. Chem. A* **2011**, *21*, 1226–1233. [CrossRef]

32. Marandi, A.; Tangestaninejad, S.; Moghadam, M.; Mirkhani, V.; Mechler, A.; Mohammadpoor-Baltork, I.; Zadehahmadi, F. Dodecatungstocobaltate heteropolyanion encapsulation into MIL-101(Cr) metal-organic framework scaffold provides a highly efficient heterogeneous catalyst for methanolysis of epoxides. *Appl. Organomet. Chem.* **2018**, *32*, 4065–4074. [CrossRef]

33. Wang, X.S.; Huang, Y.B.; Lin, Z.J.; Cao, R. Phosphotungstic acid encapsulated in the mesocages of amine-functionalized metal-organic frameworks for catalytic oxidative desulfurization. *Dalton Trans.* **2014**, *43*, 11950–11958. [CrossRef] [PubMed]

34. Abednatanzi, S.; Leus, K.; Derakhshandeh, P.G.; Nahra, F.; De Keukeleere, K.; Van Hecke, K.; Van Driessche, I.; Abbasi, A.; Nolan, S.P.; Der Voort, P.V. POM@IL-MOFs –Inclusion of POMs in ionic liquid modified MOFs to produce recyclable oxidation catalysts. *Catal. Sci. Technol.* **2017**, *7*, 1478–1487. [CrossRef]

35. Shiju, N.R.; Alberts, A.H.; Khalid, S.; Brown, D.R.; Rothenberg, G. Mesoporous silica with site-isolated amine and phosphotungstic acid groups: A solid catalyst with tunable antagonistic functions for one-pot tandem reactions. *Angew. Chem. Int. Ed.* **2011**, *50*, 9615–9619. [CrossRef] [PubMed]

36. Ramos-Fernandez, E.V.; Pieters, C.; Van der Linden, B.; Juan-Alcaniz, J.; Serra-Crespo, P.; Verhoeven, M.W.G.M.; Niemantsverdriet, H.; Gascon, J.; Kapteijn, F. Highly dispersed platinum in metal organic framework NH_2-MIL-101(Al) containing phosphotungstic acid—Characterization and catalytic performance. *J. Catal.* **2012**, *289*, 42–52. [CrossRef]

37. Bromberg, L.; Su, X.; Hatton, T.A. Heteropolyacid-functionalized aluminum 2-aminoterephthalate metal-organic frameworks as reactive aldehyde sorbents and catalysts. *ACS Appl. Mater. Interfaces* **2013**, *5*, 5468–5477. [CrossRef]

38. Bajpe, S.R.; Kirschhock, C.E.; Aerts, A.; Breynaert, E.; Absillis, G.; Parac-Vogt, T.N.; Giebeler, L.; Martens, J.A. Direct observation of molecular-level template action leading to self-assembly of a porous framework. *Chemistry* **2010**, *16*, 3926–3932. [CrossRef]

39. Liu, X.; Luo, J.; Sun, T.; Yang, S. A simple approach to the preparation of $H_6P_2W_{18}O_{62}/Cu_3(BTC)_2$ (BTC = 1,3,5-benzenetricarboxylate) and its catalytic performance in the synthesis of acetals/ketals. *React. Kinet. Mech. Catal.* **2015**, *116*, 159–171. [CrossRef]

40. Wee, L.H.; Bajpe, S.R.; Janssens, N.; Hermans, I.; Houthoofd, K.; Kirschhock, C.E.; Martens, J.A. Convenient synthesis of $Cu_3(BTC)_2$ encapsulated Keggin heteropolyacid nanomaterial for application in catalysis. *Chem. Commun.* **2010**, *46*, 8186–8193. [CrossRef]

41. Song, J.; Luo, Z.; Britt, D.K.; Furukawa, H.; Yaghi, O.M.; Hardcastle, K.I.; Hill, C.L. A multiunit catalyst with synergistic stability and reactivity: A polyoxometalate-metal organic framework for aerobic decontamination. *J. Am. Chem. Soc.* **2011**, *133*, 16839–16846. [CrossRef] [PubMed]

42. Liu, S.-M.; Zhang, Z.; Li, X.; Jia, H.; Ren, M.; Liu, S. Ti-substituted Keggin-type polyoxotungstate as proton and electron reservoir encaged into metal-organic framework for carbon dioxide photoreduction. *Adv. Mater. Interfaces* **2018**, *5*, 1801062. [CrossRef]

43. Sun, C.Y.; Liu, S.X.; Liang, D.D.; Shao, K.Z.; Ren, Y.H.; Su, Z.M. Highly stable crystalline catalysts based on a microporous metal-organic framework and polyoxometalates. *J. Am. Chem.Soc.* **2009**, *131*, 1883–1888. [CrossRef] [PubMed]

44. Huang, Z.; Yang, Z.; Hussain, M.Z.; Chen, B.; Jia, Q.; Zhu, Y.; Xia, Y. Polyoxometallates@zeolitic-imidazolate-framework derived bimetallic tungsten-cobalt sulfide/porous carbon nanocomposites as efficient bifunctional electrocatalysts for hydrogen and oxygen evolution. *Electrochim. Acta* **2020**, *330*, 135335. [CrossRef]

45. Malkar, R.S.; Yadav, G.D. Synthesis of cinnamyl benzoate over novel heteropoly acid encapsulated ZIF-8. *Appl. Catal. A: Gen.* **2018**, *560*, 54–65. [CrossRef]

46. Jeon, Y.; Chi, W.S.; Hwang, J.; Kim, D.H.; Kim, J.H.; Shul, Y.G. Core-shell nanostructured heteropoly acid-functionalized metal-organic frameworks: Bifunctional heterogeneous catalyst for efficient biodiesel production. *Appl. Catal. B Environ.* **2019**, *242*, 51–59. [CrossRef]

47. Kong, X.J.; Lin, Z.; Zhang, Z.M.; Zhang, T.; Lin, W. Hierarchical integration of photosensitizing metal-organic frameworks and nickel-containing polyoxometalates for efficient visible-light-driven hydrogen evolution. *Angew. Chem. Int. Ed.* **2016**, *55*, 6411–6416. [CrossRef]

48. Zhong, X.; Lu, Y.; Luo, F.; Liu, Y.; Li, X.; Liu, S. A nanocrystalline POM@MOFs catalyst for the degradation of phenol: Effective cooperative catalysis by metal nodes and POM guests. *Eur. J. Chem.* **2018**, *24*, 3045–3051. [CrossRef]

49. Li, G.; Zhang, K.; Li, C.; Gao, R.; Cheng, Y.; Hou, L.; Wang, Y. Solvent-free method to encapsulate polyoxometalate into metal-organic frameworks as efficient and recyclable photocatalyst for harmful sulfamethazine degrading in water. *Appl. Catal. B Environ.* **2019**, *245*, 753–759. [CrossRef]

50. Granadeiro, C.M.; Barbosa, A.D.S.; Ribeiro, S.; Santos, I.C.M.S.; De Castro, B.; Cunha-Silva, L.; Balula, S.S. Oxidative catalytic versatility of a trivacant polyoxotungstate incorporated into MIL-101(Cr). *Catal. Sci. Technol.* **2014**, *4*, 1416–1425. [CrossRef]

51. Ribeiro, S.; Barbosa, A.D.S.; Gomes, A.C.; Pillinger, M.; Gonçalves, I.S.; Cunha-Silva, L.; Balula, S.S. Catalytic oxidative desulfurization systems based on Keggin phosphotungstate and metal-organic framework MIL-101. *Fuel Process. Technol.* **2013**, *116*, 350–357. [CrossRef]

52. Hu, X.F.; Lu, Y.K.; Dai, F.N.; Liu, C.G.; Liu, Y.Q. Host-guest synthesis and encapsulation of phosphotungstic acid in MIL-101 via "bottle around ship": An effective catalyst for oxidative desulfurization. *Micropor. Mesopor. Mat.* **2013**, *170*, 36–44. [CrossRef]

53. Fazaeli, R.; Aliyan, H.; Masoudinia, M.; Heidari4, Z. Building MOF bottles (MIL-101 Family as heterogeneous single-site catalysts) around $H_3PW_{12}O_{40}$ ships: An efficient catalyst for selective oxidation of sulfides to sulfoxides and sulfones. *J. Mater. Chem. Eng.* **2014**, *2*, 46–55.

54. Sun, M.; Chen, W.-C.; Zhao, L.; Wang, X.-L.; Su, Z.-M. A PTA@MIL-101(Cr)-diatomite composite as catalyst for efficient oxidative desulfurization. *Inorg. Chem. Commun.* **2018**, *87*, 30–35. [CrossRef]

55. Julião, D.; Gomes, A.C.; Pillinger, M.; Cunha-Silva, L.; De Castro, B.; Gonçalves, I.S.; Balula, S.S. Desulfurization of model diesel by extraction/oxidation using a zinc-substituted polyoxometalate as catalyst under homogeneous and heterogeneous (MIL-101(Cr) encapsulated) conditions. *Fuel Process. Technol.* **2015**, *131*, 78–86. [CrossRef]

56. Julião, D.; Gomes, A.C.; Pillinger, M.; Valença, R.; Ribeiro, J.C.; De Castro, B.; Gonçalves, I.S.; Cunha Silva, L.; Balula, S.S. Zinc-substituted polyoxotungstate@amino-MIL-101(Al)—An efficient catalyst for the sustainable desulfurization of model and real diesels. *Eur. J. Inorg. Chem.* **2016**, *2016*, 5114–5122. [CrossRef]

57. Ribeiro, S.; Granadeiro, C.M.; Silva, P.; Almeida Paz, F.A.; De Biani, F.F.; Cunha-Silva, L.; Balula, S.S. An efficient oxidative desulfurization process using terbium-polyoxometalate@MIL-101(Cr). *Catal. Sci. Technol.* **2013**, *3*, 2404–2414. [CrossRef]

58. Granadeiro, C.M.; Nogueira, L.S.; Julião, D.; Mirante, F.; Ananias, D.; Balula, S.S.; Cunha-Silva, L. Influence of a porous MOF support on the catalytic performance of Eu-polyoxometalate based materials: Desulfurization of a model diesel. *Catal. Sci. Technol.* **2016**, *6*, 1515–1522. [CrossRef]

59. Granadeiro, C.M.; Barbosa, A.D.S.; Silva, P.; Paz, F.A.A.; Saini, V.K.; Pires, J.; De Castro, B.; Balula, S.S.; Cunha-Silva, L. Monovacant polyoxometalates incorporated into MIL-101(Cr): Novel heterogeneous catalysts for liquid phase oxidation. *Appl. Catal. A Gen.* **2013**, *453*, 316–326. [CrossRef]

60. Wang, X.-S.; Li, L.; Liang, J.; Huang, Y.-B.; Cao, R. Boosting oxidative desulfurization of model and real gasoline over phosphotungstic acid encapsulated in metal-organic frameworks: The window size matters. *ChemCatChem* **2017**, *9*, 971–979. [CrossRef]

61. Ni, X.-L.; Liu, J.; Liu, Y.-Y.; Leus, K.; Depauw, H.; Wang, A.-J.; Van Der Voort, P.; Zhang, J.; Hu, Y.-K. Synthesis, characterization and catalytic performance of Mo based metal- organic frameworks in the epoxidation of propylene by cumene hydroperoxide. *Chin. Chem. Lett.* **2017**, *28*, 1057–1061. [CrossRef]

62. Maksimchuk, N.V.; Timofeeva, M.N.; Melgunov, M.S.; Shmakov, A.N.; Chesalov, Y.A.; Dybtsev, D.N.; Fedin, V.P.; Kholdeeva, O.A. Heterogeneous selective oxidation catalysts based on coordination polymer MIL-101 and transition metal-substituted polyoxometalates. *J. Catal.* **2008**, *257*, 315–323. [CrossRef]

63. Saedi, Z.; Tangestaninejad, S.; Moghadam, M.; Mirkhani, V.; Mohammadpoor-Baltork, I. The effect of encapsulated Zn-POM on the catalytic activity of MIL-101 in the oxidation of alkenes with hydrogen peroxide. *J. Coord. Chem.* **2012**, *65*, 463–473. [CrossRef]

64. Maksimchuk, N.V.; Kovalenko, K.A.; Arzumanov, S.S.; Chesalov, Y.A.; Melgunov, M.S.; Stepanov, A.G.; Fedin, V.P.; Kholdeeva, O.A. Hybrid polyoxotungstate/MIL-101 materials: Synthesis, characterization, and catalysis of H_2O_2-based alkene epoxidation. *Inorg. Chem.* **2010**, *49*, 2920–2930. [CrossRef] [PubMed]

65. Balula, S.S.; Granadeiro, C.M.; Barbosa, A.D.S.; Santos, I.C.M.S.; Cunha-Silva, L. Multifunctional catalyst based on sandwich-type polyoxotungstate and MIL-101 for liquid phase oxidations. *Catal. Today* **2013**, *210*, 142–148. [CrossRef]

66. Tong, J.; Wang, W.; Su, L.; Li, Q.; Liu, F.; Ma, W.; Lei, Z.; Bo, L. Highly selective oxidation of cyclohexene to 2-cyclohexene-1-one over polyoxometalate/metal–organic framework hybrids with greatly improved performances. *Catal. Sci. Technol.* **2017**, *7*, 222–230. [CrossRef]

67. Ding, J.-W.; Wang, R. A new green system of HPW@MOFs catalyzed desulfurization using O_2 as oxidant. *Chin. Chem. Lett.* **2016**, *27*, 655–658. [CrossRef]

68. Rafiee, E.; Nobakht, N. Keggin type heteropoly acid, encapsulated in metal-organic framework: A heterogeneous and recyclable nanocatalyst for selective oxidation of sulfides and deep desulfurization of model fuels. *J. Mol. Catal. A Chem.* **2015**, *398*, 17–25. [CrossRef]

69. Babahydari, A.K.; Fareghi-Alamdari, R.; Hafshejani, S.M.; Rudbari, H.A.; Farsani, M.R. Heterogeneous oxidation of alcohols with hydrogen peroxide catalyzed by polyoxometalate metal–organic framework. *J. Iran. Chem. Soc.* **2016**, *13*, 1463–1470. [CrossRef]

70. Zhu, J.; Shen, M.-N.; Zhao, X.-J.; Wang, P.-C.; Lu, M. Polyoxometalate-based metal-organic frameworks as catalysts for the selective oxidation of alcohols in micellar systems. *ChemPlusChem* **2014**, *79*, 872–878. [CrossRef]

71. Wang, S.; Liu, Y.; Zhang, Z.; Li, X.; Tian, H.; Yan, T.; Zhang, X.; Liu, S.; Sun, X.; Xu, L.; et al. One-step template-free fabrication of ultrathin mixed-valence polyoxovanadate-incorporated metal organic framework nanosheets for highly efficient selective oxidation catalysis in air. *ACS Appl. Mater. Interfaces* **2019**, *11*, 12786–12796. [CrossRef] [PubMed]

72. Yang, X.L.; Qiao, L.M.; Dai, W.L. One-pot synthesis of a hierarchical microporous-mesoporous phosphotungstic acid-HKUST-1 catalyst and its application in the selective oxidation of cyclopentene to glutaraldehyde. *Chin. J. Catal.* **2015**, *36*, 1875–1885. [CrossRef]

73. Ke, F.; Guo, F.; Yu, J.; Yang, Y.Q.; He, Y.; Chang, L.Z.; Wan, X.C. Highly site-selective epoxidation of polyene catalyzed by metal-organic frameworks assisted by polyoxometalate. *J. Inorg. Organomet. Polym. Mater.* **2017**, *27*, 843–849. [CrossRef]

74. Yang, H.; Li, J.; Wang, L.Y.; Dai, W.; Lv, Y.; Gao, S. Exceptional activity for direct synthesis of phenol from benzene over PMoV@MOF with O_2. *Catal. Commun.* **2013**, *35*, 101–104. [CrossRef]

75. Liu, Y.W.; Liu, S.M.; Liu, S.X.; Liang, D.D.; Li, S.J.; Tang, Q.; Wang, X.Q.; Miao, J.; Shi, Z.; Zheng, Z.P. Facile synthesis of a nanocrystalline metal-organic framework impregnated with a phosphovanadomolybdate and its remarkable catalytic performance in ultradeep oxidative desulfurization. *ChemCatChem* **2013**, *5*, 3086–3091. [CrossRef]

76. Li, S.-W.; Gao, R.-M.; Zhao, J.-S. Deep oxidative desulfurization of fuel catalyzed by modified heteropolyacid: The comparison performance of three kinds of ionic liquids. *ACS Sustain. Chem. Eng.* **2018**, *6*, 15858–15866. [CrossRef]

77. Li, S.W.; Li, J.R.; Jin, Q.P.; Yang, Z.; Zhang, R.L.; Gao, R.M.; Zhao, J.S. Preparation of mesoporous Cs-PO7M@MOF-199@MCM-41 under two different synthetic methods for a highly oxidesulfurization of dibenzothiophene. *J. Hazard. Mater.* **2017**, *337*, 208–216. [CrossRef] [PubMed]

78. Gao, Y.; Lv, Z.; Gao, R.; Zhang, G.; Zheng, Y.; Zhao, J. Oxidative desulfurization process of model fuel under molecular oxygen by polyoxometalate loaded in hybrid material CNTs@MOF-199 as catalyst. *J. Hazard. Mater.* **2018**, *359*, 258–265. [CrossRef]

79. Yang, H.; Li, J.; Zhang, H.Y.; Lv, Y.; Gao, S. Facile synthesis of POM@MOF embedded in SBA-15 as a steady catalyst for the hydroxylation of benzene. *Micropor. Mesopor. Mat.* **2014**, *195*, 87–91. [CrossRef]

80. Zhang, Y.; Zhang, W.; Zhang, J.; Dong, Z.; Zhang, X.; Ding, S. One-pot synthesis of cup-like ZSM-5 zeolite and its excellent oxidative desulfurization performance. *RSC Adv.* **2018**, *8*, 31979–31983. [CrossRef]

81. Tian, H.R.; Liu, Y.W.; Zhang, Z.; Liu, S.M.; Dang, T.Y.; Li, X.H.; Sun, X.W.; Lu, Y.; Liu, S.X. A multicentre synergistic polyoxometalate-based metal-organic framework for one-step selective oxidative cleavage of beta-O-4 lignin model compounds. *Green Chem.* **2020**, *22*, 248–255. [CrossRef]

82. Salomon, W.; Roch-Marchal, C.; Mialane, P.; Rouschmeyer, P.; Serre, C.; Haouas, M.; Taulelle, F.; Yang, S.; Ruhlmann, L.; Dolbecq, A. Immobilization of polyoxometalates in the Zr-based metal organic framework UiO-67. *Chem. Commun.* **2015**, *51*, 2972–2975. [CrossRef] [PubMed]

83. Yang, X.-L.; Qiao, L.-M.; Dai, W.-L. Phosphotungstic acid encapsulated in metal-organic framework UiO-66: An effective catalyst for the selective oxidation of cyclopentene to glutaraldehyde. *Micropor. Mesopor. Mat.* **2015**, *211*, 73–81. [CrossRef]

84. Song, X.; Hu, D.; Yang, X.; Zhang, H.; Zhang, W.; Li, J.; Jia, M.; Yu, J. Polyoxomolybdic cobalt encapsulated within Zr-based metal–organic frameworks as efficient heterogeneous catalysts for olefins epoxidation. *ACS Sustain. Chem. Eng.* **2019**, *7*, 3624–3631. [CrossRef]

85. Buru, C.T.; Wasson, M.C.; Farha, O.K. $H_5PV_2Mo_{10}O_{40}$ polyoxometalate encapsulated in NU-1000 metal–organic framework for aerobic oxidation of a mustard gas simulant. *ACS Appl. Nano. Mater.* **2019**, *3*, 658–664. [CrossRef]

86. Buru, C.T.; Platero-Prats, A.E.; Chica, D.G.; Kanatzidis, M.G.; Chapman, K.W.; Farha, O.K. Thermally induced migration of a polyoxometalate within a metal–organic framework and its catalytic effects. *J. Mater. Chem. A* **2018**, *6*, 7389–7394. [CrossRef]

87. Buru, C.T.; Li, P.; Mehdi, B.L.; Dohnalkova, A.; Platero-Prats, A.E.; Browning, N.D.; Chapman, K.W.; Hupp, J.T.; Farha, O.K. Adsorption of a catalytically accessible polyoxometalate in a mesoporous channel-type metal–organic framework. *Chem. Mater.* **2017**, *29*, 5174–5181. [CrossRef]

88. Hao, X.L.; Ma, Y.Y.; Zang, H.Y.; Wang, Y.H.; Li, Y.G.; Wang, E.B. A polyoxometalate-encapsulating cationic metal-organic framework as a heterogeneous catalyst for desulfurization. *Eur. J. Chem.* **2015**, *21*, 3778–3784. [CrossRef]

89. Roy, S.; Vemuri, V.; Maiti, S.; Manoj, K.S.; Subbarao, U.; Peter, S.C. Two Keggin-based isostructural POMOF hybrids: Synthesis, crystal structure, and catalytic properties. *Inorg. Chem.* **2018**, *57*, 12078–12092. [CrossRef]

90. Ma, Y.Y.; Peng, H.Y.; Liu, J.N.; Wang, Y.H.; Hao, X.L.; Feng, X.J.; Khan, S.U.; Tan, H.Q.; Li, Y.G. Polyoxometalate-based metal-organic frameworks for selective oxidation of aryl alkenes to aldehydes. *Inorg. Chem.* **2018**, *57*, 4109–4116. [CrossRef]

91. Li, D.D.; Ma, X.Y.; Wang, Q.Z.; Ma, P.T.; Niu, J.Y.; Wang, J.P. Copper-containing polyoxometalate-based metal-organic frameworks as highly efficient heterogeneous catalysts toward selective oxidation of alkylbenzenes. *Inorg. Chem.* **2019**, *58*, 15832–15840. [CrossRef] [PubMed]

92. Li, D.D.; Xu, Q.F.; Li, Y.G.; Qiu, Y.T.; Ma, P.T.; Niu, J.Y.; Wang, J.P. A Stable polyoxometalate-based metal-organic framework as highly efficient heterogeneous catalyst for oxidation of alcohols. *Inorg. Chem.* **2019**, *58*, 4945–4953. [CrossRef] [PubMed]

93. Ghahramaninezhad, M.; Pakdel, F.; Niknam Shahrak, M. Boosting oxidative desulfurization of model fuel by POM-grafting ZIF-8 as a novel and efficient catalyst. *Polyhedron* **2019**, *170*, 364–372. [CrossRef]

94. Li, Y.Q.; Gao, Q.; Zhang, L.J.; Zhou, Y.S.; Zhong, Y.X.; Ying, Y.; Zhang, M.C.; Huang, C.Q.; Wang, Y.A. $H_5PV_2Mo_{10}O_{40}$ encapsulated in MIL-101(Cr): Facile synthesis and characterization of rationally designed composite materials for efficient decontamination of sulfur mustardt. *Dalton Trans.* **2018**, *47*, 6394–6403. [CrossRef]

95. Malkar, R.S.; Daly, H.; Hardacre, C.; Yadav, G.D. Aldol condensation of 5-hydroxymethylfurfural to fuel precursor over novel aluminum exchanged-DTP@ZIF-8. *ACS Sustain. Chem. Eng.* **2019**, *7*, 16215–16224. [CrossRef]

96. Saikia, M.; Bhuyan, D.; Saikia, L. Keggin type phosphotungstic acid encapsulated chromium (III) terephthalate metal organic framework as active catalyst for Biginelli condensation. *Appl. Catal. A Gen.* **2015**, *505*, 501–506. [CrossRef]

97. Deng, Q.; Nie, G.; Pan, L.; Zou, J.-J.; Zhang, X.; Wang, L. Highly selective self-condensation of cyclic ketones using MOF-encapsulating phosphotungstic acid for renewable high-density fuel. *Green Chem.* **2015**, *17*, 4473–4481. [CrossRef]

98. Bromberg, L.; Diao, Y.; Wu, H.; Speakman, S.A.; Hatton, T.A. Chromium(III) terephthalate metal organic framework (MIL-101): HF-free synthesis, structure, polyoxometalate composites, and catalytic properties. *Chem. Mater.* **2012**, *24*, 1664–1675. [CrossRef]

99. Bromberg, L.; Hatton, T.A. Aldehyde-alcohol reactions catalyzed under mild conditions by chromium(III) terephthalate metal organic framework (MIL-101) and phosphotungstic acid composites. *ACS Appl. Mater. Interfaces* **2011**, *3*, 4756–4764. [CrossRef]

100. Chen, M.; Yan, J.Q.; Tan, Y.; Li, Y.F.; Wu, Z.M.; Pan, L.S.; Liu, Y.J. Hydroxyalkylation of phenol with formaldehyde to bisphenol F catalyzed by Keggin phosphotungstic acid encapsulated in metal-organic frameworks MIL-100(Fe or Cr) and MIL-101(Fe or Cr). *Ind. Eng. Chem. Res.* **2015**, *54*, 11804–11813. [CrossRef]

101. Nobakht, N.; Faramarzi, M.A.; Shafiee, A.; Khoobi, M.; Rafiee, E. Polyoxometalate-metal organic framework-lipase: An efficient green catalyst for synthesis of benzyl cinnamate by enzymatic esterification of cinnamic acid. *Int. J. Biol. Macromol.* **2018**, *113*, 8–19. [CrossRef] [PubMed]

102. Wee, L.H.; Janssens, N.; Bajpe, S.R.; Kirschhock, C.E.A.; Martens, J.A. Heteropolyacid encapsulated in Cu$_3$(BTC)$_2$ nanocrystals: An effective esterification catalyst. *Catal. Today* **2011**, *171*, 275–280. [CrossRef]

103. Guo, T.M.; Qiu, M.; Qi, X.H. Selective conversion of biomass-derived levulinic acid to ethyl levulinate catalyzed by metal organic framework (MOF)-supported polyoxometalates. *Appl. Catal. A: Gen.* **2019**, *572*, 168–175. [CrossRef]

104. Zhang, Q.Y.; Yang, T.T.; Liu, X.F.; Yue, C.Y.; Ao, L.F.; Deng, T.L.; Zhang, Y.T. Heteropoly acid-encapsulated metal-organic framework as a stable and highly efficient nanocatalyst for esterification reaction. *RSC Adv.* **2019**, *9*, 16357–16365. [CrossRef]

105. Marandi, A.; Bahadori, M.; Tangestaninejad, S.; Moghadam, M.; Mirkhani, V.; Mohammadpoor-Baltork, I.; Frohnhoven, R.; Mathur, S.; Sandleben, A.; Klein, A. Cycloaddition of CO$_2$ with epoxides and esterification reactions using the porous redox catalyst Co-POM@MIL-101(Cr). *New J. Chem.* **2019**, *43*, 15585–15595. [CrossRef]

106. Zhu, J.; Wang, P.-C.; Lu, M. Study on the one-pot oxidative esterification of glycerol with MOF supported polyoxometalates as catalyst. *Catal. Sci. Technol.* **2015**, *5*, 3383–3393. [CrossRef]

107. Zhang, Q.Y.; Liu, X.F.; Yang, T.T.; Yue, C.Y.; Pu, Q.L.; Zhang, Y.T. Facile synthesis of polyoxometalates tethered to post Fe-BTC frameworks for esterification of free fatty acids to biodiesel. *RSC Adv.* **2019**, *9*, 8113–8120. [CrossRef]

108. Zhang, F.M.; Jin, Y.; Shi, J.; Zhong, Y.J.; Zhu, W.D.; El-Shall, M.S. Polyoxometalates confined in the mesoporous cages of metal-organic framework MIL-100(Fe): Efficient heterogeneous catalysts for esterification and acetalization reactions. *Chem. Eng. J.* **2015**, *269*, 236–244. [CrossRef]

109. Janssens, N.; Wee, L.H.; Bajpe, S.; Breynaert, E.; Kirschhock, C.E.A.; Martens, J.A. Recovery and reuse of heteropolyacid catalyst in liquid reaction medium through reversible encapsulation in Cu$_3$(BTC)$_2$ metal-organic framework. *Chem. Sci.* **2012**, *3*, 1847–1850. [CrossRef]

110. Zang, Y.D.; Shi, J.; Zhao, X.M.; Kong, L.C.; Zhang, F.M.; Zhong, Y.J. Highly stable chromium(III) terephthalate metal organic framework (MIL-101) encapsulated 12-tungstophosphoric heteropolyacid as a water-tolerant solid catalyst for hydrolysis and esterification. *React. Kinet. Mech. Cat.* **2013**, *109*, 77–89. [CrossRef]

111. Nikseresht, A.; Daniyali, A.; Ali-Mohammadi, M.; Afzalinia, A.; Mirzaie, A. Ultrasound-assisted biodiesel production by a novel composite of Fe(III)-based MOF and phosphotangestic acid as efficient and reusable catalyst. *Ultrason. Sonochem.* **2017**, *37*, 203–207. [CrossRef] [PubMed]

112. Wan, H.; Chen, C.; Wu, Z.W.; Que, Y.G.; Feng, Y.; Wang, W.; Wang, L.; Guan, G.F.; Liu, X.Q. Encapsulation of heteropolyanion-based ionic liquid within the metal-organic framework MIL-100(Fe) for biodiesel production. *ChemCatChem* **2015**, *7*, 441–449. [CrossRef]

113. Liu, Y.W.; Liu, S.M.; He, D.F.; Li, N.; Ji, Y.J.; Zheng, Z.P.; Luo, F.; Liu, S.X.; Shi, Z.; Hu, C.W. Crystal facets make a profound difference in polyoxometalate-containing metal-organic frameworks as catalysts for biodiesel production. *J. Am. Chem. Soc.* **2015**, *137*, 12697–12703. [CrossRef] [PubMed]

114. Xie, W.; Wan, F. Immobilization of polyoxometalate-based sulfonated ionic liquids on UiO-66-2COOH metal-organic frameworks for biodiesel production via one-pot transesterification-esterification of acidic vegetable oils. *Chem. Eng. J.* **2019**, *365*, 40–50. [CrossRef]

115. Chen, L.N.; Zhang, X.B.; Zhou, J.H.; Xie, Z.X.; Kuang, Q.; Zheng, L.S. A nano-reactor based on PtNi@metal-organic framework composites loaded with polyoxometalates for hydrogenation-esterification tandem reactions. *Nanoscale* **2019**, *11*, 3292–3299. [CrossRef]

116. Zhang, Y.M.; Degirmenci, V.; Li, C.; Hensen, E.J.M. Phosphotungstic acid encapsulated in metal-organic framework as catalysts for carbohydrate dehydration to 5-Hydroxymethylfurfural. *ChemSusChem* **2011**, *4*, 59–64. [CrossRef]

117. Wang, Z.H.; Chen, Q.W. Conversion of 5-hydroxymethylfurfural into 5-ethoxymethylfurfural and ethyl levulinate catalyzed by MOF-based heteropolyacid materials. *Green Chem.* **2016**, *18*, 5884–5889. [CrossRef]

118. Xie, W.L.; Wan, F. Biodiesel production from acidic oils using polyoxometalate-based sulfonated ionic liquids functionalized metal-organic frameworks. *Catal. Lett.* **2019**, *149*, 2916–2929. [CrossRef]

119. Cai, X.X.; Xu, Q.H.; Tu, G.M.; Fu, Y.H.; Zhang, F.M.; Zhu, W.D. Synergistic catalysis of ruthenium nanoparticles and polyoxometalate integrated within single UiO-66 microcrystals for boosting the efficiency of methyl levulinate to gamma-valerolactone. *Front. Chem.* **2019**, *7*, 24–34. [CrossRef]

120. Chen, J.; Wang, S.; Huang, J.; Chen, L.; Ma, L.; Huang, X. Conversion of cellulose and cellobiose into sorbitol catalyzed by ruthenium supported on a polyoxometalate/metal-organic framework hybrid. *ChemSusChem* **2013**, *6*, 2213. [CrossRef]

121. Wee, L.H.; Wiktor, C.; Turner, S.; Vanderlinden, W.; Janssens, N.; Bajpe, S.R.; Houthoofd, K.; Van Tendeloo, G.; De Feyter, S.; Kirschhock, C.E.A.; et al. Copper benzene tricarboxylate metal-organic framework with wide permanent mesopores stabilized by Keggin polyoxometallate ions. *J. Am. Chem. Soc.* **2012**, *134*, 10911–10919. [CrossRef] [PubMed]

122. Wee, L.H.; Bonino, F.; Lamberti, C.; Bordiga, S.; Martens, J.A. Cr-MIL-101 encapsulated Keggin phosphotungstic acid as active nanomaterial for catalysing the alcoholysis of styrene oxide. *Green Chem.* **2014**, *16*, 1351–1357. [CrossRef]

123. Ma, F.J.; Liu, S.X.; Sun, C.Y.; Liang, D.D.; Ren, G.J.; Wei, F.; Chen, Y.G.; Su, Z.M. A sodalite-type porous metal-organic framework with polyoxometalate templates: Adsorption and decomposition of dimethyl methylphosphonate. *J. Am. Chem. Soc.* **2011**, *133*, 4178–4181. [CrossRef] [PubMed]

124. Ma, F.J.; Liu, S.X.; Ren, G.J.; Liang, D.D.; Sha, S. A hybrid compound based on porous metal-organic frameworks and polyoxometalates: NO adsorption and decomposition. *Inorg. Chem. Commun.* **2012**, *22*, 174–177. [CrossRef]

125. Liang, D.D.; Liu, S.X.; Ma, F.J.; Wei, F.; Chen, Y.G. A crystalline catalyst based on a porous metal-organic framework and 12-tungstosilicic acid: Particle size control by hydrothermal synthesis for the formation of dimethyl ether. *Adv. Synth. Catal.* **2011**, *353*, 733–742. [CrossRef]

126. Micek-Ilnicka, A.; Gil, B. Heteropolyacid encapsulation into the MOF: Influence of acid particles distribution on ethanol conversion in hybrid nanomaterials. *Dalton Trans.* **2012**, *41*, 12624–12629. [CrossRef]

127. Juliao, D.; Barbosa, A.D.S.; Peixoto, A.F.; Freire, C.; De Castro, B.; Balula, S.S.; Cunha-Silva, L. Improved catalytic performance of porous metal-organic frameworks for the ring opening of styrene oxide. *CrystEngComm* **2017**, *19*, 4219–4226. [CrossRef]

128. Shah, W.A.; Noureen, L.; Nadeem, M.A.; Kogerler, P. Encapsulation of Keggin-type manganese-polyoxomolybdates in MIL-100 (Fe) for efficient reduction of p-nitrophenol. *J. Solid State Chem.* **2018**, *268*, 75–82. [CrossRef]

129. Ullah, L.; Zhao, G.Y.; Xu, Z.C.; He, H.Y.; Usman, M.; Zhang, S.J. 12-Tungstophosphoric acid niched in Zr-based metal-organic framework: A stable and efficient catalyst for Friedel-Crafts acylation. *Sci. China Chem.* **2018**, *61*, 402–411. [CrossRef]

130. Khder, A.S.; Hassan, H.M.A.; El-Shall, M.S. Metal-organic frameworks with high tungstophosphoric acid loading as heterogeneous acid catalysts. *Appl. Catal. A Gen.* **2014**, *487*, 110–118. [CrossRef]

131. Xie, W.L.; Yang, X.L.; Hu, P.T. $Cs_{2.5}H_{0.5}PW_{12}O_{40}$ Encapsulated in metal-organic framework UiO-66 as heterogeneous catalysts for acidolysis of soybean oil. *Catal. Lett.* **2017**, *147*, 2772–2782. [CrossRef]

132. Lu, B.B.; Yang, J.; Che, G.B.; Pei, W.Y.; Ma, J.F. Highly stable copper(I)-based metal-organic framework assembled with resorcin[4]arene and polyoxometalate for efficient heterogeneous catalysis of azide-alkyne "Click" reaction. *ACS Appl. Mater. Interfaces* **2018**, *10*, 2628–2636. [CrossRef] [PubMed]

133. Ahn, S.; Nauert, S.L.; Buru, C.T.; Rimoldi, M.; Choi, H.; Schweitzer, N.M.; Hupp, J.T.; Farha, O.K.; Notestein, J.M. Pushing the limits on metal-organic frameworks as a catalyst support: NU-1000 supported tungsten catalysts for o-xylene isomerization and disproportionation. *J. Am. Chem. Soc.* **2018**, *140*, 8535–8543. [CrossRef] [PubMed]

134. Sartipi, S.; Romero, M.J.V.; Rozhko, E.; Que, Z.Y.; Stil, H.A.; De With, J.; Kapteijn, F.; Gascon, J. Dynamic release-immobilization of a homogeneous rhodium hydroformylation catalyst by a polyoxometalate metal-organic framework composite. *ChemCatChem* **2015**, *7*, 3243–3247. [CrossRef]

135. Taherimehr, M.; Van de Voorde, B.; Wee, L.H.; Martens, J.A.; De Vos, D.E.; Pescarmona, P.P. Strategies for enhancing the catalytic performance of metal-organic frameworks in the fixation of CO_2 into cyclic carbonates. *ChemSusChem* **2017**, *10*, 1283–1291. [CrossRef]

136. Han, J.Y.; Wang, D.P.; Du, Y.H.; Xi, S.B.; Chen, Z.; Yin, S.M.; Zhou, T.H.; Xu, R. Polyoxometalate immobilized in MIL-101(Cr) as an efficient catalyst for water oxidation. *Appl. Catal. A Gen.* **2016**, *521*, 83–89. [CrossRef]

137. Mukhopadhyay, S.; Debgupta, J.; Singh, C.; Kar, A.; Das, S.K. A Keggin polyoxometalate shows water oxidation activity at neutral pH: POM@ZIF-8, an efficient and robust electrocatalyst. *Angew. Chem. Int. Ed.* **2018**, *57*, 1918–1923. [CrossRef]

138. Li, Q.Y.; Zhang, L.; Xu, Y.X.; Li, Q.; Xue, H.G.; Pang, H. Smart yolk/shell ZIF-67@POM hybrids as efficient electrocatalysts for the oxygen evolution reaction. *ACS Sustain. Chem. Eng.* **2019**, *7*, 5027–5033. [CrossRef]

139. Mukhopadhyay, S.; Basu, O.; Kar, A.; Das, S.K. Efficient electrocatalytic water oxidation by Fe(salen)-MOF composite: Effect of modified microenvironment. *Inorg. Chem.* **2020**, *59*, 472–483. [CrossRef]

140. Li, S.B.; Zhang, L.; Lan, Y.Q.; O'Halloran, K.P.; Ma, H.Y.; Pang, H.J. Polyoxometalate-encapsulated twenty-nuclear silver-tetrazole nanocage frameworks as highly active electrocatalysts for the hydrogen evolution reaction. *Chem. Commun.* **2018**, *54*, 1964–1967. [CrossRef]

141. Zhang, L.; Li, S.B.B.; Gomez-Garcia, C.J.; Ma, H.Y.; Zhang, C.J.; Pang, H.J.; Li, B.N. Two novel polyoxometalate-encapsulated metal-organic nanotube frameworks as stable and highly efficient electrocatalysts for hydrogen evolution reaction. *ACS Appl. Mater. Interfaces* **2018**, *10*, 31498–31504. [CrossRef] [PubMed]

142. Fernandes, D.M.; Barbosa, A.D.S.; Pires, J.; Balula, S.S.; Cunha-Silva, L.; Freire, C. Novel composite material polyoxovanadate@MIL-101(Cr): A highly efficient electrocatalyst for ascorbic acid oxidation. *ACS Appl. Mater. Interfaces* **2013**, *5*, 13382–13390. [CrossRef] [PubMed]

143. Yan, J.; Zhou, W.Z.; Tan, H.Q.; Feng, X.J.; Wang, Y.H.; Li, Y.G. Ultrafine Ag/polyoxometalate-doped AgCl nanoparticles in metal-organic framework as efficient photocatalysts under visible light. *CrystEngComm* **2016**, *18*, 8762–8768. [CrossRef]

144. Chen, D.M.; Liu, X.H.; Zhang, N.N.; Liu, C.S.; Du, M. Immobilization of polyoxometalate in a cage-based metal-organic framework towards enhanced stability and highly effective dye degradation. *Polyhedron* **2018**, *152*, 108–113. [CrossRef]

145. Paille, G.; Gomez-Mingot, M.; Roch-Marchal, C.; Lassalle-Kaiser, B.; Mialane, P.; Fontecave, M.; Mellot-Draznieks, C.; Dolbecq, A. A fully noble metal-free photosystem based on cobalt-polyoxometalates immobilized in a porphyrinic metal-organic framework for water oxidation. *J. Am. Chem. Soc.* **2018**, *140*, 3613–3618. [CrossRef]

146. Paille, G.; Gomez-Mingot, M.; Roch-Marchal, C.; Haouas, M.; Benseghir, Y.; Pino, T.; Ha-Thi, M.H.; Landrot, G.; Mialane, P.; Fontecave, M.; et al. Thin films of fully noble metal-free POM@MOF for photocatalytic water oxidation. *ACS Appl. Mater. Interfaces* **2019**, *11*, 47837–47845. [CrossRef]

147. Shah, W.A.; Waseem, A.; Nadeem, M.A.; Kogerler, P. Leaching-free encapsulation of cobalt-polyoxotungstates in MIL-100 (Fe) for highly reproducible photocatalytic water oxidation. *Appl. Catal. A Gen.* **2018**, *567*, 132–138. [CrossRef]

148. Liang, R.W.; Chen, R.; Jing, F.F.; Qin, N.; Wu, L. Multifunctional polyoxometalates encapsulated in MIL-100(Fe): Highly efficient photocatalysts for selective transformation under visible light. *Dalton Trans.* **2015**, *44*, 18227–18236. [CrossRef]

149. He, J.C.; Han, Q.X.; Li, J.; Shi, Z.L.; Shi, X.Y.; Niu, J.Y. Ternary supramolecular system for photocatalytic oxidation with air by consecutive photo-induced electron transfer processes. *J. Catal.* **2019**, *376*, 161–167. [CrossRef]

150. Tian, P.; He, X.; Li, W.; Zhao, L.; Fang, W.; Chen, H.; Zhang, F.; Zhang, W.; Wang, W. Zr-MOFs based on Keggin-type polyoxometalates for photocatalytic hydrogen production. *J. Mater. Sci.* **2018**, *53*, 12016–12029. [CrossRef]

151. Guo, W.; Lv, H.; Chen, Z.; Sullivan, K.P.; Lauinger, S.M.; Chi, Y.; Sumliner, J.M.; Lian, T.; Hill, C.L. Self-assembly of polyoxometalates, Pt nanoparticles and metal–organic frameworks into a hybrid material for synergistic hydrogen evolution. *J. Mater. Chem. A* **2016**, *4*, 5952–5957. [CrossRef]

152. Zhang, Z.M.; Zhang, T.; Wang, C.; Lin, Z.; Long, L.S.; Lin, W. Photosensitizing metal-organic framework enabling visible-light-driven proton reduction by a Wells-Dawson-type polyoxometalate. *J. Am. Chem. Soc.* **2015**, *137*, 3197–3200. [CrossRef] [PubMed]

153. Li, H.; Yao, S.; Wu, H.L.; Qu, J.Y.; Zhang, Z.M.; Lu, T.B.; Lin, W.B.; Wang, E.B. Charge-regulated sequential adsorption of anionic catalysts and cationic photosensitizers into metal-organic frameworks enhances photocatalytic proton reduction. *Appl. Catal. B Environ.* **2018**, *224*, 46–52. [CrossRef]

154. Tian, J.; Xu, Z.Y.; Zhang, D.W.; Wang, H.; Xie, S.H.; Xu, D.W.; Ren, Y.H.; Wang, H.; Liu, Y.; Li, Z.T. Supramolecular metal-organic frameworks that display high homogeneous and heterogeneous photocatalytic activity for H$_2$ production. *Nat. Commun.* **2016**, *7*, 11580–115808. [CrossRef] [PubMed]

155. Bu, Y.H.; Li, F.Y.; Zhang, Y.Z.; Liu, R.; Luo, X.Z.; Xu, L. Immobilizing CdS nanoparticles and MoS$_2$/RGO on Zr-based metal-organic framework 12-tungstosilicate@UiO-67 toward enhanced photocatalytic H$_2$ evolution. *RSC Adv.* **2016**, *6*, 40560–40566. [CrossRef]

156. Borges, M.E.; Diaz, L. Recent developments on heterogeneous catalysts for biodiesel production by oil esterification and transesterification reactions: A review. *Renew. Sust. Energ. Rev.* **2012**, *16*, 2839–2849. [CrossRef]

157. Sadakane, M.; Steckhan, E. Electrochemical properties of polyoxometalates as electrocatalysts. *Chem. Rev.* **1998**, *98*, 219–237. [CrossRef]

158. Gao, D.D.; Trentin, I.; Schwiedrzik, L.; Gonzalez, L.; Streb, C. The reactivity and stability of polyoxometalate water oxidation electrocatalysts. *Molecules* **2020**, *25*, 157. [CrossRef]

159. Kalimuthu, V.S.; Attias, R.; Tsur, Y. Electrochemical impedance spectra of RuO$_2$ during oxygen evolution reaction studied by the distribution function of relaxation times. *Electrochem. Commun.* **2020**, *110*, 106641. [CrossRef]

160. Rausch, B.; Symes, M.D.; Chisholm, G.; Cronin, L. Decoupled catalytic hydrogen evolution from a molecular metal oxide redox mediator in water splitting. *Science* **2014**, *345*, 1326–1330. [CrossRef]

161. Qin, J.S.; Du, D.Y.; Guan, W.; Bo, X.J.; Li, Y.F.; Guo, L.P.; Su, Z.M.; Wang, Y.Y.; Lan, Y.Q.; Zhou, H.C. Ultrastable polymolybdate-based metal organic frameworks as highly active electrocatalysts for hydrogen generation from water. *J. Am. Chem. Soc.* **2015**, *137*, 7169–7177. [CrossRef] [PubMed]

162. Huang, Z.Q.; Luo, Z.; Geletii, Y.V.; Vickers, J.W.; Yin, Q.S.; Wu, D.; Hou, Y.; Ding, Y.; Song, J.; Musaev, D.G.; et al. Efficient light-driven carbon-free cobalt-based molecular catalyst for water oxidation. *J. Am. Chem. Soc.* **2011**, *133*, 2068–2071. [CrossRef] [PubMed]

 *catalysts*

Review

Application of Various Metal-Organic Frameworks (MOFs) as Catalysts for Air and Water Pollution Environmental Remediation

Sanha Jang [1], Sehwan Song [2], Ji Hwan Lim [3], Han Seong Kim [3], Bach Thang Phan [4], Ki-Tae Ha [5], Sungkyun Park [2,*] and Kang Hyun Park [1,*]

[1] Department of Chemistry, Pusan National University, Busandaehak-ro 63beon-gil, Geumjeong-gu, Busan 46241, Korea; jangs0522@naver.com
[2] Department of Physics, Pusan National University, Busandaehak-ro 63beon-gil, Geumjeong-gu, Busan 46241, Korea; sehwan465@gmail.com
[3] Department of Organic Material Science and Engineering, Pusan National University, Geumjeong-gu, Busan 46241, Korea; wdwf455@naver.com (J.H.L.); hanseongkim@pusan.ac.kr (H.S.K.)
[4] Center for Innovative Materials and Architectures (INOMAR), Vietnam National University, Ho Chi Minh City 721337, Vietnam; pbthang@inomar.edu.vn
[5] Department of Korean Medical Science, Healthy Aging Korea Medical Research Center, Pusan National University, Yangsan 50612, Gyeongsangnam-do, Korea; hagis@pusan.ac.kr
*   Correspondence: psk@pusan.ac.kr (S.P.); chemistry@pusan.ac.kr (K.H.P.); Tel.: +82-51-510-2238 (K.H.P.)

Received: 11 December 2019; Accepted: 5 February 2020; Published: 6 February 2020

Abstract: The use of metal-organic frameworks (MOFs) to solve problems, like environmental pollution, disease, and toxicity, has received more attention and led to the rapid development of nanotechnology. In this review, we discuss the basis of the metal-organic framework as well as its application by suggesting an alternative of the present problem as catalysts. In the case of filtration, we have developed a method for preparing the membrane by electrospinning while using an eco-friendly polymer. The MOFs were usable in the environmental part of catalytic activity and may provide a great material as a catalyst to other areas in the near future.

Keywords: metal organic framework; environmental pollution; filter; gas sorption; sensor; hydrogen storage; electrospinning

---

1. Introduction

Recently, environmental pollution is increasing due to toxic waste and hazardous organic compounds [1]. The amount of poisonous compounds that are released into the environment is a serious problem for human life. In the pragmatic situation, air and organic pollutants are highly involved they can be commonly expressed as particulates, acidic substances, gases, or mixtures [2,3].

Nitroaromatic compounds are found in soil, air, and water samples due to wastewater sources from the plastic, pesticide, pharmaceutical, and dye industry [4]. In addition, the development of hydrogen and methane storage systems is necessary for the widespread use of green energy. Moreover, the separation and selective gas adsorption of poisonous gases, such as nitrogen dioxide and ammonia gas, are significant in the field of air pollution [5–9].

It is essential that sensors should be developed as the way to prevent toxic materials, one of the leading causes of environment pollution, from being released into the environment. Sulfur dioxide and nitroaromatic compounds are considered to be poisonous and harmful to human life [10]. Our previous work has reported a PdAg nanoparticle infused metal-organic framework (MOF) used for the detection of 4-nitrophenol while using an electrochemical sensor [11]. In addition, Salama et al., reported an SO_2 gas sensor while using an MOF [12].

Metal-organic frameworks (MOFs) have a wide range of interesting properties such as high specific surface area and facile modification [13–26]. MOFs are microporous materials that form three-dimensional (3D) crystalline networks, which are prepared by combining various metal ions with organic linkers in an appropriate solvent [27–32]. Over the past years, MOFs have been used as catalysts, absorbents, and filters. MOFs have many advantages due to their modifiable properties, such as high specific surface area and porosity [30,33–36]. The exceptional characteristics of MOFs have led to their possible application in a wide range of technological areas, including gas sorption, separation, storage [5–9], sensing [10–12,37], and heterogeneous catalysis [38–53].

In this review, we focus on various MOFs and their applications in different fields, such as: (1) controlled gas uptake of toxic gases, including ammonia, nitrogen dioxide, and sulfur dioxide [6,7,12]; (2) sensors using PdAg nanoparticle infused MOFs and Cd-MOFs [11,54]; (3) Hydrogen gas storage [55–57]; and, (4) filtration for air and water pollution control while using nanofibrous MOFs [58,59] (Scheme 1).

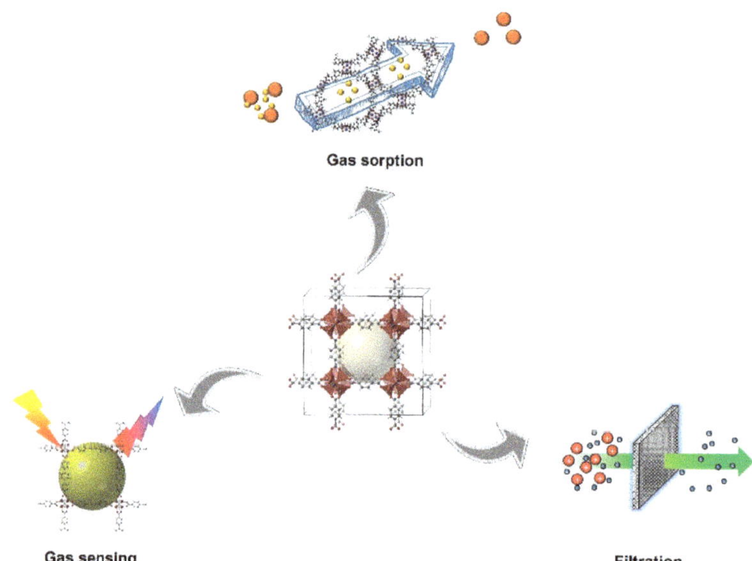

Scheme 1. Various applications by metal-organic frameworks (MOFs).

2. Basic of Stable MOFs

2.1. Characterization of MOFs

MOFs are an arising group of porous materials that were synthesized from metal ions and organic linkers [60,61]. The ever-expanding improvement in the performance of MOFs and facile control over their properties, MOFs have attracted the interest of engineers and scientists [62–67]. Earlier MOFs were synthesized from divalent metals, which showed superior properties and a diverse range of applications [60], such as gas sorption, separation, storage [5–9], sensing [10–12,37], and heterogeneous catalysis [38–53]. In particular, stable MOFs can be predicted by the strength of the metal-organic linker bond that formed in their framework. We have summarized some of the typical stable MOFs prepared from divalent, trivalent, and tetravalent metal ions in Table 1 [68].

Table 1. Abridgment of some stable MOFs [68].

MOFs [a]	Clusters/Cores	Linkers [b]	BET Surface Area ($m^2\ g^{-1}$)	Ref.
MIL-53 (Al)	$[Al(OH)(COO)_2]_n$	BDC	1181	[69]
Al-FUM	$[Al(OH)(COO)_2]_n$	FUM	1080	[70,71]
MIL-69	$[Al(OH)(COO)_2]_n$	2,6-NDC	NA	[72]
MIL-96 (Al)	$[Al_3(\mu_3\text{-}O)(COO)_6]$ $[Al(OH)(COO)_2]_n$	BTC	NA	[73]
MIL-100 (Al)	$[Al_3(\mu_3\text{-}O)(COO)_6]$	BTC	2152	[74]
MIL-101 (Al)	$[Al_3(\mu_3\text{-}O)(COO)_6]$	BDC-NH$_2$	2100	[75]
MIL-110	$[Al_8(OH)_{15}(COO)_9]$	BTC	1400	[76]
MIL-118	$[Al(OH)(COO)_2(COOH)_2]_n$	BTEC	NA	[77]
MIL-120	$[Al(OH)(COO)_2]_n$	BTEC	308	[78]
MIL-121	$[Al(OH)(COO)_2]_n$	BTEC	162	[79]
MIL-122	$[Al(OH)(COO)_2]_n$	NTC	NA	[80]
DUT-5	$[Al(OH)(COO)_2]_n$	BPDC	1613	[81]
NOTT-300	$[Al(OH)(COO)_2]_n$	BPTA	1370	[82]
CAU-1	$[Al_8(OH)_4(OCH_3)_8(COO)_{12}]$	BDC-NH$_2$	1700 c)	[83]
CAU-3-BDC	$[Al_{12}(OCH_3)_{24}(COO)_{12}]$	BDC	1550	[84]
CAU-3-BDC-NH$_2$	$[Al_{12}(OCH_3)_{24}(COO)_{12}]$	BDC-NH$_2$	1250	[84]
CAU-3-NDC	$[Al_{12}(OCH_3)_{24}(COO)_{12}]$	2,6-NDC	2320	[84]
CAU-4	$[Al(OH)(COO)_2]_n$	BTB	1520	[85]
CAU-8	$[Al(OH)(COO)_2]_n$	BeDC	600	[86]
CAU-10	$[Al(OH)(COO)_2]_n$	1,3-BDC	635	[87]
467-MOF	$[Al(OH)(COO)_2]_n$	BTTB	725	[88]
Al-PMOF	$[Al(OH)(COO)_2]_n$	TCPP	1400	[89]
PCN-333 (Al)	$[Al_3(\mu_3\text{-}O)(COO)_6]$	TATB	4000	[90]
PCN-888 (Al)	$[Al_3(\mu_3\text{-}O)(COO)_6]$	HTB	3700	[91]
Al-soc-MOF-1	$[Al_3(\mu_3\text{-}O)(COO)_6]$	TCPT	5585	[92]
MIL-53 (Cr)	$[Cr(OH)(COO)_2]_n$	BDC	NA	[93]
MIL-88A (Cr)	$[Cr_3(\mu_3\text{-}O)(COO)_6]$	FUM	NA	[94]
MIL-88B (Cr)	$[Cr_3(\mu_3\text{-}O)(COO)_6]$	BDC	NA	[94]
MIL-88C (Cr)	$[Cr_3(\mu_3\text{-}O)(COO)_6]$	2,6-NDC	NA	[94]
MIL-88D (Cr)	$[Cr_3(\mu_3\text{-}O)(COO)_6]$	BPDC	NA	[94]
MIL-96 (Cr)	$[Cr_3(\mu_3\text{-}O)(COO)_6]$ $[Cr(OH)(COO)_2]_n$	BTC	NA	[95]
MIL-100 (Cr)	$[Cr_3(\mu_3\text{-}O)(COO)_6]$	BTC	3100 c)	[96]
MIL-101 (Cr)	$[Cr_3(\mu_3\text{-}O)(COO)_6]$	BDC	4100	[63]
MIL-101-NDC (Cr)	$[Cr_3(\mu_3\text{-}O)(COO)_6]$	2,6-NDC	2100	[97]
PCN-333 (Cr)	$[Cr_3(\mu_3\text{-}O)(COO)_6]$	TATB	2548	[98]
PCN-426 (Cr)	$[Cr_3(\mu_3\text{-}O)(COO)_6]$	TMQPTC	3155	[99]
MIL-53 (Fe)	$[Fe(OH)(COO)_2]_n$	BDC	NA	[100]
MIL-68 (Fe)	$[Fe(OH)(COO)_2]_n$	BDC	665	[101]
MIL-141 (Fe)	$[Fe(OH)(COO)_2]_n$	TCPP	420	[102]
FepzTCPP (FeOH)2	$[Fe(OH)(COO)_2]_n$	Pyrazine, TCPP	760	[102]
MIL-88A (Fe)	$[Fe_3(\mu_3\text{-}O)(COO)_6]$	FUM	NA	[94]
MIL-88B (Fe)	$[Fe_3(\mu_3\text{-}O)(COO)_6]$	BDC	NA	[94]
MIL-88C (Fe)	$[Fe_3(\mu_3\text{-}O)(COO)_6]$	2,6-NDC	NA	[94]
MIL-88D (Fe)	$[Fe_3(\mu_3\text{-}O)(COO)_6]$	BPDC	NA	[94]
MIL-100 (Fe)	$[Fe_3(\mu_3\text{-}O)(COO)_6]$	BTC	2800 c)	[103]
MIL-101 (Fe)	$[Fe_3(\mu_3\text{-}O)(COO)_6]$	BDC	2823	[104]
PCN-250 (Fe)	$[Fe_3(\mu_3\text{-}O)(COO)6]$	ABDC	1486	[105]
PCN-250 (Fe$_2$Co)	$[Fe_2Co(\mu_3\text{-}O)(COO)_6]$	ABDC	1400	[105]
PCN-333 (Fe)	$[Fe_3(\mu_3\text{-}O)(COO)_6]$	TATB	2427	[90]
PCN-600 (Fe)	$[Fe_3(\mu3\text{-}O)(COO)_6]$	TCPP	2270	[106]
Tb$_2$(BDC)$_3$	$[Tb(H_2O)_2(COO)_3]n$	BDC	NA	[107]
MIL-63	$[Eu_2(\mu_3\text{-}OH)_7(COO)]n$	BTC	15	[108]
MIL-83	$[Eu(\mu_3\text{-}O)_3(COO)_3(COOH)_3]n$	1,3-ADC	NA	[109]
MIL-103	$[Tb(H_2O)(COO)_4]n$	BTB	930	[110]

Table 1. *Cont.*

MOFs [a]	Clusters/Cores	Linkers [b]	BET Surface Area (m^2 g^{-1})	Ref.
Y-BTC	$[Y(H_2O)(COO)_3]n$	BTC	1080	[111]
Tb-BTC	$[Tb(H_2O)(COO)_3]n$	BTC	786	[111]
Y-FTZB	$[Y6(\mu_3\text{-}OH)_8(COO)_6(CN_4)_6]$	FTZB	1310	[112]
Tb-FTZB	$[Tb6(\mu_3\text{-}OH)_8(COO)_6(CN_4)_6]$	FTZB	1220	[112]
Y-FUM	$[Y6(\mu_3\text{-}OH)_8(COO)_{12}]$	FUM	691	[113]
Tb-FUM	$[Tb6(\mu3\text{-}OH)8(COO)_{12}]$	FUM	503	[113]
Ce-UiO-66	$[Ce_6(\mu_3\text{-}O)_4(\mu_3\text{-}OH)_4(COO)_{12}]$	BDC	1282	[114]
Ce-UiO-66-(CH3)$_2$	$[Ce6(\mu_3\text{-}O)_4(\mu_3\text{-}OH)_4(COO)_{12}]$	BDC-$(CH_3)_2$	845	[115]
MIL-125	$[Ti_8O_8(OH)_4(COO)_{12}]$	BDC	1550	[116]
PCN-22	$[Ti_7O_6(COO)_{12}]$	TCPP	1284	[117]
COK-69	$[Ti_3O_3(COO)_6]$	CDC	NA	[118]
MOF-901	$[Ti_6O_6(OMe)_6(COO)_6]$	AB, BDA	550	[119]
MOF-902	$[Ti_6O_6(OMe)_6(COO)_6]$	AB, BPDA	400	[120]
UiO-66	$[Zr_6(\mu_3\text{-}O)_4(\mu_3\text{-}OH)_4(COO)_{12}]$	BDC	1187	[121]
UiO-67	$[Zr_6(\mu_3\text{-}O)_4(\mu_3\text{-}OH)_4(COO)_{12}]$	BPDC	3000	[121]
UiO-68	$[Zr_6(\mu_3\text{-}O)_4(\mu_3\text{-}OH)_4(COO)_{12}]$	TPDC	4170	[121]
PCN-94	$[Zr_6(\mu_3\text{-}O)_4(\mu_3\text{-}OH)_4(COO)_{12}]$	ETTC	3377	[122]
PCN-222	$[Zr_6(\mu_3\text{-}O)_4(\mu_3\text{-}OH)_4(OH)_4(H_2O)_4(COO)_8]$	TCPP	2223	[123]
PCN-223	$[Zr_6(\mu_3\text{-}O)_4(\mu_3\text{-}OH)_4(COO)_{12}]$	TCPP	1600	[124]
PCN-224	$[Zr_6(\mu_3\text{-}O)_4(\mu_3\text{-}OH)_4(OH)_6(H_2O)_6(COO)_6]$	TCPP	2600	[125]
PCN-225	$[Zr_6(\mu_3\text{-}O)_4(\mu_3\text{-}OH)_4(OH)_4(H_2O)_4(COO)_8]$	TCPP	1902	[126]
PCN-228	$[Zr_6(\mu_3\text{-}O)_4(\mu_3\text{-}OH)_4(COO)_{12}]$	TCP-1	4510	[127]
PCN-229	$[Zr_6(\mu_3\text{-}O)_4(\mu_3\text{-}OH)_4(COO)_{12}]$	TCP-2	4619	[127]
PCN-230	$[Zr_6(\mu_3\text{-}O)_4(\mu_3\text{-}OH)_4(COO)_{12}]$	TCP-3	4455	[127]
PCN-521	$[Zr_6(\mu_3\text{-}O)_4(\mu_3\text{-}OH)_4(OH)_4(H_2O)_4(COO)_8]$	MTBC	3411	[128]
PCN-700	$[Zr_6(\mu_3\text{-}O)_4(\mu_3\text{-}OH)_4(OH)_4(H_2O)_4(COO)_8]$	Me2BPDC	1807	[129]
PCN-777	$[Zr_6(\mu_3\text{-}O)_4(\mu_3\text{-}OH)_4(OH)_6(H_2O)_6(COO)_6]$	TATB	2008	[130]
PCN-133	$[Zr_6(\mu_3\text{-}O)_4(\mu_3\text{-}OH)_4(COO)_{12}]$	BTB, DCDPS	1462	[131]
PCN-134	$[Zr_6(\mu_3\text{-}O)_4(\mu_3\text{-}OH)_4(OH)_2(H_2O)_2(COO)10]$	BTB, TCPP	1946	[131]
MOF-801	$[Zr_6(\mu_3\text{-}O)_4(\mu_3\text{-}OH)_4(COO)_{12}]$	FUM	990	[132]
MOF-802	$[Zr_6(\mu_3\text{-}O)_4(\mu_3\text{-}OH)_4(OH)_2(H_2O)_2(COO)10]$	PZDC	NA	[132]
MOF-808	$[Zr_6(\mu_3\text{-}O)_4(\mu_3\text{-}OH)_4(OH)_6(H_2O)_6(COO)_6]$	BTC	2060	[132]
MOF-812	$[Zr_6(\mu_3\text{-}O)_4(\mu_3\text{-}OH)_4(COO)_{12}]$	MTB	2335	[132]
MOF-841	$[Zr_6(\mu_3\text{-}O)_4(\mu_3\text{-}OH)_4(OH)_4(H_2O)_4(COO)_8]$	MTB	1390	[132]
MOF-525	$[Zr_6(\mu_3\text{-}O)_4(\mu_3\text{-}OH)_4(COO)_{12}]$	TCPP	2620	[133]
MOF-535	$[Zr_6(\mu_3\text{-}O)_4(\mu_3\text{-}OH)_4(COO)_{12}]$	XF	1120	[133]
MOF-545	$[Zr_6(\mu_3\text{-}O)_4(\mu_3\text{-}OH)_4(OH)_4(H_2O)_4(COO)_8]$	TCPP	2260	[133]
DUT-51	$[Zr_6(\mu_3\text{-}O)_4(\mu_3\text{-}OH)_4(OH)_4(H_2O)_4(COO)_8]$	DTTDC	2335	[134]
DUT-52	$[Zr_6(\mu_3\text{-}O)_4(\mu_3\text{-}OH)_4(COO)_{12}]$	2,6-NDC	1399	[135]
DUT-84	$[Zr_6(\mu_3\text{-}O)_4(\mu_3\text{-}OH)_4(OH)_6(H_2O)_6(COO)_6]$	2,6-NDC	637	[135]
DUT-67	$[Zr_6(\mu_3\text{-}O)_4(\mu_3\text{-}OH)_4(OH)_4(H_2O)_4(COO)_8]$	TDC	1064	[136]
DUT-68	$[Zr_6(\mu_3\text{-}O)_4(\mu_3\text{-}OH)_4(OH)_4(H_2O)_4(COO)_8]$	TDC	891	[136]
DUT-69	$[Zr_6(\mu_3\text{-}O)_4(\mu_3\text{-}OH)_4(OH)_2(H_2O)_2(COO)10]$	TDC	560	[136]
NU-1000	$[Zr_6(\mu_3\text{-}O)_4(\mu_3\text{-}OH)_4(OH)_4(H_2O)_4(COO)_8]$	TBAPy	2320	[137]
NU-1100	$[Zr_6(\mu_3\text{-}O)_4(\mu_3\text{-}OH)_4(COO)_{12}]$	PTBA	4020	[138]
NU-1101	$[Zr_6(\mu_3\text{-}O)_4(\mu_3\text{-}OH)_4(COO)_{12}]$	Py-XP	4422	[139]
NU-1102	$[Zr_6(\mu_3\text{-}O)_4(\mu_3\text{-}OH)_4(COO)_{12}]$	Por-PP	4712	[139]
NU-1103	$[Zr_6(\mu_3\text{-}O)_4(\mu_3\text{-}OH)_4(COO)_{12}]$	Py-PTP	5646	[139]
NU-1104	$[Zr_6(\mu_3\text{-}O)_4(\mu_3\text{-}OH)_4(COO)_{12}]$	Por-PTP	5290	[139]
MIL-140A	$[ZrO(COO)_2]n$	BDC	415	[140]
MIL-140B	$[ZrO(COO)_2]n$	2,6-NDC	460	[140]
MIL-140C	$[ZrO(COO)_2]n$	BPDC	670	[140]
MIL-140D	$[ZrO(COO)_2]n$	Cl2ABDC	701	[140]
BUT-12	$[Zr_6(\mu_3\text{-}O)_4(\mu_3\text{-}OH)_4(OH)_4(H_2O)_4(COO)_8]$	CTTA	3387	[141]
BUT-13	$[Zr_6(\mu_3\text{-}O)_4(\mu_3\text{-}OH)_4(OH)_4(H_2O)_4(COO)_8]$	TTNA	3948	[141]
Zr-ABDC	$[Zr_6(\mu_3\text{-}O)_4(\mu_3\text{-}OH)_4(COO)_{12}]$	ABDC	3000	[142]
BUT-30	$[Zr_6(\mu_3\text{-}O)_4(\mu_3\text{-}OH)_4(COO)_{12}]$	EDDB	3940	[143]

Table 1. *Cont.*

MOFs [a]	Clusters/Cores	Linkers [b]	BET Surface Area (m^2 g^{-1})	Ref.
PIZOF	$[Zr_6(\mu_3\text{-O})_4(\mu_3\text{-OH})_4(COO)_{12}]$	PEDC	2080	[144]
Zr-BTDC	$[Zr_6(\mu_3\text{-O})_4(\mu_3\text{-OH})_4(COO)_{12}]$	BTDC	2207	[145]
Zr-BTBA	$[Zr_6(\mu_3\text{-O})_4(\mu_3\text{-OH})_4(COO)_{12}]$	BTBA	4342	[146]
Zr-PTBA	$[Zr_6(\mu_3\text{-O})_4(\mu_3\text{-OH})_4(COO)_{12}]$	PTBA	4116	[146]
Zr-BTB	$[Zr_6(\mu_3\text{-O})_4(\mu_3\text{-OH})_4(OH)_6(H_2O)_6(COO)_6]$	BTB	613	[147]
hcp UiO-67	$[Hf_{12}(\mu_3\text{-O})_8(\mu_3\text{-OH})_8(\mu_2\text{-OH})_6(COO)18]$	BPDC	1424	[148]
Zr$_{12}$-TPDC	$[Zr_{12}(\mu_3\text{-O})_8(\mu_3\text{-OH})_8(\mu_2\text{-OH})_6(COO)18]$	TPDC	1967	[149]
Hf$_{12}$-BTE	$[Hf_{12}(\mu_3\text{-O})_8(\mu_3\text{-OH})_8(\mu_2\text{-OH})_6(COO)18]$	BTE	NA	[150]
Cu-BTPP	$[Cu_3(\mu_3\text{-OH})(PZ)_3]$	BTPP	660	[151]
Ni$_3$(BTP)$_2$	$[Ni_4(PZ)_8]$	BTP	1650	[152]
Zn (1,4-BDP)	$[Zn(PZ)_2]n$	1,4-BDP	1710	[153]
Zn (1,3-BDP)	$[Zn(PZ)_2]n$	1,3-BDP	820	[153]
PCN-601	$[Ni_8(OH)_4(H_2O)_2(PZ)_{12}]$	TPP	1309	[154]
ZIF-8	$[ZnN_4]$	mIM	1947	[155]
ZIF-11	$[ZnN_4]$	bIM	1676	[155]
ZIF-67	$[CoN_4]$	mIM	1587	[156]
ZIF-90	$[ZnN_4]$	ICA	1270	[157]
ZIF-68	$[ZnN_4]$	nIM, bIM	1220	[64]
ZIF-69	$[ZnN_4]$	nIM, 5cbIM	1070	[64]
ZIF-70	$[ZnN_4]$	IM, nIM	1970	[64]

[a] These MOFs can be modified by functional organic compounds such as amino, nitro, methyl, halogen, or hydroxyl groups. These MOFs are not explained in this paper; [b] All linkers name are abbreviations [68]; [c] Langmuir surface area.

2.2. Tetravalent Metal-Carboxylate Based MOFs

Tetravalent metals, such as Ce^{4+}, Zr^{4+}, and Ti^{4+}, and carboxylate linker based MOFs are a comparatively new field of study. Lillerud et al. and Férey et al. have reported on Zr-MOFs and Ti-MOFs, respectively [116,121]. Both Zr- and Ti-MOFs have been applied in various fields because of their high stability [68]. On the other hand, Ce-MOFs are fascinating materials due to their redox properties and possible catalytic activity. For example, a Ce-MOF, which is composed of Ce^{3+} and Ce^{4+}, exhibits unique oxidase-like catalytic performance [158].

2.3. Trivalent Metal-Carboxylate Based MOFs

MOFs that are composed of trivalent metal cations and carboxylate linkers have two main secondary building units (SBUs): (1) The $[M_3(\mu_3\text{-O})(COO)_6]$ cluster, which includes a μ_3-oxo-centered trimer of MO$_6$ octahedra and (2) the $[M(OH)(COO)_2]_n$ chain, which has a μ_2-hydroxo corner sharing MO$_6$ octahedral unit [68].

2.4. Divalent Metal-Azolate Based MOFs

Another type of stable MOFs is composed of soft M^{2+} ions and azolate-based ligands while using hard soft acid base (HSAB) theory. Some of the organic reagents are in the form of azolate-based linkers (Table 2) [159]. Azoles generally release a proton to coordinate with the M ions, similar to carboxylic acids [68]. In addition, azoles display well-known coordination properties and the sp^2 nitrogen donors in pyridines and azoles are essentially alike, but different to carboxylic acids [68].

Table 2. Structure and typical coordination modes of azolates.

Linker	Structure	Typical Coordination Modes
Imidazole (HIM)		
Pyrazole (HPZ)		
Triazole (HTZ and HVTZ)		
Tetrazole (HTTZ)		

3. Toxic Gas Sensors

The detection of toxic gases is important in environmental remediation and human health problems. Accordingly, many groups have studied new sensing materials for latent modification. Schröder et al. have reported several MOFs that are used for the reversible adsorption of nitrogen dioxide [6]. The Dincǎ group have reported the use of microporous triazolate-based MOFs for the detection of ammonia gas [7]. Salma et al. have studied the synthesis of a foremost chemical sensor for the identification of sulfur dioxide at room temperature (RT) [12].

3.1. Robust Porous MOF for Reversible Adsorption of NO_2

As one of the major air pollutants, nitrogen dioxide is fatal to the environment and it causes serious health problems [75–78]. Decreasing NO_x contamination is a difficult issue due to the highly active atomic bond with oxygen and corrosive nature [79]. Therefore, various materials, including metal oxides, mesoporous silica, zeolites, and activated carbons, have been studied as NO_2 adsorbents. However, these materials show low adsorption capacities and irreversible uptake due to the disproportionation of NO^+ and NO^{3-}. MOFs have been used as solid adsorbents, but an isothermal adsorption study on NO_2 has not been conducted to date. Therefore, Han et al. studied the isothermal adsorption of MOFs and confirmed that MFM-300 (Al) can interact with highly reactive NO_2. Consequently, MFM-300 (Al) has great potential as a practical solid absorbent.

Figure 1 shows the adsorption isotherms that were obtained for MFM-300 (Al) in various gases, including NO_2, CO_2, SO_2, CO, CH_4, N_2, H_2, O_2, and Ar at room temperature and pressure. The maximum NO_2 isotherm uptake was ~4.1 mmol g^{-1} at room temperature and pressure. This value was much higher than modified Y zeolites [160], mixed oxides, such as $Ce_{1-x}Zr_xO_2$ [161], NH_3 functionalized SBA-15 [162], urea-modified mesoporous carbons [163], and activated carbons [164]. Furthermore, the crystallinity and sorption capacity were not changed after the cycling of the sorption and desorption steps.

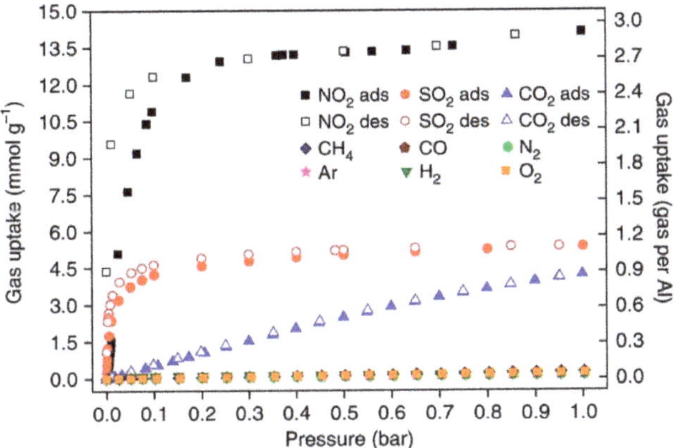

Figure 1. Gas uptake properties of MFM-300 (Al) in various gas environment such as NO_2, CO_2, SO_2, CH_4, CO, N_2, H_2, O_2, and Ar (open symbol: adsorption, solid symbol: desorption). Reproduced and adapted from Ref. [6]; Copyright (2018), Springer Nature.

3.2. Ammonia Sorption in MOFs

Ammonia (NH_3) is present in the atmosphere due to global agriculture and industry. The current industrial standard sorbents show a low affinity and limited capacity for NH_3. The heterogeneous pore size in carbon-based materials has caused a fundamental problem in studies on ammonia sorption. Recent studies have focused on sorbents containing Lewis or Brønsted acid sites that show a higher affinity toward NH_3 molecules to solve this problem. In this study, Dinca et al. showed the static and dynamic ammonia capacities of various microporous triazolate-based MOFs.

Dynamic breakthrough measurements using 0.1% NH_3 showed that Co_2Cl_2BBTA and Co_2Cl_2BTDD have a NH_3 breakthrough capacity of 8.56 and 4.78 mmol g^{-1}, respectively (Table 3). This is equal to 1.48 and 1.08 molecules of NH_3 per Co atom, respectively. The NH_3 breakthrough capacity value is reduced in 80% relative humidity (RH), regardless of the pore size due to the adsorption of water. The saturation value was 4.36 mol kg^{-1} for Co_2Cl_2BBTA and 3.38 mol kg^{-1} for Co_2Cl_2BTDD. These results indicate 0.76 and 0.77 NH_3 molecules are absorbed per open metal site in Co_2Cl_2BBTA and Co_2Cl_2BTDD, respectively.

Table 3. Saturation NH_3 breakthrough capacities value at 0.1% of MOFs.

	Dry (0% RH)	**Wet (80% RH)**
Co_2Cl_2BTDD	4.78	3.38
Co_2Cl_2BBTA	8.56	4.36
Cu_2Cl_2BBTA	7.52	5.73

3.3. Highly Performance of SO_2 MOF Sensor

Although sulfur dioxide is one of the most toxic and serious air pollutants, consumption for fossil fuel is increasing [10]. The main adverse health issues occur upon continuous exposure to SO_2, with a primary 1-h acceptable limit of 75 ppb. Thus, a sensitive sensor, which can detect even a small amount of SO_2 gas, is necessary. However, the detection of SO_2 gas by chemical reaction from CaO to $CaSO_3$ has low efficiency and irreversibility [165,166]. Therefore, it is necessary to achieve the reversible physisorption and selective interaction with SO_2. Thus far, there are many studies that have been conducted based on the metal oxide (SnO_2, WO_3, and TiO_2) that show high- sensitivity, recover time, and selectivity. However, a sensor based on metal oxide requires the high temperature (200–600 °C), which means that it requires high energy and power.

Recently, MOFs are attractive because of satisfying these requirements mentioned above. However, one of the issues using MOFs in sensing devices is directly related to the fabrication as thin films form. In this study, Salama et al. showed the fabrication of a MOF thin film on various supports and its gas-sensing properties.

Among the various MOFs, such as MFM-300 (Al), MFM-202-a, $Zn_3[Co(CN)_6]_2$, $Co_3[Co(CN)_6]_2$, Mg-MOF-74, and $Ni(bdc)(ted)_{0.5}$, they have chosen the MFM-300 (In) MOF because of its high sorption capacity. To confirm the sensing properties of MFM-300 (In) and measured the changing the capacitance. In particular, MFM-30 (In) MOF was grown on a prefunctionallized IDE with an OH-terminated self-assembled monolayer (SAM) under optimized conditions [167]. X-ray diffraction (XRD) and scanning electron microscopy (SEM) confirmed the structural properties of the film.

The MFM-300 (In) MOF sensor showed exceptional performance. In addition, it can detect SO_2 in the ppb range down to 75 ppb (Figure 2). The remarkable detection properties are related to the change in the permittivity of the thin film, depending on the adsorption of SO_2 molecules. There are two types of interaction in the adsorption process: (1) Analyte-framework interactions and (2) analyte-analyte interactions. Changing these two interactions induce a change in the capacitance of the thin film. The MOF sensor exhibited good stability over three weeks of operation.

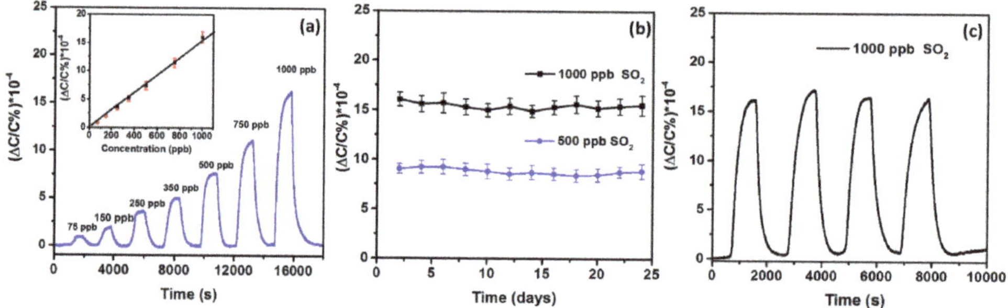

Figure 2. (**a**) Detection of SO_2 in the 75 to 1000 ppb concentration range (**b**) linear response for MFM-300 (In) MOF-sensor upon exposure to SO_2 for a 24-day (**c**) reproducibility cycles for the detection. Reproduced and adapted from Ref. [12]; Copyright (2018), RSC Journal of Materials Chemistry A.

They also measured the sensing performance with relative humidity (RH) at 350 and 1000 ppb of SO_2 gas. However, it does not show distinctive signals as compared to the "dry" condition (Figure 3a). Therefore, it confirmed that the practical applicability of MFM-300 (in) MOF as SO_2 sensor in humidity condition. The temperature dependent sensing properties from 22 °C to 80 °C showed that the performance decrease up to 35% with increasing temperature (Figure 3b). Finally, selectivity performance was conducted in various gases of MFM-300 (In) MOF. As a result, it showed good selectivity for SO_2 gas when compared to others (Figure 3c).

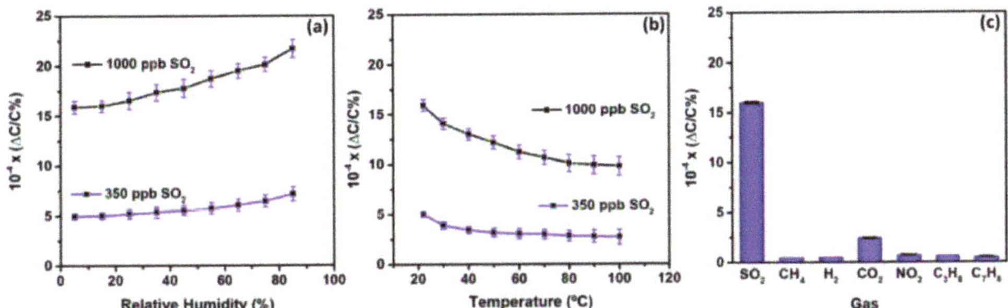

Figure 3. Environment effect of sensing performance with (**a**) relative humidity (**b**) temperature and (**c**) selectivity performance in various gases of MFM-300 (In) MOF sensor. Reproduced and adapted from Ref. [12]; Copyright (2018), RSC Journal of Materials Chemistry A.

4. Detection and Reduction of Toxic Water via MOFs

4.1. Detection of Toxic 4-Nitrophenol via AgPd Nanoparticles on Functionalized MOFs

Nitroaromatic compounds are continuous organic contaminants that originate from industrial waste. 4-Nitrophenol (4-NP) is one of the harmful phenolic pollutants found in chemical waste [54], which is due to its high polarity and subsequent high solubility in water.

4.1.1. Synthesis of UiO-66-L and AgPd Nanoparticles Embedded on UiO-66-L (L=NH$_2$ and NO$_2$)

The synthesis of UiO-66-L was previously reported in the literature [168]. ZrCl$_4$ was dispersed in DMF and the resulting mixture activated with acetic acid at 55 °C. 2-NH$_2$-H$_2$BDC and 2-NO$_2$-H$_2$BDC were used to functionalize UiO-66, respectively.

AgPd@UiO-66-L MOF was prepared via a reduction method while using sodium borohydride. UiO-66-L MOFs were homogeneously dispersed in water, and AgNO$_3$ and PdCl$_2$ were dispersed in the resulting MOFs dispersion, respectively. An aqueous solution of NaBH$_4$ was added to the Ag$^+$, Pd^{2+}, and MOFs mixture to reduce the Ag and Pd. The product was isolated via centrifugation, washing, and drying (Scheme 2).

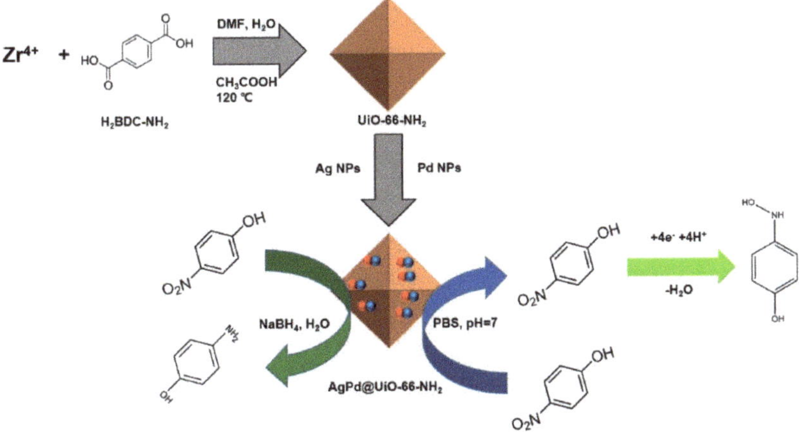

Scheme 2. Synthesis of AgPd@UiO-66-L and the electrochemical reduction mechanism of 4-nitrophenol.

4.1.2. Characterization of UiO-66-L and AgPd@UiO-66-L

In this review, the characterization of UiO-66-L and AgPd@UiO-66-L while using SEM, TEM, and XRD is described. The FE-SEM and TEM images show the good dispersion of metal NPs on the AgPd@UiO-66-NH$_2$ MOF and its octahedral morphology (Figure 4a,b). The elemental distribution of UiO-66-NH$_2$ was investigated while using HAADF and elemental mapping (Figure 4c–j), which confirmed the bimetallic AgPd nanoparticles were loaded into both the bulk MOF and on its surface.

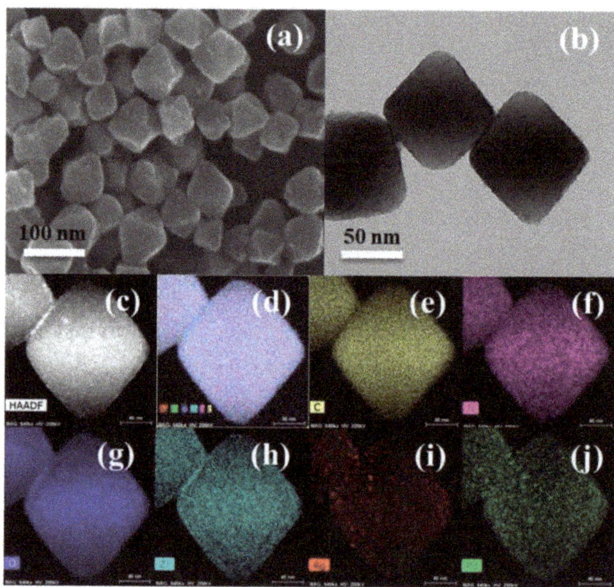

Figure 4. Scanning electron microscopy (SEM) image (**a**), TEM image (**b**), HAADF image (**c**), total element (**d**), and elemental mapping (**e–j**) of AgPd@UiO-66-NH$_2$. The bar indicate 100 nm (**a**), 50 nm (**b**), and 40 nm (**c–j**). Reproduced and adapted from Ref. [11]; Copyright (2018), Elsevier Sensors and Actuators B: Chemical.

The XRD spectra showed the crystallinity and structural aspects of the as-synthesized materials (Figure 5). The XRD spectra that were recorded for UiO-66-NH$_2$ and UiO-66-NO$_2$ were similar and in good agreement with AgPd and bare UiO-66 (Figure 5a–c). This shows that the materials crystallinity was not changed after functionalization of the -NH$_2$ and -NO$_2$ groups (Figure 5d,e). The characteristic peaks for Ag and Pd were observed at 2θ = 38.03° and 40.01°, respectively. Moreover, the existence of both Ag and Pd peaks in Figure 5f,g, respectively, show the obvious crystallinity of the AgPd alloy.

4.1.3. A Comparison of the Catalytic Performance by the Detection and Reduction of 4-Nitrophenol

In this paper, 0.5 mM of 4-nitrophenol was detected while using AgPd@UiO-66-L by cyclic voltammetry. The sharp reduction peak at –0.7 V vs. Ag/AgCl shows the reduction of 4-nitrophenol to 4-hydroxyaminophenol by AgPd@UiO-66-NH$_2$/GCE (Figure 6a). In addition, a quasi-reversible anodic peak was observed at ~0.1 V vs. Ag/AgCl, due to the oxidation of 4-hydroxyaminophenol to 4-nitrosophenol. In this result, AgPd@UiO-66-NH$_2$ showed improved performance when compared to the other materials studied. Figure 6b shows the cyclic voltammograms for the sensing of various concentrations of 4-nitrophenol (0.25–400 μm) by the AgPd@UiO-66-NH$_2$/GCE electrode in 0.1 M phosphate buffered saline (PBS) at pH 7. Increasing the concentration of 4-nitrophenol from 0.25 μM to 400 μM exhibited a linear increase in the cathodic reduction peak due to the beneficial electrocatalytic reduction of 4-nitrophenol to 4-hydroxyaminophenol.

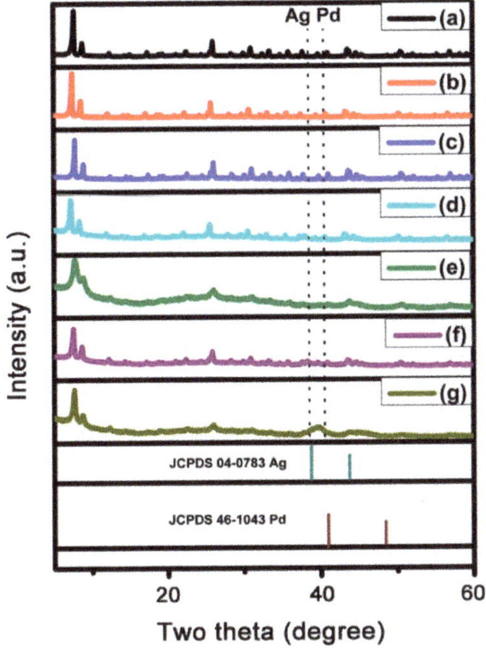

Figure 5. PXRD pattern of UiO-66 (**a**), UiO-66-NH$_2$ (**b**), UiO-66-NO$_2$ (**c**), Ag@UiO-66-NH$_2$ (**d**), Pd@UiO-66-NH$_2$ (**e**), AgPd@UiO-66-NH$_2$ (**f**), and AgPd@UiO-66-NO$_2$ (**g**). Reproduced and adapted from Ref. [11]; Copyright (2018), Elsevier Sensors and Actuators B: Chemical.

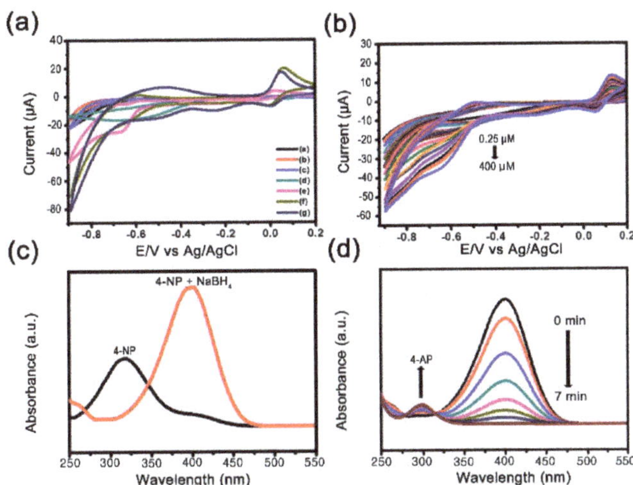

Figure 6. (**a**) CVs of the identical electrodes in 0.1 M PBS solution including 0.5 mM of 4-nitrophenol, (**b**) CVs of AgPd@UiO-66-NH$_2$/GCE in 0.1 M PBS at various concentration 4-nitrophenol (0.25 μM–400 μM), (**c**) UV-VIS absorption spectra of 4-nitrophenol before addition of NaBH$_4$ solution as compared to after, and (**d**) Intensity of absorbance peak change detection via UV-VIS spectra for the reduction of 4-nitrophenol within NaBH$_4$ in AgPd@UiO-66-NH$_2$ during the time-dependent. Reproduced and adapted from Ref. [11]; Copyright (2018), Elsevier Sensors and Actuators B: Chemical.

The catalytic activities of the as-synthesized products were also studied while using this catalytic reduction reaction using UV-Vis spectroscopy. 4-Nitrophenol shows an absorption peak at ~317 nm in an aqueous medium. Upon the addition of $NaBH_4$ powder to the solution, a new peak was detected at 400 nm (Figure 6c). The resulting thick yellow mixture and shift in the absorption peak was due to the generation of p-nitrophenolate [169]. Figure 6d showed catalytic activity by AgPd@UiO-66-NH_2 in the mixed solution of sodium borohydride and 4-nitrophenol. The color of mixed solution altered from yellow to colourless because of the activation of 4-nitrophenolate ion to 4-aminophenol.

4.2. Detection of Antibiotics in Water Using Cd-MOF as a Fluorescent Probe

Antibiotics are used to treat bacterial infections in animals and humans, but they are an important organic pollutant [170,171]. The abuse of antibiotics has led to extreme residues in subsoil water and surface water [172–177]. Therefore, luminescent MOFs have been synthesized and applied in the detection of antibiotics in water as an alternative to liquid chromatography (LC) combined with UV-Vis spectroscopy, capillary electrophoresis, Raman spectroscopy, mass spectroscopy, and ion mobility spectroscopy [141,178–193]. In this paper, the reported Cd-MOF material was used toward the detection of the antibiotic, ceftriaxone sodium (CRO).

4.2.1. Synthesis and Detection of Cd-MOF

The synthesis of Cd-MOF while using 1,4-bis(2-methyl-imidazole-1-yl)butane (bbi) was carried out while using a literature process [194]. The detection of antibiotics was achieved using UV spectroscopy in the wavelength range of 270–350 nm.

4.2.2. Characterization of Cd-MOF

The Cd-MOF was analyzed in terms of its thermal and chemical stability under harsh conditions. Figure 7a exhibited no weight loss in the temperature range of 35–300 °C due to the absence of water in the Cd-MOF. The framework started to decompose at temperatures >300 °C. Meanwhile, the Cd-MOF was stored in aqueous solutions of the antibiotic and distilled water under various pH conditions for 24 h (Figure 7b). The PXRD spectra of the treated samples are in good agreement with the base simulated data. In these results, the Cd-MOF exhibited high physical and chemical stability in alkaline, acidic, and antibiotic solutions, respectively.

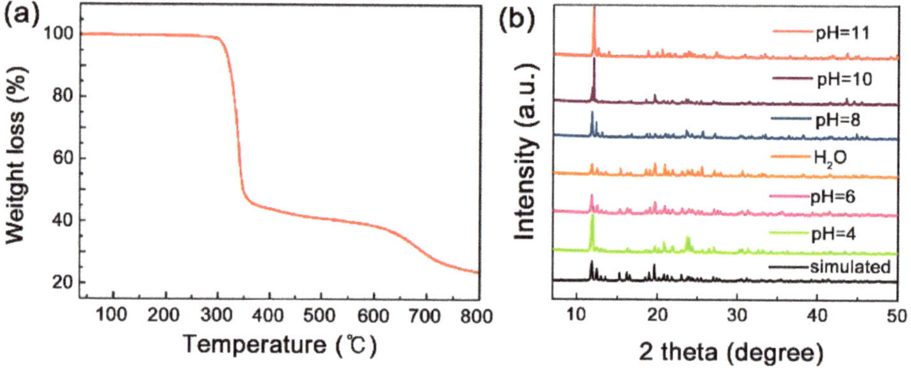

Figure 7. (a) Thermal gravity analysis graph of Cd-MOF and (b) PXRD spectrum of Cd-MOF neglected in various pH aqueous solution and water for 24 h. Reproduced and adapted from Ref. [54]; Copyright (2019), RSC Analyst.

The MOF candidates used as luminescent materials are often constructed from conjugated organic ligand linkers and d^{10} metal ions [195–200]. Therefore, the Cd-MOF shows enhanced fluorescence intensity when compared to those that are constructed from organic ligands, such as bbi and H_2L (Figure 8).

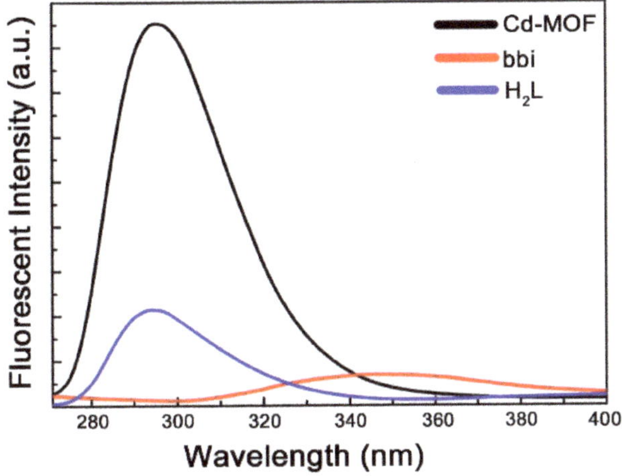

Figure 8. The fluorescence spectra of solid samples of Cd-MOF, bbi, and H_2L organic ligand. Reproduced and adapted from Ref. [54]; Copyright (2019), RSC Analyst.

4.2.3. Chemical Sensors for Antibiotic Detection

The application of Cd-MOF as a fluorescent sensor for detecting antibiotics in water was investigated because of its robust luminescence properties in water. Cd-MOF was dispersed in antibiotic solutions containing LIN, AZL, PEN, AMK, ERY, AMX, AZM, GEN, ATM, FOX, CSU, CEC, CED, CFM, MTR, SXT, and CRO (LIN: Lincomycin hydrochoride, AZL: Azithromycin latobionate, PEN: Penicillin, AMK: Amikacin, ERY: Erythromycin ethylsuccinate, AMX: Amoxicillin, AZM: Azithromycin, GEN: Gentamicin, ATM: Aztreonam, FOX: Cefoxitin, CSU: Cefathiamidine, CEC: Cefaclor, CED: Cefradine, CFM: Cefixime, MTR: metronidazole, SXT: Sulfamethoxazole, and CRO: Ceftriaxone sodium, respectively). As a result, the effective fluorescence quenching of Cd-MOF by these antibiotics follows the order of AZL < GRN < LIN < AMK < ERY < PEN < AZM < AMX < FOX < CEC < CSU < MTR < CFM < ATM < SXT < CRO (Figure 9a). In particular, the efficient detection of CRO via fluorescence quenching was ~90% in this study. Figure 9b shows the distinction of the Cd-MOF sensor for quantitative analysis while using a fluorescence titration experiment. This graph shows the straightforward and dramatic trend in the fluorescence intensity of Cd-MOF that was detected from 0 to 70 μL. Cd-MOF is an effective probe used to detect CRO. The Cd-MOF sensor in an aqueous solution of CRO exhibits highly sensitive fluorescence intensity under various pH conditions (Figure 9c, pH = 4–11). From this graph, Cd-MOF can be used under various pH conditions with no effect on the experimental results. In particular, Cd-MOF shows high performance and stability at pH = 6–7. Figure 9d shows that the rapid detection of various concentrations of CRO can be achieved when an aqueous solution of CRO was added to the Cd-MOF suspension.

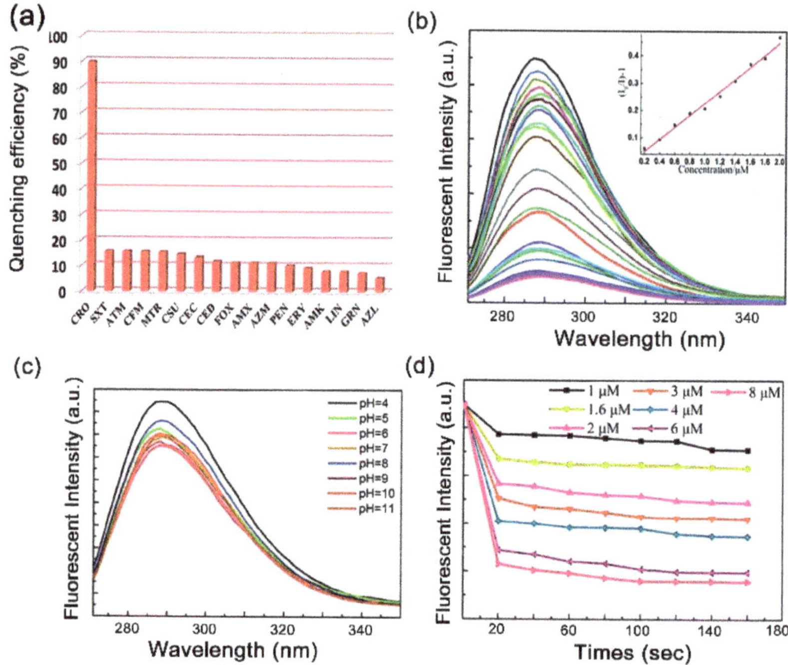

Figure 9. (**a**) Comparison of various antibiotics quenching efficiency using Cd-MOF at room temperature. (**b**) Dispersed in aqueous solution of CRO for fluorescence spectra (**c**) CRO aqueous solution tested in various pH aqueous solution using Cd-MOF, and (**d**) Fluorescence intensity of CRO solution in Cd-MOF suspensions in time dependent. Reproduced and adapted from Ref. [54]; Copyright (2019), RSC Analyst.

5. Filtration of Water and Air Pollutants Using Nanofibrous MOFs Prepared via Electrospinning Methods

5.1. Nanofiber MOF Filter for Particulate Matter

Serious threats to human health have been arisen due to the rapid development of the economy and industry, with the most dangerous of them being air pollution [201,202]. In practice, air pollutants are very diverse and are typically composed of particulate matter (PM) and toxic gases. PM is harmful to the environment affecting human health, air quality and the climate. Particulate matter whose aerodynamic diameters are <2.5 μm ($PM_{2.5}$) and <10 μm (PM_{10}) can penetrate the respiratory system and cause health problems upon prolonged exposure [203,204]. A lot of attention has been focused on researching PM filters to solve these problems. Among them, research on filters made using MOFs via electrospinning methods is a major factor.

5.1.1. Basic Theory of Electrospinning

Electrospinning is the most facile way to make nanofiber membranes containing organic and inorganic components while using polymer melts and solutions [205]. The basic principle of electrospinning is to apply a strong electric field using a high voltage power supply and drawing the fabricated fiber as they solidify (Scheme 3) [206]. It is easy to install and produce, so this method enables mass production at a low cost. Although the basic principle of electrospinning is simple, its mechanism is very complicated, because there are many factors that affect the process. Among them, the process parameters include the nozzle diameter, applied voltage, and tip-to-collector distance

(TCD) [207]. In addition, it is also influenced by the solution characteristics and physical properties, such as the concentration, viscosity, surface tension, and vapor pressure. In particular, there are many advantages that affect performance of filter, such as specific surface area, alterable fiber diameter, and pore size.

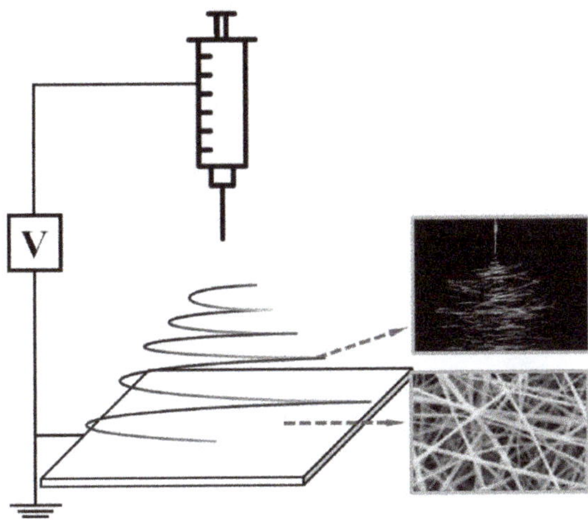

Scheme 3. Electrospinning Setup.

5.1.2. Detail Mechanism of Electrospinning

Electrospinning is the simplest way for producing micro or nanoscale fibers [208–210]. Although the basic theory of this process is uncomplicated, its detail mechanism is so complicated. Many conditions, such as solution terms and experimental environment, affect the electrospinning process.

In the case of solution term, viscosity, surface tension, vapor pressure and solution conductivity have a major influence on electrospinning. The most important of these is viscosity, which can vary greatly with electrospinning. If all conditions are the same, except for the viscosity, it will affect the thickness and formation of the fibers. If the viscosity of the solution is too high, since the drop for electrospinning is not formed, electrospinning itself might be disrupted. On the other hand, if it is considerably low, the fibers are not able to withstand the tension caused by stretching in the process of drawing fibers, so that the fiber are broken. At the optimum viscosity for electrospinning, generally, if the viscosity increases, the stretching proceeds relatively slowly, and the fibers may become thick and vice versa. Similarly, vapor pressure also affects the thickness of the fiber. The vapor pressure of the solution has an intuitive effect on the evaporation rate of the interpolation solvent, so that adjusting the vapor pressure can control the thickness of the fiber.

Surface tension is thought to influence jet formation after drop formation. If the surface tension is too high, drops are formed, but it is difficult to form a jet. This is because the surface tension of the solution is higher than the high voltage applied and, thus, the jet is not formed. In general, it is a good idea to consider the surface tension of the solution if a drop is formed but no jet is formed.

Electrospinning is basically a process for very low conductivity solutions. Electrospinning on conductive solutions is a very difficult process. The reason is that when the conductivity of the solution is high, the drop itself is not formed, and charge is directly discharged from the solution to the collector, which makes it difficult to form electrospinning. It might be possible to spinning a conductive solution by applying a very high voltage, so that the amount of charge applied is greater than that discharged, in order to spinning a conductive solution. The experimental environmental factors affecting electrospinning include TCD, voltage, nozzle size, and temperature and humidity. In the case of temperature and humidity, it is difficult to control these, which is one of the main reasons why the reproducibility of electrospinning result is different every day. Conditions that are related to electrospinning settings, such as TCD voltage and nozzle size, have a relatively small impact. During the electrospinning experiment, the TCD and voltage can be changed in real time, so it is relatively easy to know the optimal conditions for spinning. In this context, the detailed mechanisms and conditions of electrospinning are very diverse and complex, but, if the optimum conditions are found, then a fiber that has high propulsion can be obtained.

5.1.3. Characterization of MOF@PAN, PS (Polystyrene)

A high ratio of MOFs can be loaded into polymer composites without agglomeration by adjusting the morphology and particle size of dissimilar MOF crystals. Figure 10 shows the SEM images of polymer and various MOF non-woven fabrics. By controlling the electrospinning conditions, such as voltage, flow rate of the solution, and TCD, four MOFs can be formed into fiber materials. Figure 10 shows that the MOF nanoparticles are well dispersed in the polymer fibers without any apparent agglomeration, despite a high loading of 60 wt%.

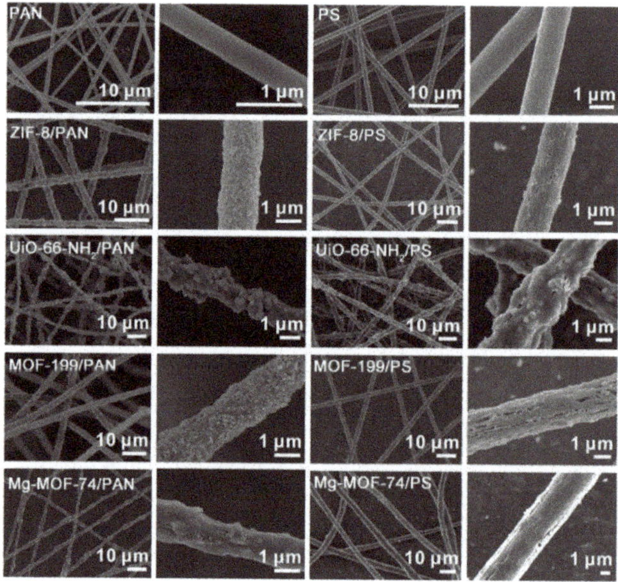

Figure 10. SEM images of polymer and various MOFs loaded nonwoven fabrics. Reproduced and adapted from Ref. [58]; Copyright (2016), ACS Journal of American Society.

5.1.4. Filtration of Particle Matter Using ZIF-8@PAN

Generally, MOF filters can capture pollutants, including particulate matter via three mechanisms: (1) Pollutants can be bound to the OMSs, (2) interaction with the functional groups in the MOF filters, and (3) electrostatic interactions with the MOF filter. Figure 11 shows the particulate matter removal efficiency that was observed for MOFs@PAN filters. Figure 11a shows that the ZIF-8@PAN filter has the highest removal efficiency for $PM_{2.5}$ and PM_{10} among the various MOFs studied. Particulate matter is very polar because of the presence of water vapor and various ions. MOFs, which have unbalanced defects and metal ions on the surface, offer positive charge. This is why the surface of PM can be polarized enhancing the electrostatic interactions. The zeta potential represents these electrostatic interactions. Among the MOFs studied, ZIF-8 exhibited the highest zeta potential of 47.5 mV. In this context, the ZIF-8@PAN filter displayed higher removal efficiency than the other filters studied and its efficiency was maintained after 48 h of exposure to polluted air (Figure 11b).

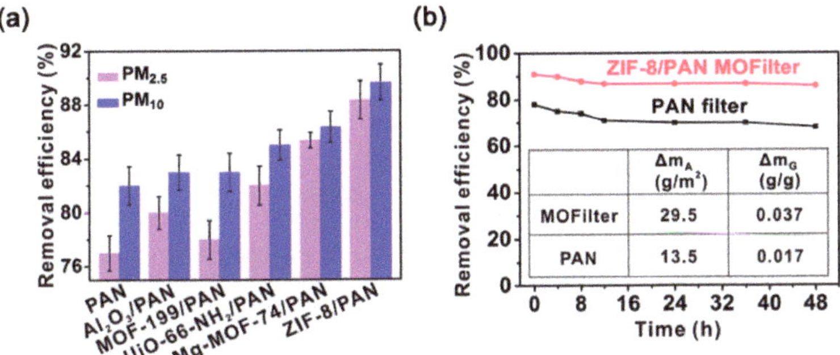

	Δm$_A$ (g/m^2)	Δm$_G$ (g/g)
MOFilter	29.5	0.037
PAN	13.5	0.017

Figure 11. (**a**) Particulate matter removal efficiencies of polyacrylonitrile (PAN) filter, Al$_2$O$_3$@PAN filter and MOF@PAN filter (**b**) Long term PM$_{2.5}$ removal efficiencies of PAN filter and ZIF-8@PAN filter. Reproduced and adapted from Ref. [58]; Copyright (2016), ACS Journal of American Society.

5.2. Nanofiber MOF Filter for Water Pollutants

Water contamination has become an important issue in environmental remediation due to the increase of urban areas and industrialization. Domestic wastewater is continuously discharged into the environment. Generally, the major pollutants in food wastewater are soluble organic food additives and insoluble organic compounds. Many methods have been used to treat these types of pollutants, such as advanced oxidation, adsorption, and photocatalytic membrane technology [211–215]. Among these methods, membrane technology is preferred due to its facile operation. However, research studies mainly concentrate on the removal of one type of pollutant, either soluble or insoluble pollutants. Recently, porous materials, such as MOFs, have been applied to water filtration to treat these contaminants [211–219]. It is necessary to use an electrospinning process to obtain superhydrophilic-underwater superoleophobic properties. In this context, an electrospun polyacrylonitrile (PAN) and MIL-100 (Fe) composite filter (PAN@ MIL-100(Fe)) have been fabricated to treat domestic wastewater.

5.2.1. Schematic of the PAN@MIL-100 (Fe) Filter

Scheme 4 shows the process that is used to prepare the PAN@MIL-100 (Fe) filter. In view of the facile electrospinning process, a H$_3$BTC/PAN electrospun fiber filter was prepared as the precursor to load MIL-100 (FE), where the PAN fiber is used as a polymer frame and trimesic acid used as the initial reaction site for MIL-100 (FE) growth. As the hydrothermal reaction proceeds, trimesic acid acts as a nucleation site for the growth of MIL-100 (Fe) on the PAN fibers.

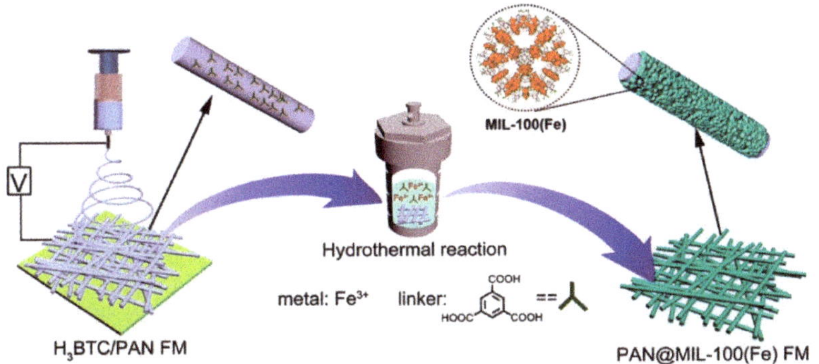

Scheme 4. (**a**) Schematic illustration of fabricating the PAN@MIL-100 (Fe) filter. Reproduced and adapted from Ref. [59]; Copyright (2019), RSC Journal of Materials Chemistry A.

5.2.2. Characterization of the PAN@MIL-100 (Fe) Filter

Figure 12a and b show that the H$_3$BTC/PAN fiber filter has a smooth surface without any beads. The average diameter is 110 nm. After the growth process, the PAN@MIL-100 (Fe) filter has a rough fiber surface with lots of particles. Many particles are covered on the PAN fibers with an average diameter of 211 nm, which is increased during the coating process. This indicates that the MOFs are successfully coated onto the PAN fibers without any aggregation.

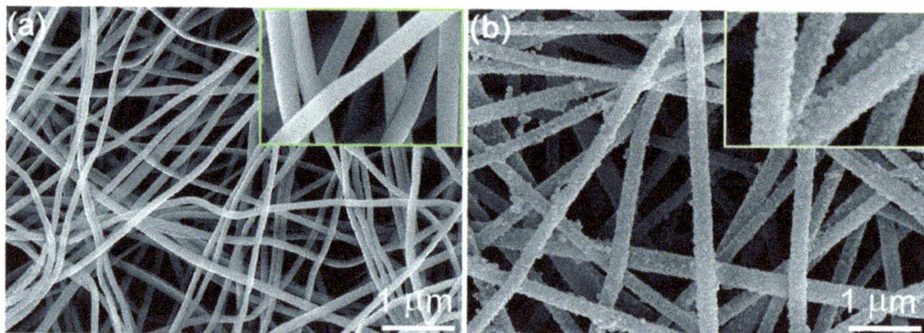

Figure 12. SEM image of (**a**) H$_3$BTC/PAN filter and (**b**) PAN@MIL-100 (Fe) filter. Reproduced and adapted from Ref. [59]; Copyright (2019), RSC Journal of Materials Chemistry A.

5.2.3. Filtration of the Wastewater with Soluble Pollutants Using PAN@MIL-100 (Fe) Filter

The electrospun fiber filter that was prepared without MOFs has macro-size pores, so it is difficult to effectively separate the pollutants. The adsorption interactions between the fibers and pollutants is major factor in the filtration performance of the filter. In this context, Figure 13 shows the results of filtering amaranth red (AR) and vanillin (VA) as soluble pollutants. AR and VA are approximately 99% removed and the removal efficiencies were both >95% after 10 adsorption-desorption cycles.

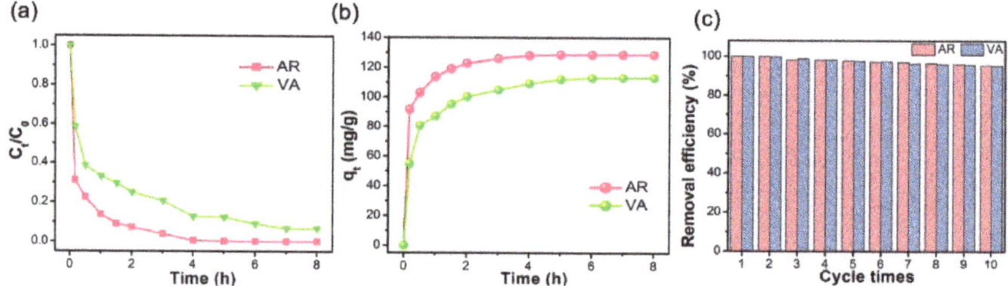

Figure 13. (**a**) Removal efficiency and (**b**) adsorption kinetic curves toward AR and VA by PAN@MIL-100 (Fe) filter. (**c**) Adsorption-desorption cycles. Reproduced and adapted from Ref. [59]; Copyright (2019), RSC Journal of Materials Chemistry A.

5.2.4. Filtration of Wastewater Containing Insoluble Pollutants Using the PAN@MIL-100 (Fe) Filter

In the removal of oil, an important parameter is the surface wettability. To treat insoluble (oil) pollutants, it is essential that the filter has the property of selective wettability (superhydrophilicity and underwater superoleophobicity), which allows for water to pass through filter, but not oil. The selective wettability is that the filter can wet both water and oil in air, but in water has only hydrophilicity. Its basic mechanism is that water around the filter acts as a barrier to prevent oil from passing through. Figure 14a,b show that the PAN@MIL-100 (Fe) filter has selective wettability and the underwater oil pollutants contact angles are 151° and 154°. After five cycles, the oil removal efficiency is only slightly changed.

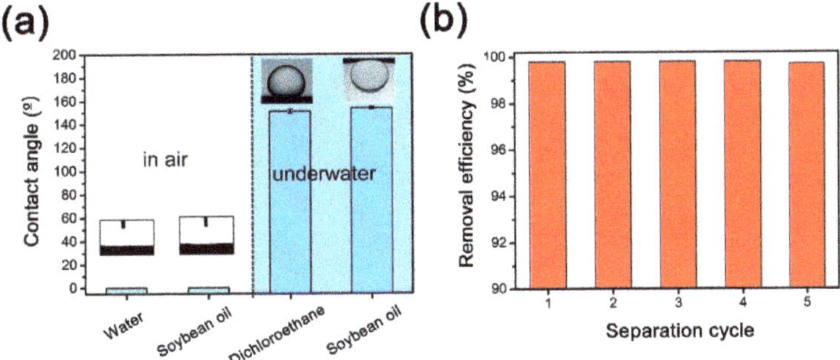

Figure 14. (**a**) Contact angles of water and oil in air and under water. (**b**) Separation efficiency versus the recycling number. Reproduced and adapted from Ref. [59]; Copyright (2019), RSC Journal of Materials Chemistry A.

6. Conclusions

The application of various MOFs materials was reviewed with a focus on toxic sensor, reduction catalyst, hydrogen storage, and filter, which act as successful functional materials. We reviewed all of the materials while using MOFs that exhibited good performance and various application. The application of MOF material for toxic sensor, such as NO_2, SO_2, and ammonia gas, showed high performance via MFM-300 (Al), Co based MOFs (Co_2Cl_2BTDD and Co_2Cl_2BBTA), and MFM-300 (In). Our previous catalyst e.g., PdAg@UiO-66-L for the detection and reduction of 4-nitrophenol also showed good catalytic activity. Moreover, a new Cd-MOF as a fluorescent detect exhibited high sensitivity and selectivity in CRO. Various MOFs for hydrogen adsorption, such as MOF-177, SNU-6, and MOF-74 composited by Co/Ni mixed-material, exhibited good performance while using physisorption analysis method. Filtration for particle matters and wastewater using organic fiber modified as MOF-based material showed good adsorption in polluted condition. In summary, MOFs can be expected one of the best candidate to solve environmental pollution and energy storage in the near future.

Author Contributions: K.H.P. provided academic direction and Discussion of the results and revising the full manuscript. S.J. collected materials and wrote the introduction, basic of stable MOFs, and conclusion part. S.S. and J.H.L. contributed to the materials and methods and results and discussion equally. H.S.K., B.T.P., K.-T.H., and S.P. participated in discussion of the results. All authors have read and agreed to the published version of the manuscript.

Funding: This research was funded by Basic Science Research Program through the National Research Foundation of Korea (NRF) funded by the Ministry of Science, ICT & Future Planning (NRF-2017R1A4A1015533, 2017R1D1A1B03036303 and 2018R1D1A1B07045663). This research was partially supported by VietNam National University Ho Chi Minh City (NCM2019-50-01).

Acknowledgments: This research was supported by PNU-RENovation (2018–2019).

Conflicts of Interest: The authors declare no conflict of interest.

Abbreviations

BDC	terephthalate
FUM	fumarate
2,6-NDC	naphthalene-2,6-dicarboxylate
BTC	benzene-1,3,5-tricarboxylate
BDC-NH_2	2-aminoterephthalate
BTEC	1,2,4,5-benzenetetracarboxylate
NTC	1,4,5,8-naphthalenetetracarboxylate

BPDC	biphenyl-4,4′-dicarboxylate
BPTA	biphenyl-3,3′,5,5′-tetracarboxylate
BTB	1,3,5-benzenetrisbenzoate
BeDC	4,4′-benzophenonedicarboxylate
1,3-BDC	isophthalate
BTTB	4,4′,4′′-[benzene-1,3,5-triyl-tris(oxy)]tribenzoate
TCPP	meso-tetrakis(4-carboxylatephenyl)porphyrin
TATB	4,4′,4′′-s-triazine-2,4,6-triyl-tribenzoate
TCPT	3,3′′,5,5′′-tetrakis(4-carboxyphenyl)-p-terphenyl
TMQPTC	2′,3′′,5′′,6′-tetramethyl-[1,1′:4′,1′′:4′′,1′′′-quaterphenyl]-3,3′′′,5,5′′′-tetracarboxylate
ABDC	4,4-azobenzenedicarboxylate
1,3-ADC	1,3-adamantanedicarboxylate
FTZB	2-fluoro-4-(tetrazol-5-yl)benzoate
CDC	trans-1,4-cyclohexanedicarboxylate
AB	4-aminobenzoate
BDA	benzene-1,4-dialdehyde
BPDA	4,4′-biphenyldicarboxaldehyde
TPDC	[1,1′:4′,1′′-terphenyl]-4,4′′-dicarboxylate
ETTC	4′,4′′,4′′′,4′′′′-(ethene-1,1,2,2-tetrayl)tetrabiphenyl-4-carboxylate
MTBC	4′,4′′,4′′′,4′′′′-methanetetrayltetrabiphenyl-4-carboxylate
PZDC	1H-pyrazole-3,5-dicarboxylate
MTB	4,4′,4′′,4′′′-methanetetrayltetrabenzoate
XF	4,4′-((1E,1′E)-(2,5-bis((4-carboxylatephenyl)ethynyl)-1,4-phenylene)bis(ethene-2,1-diyl))dibenzoate
DTTDC	dithieno[3, 2-b;2′,3′-d]-thiophene-2,6- dicarboxylate
TDC	2,5-thiophenedicarboxylate
TBAPy	1,3,6,8-tetrakis(p-benzoate)pyrene
PTBA	4-[2-[3,6,8-tris[2-(4-carboxylatephenyl)-ethynyl]-pyren-1-yl]ethynyl]-benzoate
Py-XP	4′,4′′′,4′′′′′,4′′′′′′′-(pyrene-1,3,6,8-tetrayl) tetrakis(2′,5′-dimethyl-[1,1′-biphenyl]-4-carboxylate
Por-PP	meso-tctrakis-(4-carboxylatebiphenyl)- porphyrin
Py-PTP	4,4′,4′′,4′′′-((pyrene-1,3,6,8-tetrayltetrakis(benzene-4,1-diyl))tetrakis(ethyne-2,1-diyl))tetrabenzoate
Por-PTP	meso-tetrakis-(4-((phenyl)ethynyl)benzoate)porphyrin
EDDB	4,4′-(ethyne-1,2-diyl)dibenzoate
CTTA	5′-(4-carboxyphenyl)-2′,4′,6′-trimethyl-[1,1′:3′,1′′-terphenyl]-4,4′′-dicarboxylate
TTNA	6,6′,6′′- (2,4,6-trimethylbenzene-1,3,5-triyl)tris(2-naphthoate))
PEDC	4,4′-(1,4-phenylenebis- (ethyne-2,1-diyl))dibenzoate
BTDC	2,2′-bithiophene-5,5′-dicarboxylate
BTBA	4,4′,4′′,4′′′-(biphenyl-3,3′,5,5′-tetrayltetrakis(ethyne-2,1-diyl))tetrabenzoate
PTBA	4-[2-[3,6,8-tris[2-(4-carboxylatephenyl)-ethynyl]-pyren-1-yl]ethynyl]-benzoate
BTE	4,4′,4′′-(benzene-1,3,5-triyl-tris(ethyne-2,1-diyl))tribenzoate
BTPP	1,3,5-Tris((1H-pyrazol-4-yl)phenyl)benzene
BTP	1,3,5-tris(1H-pyrazol-4-yl)benzene
1,4-BDP	1,4-benzenedi(4′-pyrazolyl)
1,3-BDP	1,3-benzenedi(4′-pyrazolyl)
TPP	10,15,20-tetra(1H-pyrazol-4-yl)-porphyrin
mIM	2-methylimidazolate
bIM	benzimidazolate
nIM	2-nitroimidazolate
5cbIM	5-chlorobenzimidazolate
ICA	imidazolate-2-carboxyaldehyde
5-mTz	5-methyltetrazolate
2-mbIM	2-methylbenzimidazolate

References

1. Sandoval, A.; Hernández-Ventura, C.; Klimova, T.E. Titanate nanotubes for removal of methylene blue dye by combined adsorption and photocatalysis. *Fuel* **2017**, *198*, 22–30. [CrossRef]

2. Kampa, M.; Castanas, E. Human health effect of air polution. *Environ. Pollut.* **2008**, *151*, 362–367. [CrossRef] [PubMed]

3. Huang, J.; Tan, S.; Lund, P.D.; Zhou, H. Impact of H_2O on organic-inorganic hybrid perovskite solar cells. *Energy Environ. Sci.* **2017**, *10*, 2284–2311. [CrossRef]

4. Liu, Z.; Yang, C.; Qiao, C. Biodegradation of p-nitrophenol and 4-chlorophenol by Stenotrophomonas sp. *FEMS Microbiol. Lett.* **2007**, *277*, 150–156. [CrossRef] [PubMed]

5. Li, H.; Wang, K.; Sun, Y.; Lollar, C.T.; Li, J.; Zhou, H.C. Recent advances in gas storage and separation using metal–organic frameworks. *Mater. Today* **2018**, *21*, 108–121. [CrossRef]

6. Han, X.; Godfrey, H.G.W.; Briggs, L.; Davies, A.J.; Cheng, Y.; Daemen, L.L.; Sheveleva, A.M.; Tuna, F.; McInnes, E.J.L.; Sun, J.; et al. Reversible adsorption of nitrogen dioxide within a robust porous metal–organic framework. *Nat. Mater.* **2018**, *17*, 691–696. [CrossRef]

7. Rieth, A.J.; Dincă, M. Controlled Gas Uptake in Metal-Organic Frameworks with Record Ammonia Sorption. *J. Am. Chem. Soc.* **2018**, *140*, 3461–3466. [CrossRef]

8. Wang, C.; Zhang, H.; Feng, C.; Gao, S.; Shang, N.; Wang, Z. Multifunctional Pd@MOF core-shell nanocomposite as highly active catalyst for p-nitrophenol reduction. *Catal. Commun.* **2015**, *72*, 29–32. [CrossRef]

9. Li, Y.Z.; Wang, H.H.; Yang, H.Y.; Hou, L.; Wang, Y.Y.; Zhu, Z. An Uncommon Carboxyl-Decorated Metal–Organic Framework with Selective Gas Adsorption and Catalytic Conversion of CO_2. *Chem. A Eur. J.* **2018**, *24*, 865–871. [CrossRef]

10. Galloway, J.N. Acid deposition: Perspectives in time and space. *Water Air Soil Pollut.* **1995**, *85*, 15–24. [CrossRef]

11. Hira, S.A.; Nallal, M.; Park, K.H. Fabrication of PdAg nanoparticle infused metal-organic framework for electrochemical and solution-chemical reduction and detection of toxic 4-nitrophenol. *Sens. Actuators B Chem.* **2019**, *298*, 126861. [CrossRef]

12. Chernikova, V.; Yassine, O.; Shekhah, O.; Eddaoudi, M.; Salama, K.N. Highly sensitive and selective SO_2 MOF sensor: The integration of MFM-300 MOF as a sensitive layer on a capacitive interdigitated electrode. *J. Mater. Chem. A* **2018**, *6*, 5550–5554. [CrossRef]

13. Yaqoob, L.; Noor, T.; Iqbal, N.; Nasir, H.; Zaman, N. Development of Nickel-BTC-MOF-derived nanocomposites with rGO towards electrocatalytic oxidation of methanol and its product analysis. *Catalysts* **2019**, *9*, 856. [CrossRef]

14. Liu, C.; Wang, W.; Liu, B.; Qiao, J.; Lv, L.; Gao, X.; Zhang, X.; Xu, D.; Liu, W.; Liu, J.; et al. Recent advances in MOF-based nanocatalysts for photo-promoted CO_2 reduction applications. *Catalysts* **2019**, *9*, 658. [CrossRef]

15. Padmanaban, S.; Yoon, S. Surface modification of a MOF-based catalyst with Lewis metal salts for improved catalytic activity in the fixation of CO_2 into polymers. *Catalysts* **2019**, *9*, 892. [CrossRef]

16. Wang, S.; Gao, Q.; Dong, X.; Wang, Q.; Niu, Y.; Chen, Y.; Jinag, H. Enhancing the Water Resistance of Mn-MOF-74 by Modification in Low Temperature NH_3-SCR. *Catalysts* **2019**, *9*, 1004. [CrossRef]

17. Xie, X.; Huang, M. Enzymatic Production of Biodiesel Using Immobilized Lipase on Core-Shell Structured Fe_3O_4@MIL-100(Fe) Composites. *Catalysts* **2019**, *9*, 850. [CrossRef]

18. Kim, A.; Muthuchamy, N.; Yoon, C.; Joo, S.H.; Park, K.H. MOF-derived Cu@Cu_2O nanocatalyst for oxygen reduction reaction and cycloaddition reaction. *Nanomaterials* **2018**, *8*, 138. [CrossRef]

19. Sun, W.; Li, X.; Sun, C.; Huang, Z.; Xu, H.; Shen, W. Insights into the pyrolysis processes of Ce-MOFs for preparing highly active catalysts of toluene combustion. *Catalysts* **2019**, *9*, 682. [CrossRef]

20. Maksimchuk, N.; Lee, J.S.; Ayupov, A.; Chang, J.S.; Kholdeeva, O. Cyclohexene oxidation with H_2O_2 over metal-organic framework MIL-125(Ti): The effect of protons on reactivity. *Catalysts* **2019**, *9*, 324. [CrossRef]

21. Ren, R.; Zhao, H.; Sui, X.; Guo, X.; Huang, X.; Wang, Y.; Dong, Q.; Chen, J. Exfoliated Molybdenum Disulfide Encapsulated in a Metal Organic Framework for Enhanced Photocatalytic Hydrogen Evolution. *Catalysts* **2019**, *9*, 89. [CrossRef]

22. Huh, S. Direct Catalytic Conversion of CO_2 to Cyclic Organic Carbonates under Mild Reaction Conditions by Metal—Organic Frameworks. *Catalysts* **2019**, *9*, 34. [CrossRef]

23. Martínez, F.M.; Albiter, E.; Alfaro, S.; Luna, A.L.; Colbeau-Justin, C.; Barrera-Andrade, J.M.; Remita, H.; Valenzuela, M.A. Hydrogen production from glycerol photoreforming on TiO_2/HKUST-1 composites: Effect of preparation method. *Catalysts* **2019**, *9*, 338. [CrossRef]

24. Stawowy, M.; Róziewicz, M.; Szczepańska, E.; Silvestre-Albero, J.; Zawadzki, M.; Musioł, M.; Łuzny, R.; Kaczmarczyk, J.; Trawczyński, J.; Łamacz, A. The impact of synthesis method on the properties and CO_2 sorption capacity of UiO-66(Ce). *Catalysts* **2019**, *9*, 309. [CrossRef]

25. Bedia, J.; Muelas-ramos, V.; Peñas-garzón, M.; Almudena, G.; Rodríguez, J.J.; Belver, C. A Review on the Synthesis and Characterization of Metal Organic Frameworks for Photocatalytic Water Purification. *Catalysts* **2019**, *9*, 52. [CrossRef]

26. Singh, H.; Zhuang, S.; Nunna, B.B.; Lee, E.S. Thermal stability and potential cycling durability of nitrogen-doped graphene modified by metal-organic framework for oxygen reduction reactions. *Catalysts* **2018**, *8*, 607. [CrossRef]

27. He, L.; Liu, Y.; Liu, J.; Xiong, Y.; Zheng, J.; Liu, Y.; Tang, Z. Core-shell noble-metal@metal-organic-framework nanoparticles with highly selective sensing property. *Angew. Chem. Int. Ed.* **2013**, *52*, 3741–3745. [CrossRef]

28. Lin, X.; Gao, G.; Zheng, L.; Chi, Y.; Chen, G. Encapsulation of strongly fluorescent carbon quantum dots in metal-organic frameworks for enhancing chemical sensing. *Anal. Chem.* **2014**, *86*, 1223–1228. [CrossRef]

29. Wu, C.M.; Rathi, M.; Ahrenkiel, S.P.; Koodali, R.T.; Wang, Z. Facile synthesis of MOF-5 confined in SBA-15 hybrid material with enhanced hydrostability. *Chem. Commun.* **2013**, *49*, 1223–1225. [CrossRef]

30. Gu, Z.G.; Li, D.J.; Zheng, C.; Kang, Y.; Wöll, C.; Zhang, J. MOF-Templated Synthesis of Ultrasmall Photoluminescent Carbon-Nanodot Arrays for Optical Applications. *Angew. Chem. Int. Ed.* **2017**, *56*, 1–7. [CrossRef]

31. Wei, Y.; Han, S.; Walker, D.A.; Fuller, P.E.; Grzybowski, B.A. Nanoparticle core/shell architectures within mof crystals synthesized by reaction diffusion. *Angew. Chem. Int. Ed.* **2012**, *51*, 7435–7439. [CrossRef] [PubMed]

32. Li, Z.; Zeng, H.C. Surface and bulk integrations of single-layered Au or Ag nanoparticles onto designated crystal planes {110} or {100} of ZIF-8. *Chem. Mater.* **2013**, *25*, 1761–1768. [CrossRef]

33. Yaghi, O.M.; O'Keeffe, M.; Ockwig, N.W.; Chae, H.K.; Eddaoudi, M.; Kim, J. Reticular synthesis and the design of new materials. *Nature* **2003**, *423*, 705–714. [CrossRef] [PubMed]

34. Kitagawa, S.; Kitaura, R.; Noro, S.I. Functional porous coordination polymers. *Angew. Chem. Int. Ed.* **2004**, *43*, 2334–2375. [CrossRef]

35. Férey, G. Hybrid porous solids: Past, present, future. *Chem. Soc. Rev.* **2008**, *37*, 191–214. [CrossRef]

36. O'Keeffe, M.; Yaghi, O.M. Deconstructing the Crystal Structures of Metal À Organic Frameworks and Related Materials into Their Underlying Nets. *Chem. Rev.* **2003**, *31*, 474–484.

37. Masih, D.; Chernikova, V.; Shekhah, O.; Eddaoudi, M.; Mohammed, O.F. Zeolite-like Metal-Organic Framework (MOF) Encaged Pt(II)-Porphyrin for Anion-Selective Sensing. *ACS Appl. Mater. Interfaces* **2018**, *10*, 11399–11405. [CrossRef]

38. Sonkar, P.K.; Prakash, K.; Yadav, M.; Ganesan, V.; Sankar, M.; Gupta, R.; Yadav, D.K. Co(II)-porphyrin-decorated carbon nanotubes as catalysts for oxygen reduction reactions: An approach for fuel cell improvement. *J. Mater. Chem. A* **2017**, *5*, 6263–6276. [CrossRef]

39. Song, E.; Shi, C.; Anson, F.C. Comparison of the behavior of several cobalt porphyrins as electrocatalysts for the reduction of O_2 at graphite electrodes. *Langmuir* **1998**, *14*, 4315–4321. [CrossRef]

40. El-Deab, M.S.; Okajima, T.; Ohsaka, T. Electrochemical reduction of oxygen on gold nanoparticle-electrodeposited glassy carbon electrodes. *J. Electrochem. Soc.* **2003**, *150*, 851–857. [CrossRef]

41. Liao, L.; Bian, X.; Xiao, J.; Liu, B.; Scanlon, M.D.; Girault, H.H. Nanoporous molybdenum carbide wires as an active electrocatalyst towards the oxygen reduction reaction. *Phys. Chem. Chem. Phys.* **2014**, *16*, 10088–10094. [CrossRef] [PubMed]

42. Sui, S.; Wang, X.; Zhou, X.; Su, Y.; Riffat, S.; Liu, C.J. A comprehensive review of Pt electrocatalysts for the oxygen reduction reaction: Nanostructure, activity, mechanism and carbon support in PEM fuel cells. *J. Mater. Chem. A* **2017**, *5*, 1808–1825. [CrossRef]

43. Ma, T.Y.; Zheng, Y.; Dai, S.; Jaroniec, M.; Qiao, S.Z. Mesoporous $MnCo_2O_4$ with abundant oxygen vacancy defects as high-performance oxygen reduction catalysts. *J. Mater. Chem. A* **2014**, *2*, 8676–8682. [CrossRef]

44. Huang, T.; Mao, S.; Zhou, G.; Wen, Z.; Huang, X.; Ci, S.; Chen, J. Hydrothermal synthesis of vanadium nitride and modulation of its catalytic performance for oxygen reduction reaction. *Nanoscale* **2014**, *6*, 9608–9613. [CrossRef] [PubMed]

45. Zhang, X.; Chen, Y.; Wang, J.; Zhong, Q. Nitrogen and fluorine dual-doped carbon black as an efficient cathode catalyst for oxygen reduction reaction in neutral medium. *ChemistrySelect* **2016**, *4*, 696–702. [CrossRef]

46. Kjaergaard, C.H.; Rossmeisl, J.; Nørskov, J.K. Enzymatic versus inorganic oxygen reduction catalysts: Comparison of the energy levels in a free-energy scheme. *Inorg. Chem.* **2010**, *49*, 3567–3572. [CrossRef]

47. Cheng, F.; Chen, J. Metal-air batteries: From oxygen reduction electrochemistry to cathode catalysts. *Chem. Soc. Rev.* **2012**, *41*, 2172–2192. [CrossRef]

48. He, G.; Qiao, M.; Li, W.; Lu, Y.; Zhao, T.; Zou, R.; Li, B.; Darr, J.A.; Hu, J.; Titirici, M.M.; et al. S, N-Co-Doped Graphene-Nickel Cobalt Sulfide Aerogel: Improved Energy Storage and Electrocatalytic Performance. *Adv. Sci.* **2017**, *4*, 1600214. [CrossRef]

49. Yang, J.; Sun, H.; Liang, H.; Ji, H.; Song, L.; Gao, C.; Xu, H. A Highly Efficient Metal-Free Oxygen Reduction Electrocatalyst Assembled from Carbon Nanotubes and Graphene. *Adv. Mater.* **2016**, *28*, 4606–4613. [CrossRef]

50. Mukerjee, S.; Srinivasan, S. Enhanced electrocatalysis of oxygen reduction on platinum alloys in proton exchange membrane fuel cells. *J. Electroanal. Chem.* **1993**, *357*, 201–224. [CrossRef]

51. Kundu, S.; Nagaiah, T.C.; Xia, W.; Wang, Y.; Van Dommele, S.; Bitter, J.H.; Santa, M.; Grundmeier, G.; Bron, M.; Schuhmann, W.; et al. Electrocatalytic activity and stability of nitrogen-containing carbon nanotubes in the oxygen reduction reaction. *J. Phys. Chem. C* **2009**, *113*, 14302–14310. [CrossRef]

52. Su, B.; Hatay, I.; Trojánek, A.; Samec, Z.; Khoury, T.; Gros, C.P.; Barbe, J.M.; Daina, A.; Carrupt, P.A.; Girault, H.H. Molecular electrocatalysis for oxygen reduction by cobalt porphyrins adsorbed at liquid/liquid interfaces. *J. Am. Chem. Soc.* **2010**, *132*, 2655–2662. [CrossRef] [PubMed]

53. Gotti, G.; Fajerwerg, K.; Evrard, D.; Gros, P. Electrodeposited gold nanoparticles on glassy carbon: Correlation between nanoparticles characteristics and oxygen reduction kinetics in neutral media. *Electrochim. Acta* **2014**, *128*, 412–419. [CrossRef]

54. Xing, P.; Wu, D.; Chen, J.; Song, J.; Mao, C.; Gao, Y.; Niu, H. A Cd-MOF as a fluorescent probe for highly selective, sensitive and stable detection of antibiotics in water. *Analyst* **2019**, *144*, 2656–2661. [CrossRef]

55. Park, H.J.; Suh, M.P. Mixed-ligand metal-organic frameworks with large pores: Gas sorption properties and single-crystal-to-single-crystal transformation on guest exchange. *Chem. A Eur. J.* **2008**, *14*, 8812–8821. [CrossRef]

56. Villajos, J.A.; Orcajo, G.; Martos, C.; Botas, J.Á.; Villacañas, J.; Calleja, G. Co/Ni mixed-metal sited MOF-74 material as hydrogen adsorbent. *Int. J. Hydrogen Energy* **2015**, *40*, 5346–5352. [CrossRef]

57. Wong-Foy, A.G.; Matzger, A.J.; Yaghi, O.M. Exceptional H_2 saturation uptake in microporous metal-organic frameworks. *J. Am. Chem. Soc.* **2006**, *128*, 3494–3495. [CrossRef]

58. Zhang, Y.; Yuan, S.; Feng, X.; Li, H.; Zhou, J.; Wang, B. Preparation of nanofibrous metal-organic framework filters for efficient air pollution control. *J. Am. Chem. Soc.* **2016**, *138*, 5785–5788. [CrossRef]

59. Zhao, R.; Tian, Y.; Li, S.; Ma, T.; Lei, H.; Zhu, G. An electrospun fiber based metal–organic framework composite membrane for fast, continuous, and simultaneous removal of insoluble and soluble contaminants from water. *J. Mater. Chem. A* **2019**, *7*, 22559–22570. [CrossRef]

60. Li, H.; Eddaoudi, M.; O'Keefe, M.; Yaghi, O.M. Design and synthesis of an exceptionally stable and highly porous metal-organic framework. *Nature* **1999**, *402*, 276–279. [CrossRef]

61. Zhou, H.C.; Long, J.R.; Yaghi, O.M. Introduction to metal-organic frameworks. *Chem. Rev.* **2012**, *112*, 673–674. [CrossRef] [PubMed]

62. Eddaoudi, M.; Kim, J.; Rosi, N.; Vodak, D.; Wachter, J.; O'Keeffe, M.; Yaghi, O.M. Systematic design of pore size and functionality in isoreticular MOFs and their application in methane storage. *Science* **2002**, *295*, 469–472. [CrossRef] [PubMed]

63. Férey, C.; Mellot-Draznieks, C.; Serre, C.; Millange, F.; Dutour, J.; Surblé, S.; Margiolaki, I. Chemistry: A chromium terephthalate-based solid with unusually large pore volumes and surface area. *Science* **2005**, *309*, 2040–2042. [CrossRef] [PubMed]

64. Banerjee, R.; Phan, A.; Knobler, C.; Furukawa, H.; O'Keeffe, M.; Yaghi, O.M. High-Throughput Synthesis of Zeolitic Imidazolate Frameworks and Application to CO_2 Capture. *Science* **2008**, *939*, 939–944. [CrossRef] [PubMed]

65. Lu, W.; Wei, Z.; Gu, Z.Y.; Liu, T.F.; Park, J.; Park, J.; Tian, J.; Zhang, M.; Zhang, Q.; Gentle, T.; et al. Tuning the structure and function of metal-organic frameworks via linker design. *Chem. Soc. Rev.* **2014**, *43*, 5561–5593. [CrossRef]

66. Stock, N.; Biswas, S. Synthesis of metal-organic frameworks (MOFs): Routes to various MOF topologies, morphologies, and composites. *Chem. Rev.* **2012**, *112*, 933–969. [CrossRef]

67. Li, M.; Li, D.; O'Keeffe, M.; Yaghi, O.M. Topological analysis of metal-organic frameworks with polytopic linkers and/or multiple building units and the minimal transitivity principle. *Chem. Rev.* **2014**, *114*, 1343–1370. [CrossRef]

68. Yuan, S.; Feng, L.; Wang, K.; Pang, J.; Bosch, M.; Lollar, C.; Sun, Y.; Qin, J.; Yang, X.; Zhang, P.; et al. Stable Metal–Organic Frameworks: Design, Synthesis, and Applications. *Adv. Mater.* **2018**, *30*, 1704303. [CrossRef]

69. Loiseau, T.; Serre, C.; Huguenard, C.; Fink, G.; Taulelle, F.; Henry, M.; Bataille, T.; Férey, G. A Rationale for the Large Breathing of the Porous Aluminum Terephthalate (MIL-53) Upon Hydration. *Chem. A Eur. J.* **2004**, *10*, 1373–1382. [CrossRef]

70. Alvarez, E.; Guillou, N.; Martineau, C.; Bueken, B.; Van De Voorde, B.; Le Guillouzer, C.; Fabry, P.; Nouar, F.; Taulelle, F.; De Vos, D.; et al. The Structure of the Aluminum Fumarate Metal–Organic Framework A520. *Angewandte* **2015**, 3664–3668. [CrossRef]

71. Gaab, M.; Trukhan, N.; Maurer, S.; Gummaraju, R.; Müller, U. The progression of Al-based metal-organic frameworks–From academic research to industrial production and applications. *Microporous Mesoporous Mater.* **2012**, *157*, 131–136. [CrossRef]

72. Loiseau, T.; Mellot-draznieks, C.; Muguerra, H.; Férey, G.; Haouas, M.; Taulelle, F. Hydrothermal synthesis and crystal structure of a new three-dimensional aluminum-organic framework MIL-69 with 2,6-naphthalenedicarboxylate(ndc),Al(OH)(ndc)·H_2O. *C. R. Chim.* **2005**, *8*, 765–772. [CrossRef]

73. Loiseau, T.; Lecroq, L.; Volkringer, C.; Marrot, J.; Fèrey, G.; Haouas, M.; Taulelle, F.; Bourrelly, S.; Llewellyn, P.L.; Latroche, M. MIL-96, a Porous Aluminum Trimesate 3D Structure Constructed from a Hexagonal Network of 18-Membered Rings and μ_3-Oxo-Centered Trinuclear Units. *J. Am. Chem. Soc.* **2006**, *128*, 10223–10230. [CrossRef] [PubMed]

74. Volkringer, C.; Popov, D.; Loiseau, T.; Fèrey, G.; Burghammer, M.; Riekel, C.; Haouas, M.; Taulelle, F. Synthesis, Single-Crystal X-ray Microdiffraction and NMR Characterizaion of the Giant Pore Metal-Organic Framework Aluminum Trimesate MIL-100. *Chem. Mater.* **2009**, *21*, 5695–5697. [CrossRef]

75. Serra-Crespo, P.; Ramos-Fernandez, E.V.; Gascon, J.; Kapteijn, F. Synthesis and characterization of an amino functionalized MIL-101(Al): Separation and catalytic properties. *Chem. Mater.* **2011**, *23*, 2565–2572. [CrossRef]

76. Volkringer, C.; Popov, D.; Loiseau, T.; Guillou, N.; Fèrey, G.; Haouas, M.; Taulelle, F.; Mellot-Draznieks, C.; Burghammer, M.; Riekel, C. A microdiffraction set-up for nanoporous metal-organic-framework-type solids. *Nat. Mater.* **2007**, *6*, 760–764. [CrossRef]

77. Volkringer, C.; Loiseau, T.; Guillou, N.; Fèrey, G.; Haouas, M.; Taulelle, F.; Audebrand, N.; Margiolaki, I.; Popov, D.; Burghammer, M.; et al. Structural transitions and flexibility during dehydration-Rehydration process in the MOF-type aluminum pyromellitate $Al_2(OH)_2[C_{10}O_8H_2]$(MIL-118). *Cryst. Growth Des.* **2009**, *9*, 2927–2936. [CrossRef]

78. Volkringer, C.; Loiseau, T.; Haouas, M.; Taulelle, F.; Popov, D.; Burghammer, M.; Riekel, C.; Zlotea, C.; Cuevas, F.; Latroche, M.; et al. Occurrence of uncommon infinite chains consisting of edge-sharing octahedra in a porous metal organic framework-type aluminum pyromellitate $Al_4(OH)_8[C_{10}O_8H_2]$(MIL-120): Synthesis, Structure, and Gas Sorption Properties. *Chem. Mater.* **2009**, *21*, 5783–5791. [CrossRef]

79. Volkringer, C.; Loiseau, T.; Guillou, N.; Férey, G.; Haouas, M.; Taulelle, F.; Elkaim, E.; Stock, N. High-throughput aided synthesis of the porous metal-organic framework-type aluminum pyromellitate, MIL-121, with extra carboxylic acid functionalization. *Inorg. Chem.* **2010**, *49*, 9852–9862. [CrossRef]

80. Volkringer, C.; Loiseau, T.; Guillou, N.; Férey, G.; Elkaïm, E. Syntheses and structures of the MOF-type series of metal 1,4,5,8,-naphthalenetetracarboxylates $M_2(OH)_2[C_{14}O_8H_4]$(Al, Ga, In) with infinite *trans*-connected M-OH-M chains (MIL-122). *Solid State Sci.* **2009**, *11*, 1507–1512. [CrossRef]

81. Senkovska, I.; Hoffmann, F.; Fröba, M.; Getzschmann, J.; Böhlmann, W.; Kaskel, S. New highly porous aluminium based metal-organic frameworks: Al(OH)(ndc) (ndc = 2,6-naphthalene dicarboxylate) and Al(OH)(bpdc) (bpdc = 4,4′-biphenyl dicarboxylate). *Microporous Mesoporous Mater.* **2009**, *122*, 93–98. [CrossRef]

82. Yang, S.; Sun, J.; Ramirez-Cuesta, A.J.; Callear, S.K.; David, W.I.; Anderson, D.P.; Newby, R.; Blake, A.J.; Parker, J.E.; Tang, C.C.; et al. Selectivity and direct visualization of carbon dioxide and sulfur dioxide in a decorated porous host. *Nat. Chem.* **2012**, *4*, 887–894. [CrossRef] [PubMed]

83. Ahnfeldt, T.; Guillou, N.; Gunzelmann, D.; Margiolaki, I.; Loiseau, T.; Férey, G.; Senker, J.; Stock, N. [Al$_4$(OH)$_2$(OCH$_3$)$_4$(H$_2$N-Bdc)$_3$]·xH$_2$O: A 12-connected porous metal-organic framework with an unprecedented aluminum-containing brick. *Angew. Chem. Int. Ed.* **2009**, *48*, 5163–5166. [CrossRef] [PubMed]

84. Reinsch, H.; Feyand, M.; Ahnfeldt, T.; Stock, N. CAU-3: A new family of porous MOFs with a novel Al-based brick: [Al$_2$(OCH$_3$)$_4$(O$_2$C-X-CO$_2$)] (X = aryl). *Dalton Trans.* **2012**, *41*, 4164–4171. [CrossRef]

85. Reinsch, H.; Krüger, M.; Wack, J.; Senker, J.; Salles, F.; Maurin, G.; Stock, N. A new aluminium-based microporous metal-organic framework: Al(BTB) (BTB = 1,3,5-benzenetrisbenzoate). *Microporous Mesoporous Mater.* **2012**, *157*, 50–55. [CrossRef]

86. Reinsch, H.; Krüger, M.; Marrot, J.; Stock, N. First keto-functionalized microporous Al-based metal-organic framework: [Al(OH)(O$_2$C-C$_6$H$_4$-CO-C$_6$H$_4$-CO$_2$)]. *Inorg. Chem.* **2013**, *52*, 1854–1859. [CrossRef]

87. Barpanda, P.; Liu, G.; Ling, C.D.; Tamaru, M.; Avdeev, M.; Chung, S.C.; Yamada, Y.; Yamada, A. Na$_2$FeP$_2$O$_7$: A safe cathode for rechargeable sodium-ion batteries. *Chem. Mater.* **2013**, *25*, 3480–3487. [CrossRef]

88. Wang, Z.W.; Chen, M.; Liu, C.S.; Wang, X.; Zhao, H.; Du, M. A Versatile Al^{III}-Based Metal-Organic Framework with High Physicochemical Stability. *Chem. Eur. J.* **2015**, *21*, 17215–17219. [CrossRef]

89. Fateeva, A.; Chater, P.A.; Ireland, C.P.; Tahir, A.A.; Khimyak, Y.Z.; Wiper, P.V.; Darwent, J.R.; Rosseinsky, M.J. A water-stable porphyrin-based metal-organic framework active for visible-light photocatalysis. *Angew. Chem. Int. Ed.* **2012**, *51*, 7440–7444. [CrossRef]

90. Feng, D.; Liu, T.F.; Su, J.; Bosch, M.; Wei, Z.; Wan, W.; Yuan, D.; Chen, Y.P.; Wang, X.; Wang, K.; et al. Stable metal-organic frameworks containing single-molecule traps for enzyme encapsulation. *Nat. Commun.* **2015**, *6*, 1–8. [CrossRef]

91. Lian, X.; Chen, Y.P.; Liu, T.F.; Zhou, H.C. Coupling two enzymes into a tandem nanoreactor utilizing a hierarchically structured MOF. *Chem. Sci.* **2016**, *7*, 6969–6973. [CrossRef]

92. Alezi, D.; Belmabkhout, Y.; Suyetin, M.; Bhatt, P.M.; Weseliński, L.J.; Solovyeva, V.; Adil, K.; Spanopoulos, I.; Trikalitis, P.N.; Emwas, A.H.; et al. MOF Crystal Chemistry Paving the Way to Gas Storage Needs: Aluminum-Based soc -MOF for CH$_4$, O$_2$, and CO$_2$ Storage. *J. Am. Chem. Soc.* **2015**, *137*, 13308–13318. [CrossRef]

93. Serre, C.; Millange, F.; Thouvenot, C.; Noguès, M.; Marsolier, G.; Louër, D.; Férey, G. Very large breathing effect in the first nanoporous chromium(III)-based solids: MIL-53 or Cr^{III}(OH)·{O$_2$C-C$_6$H$_4$-CO$_2$}·{HO$_2$C-C$_6$H$_4$-CO$_2$H}$_x$·H$_2$O$_y$. *J. Am. Chem. Soc.* **2002**, *124*, 13519–13526. [CrossRef]

94. Serre, C.; Mellot-Draznieks, C.; Surlbé, S.; Audebrand, N.; Filinchuk, Y.; Férey, G. Role of Solvent-Host Interactions That Lead to Very Large Swelling of Hybrid Frameworks. *Science* **2007**, *315*, 1828–1832. [CrossRef]

95. Long, P.; Wu, H.; Zhao, Q.; Wang, Y.; Dong, J.; Li, J. Solvent effect on the synthesis of MIL-96(Cr) and MIL-100(Cr). *Microporous Mesoporous Mater.* **2011**, *142*, 489–493. [CrossRef]

96. Férey, G.; Serre, C.; Mellot-Draznieks, C.; Millange, F.; Surblé, S.; Dutour, J.; Margiolaki, I. A hybrid solid with giant pores prepared by a combination of targeted chemistry, simulation, and powder diffraction. *Angew. Chem. Int. Ed.* **2004**, *43*, 6296–6301. [CrossRef]

97. Sonnauer, A.; Hoffmann, F.; Fröba, M.; Kienle, L.; Duppel, V.; Thommes, M.; Serre, C.; Férey, G.; Stock, N. Giant pores in a chromium 2,6-Naphthalenedicarboxylate Open- Framework structure with MIL-101 topology. *Angew. Chem. Int. Ed.* **2009**, *48*, 3791–3794. [CrossRef]

98. Lian, X.; Feng, D.; Chen, Y.P.; Liu, T.F.; Wang, X.; Zhou, H.C. The preparation of an ultrastable mesoporous Cr(III)-MOF via reductive labilization. *Chem. Sci.* **2015**, *6*, 7044–7048. [CrossRef]

99. Liu, T.F.; Zou, L.; Feng, D.; Chen, Y.P.; Fordham, S.; Wang, X.; Liu, Y.; Zhou, H.C. Stepwise synthesis of robust metal-organic frameworks via postsynthetic metathesis and oxidation of metal nodes in a single-crystal to single-crystal transformation. *J. Am. Chem. Soc.* **2014**, *136*, 7813–7816. [CrossRef]

100. Whitfield, T.R.; Wang, X.; Liu, L.; Jacobson, A.J. Metal-organic frameworks based on iron oxide octahedral chains connected by benzenedicarboxylate dianions. *Solid State Sci.* **2005**, *7*, 1096–1103. [CrossRef]

101. Fateeva, A.; Horcajada, P.; Devic, T.; Serre, C.; Marrot, J.; Grenèche, J.M.; Morcrette, M.; Tarascon, J.M.; Maurin, G.; Férey, G. Synthesis, structure, characterization, and redox properties of the porous MIL-68(Fe) solid. *Eur. J. Inorg. Chem.* **2010**, *24*, 3789–3794. [CrossRef]

102. Fateeva, A.; Clarisse, J.; Pilet, G.; Grenèche, J.M.; Nouar, F.; Abeykoon, B.K.; Guegan, F.; Goutaudier, C.; Luneau, D.; Warren, J.E.; et al. Iron and porphyrin metal-organic frameworks: Insight into structural diversity, stability, and porosity. *Cryst. Growth Des.* **2015**, *15*, 1819–1826. [CrossRef]

103. Horcajada, P.; Surblé, S.; Serre, C.; Hong, D.Y.; Seo, Y.K.; Chang, J.S.; Grenèche, J.M.; Margiolaki, I.; Férey, G. Synthesis and catalytic properties of MIL-100(Fe), an iron(III) carboxylate with large pores. *Chem. Commun.* **2007**, *27*, 2820–2822. [CrossRef]

104. Lupu, D.; Ardelean, O.; Blanita, G.; Borodi, G.; Lazar, M.D.; Biris, A.R.; Ioan, C.; Mihet, M.; Misan, I.; Popeneciu, G. Synthesis and hydrogen adsorption properties of a new iron based porous metal-organic framework. *Int. J. Hydrogen Energy* **2011**, *36*, 3586–3592. [CrossRef]

105. Feng, D.; Wang, K.; Wei, Z.; Chen, Y.P.; Simon, C.M.; Arvapally, R.K.; Martin, R.L.; Bosch, M.; Liu, T.F.; Fordham, S.; et al. Kinetically tuned dimensional augmentation as a versatile synthetic route towards robust metal-organic frameworks. *Nat. Commun.* **2014**, *5*, 1–8. [CrossRef]

106. Wang, K.; Feng, D.; Liu, T.F.; Su, J.; Yuan, S.; Chen, Y.P.; Bosch, M.; Zou, X.; Zhou, H.C. A series of highly stable mesoporous metalloporphyrin Fe-MOFs. *J. Am. Chem. Soc.* **2014**, *136*, 13983–13986. [CrossRef]

107. Reineke, T.M.; Eddaoudi, M.; Fehr, M.; Kelley, D.; Yaghi, O.M. From condensed lanthanide coordination solids to microporous frameworks having accessible metal sites. *J. Am. Chem. Soc.* **1999**, *121*, 1651–1657. [CrossRef]

108. Serre, C.; Férey, G. Hydrothermal synthesis, thermal behaviour and structure determination from powder data of a porous three-dimensional europium trimesate: $Eu_3(H_2O)(OH)_6[C_6H_3(CO_2)_3]\cdot 3H_2O$ or MIL-63. *J. Mater. Chem.* **2002**, *12*, 3053–3057. [CrossRef]

109. Millange, F.; Serre, C.; Marrot, J.; Garbdant, N.; Pellé, F.; Férey, G. Synthesis, structure and properties of a three-dimensional porous rare-earth carboxylate MIL-83(Eu): $Eu_2(O_2C\text{-}C_{10}H_{14}\text{-}CO_2)_3$. *J. Mater. Chem.* **2004**, *14*, 642–645. [CrossRef]

110. Devic, T.; Serre, C.; Audebrand, N.; Marrot, J.; Férey, G. MIL-103, a 3-D lanthanide-based metal organic framework with large one-dimensional tunnels and a high surface area. *J. Am. Chem. Soc.* **2005**, *127*, 12788–12789. [CrossRef]

111. Jiang, H.L.; Tsumori, N.; Xu, Q. A series of (6,6)-connected porous lanthanide-organic framework enantiomers with high thermostability and exposed metal sites: Scalable syntheses, structures, and sorption properties. *Inorg. Chem.* **2010**, *49*, 10001–10006. [CrossRef]

112. Xue, D.X.; Cairns, A.J.; Belmabkhout, Y.; Wojtas, L.; Liu, Y.; Alkordi, M.H.; Eddaoudi, M. Tunable rare-earth fcu-MOFs: A platform for systematic enhancement of CO_2 adsorption energetics and uptake. *J. Am. Chem. Soc.* **2013**, *135*, 7660–7667. [CrossRef]

113. Assen, A.H.; Belmabkhout, Y.; Adil, K.; Bhatt, P.M.; Xue, D.; Jiang, H.; Eddaoudi, M. Ultra-Tuning of the Rare-Earth fcu-MOF Aperture Size for Selective Molecular Exclusion of Branched Paraffins. *Angew. Chem. Int. Ed.* **2015**, *54*, 14353–14358. [CrossRef]

114. Lammert, M.; Wharmby, M.T.; Smolders, S.; Bueken, B.; Lieb, A.; Lomachenko, K.A.; Vos, D.D.; Stock, N. Cerium-based metal organic frameworks with UiO-66 architecture: Synthesis, properties and redox catalytic activity. *Chem. Commun.* **2015**, *51*, 12578–12581. [CrossRef]

115. Dalapati, R.; Sakthivel, B.; Dhakshinamoorthy, A.; Buragohain, A.; Bhunia, A.; Janiak, C.; Biswas, S. A highly stable dimethyl-functionalized Ce(IV)-based UiO-66 metal-organic framework material for gas sorption and redox catalysis. *CrystEngComm* **2016**, *18*, 7855–7864. [CrossRef]

116. Dan-Hardi, M.; Serre, C.; Frot, T.; Rozes, L.; Maurin, G.; Sanchez, C.; Férey, G. A new photoactive crystalline highly porous titanium(IV) dicarboxylate. *J. Am. Chem. Soc.* **2009**, *131*, 10857–10859. [CrossRef]

117. Yuan, S.; Liu, T.F.; Feng, D.; Tian, J.; Wang, K.; Qin, J.; Zhang, Q.; Chen, Y.P.; Bosch, M.; Zou, L.; et al. A single crystalline porphyrinic titanium metal-organic framework. *Chem. Sci.* **2015**, *6*, 3926–3930. [CrossRef]

118. Bueken, B.; Vermoortele, F.; Vanpoucke, D.E.P.; Reinsch, H.; Tsou, C.; Valvekens, P.; De Baerdemaeker, T.; Ameloot, R.; Kirschhock, C.E.A.; Van Speybroeck, V.; et al. A Flexible Photoactive Titanium Metal–Organic Framework Based on a $[Ti^{IV}_3(\mu_3\text{-}O)(O)_2(COO)_6]$ Cluster. *Angew. Chem. Int. Ed.* **2015**, *54*, 13912–13917. [CrossRef]

119. Furukawa, H.; Doan, T.L.H.; Cordova, K.E.; Yaghi, O.M. A Titanium–Organic Framework as an Exemplar of Combining the Chemistry of Metal–and Covalent–Organic Frameworks. *J. Am. Chem. Soc.* **2016**, *138*, 4330–4333.

120. Nguyen, H.L.; Vu, T.T.; Le, D.; Doan, T.L.H.; Nguyen, V.Q.; Phan, N.T.S. A Titanium–Organic Framework: Engineering of the Band-Gap Energy for Photocatalytic Property Enhancement. *ACS Catal.* **2017**, *7*, 338–342. [CrossRef]

121. Cavka, J.H.; Jakobsen, S.; Olsbye, U.; Guillou, N.; Lamberti, C.; Bordiga, S.; Lillerud, K.P. A New Zirconium Inorganic Building Brick Forming Metal Organic Frameworks with Exceptional Stability. *J. Am. Chem. Soc.* **2008**, *130*, 13850–13851. [CrossRef]

122. Wei, Z.; Gu, Z.; Arvapally, R.K.; Chen, Y.; Mcdougald, R.N.; Ivy, J.F.; Yakovenko, A.A.; Feng, D.; Omary, M.A.; Zhou, H. Rigidifying Fluorescent Linkers by Metal–Organic Framework Formation for Fluorescence Blue Shift and Quantum Yield Enhancement. *J. Am. Chem. Soc.* **2014**, *136*, 8269–8276. [CrossRef]

123. Feng, D.; Gu, Z.; Li, J.; Jiang, H.; Wei, Z.; Zhou, H. Zirconium-Metalloporphyrin PCN-222: Mesoporous Metal–Organic Frameworks with Ultrahigh Stability as Biomimetic Catalysts. *Angew. Chem. Int. Ed.* **2012**, *51*, 10307–10310. [CrossRef]

124. Feng, D.; Gu, Z.; Chen, Y.; Park, J.; Wei, Z.; Sun, Y.; Bosch, M.; Yuan, S.; Zhou, H. A Highly Stable Porphyrinic Zirconium Metal–Organic Framework with shp-a Topology. *J. Am. Chem. Soc.* **2014**, *136*, 17714–17717. [CrossRef]

125. Feng, D.; Chung, W.; Wei, Z.; Gu, Z.; Jiang, H.; Chen, Y.; Darensbourg, D.J.; Zhou, H. Construction of Ultrastable Porphyrin Zr Metal–Organic Frameworks through Linker Elimination. *J. Am. Chem. Soc.* **2013**, *135*, 17105–17110. [CrossRef]

126. Jiang, H.; Feng, D.; Wang, K.; Gu, Z.; Wei, Z.; Chen, Y.; Zhou, H. An Exceptionally Stable, Porphyrinic Zr Metal–Organic Framework Exhibiting pH-Dependent Fluorescence. *J. Am. Chem. Soc.* **2013**, *135*, 13934–13938. [CrossRef]

127. Liu, T.; Feng, D.; Chen, Y.; Zou, L.; Bosch, M.; Yuan, S.; Wei, Z.; Fordgam, S.; Wang, K.; Zhou, H. Topology-Guided Design and Syntheses of Highly Stable Mesoporous Porphyrinic Zirconium Metal–Organic Frameworks with High Surface Area. *J. Am. Chem. Soc.* **2015**, *137*, 413–419. [CrossRef]

128. Zhang, M.; Chen, Y.; Bosch, M.; Gentle, T.; Wang, K.; Feng, D.; Wang, Z.U.; Zhou, H. Symmetry-Guided Synthesis of Highly Porous Metal–Organic Frameworks with Fluorite Topology. *Angew. Chem. Int. Ed.* **2014**, *53*, 815–818. [CrossRef]

129. Yuan, S.; Lu, W.; Chen, Y.; Zhang, Q.; Liu, T.; Feng, D.; Wang, X.; Qin, J.; Zhou, H. Sequential Linker Installation: Precise Placement of Functional Groups in Multivariate Metal–Organic Frameworks. *J. Am. Chem. Soc.* **2015**, *137*, 3177–3180. [CrossRef]

130. Feng, D.; Wang, K.; Su, J.; Liu, T.; Park, J.; Wei, Z.; Bosch, M.; Yakovenko, A.; Zou, X.; Zhou, H. A Highly Stable Zeotype Mesoporous Zirconium Metal–Organic Framework with Ultralarge Pores. *Angew. Chem. Int. Ed.* **2015**, *54*, 149–154. [CrossRef]

131. Yuan, S.; Qin, J.; Zou, L.; Chen, Y.; Wang, X.; Zhang, Q.; Zhou, H. Thermodynamically Guided Synthesis of Mixed-Linker Zr-MOFs with Enhanced Tunability. *J. Am. Chem. Soc.* **2016**, *138*, 6636–6642. [CrossRef] [PubMed]

132. Furukawa, H.; Gándara, F.; Zhang, Y.; Jiang, J.; Queen, W.L.; Hudson, M.R.; Yaghi, O.M. Water Adsorption in Porous Metal–Organic Frameworks and Related Materials. *J. Am. Chem. Soc.* **2014**, *136*, 4369–4381. [CrossRef] [PubMed]

133. Morris, W.; Volosskiy, B.; Demir, S.; Gándara, F.; Mcgrier, P.L.; Furukawa, H.; Cascio, D.; Stoddart, F.; Yaghi, O.M. Synthesis, Structure, and Metalation of Two New Highly Porous Zirconium Metal–Organic Frameworks. *Inorg. Chem.* **2012**, *51*, 6443–6445. [CrossRef] [PubMed]

134. Bon, V.; Senkovskyy, V.; Senkovska, I.; Kaskel, S. Zr(IV) and Hf(IV) based metal-organic frameworks with reo-topology. *Chem. Commun.* **2012**, *48*, 8407–8409. [CrossRef]

135. Bon, V.; Senkovska, I.; Weiss, M.S.; Kaskel, S. Tailoring of network dimensionality and porosity adjustment in Zr- and Hf-based MOFs. *CrystEngComm* **2013**, *15*, 9572–9577. [CrossRef]

136. Bon, V.; Senkovska, I.; Baburin, I.A.; Kaskel, S. Zr- and Hf-Based Metal–Organic Frameworks: Tracking Down the Polymorphism. *Cryst. Growth Des.* **2013**, *13*, 1231–1237. [CrossRef]

137. Deria, P.; Mondloch, J.E.; Tylianakis, E.; Ghosh, P.; Bury, W.; Snurr, R.Q.; Hupp, J.T.; Farha, O.K. Perfl uoroalkane Functionalization of NU-1000 via Solvent-Assisted Ligand Incorporation: Synthesis and CO_2 Adsorption Studies. *J. Am. Chem. Soc.* **2013**, *135*, 16801–16804. [CrossRef]

138. Gutov, O.V.; Bury, W.; Gomez-gualdron, D.A.; Krungleviciute, V.; Fairen-Jimenez, D.; Mondloch, J.E.; Sarjeant, A.A.; Al-Juaid, S.S.; Snurr, R.Q.; Hupp, J.T.; et al. Water-Stable Zirconium-Based Metal–Organic Framework Material with High-Surface Area and Gas-Storage Capacities. *Chem. Eur. J.* **2014**, *20*, 12389–12393. [CrossRef]

139. Wang, T.C.; Bury, W.; Gómez-Gualdrón, D.A.; Vermeulen, N.A.; Mondloch, J.E.; Deria, P.; Zhang, K.; Moghadam, P.Z.; Sarjeant, A.A.; Snurr, R.Q.; et al. Ultrahigh Surface Area Zirconium MOFs and Insights into the Applicability of the BET Theory. *J. Am. Chem. Soc.* **2015**, *137*, 3585–3591. [CrossRef]

140. Guillerm, V.; Ragon, F.; Devic, T.; Vishnuvarthan, M.; Campo, B.; Vimont, A.; Clet, G.; Yang, Q.; Maurin, G.; Férey, G.; et al. A Series of Isoreticular, Highly Stable, Porous Zirconium Oxide Based Metal–Organic Frameworks. *Angew. Chem. Int. Ed.* **2012**, *51*, 9267–9271. [CrossRef]

141. Wang, B.; Lv, X.; Feng, D.; Xie, L.; Zhang, J.; Li, M.; Xie, Y.; Li, J.; Zhou, H. Highly Stable Zr(IV)-Based Metal–Organic Frameworks for the Detection and Removal of Antibiotics and Organic Explosives in Water. *J. Am. Chem. Soc.* **2016**, *138*, 6204–6216. [CrossRef] [PubMed]

142. Schaate, A.; Dühnen, S.; Platz, G.; Lilienthal, S.; Schneider, A.M.; Behrens, P. A Novel Zr-Based Porous Coordination Polymer Containing Azobenzenedicarboxylate as a Linker. *Eur. J. Inorg. Chem.* **2012**, *2012*, 790–796. [CrossRef]

143. Lv, X.; Tong, M.; Huang, H.; Wang, B.; Gan, L.; Yang, Q.; Zhong, C.; Li, J. Journal of Solid State Chemistry A high surface area Zr (IV) -based metal–organic framework showing stepwise gas adsorption and selective dye uptake. *J. Solid State Chem.* **2015**, *223*, 104–108. [CrossRef]

144. Schaate, A.; Roy, P.; Preuße, T.; Lohmeier, S.J.; Godt, A.; Behrens, P. Porous Interpenetrated Zirconium–Organic Frameworks (PIZOFs): A Chemically Versatile Family of Metal–Organic Frameworks. *Chem. Eur. J.* **2011**, *17*, 9320–9325. [CrossRef] [PubMed]

145. Yoon, M.; Moon, D. New Zr (IV) based metal-organic framework comprising a sulfur-containing ligand: Enhancement of CO_2 and H_2 storage capacity. *Microporous Mesoporous Mater.* **2015**, *215*, 116–122. [CrossRef]

146. Kalidindi, S.B.; Nayak, S.; Briggs, M.E.; Jansat, S.; Katsoulidis, A.P.; Miller, G.J.; Warren, J.E.; Antypov, D.; Corà, F.; Slater, B.; et al. Chemical and Structural Stability of Zirconium-based Metal–Organic Frameworks with Large Three-Dimensional Pores by Linker Engineering. *Angew. Chem. Int. Ed.* **2015**, *54*, 221–226. [CrossRef]

147. Wang, R.; Wang, Z.; Xu, Y.; Dai, F.; Zhang, L.; Sun, D. Porous Zirconium Metal–Organic Framework Constructed from 2D → 3D Interpenetration Based on a 3,6-Connected kgd Net. *Inorg. Chem.* **2014**, *53*, 7086–7088. [CrossRef]

148. Cliffe, M.J.; Castillo-Martinez, E.; Wu, Y.; Lee, J.; Forse, A.C.; Firth, F.C.N.; Moghadam, P.Z.; Fairen-Jimenez, D.; Gaultois, M.W.; Hill, J.A.; et al. Metal–Organic Nanosheets Formed via Defect-Mediated Transformation of a Hafnium Metal–Organic Framework. *J. Am. Chem. Soc.* **2017**, *139*, 5397–5404. [CrossRef]

149. Ji, P.; Manna, K.; Lin, Z.; Feng, X.; Urban, A.; Song, Y.; Lin, W. Single-site cobalt catalysts at new $Zr_{12}(\mu_3\text{-}O)_8(\mu_3\text{-}OH)_8(\mu_2\text{-}OH)_6$ Metal-Organic framework nodes for highly active hydrogenation of nitroarenes, nitriles, and isocyanides. *J. Am. Chem. Soc.* **2017**, *139*, 7004–7011. [CrossRef]

150. Cao, L.; Lin, Z.; Shi, W.; Wang, Z.; Zhang, C.; Hu, X.; Wang, C.; Lin, W. Exciton Migration and Amplified Quenching on Two-Dimensional Metal–Organic Layers. *J. Am. Chem. Soc.* **2017**, *139*, 7020–7029. [CrossRef]

151. Tăbăcaru, A.; Galli, S.; Pettinari, C.; Masciocchi, N.; McDonald, T.M.; Long, J.R. Nickel(II) and copper (I,II)-based metal-organic frameworks incorprating an extended tris-pyrazolate linker. *CrystEngComm* **2015**, *17*, 4992–5001. [CrossRef]

152. Colombo, V.; Galli, S.; Choi, H.J.; Han, G.D.; Maspero, A.; Palmisano, G.; Masciocchi, N.; Long, J.R. High thermal and chemical stability in pyrazolate-bridged metal–organic frameworks with exposed metal sites. *Chem. Sci.* **2011**, *2*, 1311–1319. [CrossRef]

153. Choi, H.J.; Dinc, M.; Dailly, A.; Long, J.R. Hydrogen storage in water-stable metal–organic frameworks incorporating. *Energy Environ. Sci.* **2010**, *3*, 117–123. [CrossRef]

154. Wang, K.; Lv, X.; Feng, D.; Li, J.; Chen, S.; Sun, J.; Song, L.; Xie, Y.; Li, J.; Zhou, H. Pyrazolate-Based Porphyrinic Metal–Organic Framework with Extraordinary Base-Resistance. *J. Am. Chem. Soc.* **2016**, *138*, 914–919. [CrossRef]

155. Park, K.S.; Ni, Z.; Côté, A.P.; Choi, J.Y.; Huang, R.; Uribe-Romo, F.J.; Chae, H.K.; Keeffe, M.O.; Yaghi, O.M. Exceptional chemical and thermal stability of zeolitic imidazolate frameworks. *Proc. Natl. Acad. Sci. USA* **2006**, *10*, 10186–10191. [CrossRef]

156. Wu, H.; Qian, X.; Zhu, H.; Ma, S.; Zhu, G.; Long, Y. Controlled synthesis of highly stable zeolitic imidazolate framework-67 dodecahedra and their use towards the templated formation of a hollow Co_3O_4 catalyst for CO oxidation. *RSC Adv.* **2016**, *6*, 6915–6920. [CrossRef]

157. Morris, W.; Doonan, C.J.; Furukawa, H.; Banerjee, R.; Yaghi, O.M. Crystals as Molecules: Postsynthesis Covalent Functionalization of Zeolitic Imidazolate Frameworks. *J. Am. Chem. Soc.* **2008**, *130*, 12626–12627. [CrossRef]

158. Xiong, Y.; Chen, S.; Ye, F.; Su, L.; Zhang, C.; Shen, S.; Zhao, S. Synthesis of a mixed valence state Ce-MOF as an oxidase mimetic for the colorimetric detection of biothiols. *Chem. Commun.* **2015**, *51*, 4635–4638. [CrossRef]

159. Zhang, J.P.; Zhang, Y.B.; Lin, J.B.; Chen, X.M. Metal azolate frameworks: From crystal engineering to functional materials. *Chem. Rev.* **2012**, *112*, 1001–1033. [CrossRef]

160. Goupil, J.M.; Hemidy, J.F.; Cornet, D. Adsorption of NO_2 on modified Y zeolites. *Zeolites* **1982**, *2*, 47–50. [CrossRef]

161. Levasseur, B.; Ebrahim, A.M.; Bandosz, T.J. Role of Zr^{4+} cations in NO_2 adsorption on $Ce_{1-x}Zr_xO_2$ mixed oxides at ambient conditions. *Langmuir* **2011**, *27*, 9379–9386. [CrossRef] [PubMed]

162. Levasseur, B.; Ebrahim, A.M.; Bandosz, T.J. Interactions of NO_2 with amine-functionalized SBA-15: Effects of synthesis route. *Langmuir* **2012**, *28*, 5703–5714. [CrossRef] [PubMed]

163. Florent, M.; Tocci, M.; Bandosz, T.J. NO_2 adsorption at ambient temperature on urea-modified ordered mesoporous carbon. *Carbon* **2013**, *63*, 283–293. [CrossRef]

164. Belhachemi, M.; Jeguirim, M.; Limousy, L.; Addoun, F. Comparison of NO_2 removal using date pits activated carbon and modified commercialized activated carbon via different preparation methods: Effect of porosity and surface chemistry. *Chem. Eng. J.* **2014**, *253*, 121–129. [CrossRef]

165. Ryu, H.J.; Grace, J.R.; Lim, C.J. Simultaneous CO_2/SO_2 capture characteristics of three limestones in a fluidized-bed reactor. *Energy Fuels* **2006**, *20*, 1621–1628. [CrossRef]

166. Tchalala, M.R.; Bhatt, P.M.; Chappanda, K.N.; Tavares, S.R.; Adil, K.; Belmabkhout, Y.; Shkurenko, A.; Cadiau, A.; Heymans, N.; De Weireld, G.; et al. Fluorinated MOF platform for selective removal and sensing of SO_2 from flue gas and air. *Nat. Commun.* **2019**, *10*, 1–10. [CrossRef]

167. Savage, M.; Cheng, Y.; Easun, T.L.; Eyley, J.E.; Argent, S.P.; Warren, M.R.; Lewis, W.; Murray, C.; Tang, C.C.; Frogley, M.D.; et al. Selective Adsorption of Sulfur Dioxide in a Robust Metal–Organic Framework Material. *Adv. Mater.* **2016**, *28*, 8705–8711. [CrossRef]

168. Luan, Y.; Qi, Y.; Gao, H.; Zheng, N.; Wang, G. Synthesis of an amino-functionalized metal-organic framework at a nanoscale level for gold nanoparticle deposition and catalysis. *J. Mater. Chem. A* **2014**, *2*, 20588–20596. [CrossRef]

169. Vinoth, R.; Babu, S.G.; Bahnemann, D.; Neppolian, B. Nitrogen doped reduced graphene oxide hybrid metal free catalyst for effective reduction of 4-nitrophenol. *Sci. Adv. Mater.* **2015**, *7*, 1443–1449. [CrossRef]

170. Firestone, M.; Kavlock, R.; Zenick, H.; Kramer, M. The U.S. environmental protection agency strategic plan for evaluating the toxicity of chemicals. *J. Toxicol. Environ. Health Part B* **2010**, *13*, 139–162. [CrossRef]

171. Tacconelli, E.; Carrara, E.; Savoldi, A.; Harbarth, S.; Mendelson, M.; Monnet, D.L.; Pulcini, C.; Kahlmeter, G.; Kluytmans, J.; Carmeli, Y.; et al. Discovery, research, and development of new antibiotics: The WHO priority list of antibiotic-resistant bacteria and tuberculosis. *Lancet Infect. Dis.* **2018**, *18*, 318–327. [CrossRef]

172. Yan, C.; Yang, Y.; Zhou, J.; Liu, M.; Nie, M.; Shi, H.; Gu, L. Antibiotics in the surface water of the Yangtze Estuary: Occurrence, distribution and risk assessment. *Environ. Pollut.* **2013**, *175*, 22–29. [CrossRef] [PubMed]

173. Papadopoulos, F.; Parissopoulos, G.; Papadopoulos, A.; Zdragas, A.; Ntanos, D.; Prochaska, C.; Metaxa, I. Assessment of reclaimed municipal wastewater application on rice cultivation. *Environ. Manag.* **2009**, *43*, 135–143. [CrossRef] [PubMed]

174. Pan, M.; Chu, L.M. Fate of antibiotics in soil and their uptake by edible crops. *Sci. Total Environ.* **2017**, *599–600*, 500–512. [CrossRef]

175. Chen, F.; Ying, G.G.; Kong, L.X.; Wang, L.; Zhao, J.L.; Zhou, L.J.; Zhang, L.J. Distribution and accumulation of endocrine-disrupting chemicals and pharmaceuticals in wastewater irrigated soils in Hebei, China. *Environ. Pollut.* **2011**, *159*, 1490–1498. [CrossRef]

176. Segura, P.A.; Takada, H.; Correa, J.A.; Saadi, K.E.; Koike, T.; Onwona-Agyeman, S.; Ofosu-Anim, J.; Sabi, E.B.; Wasonga, O.V.; Mghalu, J.M.; et al. Global occurrence of anti-infectives in contaminated surface waters: Impact of income inequality between countries. *Environ. Int.* **2015**, *80*, 89–97. [CrossRef]

177. Martinez, J.L. Environmental pollution by antibiotics and by antibiotic resistance determinants. *Environ. Pollut.* **2009**, *157*, 2893–2902. [CrossRef]

178. Nandi, S.; Banesh, S.; Trivedi, V.; Biswas, S. A dinitro-functionalized metal-organic framework featuring visual and fluorogenic sensing of H_2S in living cells, human blood plasma and environmental samples. *Analyst* **2018**, *143*, 1482–1491. [CrossRef]

179. Yao, Z.Q.; Li, G.Y.; Xu, J.; Hu, T.L.; Bu, X.H. A Water-Stable Luminescent Zn^{II} Metal-Organic Framework as Chemosensor for High-Efficiency Detection of Cr^{VI}-Anions ($Cr_2O_7^{2-}$ and CrO_4^{2-}) in Aqueous Solution. *Chem. A Eur. J.* **2018**, *24*, 3192–3198. [CrossRef]

180. Hou, S.L.; Dong, J.; Jiang, X.L.; Jiao, Z.H.; Wang, C.M.; Zhao, B. Interpenetration-Dependent Luminescent Probe in Indium-Organic Frameworks for Selectively Detecting Nitrofurazone in Water. *Anal. Chem.* **2018**, *90*, 1516–1519. [CrossRef]

181. Håkansson, K.; Coorey, R.V.; Zubarev, R.A.; Talrose, V.L.; Håkansson, P. Low-mass ions observed in plasma desorption mass spectrometry of high explosives. *J. Mass Spectrom.* **2000**, *35*, 337–346. [CrossRef]

182. Moros, J.; Laserna, J.J. New Raman-laser-induced breakdown spectroscopy identity of explosives using parametric data fusion on an integrated sensing platform. *Anal. Chem.* **2011**, *83*, 6275–6285. [CrossRef] [PubMed]

183. Benito-Peña, E.; Urraca, J.L.; Moreno-Bondi, M.C. Quantitative determination of penicillin V and amoxicillin in feed samples by pressurised liquid extraction and liquid chromatography with ultraviolet detection. *J. Pharm. Biomed. Anal.* **2009**, *49*, 289–294. [CrossRef] [PubMed]

184. Samanta, P.; Desai, A.V.; Sharma, S.; Chandra, P.; Ghosh, S.K. Selective Recognition of Hg^{2+} ion in Water by a Functionalized Metal-Organic Framework (MOF) Based Chemodosimeter. *Inorg. Chem.* **2018**, *57*, 2360–2364. [CrossRef]

185. Wang, C.X.; Xia, Y.P.; Yao, Z.Q.; Xu, J.; Chang, Z.; Bu, X.H. Two luminescent coordination polymers as highly selective and sensitive chemosensors for Cr^{VI}-anions in aqueous medium. *Dalton Trans.* **2019**, *48*, 387–394. [CrossRef]

186. Liu, X.J.; Zhang, Y.H.; Chang, Z.; Li, A.L.; Tian, D.; Yao, Z.Q.; Jia, Y.Y.; Bu, X.H. A Water-Stable Metal-Organic Framework with a Double-Helical Structure for Fluorescent Sensing. *Inorg. Chem.* **2016**, *55*, 7326–7328. [CrossRef]

187. Xing, K.; Fan, R.; Du, X.; Song, Y.; Chen, W.; Zhou, X.; Zheng, X.; Wang, P.; Yang, Y. A windmill-like Zn_3L_2 cage exhibiting conformational change imparted sensing for DMA and highly selective naked-eye detection of Co^{2+} ion by dynamic quenching. *Sens. Actuators B Chem.* **2018**, *257*, 68–76. [CrossRef]

188. Du, T.; Zhang, H.; Ruan, J.; Jiang, H.; Chen, H.Y.; Wang, X. Adjusting the Linear Range of Au-MOF Fluorescent Probes for Real-Time Analyzing Intracellular GSH in Living Cells. *ACS Appl. Mater. Interfaces* **2018**, *10*, 12417–12423. [CrossRef]

189. Wang, M.; Guo, L.; Cao, D. Amino-Functionalized Luminescent Metal-Organic Framework Test Paper for Rapid and Selective Sensing of SO_2 Gas and Its Derivatives by Luminescence Turn-On Effect. *Anal. Chem.* **2018**, *90*, 3608–3614. [CrossRef]

190. Tabrizchi, M.; ILbeigi, V. Detection of explosives by positive corona discharge ion mobility spectrometry. *J. Hazard. Mater.* **2010**, *176*, 692–696. [CrossRef]

191. Moreno-González, D.; Lara, F.J.; Jurgovská, N.; Gámiz-Gracia, L.; García-Campaña, A.M. Determination of aminoglycosides in honey by capillary electrophoresis tandem mass spectrometry and extraction with molecularly imprinted polymers. *Anal. Chim. Acta* **2015**, *891*, 321–328. [CrossRef] [PubMed]

192. Blasco, C.; Corcia, A.D.; Picó, Y. Determination of tetracyclines in multi-specie animal tissues by pressurized liquid extraction and liquid chromatography-tandem mass spectrometry. *Food Chem.* **2009**, *116*, 1005–1012. [CrossRef]

193. Zhou, Z.; Han, M.L.; Fu, H.R.; Ma, L.F.; Luo, F.; Li, D.S. Engineering design toward exploring the functional group substitution in 1D channels of Zn-organic frameworks upon nitro explosives and antibiotics detection. *Dalton Trans.* **2018**, *47*, 5359–5365. [CrossRef] [PubMed]

194. Huang, X.Y.; Yue, K.F.; Jin, J.C.; Liu, J.Q.; Wang, C.J.; Wang, Y.Y.; Shi, Q.Z. Three-dimensional fivefold interpenetrating microporous metal-organic framework based on mixed flexible ligands. *Inorg. Chem. Commun.* **2010**, *13*, 338–341. [CrossRef]

195. Maspoch, D.; Ruiz-Molina, D.; Veciana, J. Old materials with new tricks: Multifunctional open-framework materials. *Chem. Soc. Rev.* **2007**, *36*, 770–818. [CrossRef]

196. Allendorf, M.D.; Bauer, C.A.; Bhakta, R.K.; Houk, R.J.T. Luminescent metal-organic frameworks. *Chem. Soc. Rev.* **2009**, *38*, 1330–1352. [CrossRef]

197. Dai, J.C.; Wu, X.T.; Fu, Z.Y.; Cui, C.P.; Hu, S.M.; Du, W.X.; Wu, L.M.; Zhang, H.H.; Sun, R.Q. Synthesis, structure, and fluorescence of the novel cadmium(II)-Trimesate coordination polymers with different coordination architectures. *Inorg. Chem.* **2002**, *41*, 1391–1396. [CrossRef]

198. Tian, D.; Li, Y.; Chen, R.Y.; Chang, Z.; Wang, G.Y.; Bu, X.H. A luminescent metal-organic framework demonstrating ideal detection ability for nitroaromatic explosives. *J. Mater. Chem. A* **2014**, *2*, 1465–1470. [CrossRef]

199. Yam, V.W.-W.; Lo, K.K.-W. Luminescent polynuclear d^{10} metal complexes. *Chem. Soc. Rev.* **1999**, *28*, 323–334. [CrossRef]

200. Cui, Y.; Yue, Y.; Qian, G.; Chen, B. Luminescent functional metal-organic frameworks. *Chem. Rev.* **2012**, *112*, 1126–1162. [CrossRef]

201. Fiore, A.M.; Naik, V.; Spracklen, D.V.; Steiner, A.; Unger, N.; Prather, M.; Bergmann, D.; Cameron-Smith, P.J.; Cionni, I.; Collins, W.J.; et al. Global air quality and climate. *Chem. Soc. Rev.* **2012**, *41*, 6663–6683. [CrossRef] [PubMed]

202. Lelieveld, J.; Evans, J.S.; Fnais, M.; Giannadaki, D.; Pozzer, A. The contribution of outdoor air pollution sources to premature mortality on a global scale. *Nature* **2015**, *525*, 367–371. [CrossRef] [PubMed]

203. Hoek, G.; Krishnan, R.M.; Beelen, R.; Peters, A.; Ostro, B.; Brunekreef, B.; Kaufman, J.D. Long-term air pollution exposure and cardio-respiratory mortality: A review. *Environ. Health* **2013**, *12*, 43–57. [CrossRef] [PubMed]

204. Seaton, A.; MacNee, W.; Donaldson, K.; Godden, D. Particulate air pollution and acute health effects. *Lancet* **1995**, *345*, 176–178. [CrossRef]

205. Choi, S.; Kim, H.R.; Kim, H.S. Fabrication of superabsorbent nanofibers based on sodium polyacrylate/poly (vinyl alcohol) and their water absorption characteristics. *Polym. Int.* **2019**, *68*, 764–771. [CrossRef]

206. Carroll, C.P.; Joo, Y.L. Axisymmetric instabilities of electrically driven viscoelastic jets. *J. Non-Newton. Fluid Mech.* **2008**, *153*, 130–148. [CrossRef]

207. Hwang, Y.J.; Choi, S.; Kim, H.S. Structural deformation of PVDF nanoweb due to electrospinning behavior affected by solvent ratio. *E-Polymers* **2018**, *18*, 339–345. [CrossRef]

208. Li, D.; Xia, Y. Fabrication of titania nanofibers by electrospinning. *Nano Lett.* **2003**, *3*, 555–560. [CrossRef]

209. Kong, C.S.; Lee, S.G.; Lee, S.H.; Lee, K.Y.; Noh, H.W.; Yoo, W.S.; Kim, H.S. Electrospinning Instabilities in the Drop Formation and Multi-Jet Ejection Part I: Various Concentrations of PVA (Polyvinyl Alcohol) Polymer Solution. *J. Macromol. Sci. Part B* **2011**, *50*, 517–527. [CrossRef]

210. Kong, C.S.; Yoo, W.S.; Jo, N.G.; Kim, H.S. Electrospinning mechanism for producing nanoscale polymer fibers. *J. Macromol. Sci. Part B Phys.* **2010**, *49*, 122–131. [CrossRef]

211. Dai, X.H.; Fan, H.X.; Yi, C.Y.; Dong, B.; Yuan, S.J. Solvent-free synthesis of a 2D biochar stabilized nanoscale zerovalent iron composite for the oxidative degradation of organic pollutants. *J. Mater. Chem. A* **2019**, *7*, 6849–6858. [CrossRef]

212. Song, J.Y.; Bhadra, B.N.; Khan, N.A.; Jhung, S.H. Adsorptive removal of artificial sweeteners from water using porous carbons derived from metal azolate framework-6. *Microporous Mesoporous Mater.* **2018**, *260*, 1–8. [CrossRef]

213. Liu, N.; Chen, Y.; Zhang, W.; Qu, R.; Zhang, Q.; Feng, L.; Wei, Y. A versatile CeO_2/Co_3O_4 coated mesh for food wastewater treatment: Simultaneous oil removal and UV catalysis of food additives. *Water Res.* **2018**, *137*, 144–152. [CrossRef] [PubMed]

214. Pintor, A.M.A.; Vilar, V.J.P.; Botelho, C.M.S.; Boaventura, R.A.R. Oil and grease removal from wastewaters: Sorption treatment as an alternative to state-of-the-art technologies. A critical review. *Chem. Eng. J.* **2016**, *297*, 229–255. [CrossRef]

215. Zhu, Z.; Wang, W.; Qi, D.; Luo, Y.; Liu, Y.; Xu, Y.; Cui, F.; Wang, C.; Chen, X. Calcinable Polymer Membrane with Revivability for Efficient Oily-Water Remediation. *Adv. Mater.* **2018**, *30*, 1–8. [CrossRef] [PubMed]

216. Chatzisymeon, E.; Stypas, E.; Bousios, S.; Xekoukoulotakis, N.P.; Mantzavinos, D. Photocatalytic treatment of black table olive processing wastewater. *J. Hazard. Mater.* **2008**, *154*, 1090–1097. [CrossRef] [PubMed]

217. Dong, J.; Xu, F.; Dong, Z.; Zhao, Y.; Yan, Y.; Jin, H.; Li, Y. Fabrication of two dual-functionalized covalent organic polymers through heterostructural mixed linkers and their use as cationic dye adsorbents. *RSC Adv.* **2018**, *8*, 19075–19084. [CrossRef]

218. Zhang, L.; Sun, J.S.; Sun, F.; Chen, P.; Liu, J.; Zhu, G. Facile Synthesis of Ultrastable Porous Aromatic Frameworks by Suzuki–Miyaura Coupling Reaction for Adsorption Removal of Organic Dyes. *Chem. A Eur. J.* **2019**, *25*, 3903–3908. [CrossRef]

219. Sun, D.T.; Gasilova, N.; Yang, S.; Oveisi, E.; Queen, W.L. Rapid, Selective Extraction of Trace Amounts of Gold from Complex Water Mixtures with a Metal-Organic Framework (MOF)/Polymer Composite. *J. Am. Chem. Soc.* **2018**, *140*, 16697–16703. [CrossRef]

Review

Recent Advances in MOF-based Nanocatalysts for Photo-Promoted CO$_2$ Reduction Applications

Chang Liu [1], Wenzhi Wang [2], Bin Liu [2], Jing Qiao [1], Longfei Lv [1], Xueping Gao [1,*], Xue Zhang [1], Dongmei Xu [3], Wei Liu [3], Jiurong Liu [1,*], Yanyan Jiang [1], Zhou Wang [1], Lili Wu [1] and Fenglong Wang [1,*]

[1] School of Materials Science and Engineering, Shandong University, Jinan 250061, China
[2] School of Materials Science and Engineering, University of Jinan, Jinan 250022, China
[3] State Key Laboratory of Crystal Materials, Shandong University, Jinan 250061, China
* Correspondence: xpgao@sdu.edu.cn (X.G.); jrliu@sdu.edu.cn (J.L.); fenglong.wang@sdu.edu.cn (F.W.);
Tel.: +86-531-883-99579 (F.W.)

Received: 3 July 2019; Accepted: 30 July 2019; Published: 31 July 2019

Abstract: The conversion of CO$_2$ to valuable substances (methane, methanol, formic acid, etc.) by photocatalytic reduction has important significance for both the sustainable energy supply and clean environment technologies. This review systematically summarized recent progress in this field and pointed out the current challenges of photocatalytic CO$_2$ reduction while using metal-organic frameworks (MOFs)-based materials. Firstly, we described the unique advantages of MOFs based materials for photocatalytic reduction of CO$_2$ and its capacity to solve the existing problems. Subsequently, the latest research progress in photocatalytic CO$_2$ reduction has been documented in detail. The catalytic reaction process, conversion efficiency, as well as the product selectivity of photocatalytic CO$_2$ reduction while using MOFs based materials are thoroughly discussed. Specifically, in this review paper, we provide the catalytic mechanism of CO$_2$ reduction with the aid of electronic structure investigations. Finally, the future development trend and prospect of photocatalytic CO$_2$ reduction are anticipated.

Keywords: Metal-organic frameworks (MOFs); photocatalysis; carbon dioxide reduction; renewable energy

---

1. Introduction

Energy shortages and environment issues are global problems and challenges that are faced by human beings today [1–6]. The development of renewable energy technologies to reduce the pollutants emission has become an important research topic to maintain the sustainable development of our planet [7–12]. Artificial photosynthesis is an ideal way to effectively solve the energy and environmental problems by decomposing water to produce hydrogen or reducing CO$_2$ to high value-added chemicals or fuels [13–16]. Accordingly, searching for highly efficient materials that can convert solar energy and store it in chemicals is desired. Metal-organic frameworks (MOFs), which are known as coordination porous polymer, is a class of crystalline porous materials constructed by the coordination bond between metal ions or metal cluster nodes [17–21]. These materials have been widely used in gas separation/storage, catalysis, sensing, proton conductors, and drug delivery because of their structural diversity, design/modification, and ultra-high specific surface areas [22–25]. Based on the previous results, it is proven that the multifunctional organic ligands in the MOFs structure can play the role of "light capture antenna" [26,27]. It can effectively accept photons, generate band gap transition, and transfer electrons to metal center units. Thus, MOFs are usually used as efficient photocatalysts [28–30]. When comparing to other photocatalytic materials, MOFs exhibit big specific surface area, high porosity, and supervised capturing capability of CO$_2$ molecules, which endows them with great application prospect in the field of photocatalysis for CO$_2$ reduction. In recent years,

MOFs and their composite materials are widely used in water decomposition, hydrogen production, CO_2 reduction, and photocatalytic organic conversion [31].

Yaghi group first proposed the concept of the metal-organic frameworks in 1995, and MOFs materials were then intensively explored as new functional materials [32]. In 1997, Kitagawa group reported a three-dimensional MOF material and found its ability to adsorb gas at room temperature [33]. After that, two landmark cases of MOFs, MOF-5 and HKUST-1, were reported by Yaghi group and Williams group in 1999 [34–36]. Among them, MOF-5 is a three-dimensional skeleton that formed by coordination of $Zn_4O(CO_2)_6$ clusters and terephthalic acid ligands. Through the gas adsorption experiments, the authors found that MOFs-5 showed high specific surface area, large pore size, and a certain adsorption capacity for hydrogen. HKUST-1, as reported, is a three-dimensional skeleton that is formed by the coordination of $Cu_2(CO_2)_4$ clusters with benzotriformic acid ligands [37]. The authors found that HUKST-1 with unsaturated ligand sites can be obtained by heating water molecules that can be removed and coordinated on metal clusters [38]. Jinhee et al. report the the OCS-activation ability of chloromethanes to remove precoordinated solvent molecules from open coordination sites (OCSs) in MOFs [39]. A water molecule in HKUST-1 can easily access open metal site (OMS)with high coordination strength due to the specific coordination geometry around Cu^{2+} [40]. In particular, MOFs with OCSs have potential applications in chemical separation, molecular sorption, catalysis, ionic conduction, and sensing areas [39,41]. Since these two MOF structures were reported, the synthesis and potential applications of MOFs in gas separation, storage, catalysis, sensing, drug transportation, and so on have become hot research topics [42,43].

MOFs are extensively studied for the capture and conversion of CO_2 due to their high porosity and strong interaction with CO_2 molecules. At present, some MOFs have already been explored for their high catalytic performance in the field of photocatalysis for CO_2 reduction [44]. As photocatalysts, MOFs exhibit the following advantages. Firstly, the high specific surface area of MOFs is helpful for the gas reactants adsorption around the active site. This is beneficial to the molecule activation and catalytic transformation in the subsequent process [45,46]. Secondly, the metal-oxygen units in MOFs exhibit semiconductor-like structure due to the existence of organic ligands. MOFs with larger energy than the band gap can be excited by photons to create electron and hole pairs [47,48]. Through selectively choosing different organic ligands and metal centers, one can improve the absorption and utilization efficiency of sunlight via MOFs as light absorbing agents [49]. Besides, the separation and transfer of electrons can be promoted by changing the crystal structure, thereby which thereby inhibits the recombination of photo-induced electrons and holes [50]. In addition, MOFs, as heterogeneous catalysts, can be easily separated and recycled from the reaction system, which is beneficial for prolonging the service life of the catalyst and avoiding any pollution to the environment [51–53].

In this paper, the advances of MOFs materials for photocatalytic CO_2 reduction is systematically reviewed. This review paper starts from the research background why CO_2 reduction is important, and the mechanism studies on the photocatalytic CO_2 reduction process were then summarized. After that, the research progress of photocatalytic CO_2 reduction using MOFs were reviewed, followed by the summary of the applications of MOFs-based composite materials for photocatalytic reduction of CO_2. Finally, the current challenges and future development trend of MOFs-based materials for photocatalytic CO_2 reduction are anticipated.

2. Necessity

A large amount of fossil fuels has been combusted since the eighteenth century, so that the atmospheric CO_2 concentration increased gradually. According to the data of the National Oceanic and Atmospheric Administration (NOOA), the CO_2 concentration has exceeded 400 ppm in May, 2013, and it reached 402 ppm in May 2014 [54]. It is believed that the atmospheric CO_2 concentration will exceed 550 ppm at the end of this century [55,56]. The sudden increase of CO_2 concentration in the atmosphere can be attributed to the over-use of the fossil fuels. Currently, more than 80% of the global energy supply origins from fossil fuels, which generates a large amount of CO_2 in the atmosphere.

A hundred years ago, Arrhenius suggested that CO_2 emissions from the burning of fossil raw materials would lead to an increase in global temperatures [57,58]. Today, CO_2 has been widely accepted as the chief culprit causing global warming, climate upheaval, and many other environmental problems. Various environmental problems will become much sharper if there are no effective measures are taken to curb CO_2 emissions [59]. When the atmospheric CO_2 content rises to 450 ppm, the accompanying increase in global temperature will seriously aggravate the cessation of the hot salt circulation, and environmental problems, like melting of glaciers, will take place [60].

In the 21st century, in addition to serious environmental problems, the energy crisis is also a global issue affecting human society. In 2008, the total global energy consumption was 132,000 megawatts. According to the U.S. [61] Energy Information Administration, this number will continuously grow, and the total energy consumption in 2040 is expected to be twice of that in 2020. Although the over exploitation and use of fossil energy has caused global warming and energy crisis, we can still find some opportunities and challenges to debate these issues [62]. For example, while using a suitable method to convert CO_2 into energy materials or valuable industrial raw materials is a promising solution to close the carbon loop, and can alleviate the dependence of human beings on fossil energy and solve environmental problems that are caused by CO_2 emissions [63,64].

3. Mechanisms

Photocatalytic CO_2 reduction involves three basic processes. Under light irradiation, the electron-hole pairs could be generated in semiconductor materials upon the absorption of photons with larger energy than the forbidden band gap [65]. Subsequently, the photoexcited electron-hole pairs separate and migrate to the active sites on the surface of the semiconductor. In this process, it is necessary to reduce the bulk phase and surface recombination of photogenerated electron-hole pairs. This is the major factor limiting the efficiency of photocatalytic reduction of CO_2 [66,67]. After that, oxidation and reduction reactions occur on the surface of the semiconductor. At this time, electrons with strong enough reducing ability can reduce CO_2 molecules into hydrocarbons, such as CO, CH_4, and CH_3OH, and holes with oxidizing ability oxidize H_2O molecules to release O_2, O^{2-}, and other substances [68]. The conversion efficiency of photocatalytic CO_2 reduction depends on the capacity of the light-trapping ability of the semiconductor material, the efficiency of photo-generated carrier generation and separation, and the thermodynamic equilibrium of the surface catalytic reactions. From the kinetic point of view, the effective absorption of light, the efficient separation and migration of photo-generated electron-hole pairs, and the sufficient reactive sites on the catalyst surface are an important prerequisite for the high-efficiency photocatalytic conversion of CO_2 while using semiconductor materials [69].

The detailed mechanisms for photocatalytic CO_2 reduction process have not been discovered so far. However, mechanism studies in recent years provide valuable information to unravel this process [70]. At present, it is commonly accepted that photocatalytic CO_2 reduction is a multi-electron reduction process, as described in the Equations (2)–(8). It can be seen that the reaction process is accompanied by some unstable substances, namely intermediates. The corresponding products are different due to the specific reaction route and the number of electrons obtained during the reaction [71,72]. According to the number of electrons that were obtained by C atom, the products can be carbon monoxide, methane, formic acid, methanol, etc. [73]. In some special reaction system, some multi-carbon compounds such as ethane, acetic acid, and other compounds can also be obtained. From the perspective of Gibbs free energy, photocatalytic reduction of CO_2 is an uphill reaction, that is $\Delta G > 0$. If the reaction proceeds, a large amount of energy injection (such as incident photons) is required.

Reaction Eredox/ (V vs NHE,PH=7)

$$CO_2 + e^- \rightarrow CO^- \qquad -1.90 \tag{1}$$

$$CO_2 + H^+ + 2e^- \rightarrow HCO_2^- \qquad -0.49 \tag{2}$$

$$CO_2 + 2H^+ + 2e^- \rightarrow CO + H_2O \qquad -0.53 \tag{3}$$

$$CO_2 + 4H^+ + 4e^- \rightarrow HCHO + H_2O \qquad -0.48 \tag{4}$$

$$CO_2 + 6H^+ + 6e^- \rightarrow CH_3OH + H_2O \qquad -0.38 \tag{5}$$

$$CO_2 + 8H^+ + 8e^- \rightarrow CH_4 + 2H_2O \qquad -0.24 \tag{6}$$

$$2H^+ + 2e^- \rightarrow H_2 \qquad -0.41 \tag{7}$$

$$H_2O \rightarrow 0.5\,O_2 + 2H^+ + 2e^- \qquad 0.82 \tag{8}$$

Hendon et al. [74] elucidated the electronic structure of MIL-125 with aminated linkers through a combination of synthesis and computation. They also discussed the band gap modification of MIL-125, a TiO_2/1,4-benzenedicarboxylate (bdc) MOF, and the possible mechanism for the photocatalytic CO_2 reduction was proposed (Figure 1).

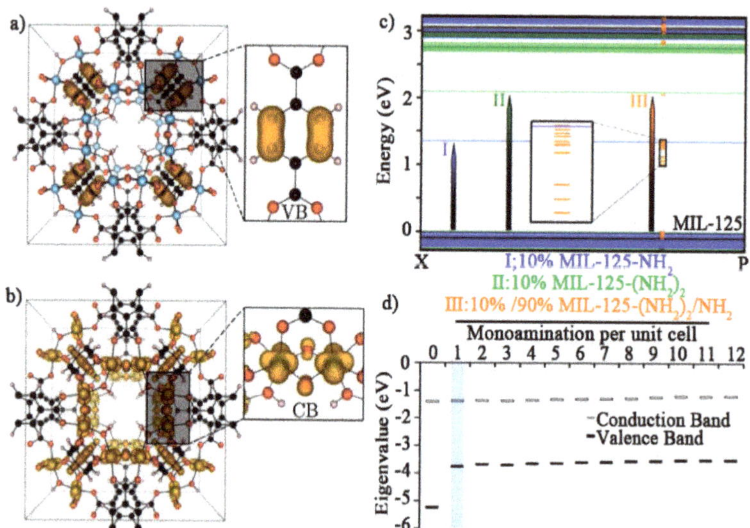

Figure 1. (**a**) the valence band is composed of the bdc C 2p orbitals (shown on the right), making these favorable for linker-based band gap modifications; (**b**) the conduction band is composed ofO 2p orbitals and Ti 3d orbitals (shown on the right). (**c**) PBEsol band structures for synthetic MIL-125 (black), 10%-MIL-125-NH2 (blue), 10%-MIL-125-(NH2)2/90%-MIL-125-NH2 (orange) and the theoretical 10%-MIL-125-(NH2)2 (green). (**d**) HSE06-calculated VB and CB energies of MIL-125-NH2 models containing increasing numbers of bdc-NH2 linkers [i.e. 0 (MIL-125) to 12 (100%-MIL-125-NH2)] per unit cell. MOFs materials for photocatalytic CO_2 reduction. Reprinted from ref. 74 with permission by the American Chemical Society.

Photocatalytic CO_2 reduction using MOFs-based materials as catalysts has drawn dramatic research interests in recent years. It is easy to design MOFs materials with accessible metal sites, specific hetero-atoms, and the ordered structure of functionalized organic ligands. This can effectively improve the efficiency of electron-hole separation and the photocatalytic performance. Porosity can make MOFs expose more active sites and channels for reactant adsorption. This can improve the charge transfer efficiency as well as improve its utilization efficiency of solar energy while inhibiting the recombination of the photo-induced electron-hole pairs in the bulk phase. Based on the above merits, people try to use different MOFs for photocatalytic CO_2 reduction. In the following text, we will introduce three typical MOFs for photocatalytic CO_2 reduction and their catalytic performances. New insights for the dominating factors on activity and selectivity of product will also be discussed. Table 1 summarizes the research progress of several typical MOF materials for photocatalytic CO_2 reduction in recent years.

Table 1. the research progress of several typical metal-organic frameworks (MOF) materials for photocatalytic CO_2 reduction.

Sample	Light Source Conditions	Product	Productivity	Ref.
$Zr_6O_4(OH)_4(bpdc)_6$	Visible light	CO	-	75
MIL-101(Fe)	Visible light	$HCOO^-$	7.375μmol/h	76
PCN-222	Visible light	$HCOO^-$	3.12μmol/h	77
NNU-28	Visible light	dicarboxylic acid	183.3μmol/h	78
$Zr_6O_4(OH)_4(L)\bullet6DMF$	Visible light	$HCOO^-$	96.2μmol/ h	79
NH_2-Uio-66(Zr)	Visible light	$HCOO^-$	1.32μmol/h	80
Ag-Ren-MOF	Visible light	CO	-	81
UiO-66-CAT	Visible light	HCOOH	9μmo/h	82
MOF-525-Co	Visible light	CO	36.67μmol/h	83
$Cd_{0.2}Zn_{0.8}S@UiO-66-NH_2$	Visible light	CH_3OH	-	84
Co-ZIF-9	Visible light	CO	28.54μmol/h	85
ZIF-67	Visible light	CO	3.89μmol/h	86
Ag@Co-ZIF-9	Visible light	CO	28.4μmol/h	87
Zn-MOF nanoliths	Visible light	CO	-	88
Zn_2GeO_4/Mg-MOF-74	Visible light	CO	1.43μmol/h	89
TiO_2-ZIF-8	Visible light	MeOH	1.21μmol/h	90
Zn/PMOF	Visible light	CH_4	10.43μmol/h	91
Co-ZIF-9/TiO_2	Visible light	CH_4	-	92
Cu-TiO_2/ZIF-8	UV-light	CO	-	93
$CsPbBr_3$@ZIFs	Visible light	CO	29.630μmol/h	94
$Ti_8O_8(OH)_4(bdc)_6$(MIL-125(Ti))	365nm UV-light	$HCOO^-$	0.814μmol/ h	95

3.1. Zr MOFs

In 2011, Wang et al. [75] chelated metal ions (such as Ir, Re, and Ru) with 4,4-biphenyldicarboxylic acid derivatives as organic ligands to construct MOFs, and the Zr-based MOF (UiO-67) systems with different metal doping were obtained. A similar synthesis strategy has also been adopted by Wang et al. who used ligand H_2L_4 for photocatalytic reduction of CO_2 to CO [76]. The total conversion number (TON) of CO_2 reduction can reach 10.9. The photocatalytic activity can be improved by doping a variety of photoactive metal nanoparticles inside MOFs. Subsequently, the authors observed a significant decrease in photocatalytic activity through a series of comparison experiments, which proved that the metal nanoparticles themselves are the real active sites that are involved in the photocatalytic reaction.

In 2015, Xu et al. [77] chose Zr-MOF (PCN-222) containing porphyrin as catalysts and found that it could be used as a stable photocatalyst to reduce CO_2 to formate ion under visible light. It was found that PCN-222 exhibited broad-spectrum absorption properties. There existed a series of extremely long lifetime electron trap states in the material, which could inhibit the recombination of photogenerated charge carriers and improve the photoreduction efficiency of CO_2. In 2016, Chen et al. [78] synthesized a new microporous stable zirconium-based metal organic skeleton (NNU-28) from 4,4'-(anthracene-9,10-bis (2,1-ethynylphenyl) dicarboxylic acid, which was used to reduce CO_2 to formate while using triethanolamine as the sacrifice agent. Under visible light irradiation, the rate of catalytic conversion of CO_2 to formate ion was 183.3 μmol/h. It was found that, in the catalytic reaction, the ligands produced about 27.3% formate ions, while the metal clusters produced about 77.7% formate ions. Under light irradiation, anthracene derivative ligands not only acted as an effective light collector, but it also sensitized Zr_6 oxygen clusters through the LMCT (linker-to-metal charge transfer) process. At the same time, the ligand itself can also be stimulated to form free radicals and produce photogenerated electrons. Figure 2 shows two catalytic pathways for the reduction of CO_2 to formate. This strategy is helpful for the design and development of MOFs materials with efficient visible light response [78].

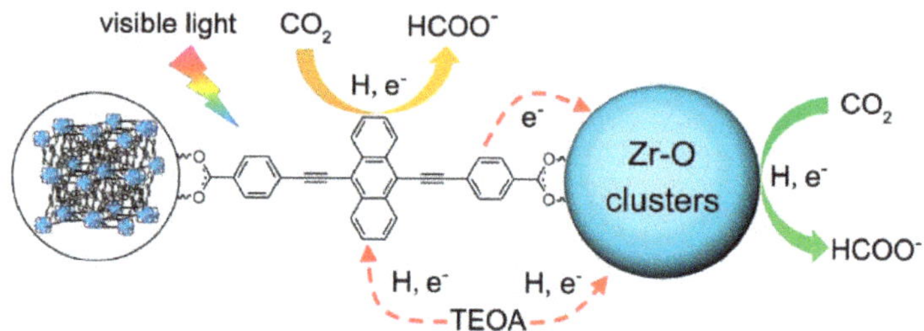

Figure 2. Two catalytic pathways for the reduction of CO_2 to formate. Reprinted from ref. 78 with permission by the Royal Society of Chemistry.

In 2018, Sun et al. [79] synthesized a porous zirconium based metal-organic framework $[(Zr_6O_4(OH)_4(L)\cdot 6DMF)$ while using dicarboxyl ligands (H_2L=2,2'-diamino -4,4'-stilbene dicarboxylic acid, DMF) with conjugated imine function. The materials showed high chemical stability and remarkable visible light absorption properties. The average rate of $HCOO^-$ formation of MOFs is about 96.2 μmol/h.

Sun et al. [80] compared the activity of NH_2-UiO-66(Zr) and NH_2-MIL-125(Ti) for photocatalytic reduction of CO_2 under visible light. The results showed that the catalytic performance of NH_2-UiO-66(Zr) was higher than that of NH_2-MIL-125(Ti) under the same reaction conditions. This is ascribed to the effective transfer of photogenerated electrons from ATA to Zr-O clusters, and made Zr-O clusters efficient photocatalytic active sites. Furthermore, some ATA ligands were replaced by 2,5-diamino terephthalic acid (DTA) and the mixed ligand NH_2-UiO-66(Zr) was obtained. It was found that the CO_2 conversion of mixed NH_2-UiO-66(Zr) was 50% higher than that of pure NH_2-UiO-66(Zr). This may be because the mixed NH_2-UiO-66(Zr) showed strong photoabsorption capacity and large CO_2 adsorption capacity, so its photocatalytic activity is obviously improved.

Choi et al. [81] reported the synthesis of composited catalysts by covalently binding $Re^I(CO)_3$(bpydc)Cl(as Re TC) to UiO-67 to Re_n-MOFs (n is the density of Re TC in the pores of MOF). Subsequently, the MOF was further modified with cubic silver nanoparticles to obtain Ag-Re_n-MOF, thus the photocatalytic activity of CO_2 conversion was significantly improved (Figure 3A, [81]). The PXRD (powder X-ray diffraction) patterns showed that the single crystal Re_3-MOF structure is preserved when different amount of Re TC is introduced into Re_n-MOF (Figure 3B, [81]). By studying the process of photocatalytic conversion of CO_2 by Re_n-MOF (Figure 3C, [81]), it was found that the catalytic activity of Re_3-MOF was the highest. In addition, under visible light irradiation, the activity of $AgRe_3$-MOF was five times higher than that of Re_3-MOF, and the conversion efficiency of CO_2 to CO was increased by seven times. This is mainly because MOF has large porosity and CO_2 adsorption capacity, which is conducive to the occurrence of catalytic reduction reaction. On the other hand, precious metals have a wide range of photo absorption and are easier to trap photogenerated electrons due to the lower Fermi levels. At the same time, their stability could be further improved due to the strong covalent bond between Re TC and MOF.

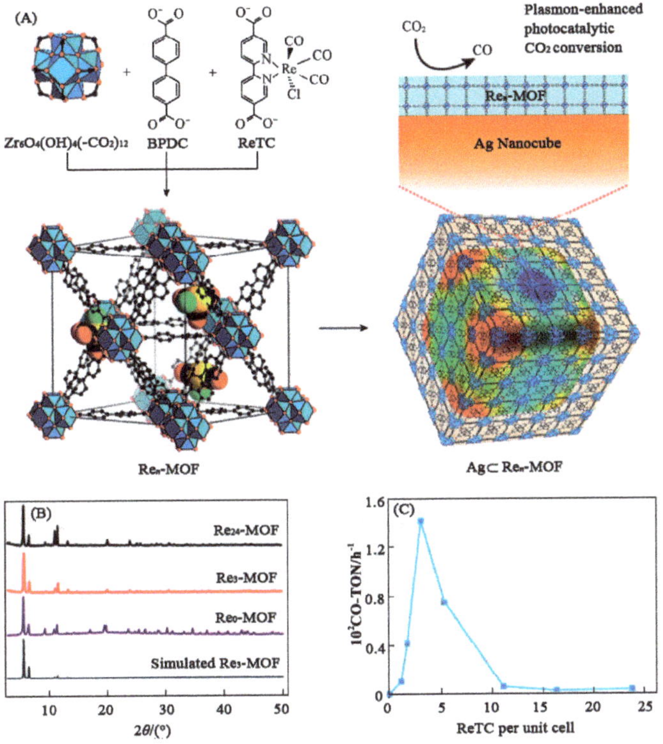

Figure 3. Structures of Ren-MOF and Ag Ren-MOF based catalysts (**A**), PXRD of Ren-MOFs (**B**), and the photocatalytic activity of Ren-MOF (**C**). Reprinted from ref. 81 with permission by the American Chemical Society.

Lee et al. [82] used UiO-66 (Zirconium 1,4-Carboxybenzene) as a precursor to obtain UiO-66-CAT with Cr^{3+} or Ga^{3+} sites as catalysts for photocatalytic CO_2 reduction. In the presence of TEOA and BNAH, the TON (turnover number) values of UiO-66-Cr CAT and UiO-66- Ga CAT are 11.22 ±0.37 and 6.14±0.22, and the amount of HCOOH that is produced by catalytic reduction of CO_2 after visible light irradiation for 6h were (51.73±2.64) and (28.78 ±2.52) μmol, respectively. The activity of UiO-66-Cr CAT is about twice higher than that of UiO-66-Ga CAT, which is mainly attributed to the fact that Cr^{3+} is more efficient than Ga^{3+} for the rapid transfer of electrons. At the same time, Cr derivatives show higher reduction efficiency than Ga derivatives due to their open shell structure.

Zhang et al. [83] reported Zr- porphyrin MOF (MOF-525-Co) as efficient catalysts for CO_2 conversion. Using TEOA as a sacrificial agent, MOF-525-Co could efficiently catalyze the reduction of CO_2 to CO and CH_4 under visible light irradiation. When compared with Zn-MOF-525 and MOF-525, MOF-525-Co showed the highest catalytic activity and CO_2 adsorption capacity. The metallized MOFs is obviously improved, and exhibited strong charge separation ability and energy conversion efficiency. The highest catalytic performance of cobalt metallized MOFs is mainly due to the fact that the introduction of monoatomic Co into MOF-525 can significantly improve the electron-hole separation efficiency in porphyrin ligands. At the same time, the photogenerated electrons rapidly migrated from the porphyrin center to the surface of the catalyst, thus the electrons with long lifetime were obtained, which effectively activated the CO_2 molecules that were adsorbed on the Co center.

Su et al. [84] prepared a series of $Cd_{0.2}Zn_{0.8}S@UiO-66-NH_2$ composites with different UiO-66-NH$_2$ content by solvothermal method, which were used for photocatalytic reduction of CO_2 to CH_3OH. The results showed that the single UiO-66-NH$_2$ showed no activity for photocatalytic CO_2 reduction, but $Cd_xZn_{1-x}S$ with adjustable composition and band gap could be efficiently excited by visible light. All of the $Cd_{0.2}Zn_{0.8}S@UiO-66-NH_2$ samples showed excellent photocatalytic activity when compared with $Cd_{0.2}Zn_{0.8}S$. When the content of UiO-66-NH$_2$ was 20% (mass fraction), the catalyst showed the best photocatalytic activity, and the formation rate of CH_3OH is 3.4 times higher than that of single structure $Cd_{0.2}Zn_{0.8}S$. This is mainly due to the effective charge separation and transfer at the interface between $Cd_{0.2}Zn_{0.8}S$ and UiO-66-NH$_2$. Thus, the photogenerated electrons that were absorbed by $Cd_{0.2}Zn_{0.8}S$ and UiO-66-NH$_2$ can be quickly transferred to the surface for CO_2 reduction. In addition, $Cd_{0.2}Zn_{0.8}S@UiO-66-NH_2$ photocatalyst showed excellent stability in the process of photocatalytic reduction of CO_2.

3.2. Zn MOFs

In 2015, Wang et al. [85] reported the establishment of CO_2 photoreduction system while using the CdS semiconductor and Co-ZIF-9 as catalyst and co-catalyst, respectively. Under mild reaction conditions, the reaction system of bipyridine and triethanolamine showed high catalytic activity when CO_2 was deoxidized to CO under visible light irradiation. Under the irradiation of monochromatic light at a wavelength of 420 nm, the quantum efficiency could reach 1.93%.

In 2018, Wang et al. [86] synthesized a series of ZIF-67 nanocrystals with a different morphology by the solvent induction method. Taking the advantages of MOF, the capture of CO_2 was controlled by controlling its morphology, and their photocatalytic performance was further improved. In the same year, Chen [87] and co-workers fabricated the Ag-Co-ZIF-9 nanocomposited materials with different Ag loading by the photo deposition method to study the effect of Ag NPs on the reaction performance of Co-ZIF-9 in CO_2 photo reduction reaction. In this study, Co-ZIF-9, with a rod structure was obtained by the reflux method, and ultra-small Ag nanoparticles (< 5 nm) were doped into Co-ZIF-9 by photodeposition. With the help of photosensitizer, the Ag@Co-ZIF-9 composite showed the catalytic performance of converting CO_2 to CO under the irradiation of visible light. With the increase of Ag nanoparticles, the formation of CO obviously increased while the amount of H_2 decreased. When compared with pure Co-ZIF-9, the photocatalytic activity of Ag@Co-ZIF-9 can be improved by two times (about 28.4 µmol CO), and selectivity about 20% (22.9 µmol H_2). The experimental results showed that Ag NPs in Co-ZIF-9 could act as an electron trap and active site for CO_2 reduction, thus the efficiency and selectivity of MOF materials in CO_2 photo reduction were improved.

Subsequently, Ye et al. [88] developed and used the ultra-thin two-dimensional Zn-MOF nanoliths to reduce CO_2 to CO. They firstly tried to establish two novel non-precious metal mixed photocatalytic systems. The catalyst showed excellent photocatalytic activity and selectivity under mild reaction conditions. It was confirmed that the Zn-MOF nanoparticles show better charge transfer ability than the Zn-MOF bulk materials via electrochemical impedance and PL (photoluminescence) spectroscopy analysis, thus stronger photocatalytic efficiency and selectivity were obtained. This provides feasibility for the application of photocatalysis in the development of various two-dimensional (2D) MOF materials.

In 2018, Zhao et al. [89] prepared Zn_2GeO_4/Mg-MOF-74 composites by the hydrothermal method (Figure 4). When the water was used as agent, the photocatalytic activity of Zn_2GeO_4/Mg-MOF-74 for CO_2 reduction reaction is higher than that of pure Zn_2Ge_4 nanorods or the physical mixture of Zn_2GeO_4 and Mg-MOF-74. This is mainly due to the stronger CO_2 adsorption performance of Mg-MOF-74, the lower recombination probability of photogenerated electron-hole pair and more alkali metal sites on the surface of Mg-MOF-74. In addition, the effect of H_2O on the reaction was also studied and the results show that H_2O is the reducing agent and hydrogen source involved in the reaction. In the process of reduction, the photogenerated electrons from the conduction band reduce CO_2 to CO and HCOOH, by the reaction of $CO_2+2e^-+2H^+\rightarrow HCOOH$ and $CO_2+2e^-+2H^+\rightarrow CO+H_2O$, in which the content of HCOOH is very small.

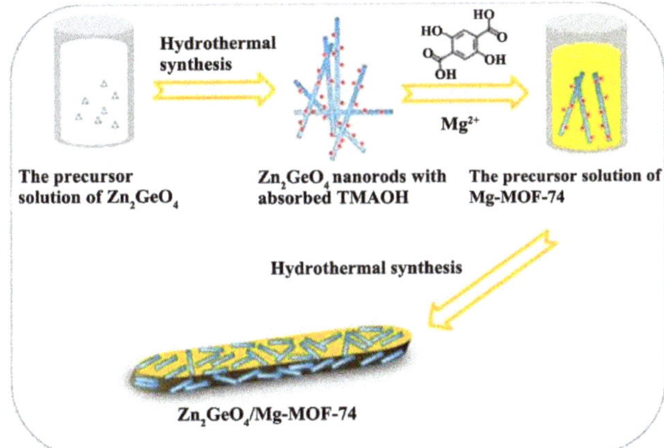

Figure 4. Schematic illustration of the synthesis of the Zn_2GeO_4/Mg-MOF-74 composites. Reprinted from ref. 89 with permission by the Royal Society of Chemistry.

In 2018, Cardoso et al. [90] modified TiO_2 nanotubes and formed a core-shell structure by layer growth of ZIF-8 nanoparticles on the surfaces. The FT-IR spectra show that the host-guest interaction depends on the pore structure and chemical properties of MOF connectors. Under UV irradiation at room temperature, CO_2 can be photocatalyzed to methanol and ethanol fuel on the electrode of composited materials. Zinc-based MOF not only provided the adsorption/activation of CO_2, but also acted as a light absorber to transfer excited electrons for photocatalytic reduction.

Sadeghi et al. [91] synthesized zinc-based porphyrin (Zn/PMOF), which could catalytically reduce CO_2 to CH_4 under light irradiation. The results showed that the yield of CH_4 was 10.43 μmol when Zn/PMOF was used as photocatalyst. After 4h irradiation, Zn/PMOF was much higher than that of CH_4 when ZnO was used as photocatalyst. At the same time, Zn/PMOF as photocatalyst showed high selectivity for CO_2 reduction, and it has better stability and repeatability when comparing to ZnO.

Yan et al. [92] loaded different amounts of TiO_2 on Co-ZIF-9 to construct Co-ZIF-9/TiO_2 nanostructure composites (ZIFx/T, x is the mass ratio of Co-ZIF-9 in the composites, T is TiO_2). The results showed that ZIF0.03/ T showed the best catalytic conversion efficiency of CO_2, and the yield of Ti/T is 2.1 times higher than that of pure TiO_2 catalyst after irradiation for 10h. Linear sweep voltammetry in CO_2 saturated solution further reveals that Co-ZIF-9 can effectively activate CO_2 and reduce the CO_2 reduction initiation potential of ZIFx/T (x ≤ 0.10). In addition, the photoluminescence spectra show that the ZIFx/T composites that were prepared by in-situ synthesis showed higher charge separation efficiency. Therefore, better CO_2 adsorption capacity and charge separation rate are beneficial to the high activity of ZIFx/T nanostructures in photocatalytic transformation.

Maina et al. [93] designed a catalytic system based on membrane reactor. The controllable encapsulation of TiO_2 and Cu^{2+} doped TiO_2 nanoparticles (Cu-TiO_2) in ZIF-8 film was realized by the rapid thermal deposition (RTD) method (Figure 5A, [93]). Under ultraviolet irradiation, the Cu-TiO_2/ZIF-8 hybrid film showed high photocatalytic activity. The results show that, when compared with the amount produced by the original ZIF-8 film alone, the yields of CO and CH_3OH increased by 188% and 50%, respectively (Figure 5B, [93]). Further studies showed that the yields of photocatalytic reduction of CO_2 to CH_3OH and CO depend on the content of Cu-TiO_2 nanoparticles that are loaded on MOF films (Figure 5C, [93]). When the loading of Cu-TiO_2 nanoparticles is 7 μg, Cu-TiO_2/ZIF-8 exhibited the best catalytic efficiency. When compared with the original ZIF-8 film, the yields of CO and CH_3OH increased by 23.3% and 70%, respectively. The sharp increase of product originated from the synergistic effect between the ability of semiconductor nanoparticles to produce photoexcited electrons under light irradiation and the high CO_2 adsorption capacity of MOF.

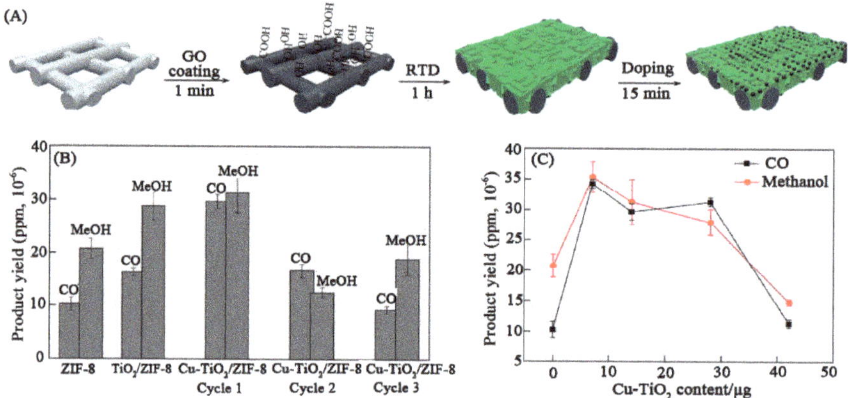

Figure 5. Fabrication of Cu-TiO$_2$/ ZIF-8 membranes (**A**), effect of membrane composition (**B**) and Cu-TiO$_2$nanoparticles loading on the product yields (**C**). Reprinted from ref. 93 with permission by the American Chemical Society.

Kong et al. [94] prepared CsPbBr$_3$@ ZIFs composites by in-situ synthesis used as CO$_2$ reduction photocatalyst with reinforcing activity (Figure 6A, [94]). The electron consumption rates of CsPbBr$_3$@ZIF-8 and CsPbBr$_3$@ZIF-67 are 15.498 and 29.630 µmol·g^{-1}·h^{-1}, which is 1.39 and 2.66 times higher than that of pure CsPbBr$_3$, respectively. The comparison of photocatalytic CO$_2$ reduction performance using CsPbBr$_3$ and CsPbBr$_3$@ZIFs showed that the ZIF coating greatly improved the catalytic activity of CsPbBr$_3$ (Figure 6B, [94]). In addition, six cycle experiments have been carried out on CsPbBr$_3$@ZIF, and it was found that the electron consumption rate suffered from negligible decrease. This indicates that it possessed good stability (Figure 6C, [94]). The synergistic effect of CsPbBr$_3$ and ZIF coating improved the stability of CsPbBr$_3$ to water molecules and enhanced the CO$_2$ capture ability and the charge separation efficiency. All of these lead to a higher conversion efficiency. Moreover, the catalytic active center Co in ZIF-67 could further accelerate the process of charge separation, activate CO$_2$ molecules, and improve the catalytic activity of CO$_2$ reduction.

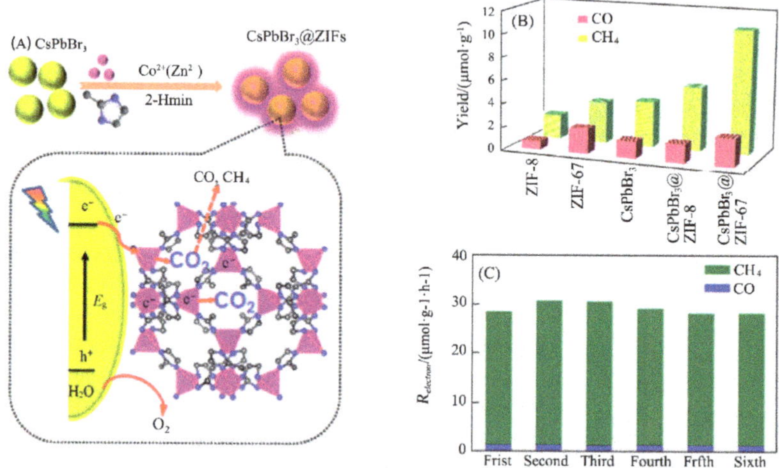

Figure 6. Schematic illustration of the fabrication process and CO$_2$ photoreduction process of CsPbBr$_3$/ZIFs (**A**) and photocatalytic CO$_2$ reduction performances of CsPbBr$_3$ and CsPbBr$_3$@ZIFs (**B,C**). Reprinted from ref. 94 with permission by the American Chemical Society.

3.3. Ti MOFs

In 2012, Fu et al. [95] reported a photosensitive MOF $Ti_8O_8(OH)_4(bdc)_6(MIL-125(Ti))$ for photocatalytic CO_2 reduction. The photocatalytic activity evaluation indicated that Ti-MOF could efficiently reduce CO_2 to $HCOO^-$ under 365 nm UV irradiation. When comparing to other MOFs, MIL-125(Ti) showed slightly higher activity. The photocatalytic results of NH_2-MIL-125 showed that the concentration of $HCOO^-$ increased in the reaction system with the extension of irradiation time, and the formation of HCOO- reached 8.14 µmol within 10 hours. On one hand, the introduction of NH_2 can promote the rapid transfer of electrons from O to Ti, in $TiO_5(OH)$ metal cluster. On the other hand, NH_2 can significantly improve the adsorption capacity of NH_2-MIL-125 (Ti) to CO_2, which is beneficial for the adsorption and activation of CO_2 in the process of photocatalytic reaction. In 2018, He [96] designed an MOF-based ternal-composite photocatalyst $(TiO_2/Cu_2O/Cu_3(BTC)_2)$ to increase the density of charge carrier and promote the activation of CO_2 molecules to improve the photoreduction capacity of CO_2. The experimental results showed that the addition of Cu_2O and $Cu_3(BTC)_2$ not only significantly improved the light conversion efficiency of CO_2, but also facilitated the formation of CH_4. The increase of charge carrier density improved the overall performance of the catalyst. The PL, XPS, and DRIFT analysis verified that the coordination of unsaturated metal sites were helpful in activating CO_2. This study provides a new way to solve the problems of low charge density and efficiency CO_2 activation, and it also provides a reasonable design for in-depth understanding of CO_2 photoreduction and other applications of mixed nanomaterials based on MOF.

4. Prospect of Photocatalytic CO_2 Reduction

The advantages of MOFs-based photocatalytic materials are obvious when comparing to conventional semiconductor materials. Thus, they have attracted more and more research attentions in photocatalysis. However, the low efficiency of this technology still hinders its wide applications in industry. The following problems should be addressed in the future. Firstly, researchers need to put forward effective strategies to improve the light absorption properties and charge separation performances. Secondly, most MOFs are not as metal oxide for semiconductor photocatalysts, especially in water or under ultraviolet light, which ultimately leads to the decreased catalyst life; hence, how to enhance their robustness is another important topic. Thirdly, there are few studies on the mechanism of photocatalytic CO_2 reduction in MOFs, especially the current understanding of the catalytic reaction path is still blurred. In addition, most of the reported photocatalytic CO_2 reduction reactions are carried out in organic solvents, requiring additional sacrificial agents. The future materials for catalytic reduction of CO_2 should be economical and environmentally friendly. Therefore, it is urgent to solve the above problems of MOFs materials for photocatalytic CO_2 reduction.

5. Conclusions

Artificial photosynthesis using catalysts to convert CO_2 to high value-added chemicals or fuels is an ideal way to effectively solve energy and environmental problems. The MOFs materials exhibit great application prospects in the field of photocatalysis, due to its ultra-high specific surface area, porous properties, modified/regulated textures, and high capture capability for CO_2 molecules. The advantages and significance of MOFs materials in CO_2 catalytic reduction are described in detail. Meanwhile, the application of typical MOFs in CO_2 photoreduction, for example, Zr-MOFs, Zn-MOFs, and Ti-MOFs, were introduced and summarized. Finally, the future development trend and prospect of photocatalytic CO_2 reduction are anticipated in this review.

Funding: This work was supported by the National Natural Science Foundation of China (No. 51572157), the Natural Science Foundation of Shandong Province (No. ZR2016BM16), Qilu Young Scholar Program of Shandong University (No. 31370088963043), and the Fundamental Research Funds of Shandong University (No. 2018JC036, 2018JC046).

Conflicts of Interest: The authors declare no conflict of interest. The funders had no role in the design of the study; in the collection, analyses, or interpretation of data; in the writing of the manuscript, or in the decision to publish the results.

References

1. Goeppert, A.; Czaun, M.; Jones, J.P.; Surya Prakash, G.K.; Olah, G.A. Recycling of carbon dioxide to methanol and derived products-closing the loop. *Chem. Soc. Rev.* **2014**, *43*, 7995–8048. [CrossRef] [PubMed]

2. Lanzafame, P.; Centi, G.; Perathoner, S. Catalysis for biomass and CO_2 use through solar energy: Opening new scenarios for a sustainable and low-carbon chemical production. *Chem. Soc. Rev.* **2014**, *43*, 7562–7580. [CrossRef] [PubMed]

3. Usubharatana, P.; McMartin, D.; Veawab, A.; Tontiwachwuthikul, P. Photocatalytic process for CO_2 emission reduction from industrial flue gas streams. *Ind. Eng. Chem. Res.* **2006**, *45*, 2558–2568. [CrossRef]

4. Habisreutinger, S.N.; Schmidt-Mende, L.; Stolarczyk, J.K. Photocatalytic reduction of CO_2 on TiO_2 and other semiconductors. *Angew. Chem. Int. Ed.* **2013**, *52*, 7372–7408. [CrossRef] [PubMed]

5. Logan, M.W.; Ayad, S.; Adamson, J.D.; Dilbeck, T.; Hanson, K.; Uribe-Romo, F.J. Systematic variation of the optical bandgap in titanium based isoreticular metal-organic frameworks for photocatalytic reduction of CO_2 under blue light. *J. Mater. Chem. A* **2017**, *5*, 11854–11863. [CrossRef]

6. White, J.L.; Baruch, M.F.; Pander, J.E.; Hu, Y.; Fortmeyer, I.C.; Park, J.E.; Zhang, T.; Liao, K.; Gu, J.; Yan, Y.; et al. Light-driven heterogeneous reduction of carbon dioxide: Photocatalysts and photoelectrodes. *Chem. Rev.* **2015**, *115*, 12888–12935. [CrossRef] [PubMed]

7. Joshi, V.V.; Meier, A.; Darsell, J.; Nachimuthu, P.; Bowden, M.; Weil, K.S. Short-term oxidation studies on nicrofer-6025HT in air at elevated temperatures for advanced coal based power plants. *Oxid. Met.* **2013**, *79*, 383–404. [CrossRef]

8. Wang, S.; Wang, X. Imidazolium ionic liquids, imidazolylidene heterocyclic carbenes, and zeolitic imidazolate frameworks for CO_2 capture and photochemical reduction. *Angew. Chem. Int. Ed.* **2015**, *55*, 2308–2320. [CrossRef] [PubMed]

9. Liu, X.; Inagaki, S.; Gong, J. Heterogeneous molecular systems for photocatalytic CO_2 reduction with water oxidation. *Angew. Chem. Int. Ed.* **2016**, *55*, 14924–14950. [CrossRef]

10. Pan, J.; Wu, X.; Wang, L.Z.; Liu, G.; Lu, M.; Cheng, H.M. Synthesis of aanatase TiO_2 rods with dominant reactive {010} facets for the photoreduction of CO_2 to CH_4 and use in dye-sensitized solar cells. *Chem. Commun.* **2011**, *47*, 8361–8363. [CrossRef]

11. Roy, S.; Varghese, O.; Paulose, M.; Grimes, C.A. Toward solar fuels: Photocatalytic conversion of carbon dioxide to hydrocarbons. *ACS Nano* **2010**, *4*, 1259–1278. [CrossRef] [PubMed]

12. Kondratenko, E.V.; Mul, G.; Baltrusaitis, J.; Larrazábal, G.O.; Ramírez, J.P. Status and perspectives of CO_2 conversion into fuels and chemicals by catalytic, photocatalytic and electrocatalytic processes. *Energ. Environ. Sci.* **2013**, *6*, 3112–3135. [CrossRef]

13. Mozia, S. Generation of useful hydrocarbons and hydrogen during photocatalytic decomposition of acetic acid on CuO/rutile photocatalysts. *Int. J. Photoenergy* **2009**, *2009*, 469069. [CrossRef]

14. Seki, T.; Kokubo, Y.; Ichikawa, S.; Suzuki, T.; Kayaki, Y.; Ikariya, T. Mesoporous silica-catalysed continuous chemical fixation of CO_2 with N, N'-dimethylethylenediamine in supercritical CO_2: The efficient synthesis of 1, 3-dimethyl-2-imidazolidinone. *Chem. Commun.* **2009**, *3*, 349–351. [CrossRef] [PubMed]

15. Anpo, M.; Yamashita, H.; Ichihashi, Y.; Ehara, S. Photocatalytic reduction of CO_2 with H_2O on various titanium oxide catalysts. *Electroanal. Chem.* **2012**, *43*, 3165–3172. [CrossRef]

16. Kazuhiko, M.; Keita, S.; Osamu, I. A polymeric-semiconductor-metal-complex hybrid photocatalyst for visible-light CO_2 reduction. *Chem. Commun.* **2013**, *49*, 10127–10129.

17. Susumu, K.; Ryo, K.; Shin-Ichiro, N. Functional porous coordination polymers. *Angew. Chem.* **2004**, *43*, 2334–2375.

18. Li, R.; Hu, J.H.; Deng, M.S.; Wang, H.L.; Wang, X.J.; Hu, Y.L.; Jiang, H.L.; Jiang, J.; Zhang, Q.; Xie, Y.; et al. Metal-organic frameworks: Integration of an inorganic semiconductor with a metal-organic framework: A platform for enhanced gaseous photocatalytic reactions. *Adv. Mater.* **2014**, *26*, 4783–4788. [CrossRef]

19. Koen, B. Lanthanide-based luminescent hybrid materials. *Chem. Rev.* **2009**, *109*, 4283–4374.

20. Yang, Q.Y.; Liu, D.H.; Zhong, C.L.; Li, J.R. Development of computational methodologies for metal-organic frameworks and their application in gas separations. *Chem. Rev.* **2013**, *113*, 8261–8323. [CrossRef]

21. Yi, F.Y.; Li, J.P.; Wu, D.; Sun, Z.M. A series of multifunctional metal-organic frameworks showing excellent luminescent sensing, sensitization, and adsorbent abilities. *Chemistry* **2015**, *21*, 11475–11482. [CrossRef] [PubMed]

22. Hu, Z.C.; Benjamin, J.D.; Li, J. Luminescent metal-organic frameworks for chemical sensing and explosive detection. *Chem. Soc. Rev.* **2014**, *43*, 5815–5840. [CrossRef] [PubMed]
23. He, X.; Gan, Z.; Fisenko, S.; Wang, D.; El-Kaderi, H.M.; Wang, W.N. Rapid formation of metal-organic frameworks (Mofs) based nanocomposites in microdroplets and their applications for CO_2 photoreduction. *ACS Appl. Mater. Interfaces* **2017**, *9*, 9688–9698. [CrossRef] [PubMed]
24. Furukawa, H.; Cordova, K.E.; O'Keeffe, M.; Yaghi, O.M. The chemistry and applications of metal-organic frameworks. *Science* **2013**, *341*, 1230444. [CrossRef] [PubMed]
25. Low, Z.X.; Yao, J.F.; Liu, Q.; He, M.; Wang, H.T. Crystal transformation in zeolitic-imidazolate framework. *Cryst. Growth Des.* **2014**, *14*, 6589–6598. [CrossRef]
26. Stassen, I.; Burtch, N.C.; Talin, A.A.; Falcaro, P.; Allendorf, M.D.; Ameloot, R. Correction: An updated roadmap for the integration of metal-organic frameworks with electronic devices and chemical sensors. *Chem. Soc. Rev.* **2017**, *46*, 3853. [CrossRef]
27. Lee, J.Y.; Farha, O.K.; Roberts, J.; Scheidt, K.A.; Nguyen, S.B.T.; Hupp, J.T. Metal-organic framework materials as catalysts. *Chem. Soc. Rev.* **2009**, *38*, 1450–1459. [CrossRef]
28. Chen, X.B.; Wang, X.Y.; Zhu, D.D.; Yan, S.J.; Lin, J. Three-component domino reaction synthesis of highly functionalized bcyclic pyrrole derivatives. *Tetrahedron* **2014**, *45*, 1047–1054. [CrossRef]
29. Liu, S.W.; Feng, C.; Li, S.T.; Peng, X.X.; Xiong, Y. Enhanced photocatalytic conversion of greenhouse gas CO_2 into solar fuels over g-C_3N_4 nanotubes with decorated transparent ZIF-8 nanoclusters. *Appl. Catal. B Environ.* **2017**, *211*, 1–10. [CrossRef]
30. Crake, A.; Christoforidis, K.C.; Kafizas, A.; Zafeiratos, S.; Petit, C. CO_2 capture and photocatalytic reduction using bifunctional TiO_2/Mofs nanocomposites under UV–Vis irradiation. *Appl. Catal. B Environ.* **2017**, *210*, 131–140. [CrossRef]
31. Long, J.R.; Yaghi, O.M. The pervasive chemistry of metal-organic frameworks. *Chem. Soc. Rev.* **2009**, *38*, 1213–1214. [CrossRef] [PubMed]
32. Li, B.Y.; Leng, K.Y.; Zhang, Y.M.; James, J.D.; Wang, J.; Hu, Y.F.; Ma, D.X.; Shi, Z.; Zhu, L.K.; Zhang, D.L. Metal-organic framework based upon the synergy of a brønsted acid framework and lewis acid centers as a highly efficient heterogeneous catalyst for fixed-bed reactions. *J. Am. Chem. Soc.* **2015**, *137*, 4243–4248. [CrossRef] [PubMed]
33. Yong, Y.; Suyetin, M.; Bichoutskaia, E.; Blake, A.J.; Allan, D.R.; Barnett, S.A.; Schroder, M. Modulating the packing of [Cu_{24}(isophthalate)$_{24}$] cuboctahedra in a triazole-containing metal-organic polyhedral framework. *Chem. Sci.* **2013**, *4*, 1731–1736.
34. Hirscher, M.; Panella, B. Hydrogen storage in metal-organic frameworks. *Chem. Rev.* **2007**, *56*, 809–835. [CrossRef]
35. Ben, V.D.V.; Bart, B.; Joeri, D.; Dirk, D.V. Adsorptive separation on metal-organic frameworks in the liquid phase. *Chem. Soc. Rev.* **2014**, *43*, 5766–5788.
36. Lee, J.Y.; Farha, O.K.; Roberts, J.; Scheidt, K.A.; Nguyen, S.B.T.; Hupp, J.T. Cheminform abstract: Metal-organic framework materials as catalysts. *Cheminform* **2010**, *40*, 1450–1459. [CrossRef]
37. Gascon, J.; Corma, A.; Kapteijn, F.; Xamena, F.X.L.I. Metal organic framework catalysis: Quo vadis? *ACS Catal.* **2014**, *4*, 361–378. [CrossRef]
38. Mercedes, A.; Esther, C.; Belén, F.; Xamena, F.X.; Llabrés, I.; Hermenegildo, G. Semiconductor behavior of a metal-organic framework (MOF). *Chem. Eur. J.* **2007**, *13*, 5106–5112.
39. Bae, J.; Lee, E.J.; Jeong, N.C. Metal coordination and metal activation abilities of commonly unreactive chloromethanes toward metal-organic frameworks. *Chem. Commun.* **2018**, *54*, 6458–6471. [CrossRef]
40. Song, D.; Bae, J.; Ji, H.; Kim, M.-B.; Bae, Y.-S.; Park, K.S.; Moon, D.; Jeong, N.C. Coordinative reduction of metal nodes enhances the hydrolytic stability of a paddlewheel metal-organic framework. *Am. Chem. Soc.* **2019**, *141*, 7853–7864. [CrossRef]
41. Howarth, A.J.; Peters, A.W.; Vermeulen, N.A.; Wang, T.C.; Hupp, J.T.; Farha, O.K. Best practices for the synthesis, activation, and characterization of metal-organ frameworks. *Chem. Mater.* **2017**, *29*, 26–39. [CrossRef]
42. Lin, R.B.; Li, F.; Liu, S.Y.; Qi, X.L.; Zhang, J.P.; Chen, X.M. A noble-metal-free porous coordination framework with exceptional sensing efficiency for oxygen. *Angew. Chem. Int. Ed.* **2013**, *52*, 13429–13433. [CrossRef]

43. Kataoka, Y.; Sato, K.; Miyazaki, Y.; Masuda, K.; Tanaka, H.; Naito, S.; Mori, W. Photocatalytic hydrogen production from water using porous material [Ru_2(ρ-BDC)$_2$]$_n$. *Energy Environ. Sci.* **2009**, *2*, 397–400. [CrossRef]

44. Hu, X.; Sun, C.I.; Qin, C.; Wang, X.; Wang, H.K.; Zhou, E.; Li, W.; Su, Z.I. Iodine-emplated assembly of unprecedented 3d–4f metal-organic frameworks as photocatalysts for hydrogen generation. *Chem. Commun.* **2013**, *49*, 3564–3566. [CrossRef]

45. Koroush, S.; Lin, Q.P.; Mao, C.Y.; Feng, P.Y. Incorporation of iron hydrogenase active sites into a highly stable metal-organic framework for photocatalytic hydrogen generation. *Chem. Commun.* **2014**, *50*, 10390–10393.

46. Feng, D.W.; Gu, Z.Y.; Li, J.R.; Jiang, H.L.; Wei, Z.W.; Zhou, H.C. Zirconium-metalloporphyrin PCN-222: Mesoporous metal-organic frameworks with ultrahigh stability as biomimetic catalysts. *Angew. Chem. Int. Ed.* **2012**, *51*, 10307–10310. [CrossRef]

47. Subhadeep, S.; Gobinda, D.; Jayshri, T.; Rahul, B. Photocatalytic metal-organic framework from Cds quantum dot incubated luminescent metallohydrogel. *J. Am. Chem. Soc.* **2014**, *136*, 14845–14851.

48. Lee, Y.; Kim, S.; Ku, K.J.; Cohen, S.M. Photocatalytic CO_2 reduction by a mixed metal (Zr/Ti), mixed ligand metal-organic framework under visible light iradiation. *Chem. Commun.* **2015**, *51*, 5735–5738. [CrossRef]

49. Sonja, P.; Fei, H.; Andreas, O.; Cohen, S.M.; Sascha, O. Enhanced photochemical hydrogen production by a molecular diiron catalyst incorporated into a metal-organic framework. *J. Am. Chem. Soc.* **2013**, *135*, 16997–17003.

50. Liu, Q.; Low, Z.X.; Li, L.; Razmjou, A.; Wang, K.; Yao, J.F.; Wang, H. ZIF-8/Zn_2GeO_4 nanorods with an enhanced CO_2 adsorption property in an aqueous medium for photocatalytic synthesis of liquid fuel. *J. Mater. Chem. A* **2013**, *1*, 11563–11569. [CrossRef]

51. Gastaldo, C.M.; Antypov, D.; Warren, J.E.; Briggs, M.E.; Chater, P.A.; Wiper, P.V.; Miller, G.J.; Khimyak, Y.Z.; Darling, G.R.; Berry, N.G. Side-Chain Control of Porosity Closure in Single and Multiple-Peptide-Based Porous Materials by Cooperative Folding. *Nat. Chem.* **2014**, *6*, 343–351. [CrossRef]

52. Zhou, J.J.; Wang, R.; Liu, X.L.; Peng, F.M.; Li, C.H.; Teng, F.; Yuan, Y.P. In situ growth of CdS nanoparticles on UiO-66 metal-organic framework octahedrons for enhanced photocatalytic hydrogen production under visible light irradiation. *Appl. Surf. Sci.* **2015**, *346*, 278–283. [CrossRef]

53. Wang, S.B.; Yao, W.S.; Lin, J.L.; Ding, Z.X.; Wang, X.C. Cobalt imidazolate metal-organic frameworks photosplit CO_2 under mild reaction conditions. *Angew. Chem. Int. Ed.* **2014**, *53*, 1034–1038. [CrossRef]

54. Kim, D.; Kelsey, K.S.; Hong, D.; Yang, P.D. Artificial photosynthesis for sustainable fuel and chemical production. *Angew. Chem. Int. Ed.* **2015**, *54*, 3259–3266. [CrossRef]

55. Serena, B.; Samuel, D.; Laia, F.; Carolina, G.S.; Miguel, G.; Craig, R.; Thibaut, S.; Antoni, L. Molecular artificial photosynthesis. *Chem. Soc. Rev.* **2014**, *43*, 7501–7519.

56. Kumar, B.; Llorente, M.; Froehlich, J.; Dang, T.; Sathrum, A.; Kubiak, C.P. Photochemical and photoelectrochemical reduction of CO_2. *Annu. Rev. Phys. Chem.* **2012**, *63*, 541–569. [CrossRef]

57. Tao, A.; Prasert, S.; Yang, P.D. Polyhedral silver nanocrystals with distinct scattering signatures. *Angew. Chem. Int. Ed.* **2006**, *45*, 4597–4601. [CrossRef]

58. Chambers, M.B.; Wang, X.; Elgrishi, N.; Christopher, H.H.; Aron, W.; Jonathan, B.; Canivet, J.; Alessandra, Q.E.; David, F.; Caroline, M.D. Photocatalytic carbon dioxide reduction with rhodium-based catalysts in solution and heterogenized within metal-organic frameworks. *ChemSusChem* **2015**, *8*, 603–608. [CrossRef]

59. Hou, W.; Wei, H.H.; Pavaskar, P.; Goeppert, A.; Aykol, M.; Cronin, S.B. Photocatalytic conversion of CO_2 to hydrocarbon fuels via plasmon-enhanced absorption and metallic interband transitions. *ACS Catal.* **2011**, *1*, 929–936. [CrossRef]

60. Meister, S.; Reithmeier, R.O.; Tschurl, M.; Heiz, U.; Rieger, B. Unraveling side reactions in the photocatalytic reduction of CO_2: Evidence for light-induced deactivation processes in homogeneous photocatalysis. *ChemSusChem* **2015**, *7*, 690–697. [CrossRef]

61. Suljo, L.; Phillip, C.; Ingram, D.B. Plasmonic-metal nanostructures for efficient conversion of solar to chemical energy. *Nat. Mater.* **2011**, *10*, 911–921.

62. Hou, W.; Cronin, S.B. A review of surface plasmon resonance-enhanced photocatalysis. *Adv. Funct. Mater.* **2013**, *23*, 1612–1619. [CrossRef]

63. Tu, W.; Zhou, Y.; Li, H.; Li, P.; Zou, Z. Au@TiO_2 yolk-shell hollow spheres for plasmonInduced photocatalytic reduction of CO_2 into solar fuel via local electromagnetic field. *Nanoscale* **2015**, *7*, 14232–14236. [CrossRef]

64. Gao, S.T.; Liu, W.H.; Shang, N.Z.; Feng, C.; Wu, Q.H.; Wang, Z.; Wang, C. Integration of a plasmonic semiconductor with a metal-organic framework: A case of Ag/AgCl@ZIF-8 with enhanced visible light photocatalytic activity. *RSC Adv.* **2014**, *4*, 61736–61742. [CrossRef]

65. Yuan, X.Z.; Hou, W.; Yan, W.; Zeng, G.M.; Chen, X.H.; Leng, L.J.; Wu, Z.B.; Hui, L. One-pot self-assembly and photoreduction synthesis of silver nanoparticle-decorated reduced graphene oxide/MIL-125(Ti) photocatalyst with improved visible light photocatalytic activity. *Appl. Organomet. Chem.* **2016**, *30*, 289–296. [CrossRef]

66. Wheeler, D.A.; Zhang, Z.J. Exciton dynamics in semiconductor nanocrystals. *Adv. Mater.* **2013**, *25*, 2878–2896. [CrossRef]

67. Easun, T.L.; Jia, J.H.; James, A.C.; Danielle, L.B.; Stapleton, C.S.; Vuong, K.Q.; Champness, N.R.; George, M.W. Photochemistry in a 3D metal-organic framework (MOF): Monitoring intermediates and reactivity of the fac-to-mer photoisomerization of Re(Diimine)(CO)$_3$Cl incorporated in a MOF. *Inorg. Chem.* **2014**, *53*, 2606–2612. [CrossRef]

68. Matsuoka, S.; Kohzuki, T.; Pac, C.; Ishida, A.; Takamuku, S.; Kusaba, M.; Nakashima, N.; Yanagida, S. Photocatalysis of oligo(P-phenylenes): Photochemical reduction of carbon dioxide with triethylamine. *J. Phys. Chem.* **1992**, *96*, 4437–4442. [CrossRef]

69. Smieja, J.M.; Benson, E.E.; Bhupendra, K.; Grice, K.A.; Seu, C.S.; Miller, A.J.M.; Mayer, J.M.; Kubiak, C.P. Kinetic and structural studies, origins of selectivity, and interfacial charge transfer in the artificial photosynthesis of Co. *Proc. Natl. Acad. Sci. USA* **2012**, *109*, 15646–15650. [CrossRef]

70. Yuan, Y.P.; Ruan, L.W.; Barber, J.; Loo, J.; Xue, C. Hetero-nanostructured suspended photocatalysts for solar-to-fuelconversion. *Energy Environ. Sci.* **2014**, *7*, 3934–3951. [CrossRef]

71. Hod, I.; Sampson, M.D.; Deria, P.; Kubiak, C.P.; Farha, O.K.; Hupp, J.T. Fe-porphyrin-based metal-organic framework films as high-surface concentration, heterogeneous catalysts for electrochemical reduction of CO$_2$. *ACS Catal.* **2015**, *5*, 6302–6309. [CrossRef]

72. Benson, E.E.; Kubiak, C.P. Structural investigations into the deactivation pathway of the CO$_2$ reduction electrocatalyst Re(Bpy)(CO)$_3$Cl. *Chem. Commun.* **2012**, *48*, 7374–7376. [CrossRef]

73. Cavka, J.H.; Jakobsen, S.; Olsbye, U.; Guillou, N.; Lamberti, C.; Bordiga, S.; Lillerud, K.P. A new zirconium inorganic building brick forming metal organic frameworks with exceptional stability. *J. Am. Chem. Soc.* **2008**, *130*, 13850–13851. [CrossRef]

74. Hendon, C.H.; Tiana, D.; Fontecave, M.; Sanchez, C.; D'arras, L.; Sassoye, C.; Rozes, L.; Mollot-Draznieks, C.; Walsh, A. Engineering the optical response of the titanium-MIL-125 metal-organic framework through ligand functionalization. *J. Am. Chem. Soc.* **2013**, *135*, 10942–10945. [CrossRef]

75. Wang, C. Doping metal-organic frameworks for water oxidation, carbon dioxide reduction, and organic photocatalysis. *J. Am. Chem. Soc.* **2011**, *133*, 13445–13454. [CrossRef]

76. Wang, D.K.; Huang, R.K.; Liu, W.J.; Sun, D.R.; Li, Z.H. Fe-based Mofs for photocatalytic CO$_2$ reduction: Role of coordination unsaturated sites and dual excitation pathways. *ACS Catal.* **2014**, *4*, 4254–4260. [CrossRef]

77. Xu, H.Q.; Hu, J.; Wang, D.; Li, Z.; Zhang, Q.; Luo, Y.; Yu, S.H.; Jiang, H.L. Visible-light photoreduction of CO$_2$ in a metal-organic framework: Boosting electron-hole separation via electron trap states. *J. Am. Chem. Soc.* **2015**, *137*, 13440–13443. [CrossRef]

78. Chen, D.S.; Xing, H.Z.; Wang, C.G.; Su, Z.M. Highly efficient visible-light-driven CO$_2$ reduction to formate by a new anthracene-based zirconium Mof via dual catalytic routes. *J. Mater. Chem. A* **2016**, *4*, 2657–2662. [CrossRef]

79. Sun, M.; Yan, S.; Sun, Y.; Yang, X.; Guo, Z.; Du, J.; Chen, D.; Chen, P.; Xing, H. Enhancement of visible-light-driven CO$_2$ reduction performance using an amine-functionalized zirconium metal-organic framework. *Dalton Trans.* **2018**, *47*, 909–915. [CrossRef]

80. Sun, D.; Fu, Y.; Liu, W.; Ye, L.; Wang, D.; Yang, L.; Fu, X.; Li, Z.D. Studies on photocatalytic CO$_2$ reduction over NH$_2$-Uio-66 (Zr) and its derivatives: Towards a better understanding of photocatalysis on metal-organic frameworks. *Chem. Eur. J.* **2013**, *42*, 14279–14285. [CrossRef]

81. Choi, K.M.; Kim, D.; Rungtaweevoranit, B.; Trickett, C.A.; Barmanbek, J.T.; Alshammari, A.S.; Yang, P.; Yaghi, O.M. Plasmon-enhanced photocatalytic CO$_2$ conversion within metal-organic frameworks under visible light. *J. Am. Chem. Soc.* **2017**, *139*, 356–362. [CrossRef]

82. Lee, Y.; Kim, S.; Fei, H.; Kang, J.K.; Cohen, S.M. Photocatalytic CO$_2$ reduction using visible light by metal-monocatecholato species in a metal-organic framework. *Chem. Commun.* **2015**, *92*, 16549–16552. [CrossRef]

83. Zhang, H.; Wei, J.; Dong, J.; Liu, G.; Shi, L.; An, P.; Zhao, G.; Kong, J.; Wang, X.; Meng, X.; et al. Efficient visible-light-driven carbon dioxide reduction by a single-atom implanted metal-organic framework. *Angew. Chem.* **2016**, *128*, 14310–14314. [CrossRef]

84. Su, Y.; Zhang, Z.; Liu, H.; Wang, Y. $Cd_{0.2}Zn_{0.8}S$@Uio-66-NH_2 nanocomposites as efficient and stable visible-light-driven photocatalyst for H_2 evolution and CO_2 reduction. *Appl. Catal. B Environ.* **2017**, *200*, 448–457. [CrossRef]

85. Wang, S.B.; Wang, X.C. Photocatalytic CO_2 reduction by CdS promoted with a zeolitic imidazolate framework. *Appl. Catal. B Environ.* **2015**, *162*, 494–500. [CrossRef]

86. Wang, M.; Liu, J.X.; Guo, C.M.; Gao, X.S.; Gong, C.H.; Wang, Y.; Liu, B.; Li, X.X.; Gurzadyan, G.G.; Sun, L.C. Metal-organic frameworks (ZIF-67) as efficient cocatalysts for photocatalytic reduction of CO_2: The role of the morphology effect. *J. Mater. Chem. A* **2018**, *6*, 4768–4775. [CrossRef]

87. Chen, M.; Han, L.; Zhou, J.; Sun, C.; Hu, C.; Wang, X.; Su, Z. Photoreduction of carbon dioxide under visible light by ultra-small Ag nanoparticles doped into Co-Zif-9. *Nanotechnology* **2018**, *29*, 284003. [CrossRef]

88. Ye, L.; Gao, Y.; Cao, S.Y.; Chen, H.; Yao, Y.A.; Hou, J.G.; Sun, L.C. Assembly of highly efficient photocatalytic CO_2 conversion systems with ultrathin two-dimensional metal-organic framework nanosheets. *Appl. Catal. B Environ.* **2018**, *227*, 54–60. [CrossRef]

89. Zhao, H.; Wang, X.S.; Feng, J.F.; Chen, Y.N.; Yang, X.; Gao, S.Y.; Cao, R. Synthesis and characterization of Zn_2GeO_4/Mg-MOF-74 composites with enhanced photocatalytic activity for CO_2 reduction. *Catal. Sci. Technol.* **2018**, *8*, 1288–1295. [CrossRef]

90. Cardoso, J.C.; Stulp, S.; de Brito, J.F.; Flor, J.B.S.; Frem, R.C.G.; Zanoni, M.V.B. MOFs based on ZIF-8 deposited on TiO_2 nanotubes increase the surface adsorption of CO_2 and its photoelectrocatalytic reduction to alcohols in aqueous media. *Appl. Catal. B Environ.* **2018**, *225*, 563–573. [CrossRef]

91. Sadeghi, N.; Sharifnia, S.; Sheikh Arabi, M. A porphyrin-based metal organic framework for high rate photoreduction of CO_2 to CH_4 in gas phase. *J. CO_2 Util.* **2016**, *16*, 450–457. [CrossRef]

92. Yan, S.C.; Ouyang, S.X.; Gao, J.; Yang, M.; Feng, J.Y.; Fan, X.X.; Wan, L.J.; Li, Z.S.; Ye, J.H.; Zhou, Y.; et al. A room-temperature reactive-template route to mesoporous $ZnGa_2O_4$ with improved photocatalytic activity in reduction of CO_2. *Angew. Chem. Int. Ed.* **2010**, *49*, 6400–6404. [CrossRef]

93. Maina, J.W.; Schütz, J.A.; Grundy, L.; Ligneris, E.D.; Yi, Z.F.; Kong, L.X.; Pozo-Gonzalo, C.; Ionescu, M.; Dumée, L.F. Inorganic nanoparticles/metal organic framework hybrid membrane reactors for efficient photocatalytic conversion of CO_2. *ACS Appl. Mater. Interfaces* **2017**, *9*, 35010–35017. [CrossRef]

94. Kong, Z.C.; Liao, J.F.; Dong, Y.J.; Xu, Y.F.; Chen, H.Y.; Kuang, D.B.; Su, C.Y. Core@shell $CsPbBr_3$@Zeolitic imidazolate framework nanocomposite for efficient photocatalytic CO_2 reduction. *ACS Energy Lett.* **2018**, *3*, 2656–2662. [CrossRef]

95. Fu, Y.H.; Sun, D.R.; Chen, Y.J.; Huang, R.K.; Ding, Z.X.; Fu, X.Z.; Li, Z.H. An amine-functionalized titanium metal-organic framework photocatalyst with visible-light-induced activity for CO_2 reduction. *Angew. Chem. Int. Ed.* **2012**, *51*, 3364–3367. [CrossRef]

96. He, X.; Wang, W.N. MOF-based ternary nanocomposites for better CO_2 photoreduction: Roles of heterojunctions and coordinatively unsaturated metal sites. *J. Mater. Chem. A* **2018**, *6*, 932–940. [CrossRef]

MDPI

St. Alban-Anlage 66

4052 Basel

Switzerland

Tel. +41 61 683 77 34

Fax +41 61 302 89 18

www.mdpi.com

Catalysts Editorial Office

E-mail: catalysts@mdpi.com

www.mdpi.com/journal/catalysts